T0177156

DO WE REALLY UNDERSTAND QUANTUM MECHANICS?
Second Edition

Quantum mechanics impacts on many areas of physics from pure theory to applications. However, it is very delicate to interpret, and philosophical difficulties as well as counter-intuitive results are apparent at a fundamental level. This book presents our current understanding of the theory, providing a historical introduction and discussing many of its interpretations.

Fully revised from the first edition, this book contains state-of-the-art research such as loophole-free experimental Bell tests; theorems on the reality of the wave function including the PBR theorem; and a new section on quantum simulation. More interpretations are now incorporated, described, and compared, with discussion of their successes and difficulties. Other sections, such as the quantum error correction codes and references, have been expanded. This book is ideal for researchers in physics and maths, and philosophers of science interested in quantum physics and its foundations.

FRANCK LALOË is Researcher Emeritus at the National Center for Scientific Research (CNRS) and is affiliated to the Laboratoire Kastler Brossel at the École Normale Supérieure. He is co-author of *Quantum Mechanics* (with Claude Cohen-Tannoudji and Bernard Diu), one of the most well-known textbooks on quantum mechanics.

DO WE REALLY UNDERSTAND QUANTUM MECHANICS?

Second Edition

FRANCK LALOË

Laboratoire Kastler Brossel, École Normale Supérieure, Paris and
Centre National de la Recherche Scientifique (CNRS), France

CAMBRIDGE
UNIVERSITY PRESS

CAMBRIDGE
UNIVERSITY PRESS

University Printing House, Cambridge CB2 8BS, United Kingdom

One Liberty Plaza, 20th Floor, New York, NY 10006, USA

477 Williamstown Road, Port Melbourne, VIC 3207, Australia

314–321, 3rd Floor, Plot 3, Splendor Forum, Jasola District Centre, New Delhi – 110025, India

79 Anson Road, #06–04/06, Singapore 079906

Cambridge University Press is part of the University of Cambridge.

It furthers the University's mission by disseminating knowledge in the pursuit of education, learning, and research at the highest international levels of excellence.

www.cambridge.org
Information on this title: www.cambridge.org/9781108477000
DOI: 10.1017/9781108569361

First published 2012
Second edition 2019

Printed and bound in Great Britain by Clays Ltd, Elcograf S.p.A.

A catalogue record for this publication is available from the British Library.

Library of Congress Cataloging-in-Publication Data
Names: Laloe, Franck, 1940– author.
Title: Do we really understand quantum mechanics? /
Franck Laloë (École Normale Supérieure, Paris).
Description: Second edition. | Cambridge ; New York, NY : Cambridge
University Press, 2019. | Includes bibliographical references and index.
Identifiers: LCCN 2018029216 | ISBN 9781108477000 (hardback)
Subjects: LCSH: Quantum theory. | Science–Philosophy.
Classification: LCC QC174.12 .L335 2019 | DDC 530.12–dc23
LC record available at https://lccn.loc.gov/2018029216

ISBN 978-1-108-47700-0 Hardback

Contents

Foreword [1]

Claude Cohen-Tannoudji

Quantum mechanics is an essential topic in today's physics curriculum at both the undergraduate and graduate levels. Quantum mechanics can explain the microscopic world with fantastic accuracy; the fruits from its insights have created technologies that have revolutionized the world. Computers, lasers, mobile telephones, and optical communications are but a few examples. The language of quantum mechanics is now an accepted part of the language of physics, and day-to-day usage of this language provides physicists with the intuition that is essential for achieving meaningful results. Nevertheless, most physicists acknowledge that, at least once in their scientific career, they have had difficulties understanding the foundations of quantum theory, perhaps even the impression that a really satisfactory and convincing formulation of the theory is still lacking.

Numerous quantum mechanics textbooks are available for explaining the quantum formalism and applying it to understand problems such as the properties of atoms, molecules, liquids, and solids; the interactions between matter and radiation; and more generally to understand the physical world that surrounds us. Other texts are available for elucidating the historical development of this discipline and describing the steps through which it went before quantum mechanics reached its modern formulation. In contrast, books are rare that review the conceptual difficulties of the theory and then provide a comprehensive overview of the various attempts to reformulate quantum mechanics in order to solve these difficulties. The present text by Franck Laloë does precisely this. It introduces and discusses in detail results and concepts such as the Einstein–Podolsky–Rosen theorem, Bell's theorem, and quantum entanglement that clearly illustrate the strange character of quantum behavior. Within the last few decades, impressive experimental progress

---

[1] Translated by D. Kleppner

has made it possible to carry out experiments that the founding fathers of quantum mechanics considered only as "thought experiments". For instance, it is now possible to follow the evolution of a single atom in real time. These experiments are briefly reviewed, providing an updated view of earlier results such as convincing violations of the Bell inequalities.

This book provides a clear and objective presentation of the alternative formulations that have been proposed to replace the traditional "orthodox" theory. The internal logic and consistency of these interpretations is carefully explained so as to provide the reader with a clear view of the formulations and a broad view of the state of the discipline. At a time when research is becoming more and more specialized, I think that it is crucial to keep some time for personal thought, to step back and ask oneself questions about the deep significance of the concepts that we employ routinely. In this text, I see the qualities of clarity, intellectual rigor, and deep analysis that I have always noticed and appreciated in the work of the author during many years of friendly collaboration. I wish the book a well-deserved great success!

Preface

This second edition has benefited from many discussions with colleagues, seminars, etc. at various places; this stimulated various improvements and additions to the first edition. I am especially grateful to Philip Stamp, who read the first edition of this book very thoroughly and provided an impressive list of excellent suggestions. My dear old friend Bill Mullin also helped a lot with many excellent remarks on some chapters. I greatly benefited from the deep understanding of Johannes Kofler concerning the domain of validity of the various Bell inequalities, and the various possible "loopholes" in the interpretation of the experiments. Philippe Grangier, Patrick Peter, Jean Bricmont, Ward Struyve, Roger Balian, Julia Kempe, and Michel le Bellac provided many excellent suggestions and remarks to improve the text.

Finally, a wonderful feature of Internet is that the author of a book can receive by mail many reactions from readers all over the world. In particular, I wish to thank Paul Slater, who, from California, sent me by email a series of particularly useful suggestions, but alas I cannot give here the long list of the many other colleagues who also helped in various ways with their suggestions.

Preface to the First Edition

In many ways, quantum mechanics is a surprising theory. It is known to be nonintuitive, and leads to representations of physical phenomena that are very different from what our daily experience could suggest. But it is also very surprising because it creates a big contrast between its triumphs and difficulties.

On the one hand, among all theories, quantum mechanics is probably one of the most successful achievements of science. It was initially invented in the context of atomic physics, but it has now expanded into many domains of physics, giving access to an enormous number of results in optics, solid-state physics, astrophysics, etc. It has actually now become a general method, a framework in which many theories can be developed, for instance to understand the properties of fluids and solids, fields, elementary particles, and leading to a unification of interactions in physics. Its range extends much further than the initial objectives of its inventors and, what is remarkable, this turned out to be possible without changing the general principles of the theory. The applications of quantum mechanics are everywhere in our twenty-first century environment, with all sorts of devices that would have been unthinkable 50 years ago.

On the other hand, conceptually this theory remains relatively fragile because of its delicate interpretation – fortunately, this fragility has little consequence for its efficiency. The reason why difficulties persist is certainly not that physicists have tried to ignore them or put them under the rug! Actually, a large number of interpretations have been proposed over the decades, involving various methods and mathematical techniques. We have a rare situation in the history of sciences: consensus exists concerning a systematic approach to physical phenomena, involving calculation methods having an extraordinary predictive power; nevertheless, almost a century after the introduction of these methods, the same consensus is far from

being reached concerning the interpretation of the theory and its foundations. This
is reminiscent of the colossus with feet of clay.

The difficulties of quantum mechanics originate from the object it uses to de-
scribe physical systems, the state vector $|\Psi\rangle$. While classical mechanics describes a
system by directly specifying the positions and velocities of its components, quan-
tum mechanics replaces them by a complex mathematical object $|\Psi\rangle$, providing a
relatively indirect description. This is an enormous change, not only mathemat-
ically, but also conceptually. The relations between $|\Psi\rangle$ and physical properties
leave much more room for discussions about the interpretation of the theory than
in classical physics. Actually, many difficulties encountered by those who tried (or
are still trying) to "really understand" quantum mechanics are related to questions
pertaining to the exact status of $|\Psi\rangle$. For instance, does it describe the physical re-
ality itself, or only some (partial) knowledge that we might have of this reality?
Does it describe ensembles of systems only (statistical description), or one single
system as well (single events)? Assume that, indeed, $|\Psi\rangle$ is affected by an imperfect
knowledge of the system; is it then not natural to expect that a better description
should exist, at least in principle? If so, what would be this deeper and more precise
description of the reality?

Another confusing feature of $|\Psi\rangle$ is that, for systems extended in space (for in-
stance, a system made of two particles at very different locations), it gives an overall description of all its physical properties in a single block, from which the notion
of space seems to have disappeared; in some cases, the properties of the two remote
particles are completely "entangled" in a way where the usual notions of space-time
and of events taking place in it seem to become diluted. It then becomes difficult,
or even impossible, to find a spatio-temporal description of their correlations that
remains compatible with relativity. All this is of course very different from the
usual concepts of classical physics, where one attributes local properties to physical systems by specifying the density, the value of fields, etc. at each point of space.
In quantum mechanics, this separability between the physical content of different
points of space is no longer possible in general. Of course, one could think that this
loss of a local description is just an innocent feature of the formalism with no special consequence. For instance, in classical electromagnetism, it is often convenient
to introduce a choice of gauge for describing the fields in an intermediate step; in
the Coulomb gauge, the potential propagates instantaneously, while Einstein relativity forbids any communication that is faster than light. But this instantaneous
propagation is just a mathematical artifact: when a complete calculation is made,
proper cancellations of the instantaneous propagation take place so that, at the end,
the relativistic limitation is perfectly obeyed. But, and as we will see, it turns out
that the situation is much less simple in quantum mechanics: in fact, a mathemat-
ical entanglement in $|\Psi\rangle$ can indeed have important physical consequences for the

result of experiments, and even lead to predictions that are, in a sense, contradictory with locality. Without any doubt, the state vector is a curious object for describing reality!

It is therefore not surprising that quantum mechanics should have given rise to so many interpretations. Their very diversity makes them interesting. Each of them introduces its own conceptual frame and view of physics, sometimes attributing to it a special status among the other natural sciences. Moreover, these interpretations may provide complementary views on the theory, shedding light onto some interesting features that, otherwise, would have gone unnoticed. The best-known example is Bohm's theory, from which Bell started to obtain a theorem illustrating general properties of quantum mechanics and entanglement, with applications ranging outside the Bohmian theory. Other examples exist, such as the use of stochastic Schrödinger dynamics to better understand the evolution of a quantum subsystem, the history interpretation and its view of complementarity, etc.

This book is intended for the curious reader who wishes to get a broad view on the general situation of quantum physics, including the various interpretations that have been elaborated, and without putting aside the difficulties when they occur. It is not a textbook designed for a first contact with quantum mechanics; there already exist many excellent reference books for students. In fact, from Chapter 1, the text assumes some familiarity with quantum mechanics and its formalism (Dirac notation, the notion of wave function, etc.). Any student who has already studied quantum mechanics for a year should have no difficulty in following the equations. Actually, there are relatively few in this book, which focuses, not on technical, but on logical and conceptual difficulties. Moreover, a chapter is inserted as an annex at the end of the book in order to help those who are not used to the quantum formalism. It offers a first contact with the notation; the reader may, while he/she progresses in the other chapters, choose a section of this chapter to clarify his/her ideas on such or such a technical point.

Chapters 1 and 2 recall the historical context, from the origin of quantum mechanics to the present situation, including the successive steps from which the present status of $|\Psi\rangle$ emerged. Paying attention to history is not inappropriate in a field where the same recurrent ideas are so often rediscovered; they appear again and again, sometimes almost identical over the years, sometimes remodelled or rephrased with new words, but in fact more or less unchanged. Therefore, a look at the past is not necessarily a waste of time! Chapters 3 and 4 discuss two important theorems that form a logical chain, the EPR (Einstein, Podolsky, and Rosen) theorem and the Bell theorem; both give rise to various forms, several of which will be described. Chapters 5 and 6 introduce more recent theorems that follow similar lines, as well as possible consequences on the status of the state vector. Chapter 7 gives a more general view on quantum entanglement, and Chapter 8 illustrates the notion with various processes that make use of it, such as quantum cryptography and teleportation. Chapter 9 discusses quantum measurement, in particular weak and continuous measurements. A few experiments are described in Chapter 10; among the huge crowd of those illustrating quantum mechanics in various circumstances, we have chosen a small fraction of them – those where state vector reduction is "seen in real time". Finally, Chapter 11, the longest of all chapters, gives an introduction and some discussion of the various interpretations of quantum mechanics. The chapters are relatively independent and the reader may probably use them in almost any order. Needless to say, no attempt was made to cover all subjects related to the foundations of quantum mechanics. A selection was unavoidable; it resulted in a list of subjects that the author considers as particularly relevant, but of course this personal choice remains somewhat arbitrary.

The motivation of this book is not to express preference for any given interpretations, as has already been done in many reference articles or monographs (we will quote several of them). It is even less to propose a new interpretation that would miraculously solve all problems. The objective is, rather, to review the various interpretations and to obtain a general perspective on the way they are related, their differences and common features, and their individual consistency. Indeed, each of these interpretations has its own logic, and it is important to remember it; a classical mistake is to mix various interpretations together. For instance, the Bohmian interpretation has sometimes been criticized by elaborating constructions that retain some elements of this interpretation, but not all, or by inserting elements that do not belong to the interpretation; one then obtains contradictions, but this says very little on the Bohmian interpretation itself. The necessity for logical consistency is general in the context of the foundations of quantum mechanics. Sometimes, the EPR argument or the Bell theorem has been misunderstood because of a confusion between their assumptions and conclusions. We will note in passing a few occasions where such mistakes are possible in order to help avoiding them. We should also mention that it is out of the question to give an exhaustive description of all interpretations of quantum mechanics here! They may be associated in many different ways, so that it is impossible to account for all possible combinations or nuances. A relatively abundant bibliography is proposed to the reader, but, in this case also, reaching any exhaustiveness is impossible; some choices have been made, sometimes arbitrary, in order to keep the total volume within reasonable limits.

To summarize, the main purpose of this book is an attempt to provide a balanced view on the conceptual situation of a theory that is undoubtedly one of the most remarkable achievements of the human mind, quantum mechanics, without hiding either difficulties or successes. As we already mentioned, its predictive power constantly obtains marvelous results in new domains, sometimes in a totally unpredictable way; nevertheless, this intellectual edifice remains the object of discussions or even controversy concerning its foundations. No one would think of discussing classical mechanics or the Maxwell equations in the same way. Maybe this signals that the final and optimum version of the theory has not yet been obtained?

Acknowledgments

Many colleagues played an important role in the elaboration of this book. The first is certainly Claude Cohen-Tannoudji, to whom I owe a lot. Over the years I benefited, as many others did, from his unique and deep way of using quantum mechanics (and even to think quantum-mechanically); more than 40 years of friendship (and of common writing) and uncountable stimulating discussions were priceless for me. Alain Aspect is another friend with whom, from the beginning of his thesis in the seventies, a constant exchange of ideas took place (it still does!). At the time, the foundations of quantum theory were not very well considered among mainstream physicists, even sometimes perceived as passé or mediocre physics; Alain and I could comfort each other and make progress together in a domain that we both found fascinating, with the encouragement of Bernard d'Espagnat. Jean Dalibard and Philippe Grangier have been other wonderful discussion partners, always open-minded with extreme intellectual clarity; I wish to thank them warmly. The title *"Do we Really Understand Quantum Mechanics?"* was suggested to me long ago by Pierre Fayet, on the occasion of two seminars on this subject he was asking me to give. This book arose from a first version of a text published in 2001 as an article in the *American Journal of Physics*, initiated during a visit at the Theoretical Physics Institute at the California University of Santa Barbara. During a session on Bose–Einstein condensation, I was lucky enough to discuss several aspects of quantum mechanics with its organizer, Antony Leggett; another lucky event favoring exchanges was to share Wojciech Zurek's office! A little later, a visit to the Lorentz Institute of Leiden was also very stimulating, in particular with the help of Stig Stenholm. As for Abner Shimony, he guided me with much useful advice and encouraged the writing of the first version of this text.

Among those who helped much with the present version of the text, Michel Le Bellac played an important role by reading the whole text and giving useful advice, which helped to improve the text. He and Michèle Leduc chose a wonderful (anonymous) reviewer who made many very relevant remarks; I am grateful to all three of them. Among the other friends who also helped efficiently on various aspects and made interesting suggestions are Roger Balian, Serge Reynaud, William Mullin, Olivier Darrigol, Bernard d'Espagnat, and Catherine Chevalley; I am very grateful for many comments, advice, questions, etc. Markus Holzmann kindly read the whole manuscript when it was completed and made many interesting suggestions. The careful editing work of Anne Rix has been invaluable for improving the homogeneity and the quality of the text.

Last but not least, concerning the chapter describing the various interpretations of quantum mechanics, I asked specialists of each of these interpretations to be kind enough to check what I had written. I thank Sheldon Goldstein for reading and commenting on the part concerning the Bohmian theory, Philip Pearle and Giancarlo Ghirardi for their advice concerning modified Schrödinger dynamics, Robert Griffiths and Roland Omnès for their comments on the history interpretation, Bernard d'Espagnat for clarifying remarks on veiled reality, Richard Healey for his help on the modal interpretation, Carlo Rovelli for his comments and suggestions on the relational interpretation, Alexei Grinbaum for illuminating comments concerning quantum logic and formal theories, and Thibault Damour for his helpful reading of my presentation of the Everett interpretation. According to the tradition it should be clear that, if nevertheless errors still subsist, the responsibility is completely the author's. Finally, without the exceptional atmosphere of my laboratory, LKB (Laboratoire Kastler Brossel), without the constant interaction with its members, and without the intellectual environment of ENS (Ecole Normale Supérieure), nothing would have been possible.

1
Historical Perspective

The founding fathers of quantum mechanics had already perceived the essence of the difficulties of quantum mechanics; today, after almost a century, the discussions are still lively and, if some very interesting new aspects have emerged, at a deeper level the questions have not changed so much. What is more recent, nevertheless, is a general change of attitude among physicists: until about 1970 or 1980, most physicists thought that the essential questions had been settled, and that "Bohr was right and proved his opponents to be wrong". This was probably a consequence of the famous discussions among Bohr, Einstein, Schrödinger, Heisenberg, Pauli, de Broglie, and others (in particular at the Solvay meetings [1–3], where Bohr's point of view had successfully resisted Einstein's extremely clever attacks). The majority of physicists did not know the details of the arguments. They nevertheless thought that the standard "Copenhagen interpretation" had clearly emerged from the infancy of quantum mechanics as the only sensible attitude for good scientists. This interpretation includes the idea that modern physics must contain indeterminacy as an essential ingredient: it is fundamentally impossible to predict the outcome of single microscopical events; it is impossible to go beyond the formalism of the wave function (or its generalization, the state vector $|\Psi\rangle$) and complete it. For some physicists, the Copenhagen interpretation also includes the difficult notion of "complementarity" – even if it is true that, depending on the context, complementarity comes in many varieties and has been interpreted in many different ways! By and large, the impression of the vast majority was that Bohr had eventually won the debate against Einstein, so that discussing again the foundations of quantum mechanics after these giants was pretentious, passé, and maybe even in bad taste.

Nowadays, the attitude of physicists is more open concerning these matters. One first reason is probably that the nonrelevance of the "impossibility theorems" put forward by the defenders of the standard interpretation, in particular by von Neumann [4], has now been better realized by the scientific community – see [5–7] and [8], as well as the discussion given in [9]). Another reason is, of course, the

great impact of the discoveries and ideas of J.S. Bell [6] in 1964. At the beginning of a new century, it is probably fair to say that we are no longer sure that the
Copenhagen interpretation is the only possible consistent attitude for physicists –
see for instance the doubts expressed by Shimony in [10]. Alternative points of
view are considered with interest: we have theories including additional variables
(or "hidden variables"[1]) [11, 12]; other theories are modified dynamics of the state
vector [7, 13–15] (nonlinear and/or stochastic evolution); at the other extreme, we
have points of view such as the so-called "many worlds interpretation" (or "many
minds interpretation", or "multibranched universe") [16]; more recently, other interpretations such as that of "decoherent histories" [17] have been put forward (this
list is nonexhaustive). These interpretations and several others will be discussed in
Chapter 11. For a recent review containing many references, see [18], which emphasizes additional variables, but which is also characteristic of the variety of positions among contemporary scientists[2]. See also an older but very interesting debate
published in *Physics Today* [19]; another very useful source of older references is
the 1971 *American Journal of Physics* "Resource Letter" [20]. But this variety of
possible alternative interpretations should not be the source of misunderstandings!
It should also be emphasized very clearly that, until now, no new fact whatsoever
(or new reasoning) has appeared that has made the Copenhagen interpretation ob-
solete in any sense.

1.1 Three periods

Three successive periods may be distinguished in the history of the elaboration of
the fundamental quantum concepts; they have resulted in the point of view that is
called "the Copenhagen interpretation", or "orthodox", or again "standard" interpretation. Actually, these terms may group different variants of the general interpretation, as we see in more detail later in this book (in particular in Chapter 11).
Here we give only a brief historical summary; we refer the reader who would like to
know more about the history of the conceptual development of quantum mechanics
to the book of Jammer [21] – see also [22] and [23]. For detailed discussions of
fundamental problems in quantum mechanics, one could also read [10, 24, 25] as
well as the references contained, or those given in [20].

---

[1] As we discuss in more detail in §11.8, we prefer to use the words "additional variables" since they are not
hidden, but actually appear directly in the results of measurements.
[2] For instance, the contrast between the titles of [10] and [18] is interesting.

1.1.1 Prehistory

Planck's name is obviously the first that comes to mind when one thinks about the birth of quantum mechanics: in 1900, he was the one who introduced the famous constant h, which now bears his name. His method was phenomenological, and his motivation was actually to explain the properties of the radiation in thermal equilibrium (blackbody radiation) by introducing the notion of finite grains of energy in the calculation of the entropy [26]. Later he interpreted them as resulting from discontinuous exchange between radiation and matter. It is Einstein who, still later (in 1905), took the idea more seriously and really introduced the notion of quantum of light (which would be named "photon" only much later, in 1926 [27]) in order to explain the wavelength dependence of the photoelectric effect – for a general discussion of the many contributions of Einstein to quantum theory, see [28].

One should nevertheless realize that the most important and urgent question at the time was not so much to explain the fine details of the properties of interactions between radiation and matter, or the peculiarities of the blackbody radiation. It was more general: to understand the origin of the stability of atoms, that is, of all matter which surrounds us and of which we are made! According to the laws of classical electromagnetism, negatively charged electrons orbiting around a positively charged nucleus should constantly radiate energy, and therefore rapidly fall onto the nucleus. Despite several attempts, explaining why atoms do not collapse but keep fixed sizes was still a complete challenge for physics[3]. One had to wait a little bit longer, until Bohr introduced his celebrated atomic model (1913), to see the appearance of the first ideas allowing the question to be tackled. He proposed the notion of "quantized permitted orbits" for electrons, as well as of "quantum jumps", to describe how they would go from one orbit to another, for instance during radiation emission processes. To be fair, we must concede that these notions have now almost disappeared from modern physics, at least in their initial forms; quantum jumps are replaced by a much more precise and powerful theory of spontaneous emission in quantum electrodynamics. But, on the other hand, one may also see a resurgence of the old quantum jumps in the modern use of the postulate of the wave packet (or state vector) reduction (§1.2.2.a). After Bohr, came Heisenberg, who, in 1925, introduced the theory that is now known as "matrix mechanics"[4], an abstract intellectual construction with a strong philosophical component, sometimes close to positivism; the classical physical quantities are replaced by "observables", corresponding mathematically to matrices, defined by suitable postulates without much help of intuition. Nevertheless, matrix mechanics contained

---

[3] For a review of the problem in the context of contemporary quantum mechanics, see [29].

[4] The names of Born and Jordan are also associated with the introduction of this theory, since they immediately made the connexion between Heisenberg's rules of calculation and those of matrices in mathematics.

many elements that turned out to be essential building blocks of modern quantum mechanics!

In retrospect, one can be struck by the very abstract and somewhat mysterious character of atomic theory at this period of history; why should electrons obey such rules, which forbid them to leave a restricted class of orbits, as if they were miraculously guided on simple trajectories? What was the origin of these quantum jumps, which were supposed to have no duration at all, so that it would make no sense to ask what were the intermediate states of the electrons during such a jump? Why should matrices appear in physics in such an abstract way, with no apparent relation with the classical description of the motion of a particle? One can guess how relieved physicists probably felt when another point of view emerged, a point of view that looked at the same time much simpler and more in the tradition of the physics of the nineteenth century: the undulatory (or wave) theory.

1.1.2 The undulatory period

The idea of associating a wave with every material particle was first introduced by de Broglie in his thesis (1924) [30]. A few years later (1927), the idea was confirmed experimentally by Davisson and Germer in their famous electron diffraction experiment [31]. For some reason, at that time de Broglie did not proceed much further in the mathematical study of this wave, so that only part of the veil of mystery was raised by him (see for instance the discussion in [32]). It is sometimes said that Debye was the first, after hearing about de Broglie's ideas, to remark that in physics a wave generally has a wave equation: the next step would then be to try and propose an equation for this new wave. The story adds that the remark was made in the presence of Schrödinger, who soon started to work on this program; he successfully and rapidly completed it by proposing the equation that now bears his name, one of the most basic equations of all physics. Amusingly, Debye himself does not seem to have remembered the event. The anecdote may be inaccurate; in fact, different reports about the discovery of this equation have been given, and we will probably never know exactly what happened. What remains clear is that the introduction in 1926 of the Schrödinger equation for the wave function[5] [33] is one of the essential milestones in the history of physics. Initially, it allowed one to understand the energy spectrum of the hydrogen atom, but it was soon extended and gave successful predictions for other atoms, then molecules and ions, solids (the theory of bands, for instance), etc. It is at present the major basic tool of many branches of modern physics and chemistry.

Conceptually, at the time of its introduction, the undulatory theory was welcomed as an enormous simplification of the new mechanics. This is particularly

[5] See footnote [12] for the relation between the state vector and the wave function.

true because Schrödinger and others (Dirac, Heisenberg) promptly showed how it could be used to recover the predictions of matrix mechanics from more intuitive considerations, using the properties of the newly introduced "wave function" – the solution of the Schrödinger equation. The natural hope was then to extend this success, and to simplify all problems raised by the mechanics of atomic particles: one would replace it by a mechanics of waves, which would be analogous to electromagnetic or sound waves. For instance, Schrödinger initially thought that all particles in the universe looked to us like point particles just because we observe them at a scale that is too large; in fact, they are tiny "wave packets" that remain localized in small regions of space. He had even shown that these wave packets remain small (they do not spread in space) when the system under study is a harmonic oscillator – alas, we now know that this is a very special case; in general, the wave packets constantly spread in space!

1.1.3 Emergence of the Copenhagen interpretation

It did not take long before it became clear that the completely undulatory theory of matter also suffered from very serious difficulties, actually so serious that physicists were soon led to abandon it. A first example of difficulty is provided by a collision between particles, where the Schrödinger wave spreads in all directions, like a circular wave in water stirred by a stone thrown into it; but, in all collision experiments, particles are observed to follow well-defined trajectories and remain localized, going in some precise direction. For instance, every photograph taken in the collision chamber of a particle accelerator shows very clearly that particles never get "diluted" in all space! This stimulated the introduction, by Born in 1926, of the probabilistic interpretation of the wave function [34]: quantum processes are fundamentally nondeterministic; the only thing that can be calculated is probabilities, given by the square of the modulus of the wave function.

Another difficulty arises as soon as one considers systems made of more than one single particle: then, the Schrödinger wave is no longer an ordinary wave since, instead of propagating in normal space, it propagates in the so-called "configuration space" of the system, a space that has $3N$ dimensions for a system made of N particles! For instance, already for the simplest of all atoms, the hydrogen atom, the wave propagates in six dimensions[6]. For a collection of atoms, the dimension grows rapidly, and becomes an astronomical number for the ensemble of atoms contained in a macroscopic sample. Clearly, the new wave was not at all similar to

---

[6] This is true if spins are ignored; if they are taken into account, four such waves propagate in six dimensions.

classical waves, which propagate in ordinary space; this deep difference will be a sort of Leitmotiv in this text[7], reappearing under various aspects here and there[8].

In passing, it is interesting to notice that the recent observation of the phenomenon of Bose–Einstein condensation in dilute gases [35] can be seen, in a sense, as a sort of realization of the initial hope of Schrödinger: this condensation provides a case where a many-particle matter wave does propagate in ordinary space. Before condensation takes place, we have the usual situation: the gas has to be described by wave functions defined in a huge configuration space. But, when the atoms are completely condensed into a single-particle wave function, they are restricted to a much simpler many-particle state built with the same ordinary wave function, as for a single particle. The matter wave then becomes similar to a classical field with two components (the real part and the imaginary part of the wave function), resembling for instance an ordinary sound wave. This illustrates why, somewhat paradoxically, the "exciting new states of matter" provided by Bose–Einstein condensates are not an example of an extreme quantum situation; in a sense, they are actually more classical than the gases from which they originate (in terms of quantum description, interparticle correlations, etc.). Conceptually, of course, this remains a very special case and does not solve the general problem associated with a naive view of the Schrödinger waves as real waves.

The purely undulatory description of particles has now disappeared from modern quantum mechanics in its standard form[9]. In addition to Born and Bohr, Heisenberg [36], Jordan [37, 38], Dirac [39], and others played an essential role in the appearance of a new formulation of quantum mechanics [23], where probabilistic and undulatory notions are incorporated in a single complex logical edifice. The probabilistic component of the theory is that, when a system undergoes a measurement, the result is fundamentally random; the theory provides only the probabilities of the different possible outcomes. The wave component of the theory is that, when no measurements are performed, the Schrödinger equation is valid. The wave function is no longer considered as a direct physical description of the system itself; it is only a mathematical object that provides the probabilities of the different results[10] – we come back to this point in more detail in §1.2.3.

The first version of the Copenhagen interpretation was completed around 1927,

[7] For instance, the nonlocality effects occurring with two correlated particles can be seen as a consequence of the fact that the wave function propagates locally, but in a six-dimensional space, while the usual definition of locality refers to ordinary space which has three dimensions.

[8] Quantum mechanics can also be formulated in a way that does not involve the configuration space, but just the ordinary space: using the formalism of field operators (sometimes called second quantization, for historical reasons) - *cf.* §1.3. One can write these operators in a form that is similar to a wave function. Nevertheless, since they are quantum operators, their analogy with a classical field is even less valid.

[9] See also §11.10 for the discussion of a nonstandard quantum theory based on such a description.

[10] In the literature, one often finds the word "ontological" to describe Schrödinger's initial point of view on the wave function, as opposed to "epistemological" to describe the probabilistic interpretation.

the year of the fifth Solvay conference [3]. Almost immediately, theorists started to extend the range of quantum mechanics from particle to fields. At that time, the interest was focused only on the electromagnetic field, associated with the photon, but the ideas were later generalized to fields associated with a wide range of particles (electrons, muons, quarks, etc.). Quantum field theory has now enormously expanded and become a fundamental tool in particle physics, within a relativistic formalism (the Schrödinger equation itself does not satisfy Lorentz invariance). A generalization of the ideas of gauge invariance of electromagnetism has led to various forms of gauge theories; some are at the root of our present understanding of the role in physics of the fundamental interactions (electromagnetic, weak, strong[11]) and led to the successful prediction of new particles. Nevertheless, despite all these remarkable successes, field theory remains, conceptually, on the same fundamental level as the theory of a single nonrelativistic particle treated with the Schrödinger equation. We come back briefly on this subject in §1.3 but, since this text is concerned mostly with conceptual issues, we will not discuss field theory beyond this chapter.

1.2 The state vector

Many discussions concerning the foundations of quantum mechanics are related to the status and physical meaning of the state vector. In §§1.2.1 and 1.2.2, we begin by first recalling its definition and use in quantum mechanics (the reader familiar with the quantum formalism might wish to skip these two sections); then, in §1.2.3, we discuss the status of the state vector in standard quantum mechanics.

1.2.1 Definition, Schrödinger evolution, Born rule

We briefly summarize how the state vector is used in quantum mechanics and its equations; more details are given in §12.1.1 and following.

1.2.1.a Definition

Consider a physical system made of N particles with mass, each propagating in ordinary space with three dimensions; the state vector $|\Psi\rangle$ (or the associated wave function[12]) replaces in quantum mechanics the N positions and N velocities which, in classical mechanics, would be used to describe the state of the system. It is often

---

[11] There is a fourth fundamental interaction in physics, gravitation. The "standard model" of field theory unifies the first three interactions, but leaves gravitation aside. Other theories unify the four fundamental interactions, but for the moment they are not considered standard.

[12] For a system of spinless particles with masses, the state vector $|\Psi\rangle$ is equivalent to a wave function, but for more complicated systems this is not the case. Nevertheless, conceptually they play the same role and are used in the same way in the theory, so that we do not need to make a distinction here.

convenient to group all these positions and velocities within the $6N$ components of a single vector $\mathbf{V}$ belonging to a real vector space with $6N$ dimensions, called "phase space"[13]; formally, one can merely consider that the state vector $|\Psi\rangle$ is the quantum equivalent of this classical vector $\mathbf{V}$. It nevertheless belongs to a space that is completely different from the phase space, a complex vector space called "space of states" (or, sometimes, the "Hilbert space" for historical reasons) with infinite dimension. The calculations in this space are often made with the help of the Dirac notation [39], which actually we will use here, and where the vectors belonging to the space of states are often called "kets".

Because the state vector belongs to a linear space, any combination of two arbitrary state vectors $|\Psi_1\rangle$ and $|\Psi_2\rangle$ belonging to the space of states:

$$|\Psi\rangle = \alpha\,|\Psi_1\rangle + \beta\,|\Psi_2\rangle \tag{1.1}$$

(where α and β are arbitrary complex numbers) is also a possible state for the system. This is called the "superposition principle" of quantum mechanics, and has many consequences.

Moreover, the formalism of quantum mechanics associates with each physical observable of the system, position(s), momentum (or momenta), energy, angular momentum, etc., a linear operator acting in the space of states, and provides rules for constructing these operators. For historical reasons (§1.1.1), each of these operators is often called an "observable"; they belong to the category of mathematical operators called "linear Hermitian operators".

1.2.1.b Schrödinger evolution

The evolution of the state vector $|\Psi(t)\rangle$ between time t_0 and t_1 is given by the Schrödinger equation:

$$i\hbar\frac{d}{dt}\,|\Psi(t)\rangle = H(t)\,|\Psi(t)\rangle \tag{1.2}$$

where $H(t)$ is the Hamiltonian evolution of the system (including the internal interactions of this system as well as the effects of classical external fields applied to it, for instance static or time-dependent magnetic fields). The Schrödinger equation is a linear differential equation, similar to many other such equations in physics. It leads to a progressive evolution of the state vector, without any quantum jump or discontinuity. It is as general as the Newton or Lagrange equations in classical mechanics, and can be applied to all physical situations, provided of course the system is well defined with a known Hamiltonian.

In particular, the Schrödinger equation can also be applied to a situation where the physical system interacts with a measurement apparatus (a spin 1/2 particle

---

[13] The phase space therefore has twice as many dimensions as the aforementioned configuration space.

entering the magnetic field gradient created by a Stern–Gerlach apparatus, for instance); it then does not select precise experimental results, but keeps all of them as potentialities (within a so-called "coherent superposition"). One more ingredient is then introduced into the theory, the Born probability rule.

1.2.1.c Born probability rule

We assume that, at time t_1, when the solution $|\Psi(t)\rangle$ of equation (1.2) takes the value $|\Psi(t_1)\rangle$, the system undergoes a measurement, associated with an operator M (observable) acting in the space of states. We note $|\ m_i >$ the eigenvectors of M associated with eigenvalues m_i ($i = 1, 2, ...$); if some eigenvalues are degenerate, several consecutive values in the series of m_i are equal, but associated with different vectors $|\ m_i >$. Since M is an Hermitian operator, the $|\ m_i >$ can be chosen as an orthonormal basis of the space of states.

The Born probability rule then states that, in an ideal measurement, the following are true:

(i) The result of a measurement associated with M can only be one of the m_i's; other results are never obtained.

(ii) If a particular eigenvalue m_i is nondegenerate, the probability $\mathcal{P}_i$ of obtaining result m_i is given by the square modulus of the scalar product of $|\Psi(t_1)\rangle$ by $|m_i\rangle$:

$$\mathcal{P}_i = |\langle m_i\ |\Psi(t_1)\rangle|^2 \tag{1.3}$$

(iii) The probability of measuring a degenerate eigenvalue is the sum of the probabilities (1.3) corresponding to all the orthonormal eigenvectors associated with this eigenvalue[14].

Rules (ii) and (iii) may be grouped in a simple form, which will be useful in what follows. If the result corresponds to an eigenvalue m that is p times degenerate, the series of p numbers $m_i, m_{i+1}, ..., m_{i+p}$ have the same value m. We can then introduce the sum of the projectors (§12.1.3) over the corresponding eigenvectors:

$$P_M(m) = |m_i\rangle\langle m_i| + |m_{i+1}\rangle\langle m_{i+1}| + ... + |m_{i+p}\rangle\langle m_{i+p}| \tag{1.4}$$

This operator is also a projector (it is equal to its square), which can be applied to the state vector $|\Psi(t_1)\rangle$ before the measurement:

$$P_M(m)|\Psi(t_1)\rangle = |\Psi'_m\rangle \tag{1.5}$$

The probability of obtaining result m in the measurement is then nothing but the square of the norm of $|\Psi'\rangle$:

$$\mathcal{P}_m = \langle\Psi'_m|\Psi'_m\rangle = \langle\Psi(t_1)|P_M(m)|\Psi(t_1)\rangle \tag{1.6}$$

[14] Similarly, in the classical theory of probabilities, if an event E can be obtained either as event e_1, or e_2, ... , or e_i, ... , and if all events e_i are exclusive, the probability of E is the sum of the probabilities of the e_i.

1.2.2 Measurement processes

The standard interpretation of quantum mechanics contains the continuous, deterministic, evolution of the wave function/state due to the Schrödinger equation. Usually, one also includes in this interpretation a second postulate of evolution; this postulate is associated with the process of measurement, and completely different from the Schrödinger evolution, since it is discontinuous. It is often called the "wave packet reduction", or "wave function collapse", or again "state vector reduction", and was introduced by von Neumann in his famous treatise (chapter VI of [4]). Most textbooks include this postulate, probably because it provides a more detailed and intuitive view of the evolution of a quantum system during measurement, but this is not indispensable. Bohr himself preferred another point of view where state vector reduction is not used[15]; we discuss this point of view in § 1.2.2.b (there exist also other interpretations of quantum mechanics that do not make use of state vector reduction, as discussed in Chapter 11; see for instance §§11.1.2, 11.8 or 11.12).

1.2.2.a Von Neumann reduction (collapse)

Suppose now that the system we study is prepared at time t_0, evolves freely (without being measured) until time t_1, where it undergoes a first measurement, and then evolves freely again until time t_2, where a second measurement is performed. Just after the first measurement at time t_1, when the corresponding result of measurement is known, it is very natural to consider that both the initial preparation and the first measurement are part of a single preparation process of the system. One then associates to this preparation a state vector that includes the information of the first result; in other words, the state vector is updated to include the interaction with the first measurement apparatus as well as the information acquired during the corresponding measurement (update of information). This is precisely what the state vector reduction (or state collapse) postulate does. The new "reduced" state vector can then be used as an initial state to calculate the probabilities of the different results corresponding to the second measurement, at time t_2.

Dirac also takes this point of view when he writes (page 9 of "Quantum Mechanics" [39]): "There are, however, two cases when we are in general obliged to consider the disturbance as causing a change in state of the system, namely, when the disturbance is an observation and when it consists in preparing the system so as to be in a given state".

---

[15] As stated in [40]: "Most importantly, Bohr's complementarity interpretation makes no mention of wave packet collapse ... or a privileged role for the subjective consciousness of the observer. Bohr was also in no way a positivist. Much of what passes for the Copenhagen interpretation is found in the writings of Werner Heisenberg, but not in Bohr" (see also footnote [22]).

We assume that the measurement is ideal[16] – it preserves the integrity of the system, as opposed to destructive measurements such as the absorption of a photon in a detector. Then, after the measurement associated with M has provided result m_i corresponding to a nondegenerate eigenvalue (and therefore to a single $|m_i\rangle$), the reduced state vector is as follows:

$$|\Psi'_{m_i}\rangle = |m_i\rangle \tag{1.7}$$

In other words, at time t_1 when the first measurement is performed and provides result m_i, the state vector jumps discontinuously from $|\Psi(t_1)\rangle$ to $|m_i\rangle$. If the same measurement is repeated very shortly after, by applying the Born rule (1.3) to state $|\Psi'\rangle$, one finds that all probabilities are zero but one. The result is then certain: one obtains m_i again.

The generalization of (1.7) to a degenerate eigenvalue is given by (1.5):

$$|\Psi'_m\rangle = c P_M(m) |\Psi(t_1)\rangle \tag{1.8}$$

where a normalization coefficient c has been added[17].

The rule can easily be generalized to more than two successive measurements. Each time a measurement is performed and a result obtained, the state vector jumps to a new value that includes this new information (but may also erase some previous information). One exception occurs if the same measurement is performed repeatedly, at times that are sufficiently close to avoid any Schrödinger evolution of the system between the measurements. Then all results are necessarily the same and, after the first measurement, the state vector reduction has no effect (but it becomes effective again as soon as a different observable is measured).

Clearly, state vector reduction is closely related to the Born probability rule. Actually, if one generalizes this rule to multi-time measurements (§11.1.2.a and Appendix G), one can derive the reduction of the state vector as a convenient rule to calculate probabilities. From this point of view, the state reduction no longer appears as a postulate, but just as a convenient way to make calculations that can be derived from the generalized Born rule.

To summarize, the general scheme then includes different stages in the evolution of the state vector. Between preparation and measurements, it evolves continuously according to the Schrödinger equation. When the system undergoes a measurement, it interacts with a measurement apparatus. The probabilities of the various outcomes can be calculated from the state vector in a perfectly well defined way from this equation and from the Born rule, but only the probabilities: the Schrödinger equation itself does not select a single result. The uniqueness of the

---

[16] We come back in more detail on the von Neumann model of measurement in §9.1.1 and on the notion of QND (quantum nondemolition) measurement.

[17] This coefficient is the inverse of the square root of $\langle \Psi(t_1)| P_M(m) |\Psi(t_1)\rangle$.

outcome in the quantum description of the system is associated with state vector reduction, a process that makes the state vector jump discontinuously (and randomly) to a new value. The emergence of a single result is then obtained (one could say "forced") explicitly in the state vector by retaining only the component corresponding to the observed outcome; all components of the state vector associated with the other results are put to zero, hence the name "reduction" (we will come back to this question in §2.1). The reduction process is discontinuous and irreversible. In this general scheme, separate rules and equations are therefore introduced, one for the "natural" continuous evolution of the system between measurements, another for the measurements performed on it. The difficulty is then to understand precisely how to avoid possible conflicts between these two different postulates.

1.2.2.b Bohr

Bohr does not use the notion of state vector reduction. He prefers to see the ensemble of all measurements performed at different times as being part of one single big experiment, which includes all experimental devices necessary to perform the series of measurements (and presumably also those involved in the preparation stage of the studied system as well). The rules of quantum mechanics then provide the probabilities corresponding to all possible series of results. In this view, one should not ask (as we just did in §1.2.2.a) what is the quantum state of the measured system between the first measurement and the second: separating the system from the whole experimental apparatus has no meaning in Bohr's interpretation of quantum mechanics (nonseparability, §3.3.3.c).

Any possible conflict between two different evolution postulates then disappears. It is nevertheless replaced by another difficulty since, in each experiment, one has to distinguish between two different parts: observed system(s) and measurement apparatus(es). Only the latter are directly accessible to human experience and can be described with ordinary language, as in classical physics. The measurement apparatuses and observers then have a very specific role. They provide the results of measurements to observers, so that they are at the origin of our perception of the physical world; but, at the same time, they also introduce an irreducible nondeterministic component into the theory and into the evolution of physical systems. The difficulty is then to decide where exactly to put the frontier between the two different parts. For instance, if the distinction is to be made in terms of size of the systems, one could ask from what size a physical system is sufficiently macroscopic to be considered as directly accessible to human experience, and will behave as a measurement apparatus. If the distinction is made in another way than size, then more elaborate rules should be specified to clarify this concept. For Bohr, it is not so much the size of the apparatus than its measuring function that matters; one could imagine situations where the same physical system is measured in one experiment, and then plays the role of the measurement apparatus (or some part of it) in another experiment. In [41], Bohr writes[18]: "This necessity of discriminating in each experimental arrangement between those parts of the physical system considered which are to be treated as measuring instruments and those which constitute the objects under investigation may indeed be said to form a *principal distinction between the classical and quantum-mechanical description of physical phenomena*".

1.2.3 Status

Under these conditions, what is the status of the state vector (or wave function) in standard quantum mechanics?

1.2.3.a Two extremes

When discussing the status of the state vector in standard quantum mechanics, two opposite mistakes should be avoided, since both "miss the target" on different sides. The first is to endorse the initial hopes of Schrödinger and to decide that the wave function directly describes the physical properties of the system, even if it propagates in the configuration space, which is in general distinct from (and of larger dimension than) ordinary space. In such a purely undulatory view, the classical positions and velocities of particles are replaced by the amplitude of a complex wave, and the very notion of point particle becomes diluted. Nevertheless, the difficulties introduced by this view are now so well known – see discussion in the preceding section – that nowadays few physicists seem to be tempted to support it, at least within standard theory[19]. At the extreme opposite, one considers that the wave function does not attempt to describe the physical properties of the system itself, but just the information that some observer has on it. It then becomes perfectly analogous to a classical probability distribution in usual statistical theory; the wave function should then get a relative (or contextual) status depending on the observer. Of course, at first sight, this seems to bring an elementary solution to the difficulties introduced into quantum mechanics by the state vector reduction and its discontinuities (§1.2.2): we all know that classical probabilities may undergo sudden jumps, and nobody considers this as a special problem. For instance, as soon as new information on an (possibly remote) event becomes available to us, the probability distribution that we associate with it undergoes a sudden and complete change; by analogy, is this not the obvious way to explain the sudden state vector reduction?

One first difficulty of this point of view is that it would naturally lead to a relative

---

[18] See also Bell's quotation in §2.5

[19] Within theories introducing modified Schrödinger dynamics (nonstandard versions requiring a modification of the Schrödinger equation; see §11.10), a purely undulatory point of view can indeed be reintroduced.

character of the wave function: if two observers had different information on the same system, should they use different wave functions to describe the same system[20]? In classical probability theory, distributions of probabilities that undergo sudden jumps are inherently "observer-dependent" and thus "subjective": the observer who has more information describes the phenomenon with a distribution that is narrower than that of another, less informed, observer; in principle, one can always imagine an observer with perfect knowledge of the phenomenon for whom the probability is perfectly peaked at one value (the process is then deterministic for this observer). But standard quantum mechanics rejects the possibility of such a perfect description of all properties of a system, even for an observer who has all the possible information. Most authors do not attribute such a relative character to the wave function or state vector (a different, nonstandard, point of view will nevertheless be mentioned in §11.3.2.b of Chapter 11). One should therefore consider the information contained in the state vector as universal, the maximum information on the physical system available to any human observer.

Moreover, when in ordinary probability theory a distribution undergoes a sudden "jump" to a more precise distribution, the reason is simply that more precise values of the variables already exist – they actually existed before the jump. In other words, the very fact that the probability distribution reflected our imperfect knowledge implies the possibility of a jump to a more precise description, closer to the reality of the system itself[21]. But this is in complete opposition with orthodox quantum mechanics, which rejects the notion of any better description of the reality (§3.3.2). In this theory, the random character of the measurement process is absolute; it is vain to try and explain a particular result of measurement by causes, since they do not exist. Ignorance cannot be invoked to explain the jumps of the state vector in standard quantum mechanics.

Classical probabilities can also be defined in the absence of observers and then called "objective": one considers a physical event taken among a whole ensemble of possible events, which all have well-defined but different dynamics because some of their initial physical properties differ, or because they undergo uncon-

[20] We assume that the two observers use the same reference frame for space-time; they differ only by the amount of information they have. If they used different Galilean reference frames, one should apply simple mathematical transformations to go from one state vector to the other. But this has no more conceptual impact than the usual transformations that allow us, in classical mechanics, to transform positions and conjugate momenta when changing the frame of reference.

For completeness, we should add that there is also room in quantum mechanics for classical uncertainties arising from an imperfect knowledge of the system, in particular with the formalism of the density operator (§12.1.6). Here, we limit ourselves to the discussion of state vectors and wave functions (pure states).

Finally, let us mention that, in a more elaborate discussion, one would introduce other elements, for instance the notion of intersubjectivity, etc. [10, 24].

[21] Normally, information is about something! An information, or a probability, refers to an object or an event that is supposed to have its own reality, independently of the information, and of the fact that someone is acquiring it or not – see for instance §VII of [42]. It would be very unusual to define the object by the information itself, raising questions about logical circularity.

trolled perturbations. To introduce probabilities, one then postulates that all these distinct events have the same "weight"; one then performs statistical averages on their ensemble. By contrast, standard quantum mechanics introduces probabilities only if observers make measurements; it rejects the idea of different initial properties, or perturbations, explaining the appearance of these probabilities.

Introducing the notion of preexisting values is precisely the basis of "unorthodox" theories with additional variables (often called hidden variables – see §11.8 and footnote [1]); uncontrolled perturbations are invoked in the theories with modified Schrödinger dynamics (§11.10). So, advocating this "information interpretation" for a straightforward explanation of state vector reduction amounts to advocating nonorthodox approaches to quantum mechanics. It is therefore important to keep in mind that, in the standard interpretation of quantum mechanics, the wave function (or state vector) gives *the* ultimate physical description of the system, with all its physical properties; it is neither contextual, nor observer dependent (subjective), but absolute; if it gives probabilistic predictions on the result of future measurements, it nevertheless remains inherently completely different from an ordinary classical distribution of probabilities.

The conclusion is that, if the content of the quantum state vector is defined as information, it must be realized that this is a new concept (a new definition of the word): this information is of different nature than what is usually meant by this word in the rest of physics. It is probabilistic, but the associated probabilities are not defined in relation to the knowledge of some specific observer; what is relevant is the universal information that is accessible to all human beings doing experiments in physics (intersubjectivity).

1.2.3.b The Copenhagen (orthodox) point of view

If none of these extremes is correct, how should we combine them to obtain the status of the state vector in orthodox quantum mechanics? To what extent should we consider that the wave function describes the physical system itself (realist interpretation), or rather that it contains only the information that we may have on it (positivistic interpretation, for instance), presumably in some sense that is more subtle than a classical distribution function? In orthodox quantum theory, the state vector has a really nontrivial status – actually, it has no equivalent in all the rest of physics.

The Copenhagen/orthodox interpretation is not defined in exactly the same way by all authors[22]; nuances exist between the different definitions of what is a quantum state. It may then be more appropriate to speak of the "standard" definition,

---

[22] For instance, Howard writes [40]: "Much of what passes for the Copenhagen interpretation is found in the writings of Werner Heisenberg, but not in Bohr. Indeed, Bohr and Heisenberg disagreed for decades in deep and important ways. The idea that there was a unitary Copenhagen point of view on interpretation was, it shall be argued, a post-war invention, for which Heisenberg was chiefly responsible. Many other physicists

namely that found in most textbooks. The standard view is that the state vector (or wave function) is associated with a preparation procedure of the physical system under consideration – it is then convenient to use the Heisenberg picture (§ 12.1.5), where the state vector is independent of time (the time dependence is transposed to the observables/operators in this picture). Dirac, in chapter I of [39], writes: "We must first generalize the meaning of a 'state' so that it can apply to an atomic system... The method of preparation may then be taken as the specification of the state". Or Stapp , when introducing the Copenhagen interpretation [43]: "The spec-ifications A on the manner of preparation of the physical system are first transcribed into a wave function $\Psi_A(x)$". Similarly, Peres [44] writes : "a state vector is not a property of a physical system, but rather represents an experimental procedure for preparing or testing one or more physical systems". He also makes the general com-ment: "quantum theory is incompatible with the proposition that measurements are processes by which we discover some unknown and pre-existing property". In this view, a wave function is an objective representation (independent of the observer) of a preparation procedure, rather than of the isolated physical system itself. The preparation procedure may imply the measurement of a physical quantity by some observer; for instance, one can prepare the state of a spin by sending the particle through a Stern–Gerlach magnet, and measuring the position of the particle at the output. But one can also prepare a system by letting it reach thermal equilibrium in interaction with a thermostat, as for instance is the case in most magnetic resonance experiments; then, the description of the system requires, not a single state vector, but many, which can all be summarized in a "density operator" (§12.1.6). In any case, it seems safe to associate the standard interpretation of the state vector (or density operator) with a physical preparation procedure of the system under study.

This does not prevent the state vector from containing some information about the properties of the system itself. Indeed, when a given physical quantity has been measured, and when a given result has been obtained, quantum mechanics predicts that the same result is always obtained if the same measurement is repeated just after. It is then natural to think that, between the measurements, the system had a physical property related to the certainty of the result of the second measurement – see Chapter 3. The result of measurement has then become a property of the sys-tem. This idea can be expressed mathematically: all operators acting in the space of states that take $|\Psi\rangle$ as one of their eigenvectors correspond to physical quantities that have precise values in this state, and can be associated with properties of the system. But, for a quantum system in a given state $|\Psi\rangle$, most operators are not in this case; only a very small proportion of the properties that could be defined in classical mechanics can still be defined quantum mechanically; moreover, the list

and philosophers, each with his own agenda, contributed to the promotion of this invention for polemical or rhetorical purposes".

of those that are defined is not fixed, but depends on $|\Psi\rangle$. Hartle proposes the following definition [45]: "The state of an individual system in quantum mechanics is, therefore, defined as the list of all propositions (concerning the individual system) together with their truth values – true, false, or indefinite".

At the end, we reach a sort of intermediate situation where neither the pure continuous undulatory interpretation nor the pure probabilistic interpretation is correct, while elements of both points of view are retained. A preparation procedure for a physical system is indeed an objective fact, and if the state vector describes such a procedure, it also necessarily has some objective component – it cannot be entirely mental and observer dependent. Sometimes some properties of the system exist, sometimes others, but most do not exist, so that their measurement will provide random results. Both interpretations are combined in a way that emphasizes the role of the whole experimental setup. Bohr described the situation by using the general concept "complementarity"; for instance, in [46], he writes "the viewpoint of complementarity presents itself as a rational generalization of the very idea of causality" (how to relate in a causal way a preparation procedure to an observation procedure).

Concerning probabilities, standard quantum mechanics makes use of a notion of probability that does not refer to the knowledge of some particular observer, but to some universal information that is shared by all physical observers, independently of their personal identity. These probabilites are not subjective, but "intersubjective". The sudden discontinuous jumps of these probabilites during measurements are of a different nature than the sudden jumps of a classical distribution of probability, when new information is acquired.

1.3 Other formalisms, field theory, path integrals

Quantum mechanics can be expressed in various forms and with different equations; these forms are basically equivalent, but may turn out to be more or less convenient, depending on the physical system under study. Since the initial introduction of the Schrödinger equation in 1926, many developments and generalizations have been proposed. It is of course completely out of the question to describe all of them here; the purpose of this brief section is only to mention a few of them that are especially important.

A crucial step in the development of quantum mechanics was the inclusion of the notion of identical particles. They can be either bosons (photons, helium 4 atoms, etc.) or fermions (electrons, protons, etc.), and obey different rules for the symmetry of their state vector under exchange of identical particles: bosons have a completely symmetric state, while fermions have an completely antisymmetric state vector. A direct consequence is that two fermions can never occupy the same individual quantum state, as postulated by Pauli in 1925 with his exclusion principle (the principle was initially introduced for electrons in atoms [47], but then generalized to all fermions). Later, Dirac [48], Fock [49], and Jordan [38, 50] proposed a very convenient treatment of identical particles with the use of creation and annihilation operators, which automatically take into account these symmetry rules, for both fermions and bosons; this method is sometimes called "second quantization" for historical reasons.

Dirac's initial motivation was the quantization of the electromagnetic field and the derivation of Einstein's coefficient of absorption and emission within quantum theory. Since then, the formalism of creation and annihilation operators has become the basic tool of a whole branch of quantum mechanics called quantum field theory, which applies to any field (not only the electromagnetic field). We have already mentioned at the end of §1.1.3 that quantum field theory is very successful and has many applications, especially in elementary particle theory, cosmology, etc. Quantum field theory can be made fully relativistic (compatible with Einstein's special relativity), as opposed to the usual Schrödinger equation (which assumes an absolute time, as in Galileo relativity). In field theory, one often prefers to take a point of view where the state vector remains constant, while the time evolution is transferred to the field operators, using what is called the "Heisenberg picture". In this way, one obtains propagation equations for the field operators that look similar to those of a classical field, except of course that the functions appearing in the equations are operators instead of ordinary functions.

Nevertheless when, at the end of the calculation, one needs to obtain a probability, the state vector is still needed to write the average value of a projector; the state vector therefore remains an essential ingredient in this version of quantum mechanics. Moreover, in many branches of atomic, molecular, and condensed matter physics, relativistic effects are actually negligible, so that the basic tool of quantum mechanics is the Schrödinger equation in these domains. From a fundamental point of view, using the Schrödinger or the Heisenberg picture does not change much to the interpretation of quantum mechanics and its conceptual difficulties (see also the discussion of § 5.3.2). One can even say that quantum field theory has as many, if not more, problems of interpretation as elementary quantum mechanics. This is especially true if one wishes to include gravity, which actually adds several specific problems (treatment of quantum space-time).

Feynman has extended Dirac's ideas to introduce another quantization procedure, called the "Feynman path integral" [51, 52]. In classical mechanics, a Lagrangian describing the motion of a physical system can be used to define the conjugate variable of each configuration variable, and then to introduce a Hamiltonian. The standard quantization procedure starts from this Hamiltonian and applies quantization rules to all pairs of conjugate variables, which become noncommuting operators. Nevertheless, Feynman showed that another method can be used: it is also possible to directly start from the Lagrangian, calculate classical integrals of this Lagrangian along various paths called "actions", and to obtain quantum probability amplitudes by summing over all possible classical paths. He showed that this procedure leads to results that are equivalent to those obtained from the standard Schrödinger equation. In this point of view, one directly obtains probability amplitudes connecting various points of configuration space at different times; one does not have to apply the Born rule to calculate the product between a bra and a ket, to use operators and diagonalize them, etc. This approach provides an interesting illustration of the relation between classical and quantum mechanics, which is analogous to the relation between geometrical and wave optics. Moreover, it turns out to be more useful in some cases, for instance when the available Lagrangian does not allow one to define conjugate variables for all configuration variables, so that no Hamiltonian can be built. It really provides a new quantization procedure that can be used in many situations, for instance in attempts to combine quantum mechanics and gravity (general relativity).

2

Present Situation, Remaining Conceptual Difficulties

Conceptual difficulties still remain in quantum mechanics, even if they had already been identified by its inventors. This does not mean that the theory is not successful! The reality is quite the opposite: in fact, independently of these difficulties, quantum mechanics is probably the most successful theory of all science. One can even consider that its ability to adapt to new situations is one of its most remarkable features. It continues to give efficient and accurate predictions while new experiments are performed, even in situations that the founding fathers had no way to imagine. Actually, there is no other theory that has been verified with the same accuracy in so many situations. Nevertheless, it remains true that conceptual difficulties subsist, and their discussion is the object of this chapter. As we will see, most of them relate to the process of quantum measurement, in particular to the very nature of the random process that takes place on this occasion.

We have seen that, in most cases, the wave function evolves gently, in a perfectly predictable and continuous way, according to the Schrödinger equation; in some cases only, when new information is obtained by measurement, unpredictable changes of the state vector take place according to the von Neumann postulate of state vector reduction. Obviously, having two different postulates for the evolution of the same mathematical object is very unusual in physics[1]. The notion was a complete novelty when it was introduced, and still remains unique, but also the source of difficulties – in particular, logical difficulties related to the compatibility between the two different postulates. Normally, one would tend to see a measurement process, not as very special, but just as an interaction process between a (possibly

---

[1] Instead of the "Schrödinger picture", it is equivalent to use the "Heisenberg picture" (§12.1.5). The Schrödinger evolution is then transferred to the observables of the system (operators acting in its space of states), while the state vector remains constant in the absence of measurement. The observables always evolve continuously, whether or not measurements are performed. But two different rules are still obeyed by the state vector: most of the time, it remains constant, but when a measurement occurs it undergoes a discontinuous jump to a new value accounting for the new information. The acquisition of information is then still considered as a completely different physical process than "normal" evolution, creating the need for a precise definition of the border between the two processes.

microscopic) measured system and a macroscopic measurement apparatus. The latter should be treated within the ordinary laws of physics: indeed, a theory where no distinction is necessary between normal evolution and measurements would seem much more general, and therefore preferable. In other words, why give such a special character to measurements so that two separate postulates become necessary? Where exactly does the range of application of the first postulate stop in favor of the second? More precisely, among all the interactions – or perturbations – that a physical system can undergo, which ones should be considered as normal (Schrödinger evolution), and which ones are a measurement (state vector reduction)? We need to understand better the reason that physics needs to introduce a split between two different worlds, and the nature of this split: Schrödinger's world, which is perfectly continuous and deterministic, and Born's measurement world, which is very different since the continuous evolution stops to be replaced by nondeterministic and discontinuous processes.

From Bohr's point of view also, measurement processes and apparatuses are clearly "discriminated" from other processes and physical systems. In Bohr's universe, in the absence of measurement, a general evolution takes place in a continuous and deterministic way according to the Schrödinger equation. But, in the particular case of events involving the interaction between a microscopic quantum system and a setup especially designed to transfer information to a macroscopic observer, an inherent randomness appears in the evolution. These measurement processes are, so to say, considered as "closed bubbles" inserted within this general evolution, closed events extending over a whole region of space-time, from their beginning to their end. They cannot be decomposed into more detailed relativistic events, and are fundamentally characterized by the fact that an intelligent human being is asking a question to Nature; the outcome is a unique answer, but nondeterministic.

Whether we prefer von Neumann's or Bohr's point of view, we are faced with a logical problem that did not exist before, in classical mechanics, when nobody thought that measurements providing information should be treated as special processes in physics. We learn from Bohr that we should not try to transpose our experience of the everyday world to microscopic systems; then, for each experiment, where exactly is the limit between these two worlds? In von Neumann's approach also, it becomes necessary to introduce a "split" between the measured system and the measurement apparatus, or between a standard quantum evolution and a measurement process. It has often been argued that the precise position of this split is irrelevant in practice, since it does not affect the physical predictions. It remains nevertheless true that, under these conditions, the theory is not perfectly well defined by its postulates. Is it really sufficient to remark that the distance between microscopic and macroscopic is so huge that the exact position of the frontier between them is unimportant? Bell (and others) complained about the shifty character of this division [53]; in the words of Mermin [54], "Bell deplored a shifty split that haunts quantum mechanics". As we will see in Chapter 11, the purpose of some other interpretations of quantum mechanics is to suppress the necessity of this split.

Moreover, it may look very surprising that, in modern physics, the "observer" and her/his acquisition of information should play such a central role, giving to the theory an unexpected anthropocentric foundation, as in astronomy in the Middle Ages. Should we really reject the description of isolated systems as unscientific, just because we are not observing them? If observers are so important, how precisely is an observer defined? For instance, can an animal perform a measurement and reduce the state vector, or is a human being required? Bell once asked with humour [55]: "Was the world wave function waiting to jump for thousand of millions of years until a single-celled living creature appeared? Or did it have to wait a little longer for some highly qualified measurer, – with a Ph.D.?" – see also [56], in particular its title. These general questions are difficult, somewhat philosophical, and we will come back to them in Chapter 11 where, for instance, we mention that London and Bauer [57] have suggested that it is the faculty of introspection that qualifies a living creature to be an observer in quantum theory; see also §11.3.2.b.

Another difficulty is related to the random character of the predictions of quantum mechanics, which provides only probabilities for the results of measurements. In itself, this situation is relatively usual in physics, where for instance statistical classical mechanics makes constant use of probabilities. But the probabilities are then associated with the notion of statistical ensemble: one assumes that the system under study is chosen randomly among many similar systems belonging to the same ensemble. Indeed, they share common properties contained in the statistical description, but a more accurate description can reveal that, at some finer level, they also differ by other individual properties. In other words, every single system has more physical properties than those specified for the whole ensemble. The same question can immediately be transposed to quantum mechanics: if its predictions are only probabilities, does it mean that it describes ensembles of systems only? Or should we consider that the state vector provides the most accurate possible description of a single system? We will often come back to this question in what follows, in particular in Chapters 3 and 11.

A general question is: should we accept as perfectly valid a physical theory that is predictive but not descriptive (a theory that makes good predictions, but provides no description of the physical events)? Many such questions have been discussed, and it is probably impossible to summarize all that has been written on the exact behavior and role of the state vector in a reasonable number of pages. As illustrations, we will discuss a few examples in this chapter: von Neumann's infinite regress (§2.1), Schrödinger's cat (§2.2), Wigner's friend (§2.3), and negative measurements (§2.4). We will then recount a number of quotations (§2.5), which give an idea of the variety of possible positions on quantum mechanics, and finally mention a few points where errors have been made in the past (§2.6).

2.1 Von Neumann's infinite regress/chain

Von Neumann, in a treatise published in 1932, introduced an explicit theory of quantum measurement (chapters 4–6 of [4]). In contrast with Bohr's approach, von Neumann considers the measuring apparatus as a quantum system. He assumes that the measured system S is put into contact with a measurement apparatus M and interacts with it for some time. M contains a macroscopic "pointer" P that, after the interaction has finished, reaches a position that depends on the initial state of S. In this chapter, we introduce the general ideas related to the von Neumann infinite chain, without writing equations; we come back in §9.1 to von Neumann's theory of measurement with more detail – see also, for instance, §9.2 of [21] and §11.2 of [58].

Assume that, initially, this system was in an eigenstate of the measured observable A characterized by one of its eigenvalues[2] a. After the end of the interaction, the pointer reaches a specific position that depends on the eigenstate; one can consider that observing the macroscopic position of P amounts to measuring the observable and obtaining a as a result.

Now, in general S is not initially in an eigenstate of the measured observable A, but in a superposition of such states. The linear Schrödinger equation then predicts that the whole physical system S+M reaches a linear superposition[3] of states after the interaction. In the different components of this superposition, the pointer has several different positions, each corresponding to a different result of measurement; this situation is described as quantum entanglement (Chapter 7). Indeed, because the Schrödinger equation is linear, it cannot make a selection between these results: it can only lead to a superposition of all possible outcomes, as if no macroscopic result at all had emerged from the measurement!

To solve this problem, one can then try to add a second stage, and use another measurement apparatus M′ in order to determine the position of the pointer of M. But the process repeats itself, and the linearity of the Schrödinger equation leads to an even stranger superposition, this time containing S+M+M′ and different positions of the pointers of M and M′. By recurrence, if one adds more and more measurement apparatuses, M″ etc., one creates a longer and longer chain of correlated systems, without ever selecting one single outcome for the measurement.

---

[2] Here, for the sake of simplicity, we assume that the eigenvalue is not degenerate, but the generalization to degenerate eigenvalues is possible.

[3] This linear superposition is written explicitly in (9.7).

This recurrent process M, M', M" is sometimes called the von Neumann infinite regress, or chain. One can summarize this discussion by saying that "Uniqueness of the results cannot emerge from the Schrödinger equation only; in fact the equation creates a chain of coherent superpositions that propagates without any end".

A simple example may be useful to illustrate the process in a more concrete way. Assume for instance that we consider a spin 1/2 atom, which enters into a Stern–Gerlach spin analyzer. If the initial state of the spin is parallel (or antiparallel) to the direction of the analyzer (the direction of the magnetic field that defines the eigenstates associated with the apparatus), the wave function of the atom is deflected upwards or downwards, depending on the initial eigenvalue of the longitudinal spin. But if the initial direction of the spin is transverse, the wave function splits into two different wave packets, one pulled upwards, the other pushed downwards; again, this is an elementary consequence of the linearity of the Schrödinger equation. Propagating further, each of the two wave packets may strike a detector, with which they interact by modifying its state (as well as theirs). For instance, the incoming spin 1/2 atoms are ionized and produce electrons; as a consequence, the initial coherent superposition now encompasses new particles. Moreover, when a whole cascade of electrons is produced, as happens in a photomultiplier, all these additional electrons also become part of the superposition. In fact, there is no intrinsic limit in what soon becomes an almost infinite chain: rapidly, the linearity of the Schrödinger equation leads to a coherent superposition of states that includes a macroscopic number of particles, macroscopic currents, and, maybe pointers or recorders printing macroscopic zeros or ones on a piece of paper! If we stick to the Schrödinger equation, there is nothing to stop this infinite regress, which has its seed in the microscopic world but rapidly develops into a big macroscopic phenomenon. Can we for instance accept the idea that, at the end, it is the brain of the experimenter (who becomes aware of the results) and therefore a human being with consciousness who enters into such a superposition?

The very notion of a brain, or consciousness, in a macroscopic superposition is neither very intuitive nor clear; no one has ever observed two contradictory results at the same time. Would this strange situation correspond to an experimental result printed on paper and looking more or less like two superimposed slides, or a double exposure of a photograph? In practice, we know that we always observe only one single result in a single experiment; linear superpositions somehow resolve themselves before they reach us, and presumably before they become sufficiently macroscopic to involve measurement apparatuses. It therefore seems obvious[4] that a proper theory should break the von Neumann chain at some point, and stop the regress when (or maybe before) it reaches the macroscopic world. This operation

---

[4] Maybe not so obvious after all? There is an interpretation of quantum mechanics that precisely rests on the idea of never breaking this chain: the Everett interpretation, which will be discussed in §11.12.

is often called the "Heisenberg cut". But when exactly and how precisely should this cut be made?

Wigner [59] draws the following conclusion from this analysis: "it is not possible to formulate the laws of quantum mechanics in a complete and consistent way without reference to human consciousness". Indeed, Von Neumann considers that the emergence of the perception of a single result of measurement is an irreducible element of the theory[5]. The solution is the introduction of a special postulate: the "projection postulate" of the state vector (§1.2.2), which forces the emergence of a single result as soon as an experimenter becomes aware of the result.

The notion of the von Neumann regress is also the source of the phenomenon of decoherence (§7.3.3). The word *decoherence* is usually used to refer to the initial stage of the chain, when the number of degrees of freedom involved in the process is still relatively limited. The von Neumann chain is more general and includes this initial stage as well as its continuation, which goes on until it reaches the other extreme where it really becomes paradoxical: the Schrödinger cat.

2.2 Schrödinger's cat, measurements

The famous story of the Schrödinger cat (1935) illustrates the same problem in a different way.

2.2.1 The argument

The cat appears in a few lines only in the context of a more general discussion written by Schrödinger [60] and entitled "The Present Situation of Quantum Mechanics". His words (after translation into English [61]) are: "One can even set up quite ridiculous cases. A cat is penned up in a steel chamber, along with the following device (which must be secured against direct interference by the cat): in a Geiger counter, there is a tiny bit of radioactive substance, so small that perhaps in the course of the hour, one of the atoms decays, but also, with equal probability, perhaps none; if it happens, the counter tube discharges, and through a relay releases a hammer that shatters a small flask of hydrocyanic acid. If one has left this entire system to itself for an hour, one would say that the cat still lives if meanwhile no atom has decayed. The first atomic decay would have poisoned it. The

---

[5] In his words ([4], pages 418-421): "... the measurement or the related process of the subjective perception is a new entity relative to the physical environment and is not reducible to the latter. Indeed, subjective perception leads us into the intellectual inner life of the individual, which is extra-observational by its very nature ... Nevertheless, it is a fundamental requirement of the scientific viewpoint – the so-called principle of the psycho-physical parallelism – that it must be possible so to describe the extra-physical process of the subjective perception as if it were in reality in the physical world – i.e., to assign to its parts equivalent physical processes in the objective environment, in ordinary space".

psi-function of the entire system would express this by having in it the living and dead cat (pardon the expression) mixed or smeared out in equal parts.

It is typical of these cases that an indeterminacy originally restricted to the atomic domain becomes transformed into macroscopic indeterminacy, which can then be resolved by direct observation. That prevents us from so naively accepting as valid a 'blurred model' for representing reality. In itself, it would not embody anything unclear or contradictory. There is a difference between a shaky or out-of-focus photograph and a snapshot of clouds and fog banks."

In other words, Schrödinger considers a von Neumann chain starting from one (or a few) radioactive atomic nucleus. The device is designed so that, if the nucleus emits a photon, this photon is detected by a gamma ray detector, which provides an electric signal, which then undergoes an electronic amplification, and triggers a macroscopic mechanical system that automatically opens a bottle of poison and eventually kills the cat (Figure 2.1). This is what happens in the branch of the state vector where a photon has been emitted; but none of these events takes place in the branch where no photon is emitted. When the probability that a photon has been emitted is about 1/2, the system reaches a state with two components of equal weight, one where the cat is alive and one where it is dead. The equations of evolution therefore predicts that the cat is at the same time alive and dead, instead of being alive or dead ("and-or question"; see note [11]). Schrödinger considers this coexistence of totally different states of the cat as an obvious impossibility (a quite ridiculous case), and concludes, therefore, that something must have happened to stop the von Neumann chain before it went too far. Again, the challenge is to explain macroscopic uniqueness: why, at a macroscopic level, a unique result (the alive, or dead, cat) emerges, while this does not happen within the linear Schrödinger equation.

The cat is of course a symbol of any macroscopic object – Einstein, in a letter to Schrödinger the same year [62], used the image of the macroscopic explosion of a barrel of powder[6]. Such an object can obviously never be in a "blurred" state containing possibilities that are contradictory (an open and closed bottle, a dead and alive cat, etc.). Therefore, as Schrödinger points out, his equation should not be pushed too far and include macroscopic objects. Standard quantum mechanics is not only incapable of avoiding these paradoxical cases, it actually provides a general recipe for creating them. The logical conclusion is that some additional ingredients are needed in the theory in order to select one of the branches of the superposition, and avoid such stupid macroscopic superpositions [7]. Needless to

---

[6] Einstein writes "No clever interpretation will be able to transform this function Ψ into an adequate description of real things; in reality, there is nothing between exploded and nonexploded".

[7] It is amusing to note that, historically, Schrödinger's name is associated to two somewhat opposite concepts. One is contained in a very powerful continuous equation of evolution that applies to all systems; the other is the cat, which is the symbol of a limit that the same equation should never reach.

$$|\Psi\rangle$$

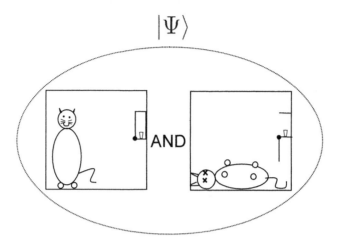

AND

Figure 2.1 The story of the Schrödinger cat illustrates how the rules of standard quantum mechanics may lead to very strange situations, where macroscopic objects are at the same time in completely different physical states. Schrödinger imagines a cat in a box where a device is triggered by the emission of a particle by a radioactive nucleus. If the particle is emitted, the device starts and opens (or breaks) a bottle of poison, which kills the cat. If the particle is not emitted, the cat remains alive. After some time, when the probability of decay of the nucleus is $1/2$, the state $|\Psi\rangle$ of the whole system contains components associated with completely different macroscopic situations, both realized at the same time. Uniqueness of macroscopic reality would require that the "AND" in the middle of the figure should be replaced by "OR", which cannot be explained within the Schrödinger equation. In the words of Schrödinger, this is a "ridiculous" situation.

say, the limit of validity of the linear equation does not have to be related to the macroscopic object itself: the branch selection process may perfectly take place before the linear superposition reaches the cat. But the difficult question is exactly where and when the process takes place.

The usual interpretation of quantum mechanics postulates that the only way to stop the linear propagation of the Schrödinger equation and break the chain is to perform an act of measurement. The question then becomes: does an elaborate animal such as a cat, or a primitive living creature such as a bacterium, have the intellectual abilities that are necessary to perform a measurement and resolve several von Neumann branches into one? At what stage of evolution can a living creature perceive its own state, projecting itself onto one of the alive or dead states? Or do humans only have access to a sufficient level of introspection to become conscious of their own observations, and to reduce the state vector? Some theories take this point of view, a case in which, when the wave function includes a cat component, the animal could remain simultaneously dead and alive for an arbitrarily long period of time, a paradoxical situation indeed.

The last sentence of Schrödinger's quotation concerning photographs has often been considered as obscure. Schrödinger probably wishes to emphasize the difference between an incomplete knowledge (an out-of-focus photograph) of a well-defined object, and an object that is inherently not sharply defined in space (a cloud) – between an indeterminacy that is related to lack of information or an inherent indeterminacy. In other words, he is already questioning the complete character of quantum mechanics (Chapter 3).

2.2.2 Misconceptions

A common misconception is that the paradox is easily solved by just invoking decoherence (§7.3.3), which explains why it is impossible in practice to observe quantum interferences between states where a cat is alive or dead. We come back to this point in §7.3.3.b in more detail, and discuss it only briefly here. Actually, (de)coherence is irrelevant in Schrödinger's argument: the cat is actually a symbol of the absurdity of a quantum state that encompasses two incompatible possibilities in ordinary life, coherent or not. It does not change the absurdity of the final situation whether the state in question is a pure state (sensitive to decoherence) or a statistical mixture (insensitive to decoherence). Actually, the standard evolution of the state vector, including decoherence, does not change the norms of any of the two components (the components where the cat is alive or dead): it only creates more and more ramifications inside both these components, without ever changing any of the two norms, which give the probability of survival of the cat. Moreover, the cat itself is already part of the environment of the radioactive atom (the detector and the bottle of poison are also in this case)[8]. The chain (the tree of possibilities) starts to propagate at a microscopic level (from the radioactive atom) and continues further and further without apparent limits; the real difficulty is to stop it from reaching the macroscopic world. It certainly does not help to remark that the chain propagates even further than the cat; invoking decoherence is not answering the question, it is repeating it.

The real problem addressed by Schrödinger is how to explain the uniqueness

---

[8] The cat itself is never in a coherent superposition of alive and dead states. When the curious state is created, the cat is already correlated (entangled) with the radioactive source, the mechanical system, the bottle of poison (open or closed), the gas in the box, etc. All these components already act as an environment and produce complete decoherence. Restoring this coherence would imply an operation of putting back all of them into the same quantum state, a clearly impossible task. The propagation of decoherence further into the environment does not add anything new to the argument.

of the macroscopic world. How can something that was indeterminate become determinate, and what is the process responsible for this? Until what point exactly does the theory remain deterministic, to become stochastic beyond this point? This question is not trivial and much more difficult to answer than just explaining the absence of macroscopic coherence. The impossibility of observing interferences between dead and alive cat states is, of course, a necessary condition for macroscopic uniqueness; but it is also far from sufficient. Schrödinger was well aware of the properties of "entanglement" in quantum mechanics (Chapter 7), a word that he introduced (and uses explicitly in the article on the cat; see §3.3.3.c). He was certainly not sufficiently naive to think that dead and alive cats can give rise to interferences. In [60], he actually never mentions the coherent or incoherent character of the superposition (the words do not appear even once in his article): for him, the question is not to avoid a coherent superposition of macroscopically different states, it is to have no superposition at all of such states[9]!

In 1952, he repeated his opinion [63] by giving a picturesque description of the situation in which a theoretical physicist would find himself in the absence of any limit put on the Schrödinger equation: "Nearly every result [a quantum theorist] pronounces is about the probability of this *or* that *or* that ... happening with usually a great many alternatives. The idea that they be not alternatives but all really happen simultaneously seems lunatic to him, just *impossible*. He thinks that if the laws of nature took *this* form for, let us say, a quarter of an hour, we should find our surroundings rapidly turning into a quagmire, or sort of a featureless jelly or plasma, all contours becoming blurred, we ourselves probably becoming jelly fish ". For Schrödinger, no sensible theorist could take this possibility seriously.

2.2.3 Relation with the measurement problem

The Schrödinger cat paradox is closely related to what is often called "the measurement problem" in quantum mechanics. In the introduction of this chapter, we have already pointed out the difficulties arising from an ill-defined border between two postulates of evolution of the state vector. The difficulty arises because the Schrödinger equation predicts that the pointer of the measurement apparatus, which is supposed to indicate the result, does not reach a well-defined position after a measurement. In general[10], the pointer reaches several positions at the same time, and macroscopic uniqueness is not obtained, exactly as for the cat. The situation is reminiscent of a river dividing into several arms: water flows at the same time in

---

[9] This is for instance the purpose of theories with a modified nonlinear Schrödinger dynamics (§11.10): they provide equations of motion where, at a macroscopic level, all probabilities dynamically go to zero, except one that goes to 1.

[10] A single position is nevertheless reached in the special case where the measured system is initially in an eigenstate of the measurement (§12.1.3); in this case, the result is certain, and no random process takes place.

one arm *and* another, not in one arm *or* another[11]. Of course, if several positions of the pointer are simultaneously realized, no well-defined result emerges from the experiment, which loses its usual meaning.

To obtain a theory where well-defined results emerge at a macroscopic level from the experiments, it is therefore necessary to stop the linear Schrödinger dynamics at some point. Among all possibilities that are predicted by the equation, one needs to make a selection by introducing a postulate, or a mechanism, that makes them exclusive instead of simultaneous. Generally speaking, probabilities are indeed defined as numbers characterizing exclusive events: if a die is thrown, it shows a single number between 1 and 6 when it stops, not several results at the same time; otherwise, the notion of probability of occurrence of the results would lose its meaning. Similarly, in quantum mechanics, to interpret the numbers obtained simultaneously from the Born rule into physical probabilities, one needs a way to introduce exclusiveness. As we will see in Chapter 11, various interpretations of quantum mechanics use different methods for this purpose.

2.2.4 Modern cats

The meaning of words sometimes changes in physics. In the recent literature in quantum electronics and optics, it has become more and more frequent to use the words "Schrödinger cat (SC)" with a different meaning. Initially, the cat was the symbol of an impossibility, an animal that can never exist (a Schrödinger gargoyle?); it was the final step of a "reductio ad absurdum" reasoning, in short, something that obviously has never been (and never will be) observed. But, nowadays, the same words are often used to describe states that are actually perfectly accessible physically, namely any coherent superposition of states that are more or less macroscopically distinguishable (the coherent character is then essential). In this new sense, Schrödinger cats have been predicted and observed in a variety of systems, for instance a large molecule propagating at the same time in different arms of an interference experiment [64], or with an ion located in two different places in a trap. Of course, decoherence may happen (§7.3.3) through correlation to the environment, often very rapidly. Theoretical calculations of this decoherence are possible within the Schrödinger equation, which is used to calculate how the initial stages of the von Neumann chain take place, and how rapidly the state vector tends to ramify into branches containing the environment.

To summarize §2.2, the paradox addresses the core of most of our difficulties with quantum mechanics: as Wigner writes [65] "measurements which leave the

---

[11] D'Espagnat speaks of "the and-or problem" to emphasize the difference between a coexistence and an alternative (end of §10-1 of Reference [25]). Bell expresses a similar idea in the quotation given in footnote [20] of page 211.

system object-plus-apparatus in one of the states with a definite position of the pointer cannot be described by the linear laws of quantum mechanics". The question is then: what is exactly the process that forces Nature to break the linearity and to make its choice among the various possibilities for the results of experiments? Indeed, the emergence of a single result in a single experiment is a major issue. As Pearle expresses it concisely [14], the problem is to "explain why events occur"!

2.3 Wigner's friend

In a theory such as standard quantum mechanics, where the observer plays such an essential role, who is entitled to play it? Wigner discusses the role of a friend, who has been asked to perform an experiment, a Stern–Gerlach measurement for instance [59]. The friend is working in a closed laboratory, so that an outside observer cannot be aware of the result before he/she opens the door (Figure 2.2). What happens just after the particle has emerged from the analyzer and when its position has been observed inside the laboratory, but is not yet known outside? For the outside observer, it is natural to consider the whole ensemble of the closed laboratory, containing both the experiment and his friend, as the "system" to be described by a big wave function. As long as the door of the laboratory remains closed and the result of the measurement unknown, this state vector will continue to contain a superposition of the two possible results; it is only later, when the result becomes known outside, that the state vector reduction should be applied. But, clearly, for Wigner's friend, who is inside the laboratory, this reasoning is just absurd! He/she will much prefer to consider that the state vector is reduced as soon as the result of the experiment is observed inside the laboratory. We are then back to a point that we have already discussed (§1.2.3), the absolute/relative character of the wave function: does this contradiction mean that we should consider two state vectors, one reduced, one not reduced, during the intermediate period of the experiment[12]? For a discussion by Wigner of the problem of the measurement, see [65].

An interpretation, sometimes associated with Wigner's name[13], assumes that the reduction of the state is a real effect taking place when a human mind interacts with the surrounding physical world and acquires some consciousness of its state; we will come back to it in §11.1.1.b. In other words, the electrical currents in the human brain may be associated with a reduction of the state vector of measured objects by some as yet unknown physical process. Of course, in this view, the re-

---

[12] Hartle considers that the answer to this question is yes [45]; see also the relational interpretation of quantum mechanics (§11.3.1).

[13] The title of [59] is indeed suggestive of this sort of interpretation; moreover, Wigner writes in this reference that "it follows (from the Wigner friend argument) that the quantum description of objects is influenced by impressions entering my consciousness". At the end of the article, he also discusses the influence of nonlinearities which would put a limit on the validity of the Schrödinger equation, and be indications of life.

Figure 2.2 Wigner assumes that a physicist outside of a closed laboratory has a friend inside who performs a quantum experiment providing results $A = \pm1$. When the friend observes a result, for instance $A = +1$, he describes the physical system with a state vector that includes this information and therefore has undergone state vector reduction. Nevertheless, as long as the door of the laboratory remains closed, the physicist outside should describe the ensemble of the experiment and of his friend with a state vector still containing all different possible results; for him, state vector reduction occurs only when he opens the door and becomes aware of the result. One then reaches a situation where the same physical reality is described at the same time by two different state vectors. To avoid this, one may assume that the first perception of the result by the mind of any observer (the friend in this case) reduces the state vector; the system then has a single state vector even if for some time it is not known by the experimenter outside.

duction takes place under the influence of the experimentalist inside the laboratory, and the question of the preceding paragraph is settled. But, even if one accepts the somewhat provocative idea of possible action of the mind (or consciousness) on the environment, this point of view does not suppress all logical difficulties: what is a human mind; what level of consciousness[14] is necessary to reduce the state, how do the electric currents in the brain act, etc.?

---

[14] See in §11.1.1 for a relation between consciousness and introspection (London and Bauer).

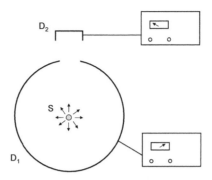

Figure 2.3 A source S surrounded by a large detector D_1 emits a particle. This detector captures the particle and provides a signal to a first recording device, except if the particle escapes through a hole corresponding to a small solid angle (in the upper direction in the figure). In this case, the particle is necessarily detected by the second detector D_2 sitting at a larger distance and providing a signal to a second recording device. In the discussion, it is assumed that the detectors are ideal and have 100% efficiency.

2.4 Negative and "interaction-free" measurements

Other paradoxical situations occur with "negative measurements" or "interaction-free measurements" in quantum mechanics [66–69]. These concepts may be illustrated with the following examples.

2.4.1 *Einstein's paradox*

Already at the Solvay Congress of 1927, Einstein proposed a Gedanken Experiment (thought experiment) that involves interaction free and nonlocal influences (page 116 of [58]). Consider a source that emits one particle with a spherical wave function (its values are independent of the direction in space). The source is surrounded by two detectors, as shown in Figure 2.3; one of the detectors, D_1, captures the particle emitted in almost all directions, except a small solid angle Ω; the second detector D_2 captures the particle inside this solid angle, but at a larger distance. We assume a perfect situation where both detectors have 100% efficiency.

Suppose that the wave packet describing the wave function of the particle reaches the first detector. Two possibilities may occur:

• Either the particle is detected by D_1. It then disappears, and the state vector is projected onto a state containing no particle and an excited detector (the first); later, the second detector D_2 will never record a particle. This first possibility occurs in most realizations of the experiment.

• Or the particle is not detected. This event occurs with a small probability, given by the solid angle subtended by the hole in the first detector seen from S, divided by 4π (we assume that the source is isotropic), but it sometimes does occur. Then only the fact that the first detector has *not* recorded the particle implies a reduction of the wave function to its component contained within Ω, implying that the second detector will *always* detect the particle later. We then have a curious situation where the probability of detection at this second detector has been greatly enhanced by a sort of "nonevent" at the first detector, that is, without any interaction between the particle and the first measurement apparatus. This illustrates that the essence of quantum measurement is something much more subtle than the often invoked "unavoidable perturbations of the measurement apparatus" (Heisenberg microscope, etc.); state vector reduction may take place even when the interactions play no role in the process.

Of course, if one assumes that, initially, the emitted particle already had a well-defined direction of emission, then the interpretation of this thought experiment becomes straightforward: detector D_2 records only the fraction of the particles that were emitted from the beginning in its direction. It is then not surprising that no interaction at all occurred with detector D_1. But standard quantum mechanics postulates that this well-defined direction of emission does not exist before any measurement. Assuming that it does amounts to the introduction of additional variables to quantum mechanics, a subject that we will discuss in much more detail below (Chapter 3 and §11.8), but which does not belong to the standard interpretation of quantum mechanics.

2.4.2 Single localization of a single particle

A similar experiment is shown in Figure 2.4: a source S emits particles one by one, each described by a Schrödinger wave propagating to the right. This wave interacts with a beam splitter device that split it into two parts (photons, neutrons, and other particles can be split in this way with appropriate devices). After propagating over some distance L_1 or L_2, each of the waves crosses a detector, D_1 in one direction, D_2 in the other. In each realization of the experiment, one single particle is emitted by S and detected, either at D_1 or D_2. In the pure Schrödinger wave description of the experiment, the situation remains perfectly symmetric: at both detectors, a von Neumann chain occurs, including states of the detectors where the particle has or has not been registered. More precisely, the state of the whole system is the sum of two kets, each containing a chain: the first contains the chain associated with a detection of the particle at D_1 while nothing happens at D_2, and the other the chain

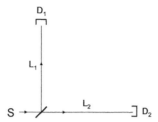

Figure 2.4 Particles are emitted one by one by a source S. The Schrödinger waves describing the particles are separated into two parts by a beam splitter; each part propagates to a detector, D_1 or D_2. In an ideal experiment, in each run of the experiment, the particle is always detected, either at D_1 or at D_2 in a perfect random way; the particle is never detected at both places.

associated with the reverse situation. Quantum mechanics predicts that a single result is observed: for a given realization, the particle is seen randomly either at D_1 or at D_2, but never at both places. If $L_1 < L_2$, the measurement at D_1 occurs first, and we have an effect of negative measurement again: if D_1 does not detect the particle, it is always detected at D_2. Conversely, when it is detected at D_1, it is certain that it will never be detected at D_2. Here we get a first flavor of quantum nonlocality: a detection at D_1 makes a detection at D_2 immediately impossible, even if the distance between the detectors is arbitrarily large (so that no information propagating at the speed of light has the time to propagate from one to the other). The resolution of the von Neumann chain into one single branch at one of the detectors (state vector reduction if one prefers) is a nonlocal phenomenon; it has instantaneous consequences on measurements performed at an arbitrary distance.

As before, the same results could easily be explained by a local model: it is sufficient to assume that, in each realization, the particle "chooses" a single direction when it crosses the beam splitter. The difficulty then seem to arise only because of the unusual description of phenomena given by quantum mechanics. But we will discuss in Chapter 4 situations where no local model can be elaborated to reproduce the quantum predictions.

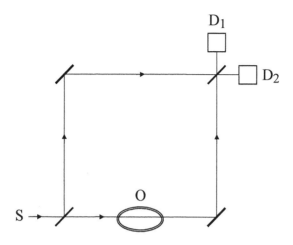

Figure 2.5 A source S emits a series of particles, one by one, into an interferometer. The path differences are adjusted in such a way that all particles reach detector
D_1 and no particle reaches D_2. When an opaque object O is inserted to block one
path for the particle, the destructive interference does not occur anymore and particles are sometimes recorded at D_2. In such an event, it seems that the presence
of the object has been measured with certainty, while no interaction whatsoever
with the object is involved since the detected particle necessarily went through
the upper path of the interferometer (otherwise, it would have been absorbed).

2.4.3 Perturbed Mach–Zehnder interferometer

Consider now Figure 2.5, with a Mach–Zehnder interferometer into which a source
emits a series of particles, one by one, which are registered at detectors D_1 and D_2.
Following [69], we assume that the path differences are adjusted to create a destructive interference effect at detector D_2; no particle can then reach this detector,
and all particles are necessarily detected in D_1 (again, we assume ideal detectors
with 100% efficiency). Now, an opaque object O is inserted in the lower path of the
interferometer, which cancels the destructive interference effect and allows some
particles to reach D_2. If we assume that the two input and output beam splitters
have a 50 % reflectivity, this event happens one time out of four. What happened
in such an event? Since the particle was not blocked by the object, it seems that
it necessarily went through the upper path of the interferometer, meaning that it
never interacted with the object. But, if the object had not been inserted, the par-

ticle could never have reached D_2! The net result is that the sole observation of a detection at D_2 reveals the presence of an object, during an event that apparently excludes any interaction with it since the observed particle is not absorbed. This phenomenon is called "interaction-free measurement" process[15]. In rough terms, one could summarize the event by saying that the object has absorbed part of the wave associated with the particle, but not the particle itself. The curious thing, then, is that the particle and one component of the wave seem to be completely dissociated in the process, since they propagate in two different arms at an arbitrary distance from each other.

Actually, what this experiment illustrates is rather that, as long as the localization of the particle has not been measured, one should not attribute any position or single trajectory to the particle in standard quantum mechanics. The correct quantum description must be made in terms of the state vector of the whole system particle + object and their correlations (this is a case of what we will call "quantum entanglement" in Chapter 7). For a while, the whole system remains in a quantum superposition of two states, one where the particle takes the lower path and perturbs the object, and another where it takes the upper path and does not perturb it ; the object is therefore potentially perturbed. But the positive measurement in D_2 suppresses one of these two states[16]; this "cancels at a distance", so to say, the effects of the interaction between the particle and the object, which ends up totally unaffected in this particular experiment.

For refinements of the ideas and experiments related to interaction-free measurements, see [70, 71]. Hardy has proposed a version involving a double interferometer, one for an electron and one for a positron, and the mutual annihilation of these particles [72]. A pedagogical introduction to interaction-free measurements is given in [73]. A possible application of these ideas is "interaction-free imaging" [74], a technique where images of objects can be obtained without perturbing them. It has been proposed to apply interaction-free measurement to electron microscopy, in order to minimize the radiation dose of biological samples and reduce the damage created by the electron beam [75–77]. Interaction-free measurements can also be used to obtain tests of macrorealism with the help of the Leggett–Garg inequalities (§4.2.2.c) [78]; experiments have been performed with an atom trapped in an optical lattice [79].

---

[15] Of course, the effect requires that the test particles *may* interact with the object, since clearly the effect would not take place if the object was completely transparent for the particle. Alain Aspect has proposed the more appropriate name "absorption free " to describe the effect.

[16] The measurement process selects the component of the state vector where the particle has propagated in the upper path of the interferometer only. The observation of the measurement result then has two effects that seem to be completely contradictory: it signals the existence of an interaction in one component of the state vector (because the detection would have been impossible if this component had not been absorbed), but at the same time it completely cancels this component according to the projection postulate.

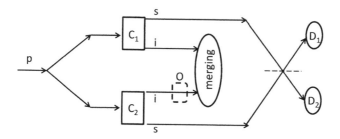

Figure 2.6 The wave of a pump photon p is split into two parts, which are then sent separately onto two nonlinear crystals C_1 and C_2. Each of these crystals converts the photon into two photons, a signal photon s and an idler photon i. This puts a two photon system into a quantum superposition of two states, one with s and i originating from C_1, and another state with s and i originating from C_2. Since the two photons are then entangled, each of them contains information on the path taken by the other. No interference is therefore observed if the two paths of the signal photon are combined on a beam splitter (the dashed line in the right of the figure) followed by two detectors D_1 and D_2. Nevertheless, if the paths of the idler photon are merged, this information disappears and interference is restored.

Now, if an object O is inserted on the path of one idler photon, the interference between the two paths of the signal photon disappears again. Observing this disappearance thus provides a way to detect the presence of object O (and even to image it) without ever detecting the photon that interacts with O. This figure illustrates the principle of the experiment, but the actual geometry of the experimental setup is significantly different.

2.4.4 Imaging with undetected photons

There exists another interesting quantum optical detection method of objects where the detected photons remain completely interaction free. The method involves pairs of photons created by the parametric down conversion of laser light in nonlinear crystals [80, 81]. In such a process, a photon of the pump laser is converted into two photons by a nonlinear optical effect; one of these photons is called the "signal", the other the "idler" (the conservation of energy implies that the sum of their frequencies is equal to the frequency of the pump photon).

The principle of the experiment discussed in [82] is illustrated in Figure 2.6. The method has some similarity with interaction-free detection, but it involves interference effects for pairs of photons, not a single particle. The general idea is to detect an interference effect between two paths followed by a quantum system made of two photons: one path where the pair of photons is created in the first crystal, and the other where it is created in the second crystal. To obtain this situation, the wave of a pump photon p is split into two equal parts with a beam splitter; one part excites a nonlinear crystal C_1, which absorbs the pump photon to create two new photons, the signal photon and the idler photon propagating in beams shown at the top of the figure; similarly, crystal C_2 creates the signal and idler photons into beams that are shown at the bottom of the figure. The two–photon quantum state is then:

$$|\Psi\rangle = \frac{1}{\sqrt{2}} [|s : 1\rangle|i : 1\rangle + |s : 2\rangle|i : 2\rangle] \qquad (2.1)$$

where $|s : 1, 2\rangle$ are the states where the signal photon propagates in the upper/lower beam, and $|i : 1, 2\rangle$ are the similar states for the idler photon. This quantum state is entangled (*cf.* Chapter 7): the state of the idler photon contains a which-way-information on the path taken by the signal photon. Therefore, if the two beams of the signal photon are recombined on a semireflecting plate (the dashed line in the right of the figure), the two beams are incoherent and give rise to no interference. Each of the detectors D_1 and D_2 then just registers the sum of two intensities (reflected and transmitted by the plate).

One can nevertheless restore interference by merging the two beams of the idler photon into a single beam, which destroys the which-way-information. In this case, the signals given by the two detectors D_1 and D_2 depend on the optical path difference, which can for instance be adjusted so that D_1 sees no signal.

Finally, if an absorbing object O is inserted into one of the beams of the idler photon before it merges with the other idler beam, the interference effect disappears for the signal photon, and D_1 sees a signal again. One can in this way detect the absorption of the idler photon by object O by observing the signal photon at D_1, that is a photon that never interacted with O. Actually, the object O does not even have to be absorbing: if it is just dephasing, its presence will also be observed as a change of phase of the signal interference. Moreover, if the experiment is repeated a large number of times with the detectors replaced by imaging devices, the correlations between the spatial propagations of the two photons can be used to obtain an image of the object, as illustrated by the image of a cat in [82].

To summarize, the signal photon that is detected gives information about the absorption and dispersion of the object without ever interacting with it; the interaction is mediated by the other photon. The perturbations acting on the idler photon (in particular, the fact that the two modes from the two crystals have been made to overlap) completely change the behavior of the signal photon, at an arbitrarily large distance. The method can have practical applications: one can for instance choose a low frequency for the idler photon in order to minimize a possible damage to the observed object, and a high frequency for the signal photon in order to optimize detection efficiency. Here again, we have a situation where the quantum effects seem to be delocalized in the usual space. Indeed, the interference effect does not occur between two classical fields propagating in three-dimensional space, but between quantum probability amplitudes associated with a pair of photons that propagate in six-dimensional configuration space.

2.4.5 Cryptography

It has also been pointed out that negative measurements can be useful in the context of quantum cryptography (§8.2). The idea proposed in [83] is that two remote partners, Alice and Bob, make random choices between two orthogonal photon polarizations; Alice sends a photon with the polarization chosen locally, which Bob will send back with a mirror to Alice only if its polarization is different from his own local random choice. The whole setup contains an interferometer that is designed so that, if the photon comes back from Bob to Alice, the probability that Alice will observe a photon at detector D_1 vanishes due to a destructive interference effect similar to that used in the example discussed in § 2.4.3. Now, by selecting only the events where Alice observes the photon at D_1 (then necessarily Bob does not detect the particle), one automatically selects events where the two remote random choices made by Alice and Bob turn out to be the same. If Alice and Bob communicate to each other the result of each run of the experiment (which detector has clicked or not), but keep secret their choice of polarization, by writing down their binary local choices for the selected events they progressively build a shared secret key. The remarkable feature of this scheme is that the events used for secret transmission are actually those where the photon has remained confined within Alice's apparatus[17]: it has not been transmitted from Alice to Bob, in perfect analogy with the particle in example discussed in § 2.4.2 (since the detection of the particle meant that the particle had not taken the path containing the object). Because the particle does not follow the transmission line between the two sites in the events retained for building the secret key, the flux of particles along this line contains no information whatsoever on the random choices made by Alice and Bob; hence an enhanced security.

2.5 A variety of points of view

The following quotations may be useful to get an idea[18] of the variety of interesting points of view that have been expressed since the appearance of quantum mechanics.

---

[17] The useful events for building the key are those where Bob did not return the Schrödinger wave to Alice's location, and did this without absorbing the particle. After Alice and Bob choose identical polarizations, the state vector develops a component where the particle propagates to Bob's site, so that it can be measured along the transmission line or at Bob's site, but this component vanishes when Alice observes the particle at D_1.

[18] With, of course, the usual proviso: short quotations taken out of their context may, sometimes, give a superficial view on the position of their authors.

Copenhagen interpretation:

(i) Bohr ([21], 2nd edition, page 204, and [84]): "There is no quantum world. There is only an abstract physical description. It is wrong to think that the task of physics is to find out how Nature is. Physics concerns what we can say about Nature". Or, similarly: "there is no quantum concept" [85].

Concerning physical phenomena: "one may strongly advocate limitation of the use of the word *phenomenon* to refer exclusively to observations obtained under specified circumstances, including an account of the whole experiment" [46].

He also defines the purpose of physics in the following way [86, 87]: "Physics is to be regarded not so much as the study of something a priori given, but rather as the development of methods of ordering and surveying human experience. In this respect our task must be to account for such experience in a manner independent of individual subjective judgement and therefore objective in the sense that it can be unambiguously communicated in ordinary human language".

While quantum mechanics is often (including nowadays) considered as a theory that is completely new with respect to classical physics (a revolution), Bohr prefers to see it as its logical continuation. He considers [88, 89] that, when introducing quantum mechanics, "the problem with which the physicists were confronted was to develop a rational generalization of classical physics[19], which would permit the harmonious incorporation of the quantum of action". For this purpose, the "principle of correspondence" should be used [90]: "The correspondence principle expresses the tendency to utilize in the systematic development of the quantum theory every feature of the classical theories in rational transcription appropriate to the fundamental contrast between the postulates and the classical theories". It is therefore natural that, on many occasions, he emphasized the importance of classical concepts to give a proper meaning to the formalism of quantum mechanics; in particular, classical concepts are indispensable to describe any process of measurement.

Bohr actually wrote many texts on quantum mechanics, sometimes rather philosophically oriented [1, 91]; more of his quotations are given in §§3.3.2 and 5.3.1, which in particular illustrate Bohr's views on space-time. In his famous Como lecture in September 1927 [92], he introduced the notion "complementarity", the relation between contradictory attributes of the same object in quantum mechanics, and then extended it far beyond, including other sciences than physics (see [91]

---

[19] Similarly, Bohr would probably have seen Einsteinian relativity as a rational generalization of classical electromagnetism (Maxwell equations).

or for instance [46], where he mentions biology[20], sociology, and psychology; see also §7.2 of [21] or Chapter 4 of [58]).

(ii) Born in 1926 (page 804 of [34]): "The motion of particles conforms to the laws of probability, but the probability itself is propagated in accordance with the laws of causality".

Or, shortly later in [93]: "The quantum theoretical description ... does not answer ... the question of where a certain particle is at a given time ... This suggests that quantum mechanics only answers properly put statistical questions, and says nothing about the course of individual phenomena. It would then be a singular fusion of mechanics and statistics".

(iii) Heisenberg [21, 94]: "But the atoms or the elementary particles are not real; they form a world of potentialities or possibilities rather than one of things and facts".

In *Physics and Philosophy* [94] (chapter V): "Natural science does not simply describe and explain nature; it is a part of the interplay between nature and ourselves; it describes nature as exposed to our method of questioning".

Chapter III of [94] is entitled "The Copenhagen Interpretation of Quantum Theory", and he writes in this chapter: "We cannot completely objectify the result of an observation, we cannot describe what 'happens' between this observation and the next". Later, he adds: "Therefore, the transition from the 'possible' to the 'actual' takes place during the act of observation. If we want to describe what happens during an atomic event, we have to realize that the word 'happens' can apply only to the observation, not to the state of affairs between two observations".

He concludes this chapter with the following: "... the measuring device has been constructed by the observer, and we have to remember that what we observe is not nature in itself but nature exposed to our method of questioning. Our scientific work in physics consists in asking questions about nature in the language that we possess and trying to get an answer from experiment by the means that are at our disposal. In this way, quantum theory reminds us, as Bohr has put it, of the old wisdom that when searching for harmony in life one must never forget that in the drama of existence we are ourselves both players and spectators. It is understandable that in our scientific relation to nature our own activity becomes very important when we have to deal with parts of nature into which we can penetrate only by using the most elaborate tools".

(iv) Jordan (as quoted by Bell in [95]): "observations not only disturb what has to be measured, they *produce* it. In a measurement of position, the electron is forced

---

[20] At the end of this article, he discusses the "complementary mode of description", and illustrates its generality by writing: "An example is offered by biology where mechanistic and vitalistic arguments are used in a typically complementary manner. In sociology too such dialectics may be often useful, particularly in problems confronting us with the study and comparison of human cultures ...".

to a decision. We compel it to assume a definite position; previously it was neither here nor there, it had not yet made its decision for a definite position...".

(v) Landau and Lifshitz, in the first section of their book on quantum mechanics [96]: "The possibility of a quantitative description of the motion of an electron requires the presence also of physical objects which obey classical mechanics to a sufficient degree of accuracy. If an electron interacts with such a 'classical object' the state of the latter is, generally speaking, altered ... In this connection the 'classical object' is usually called *apparatus*, and its interaction with the electron is spoken of as *measurement*. However, it must be most decidedly emphasized that we are here not discussing a process of measurement in which the physicist-observer takes part. By *measurement* in quantum mechanics, we understand any process of interaction between classical and quantum objects, occurring apart from and independently of any observer. The importance of the concept of measurement in quantum mechanics was elucidated by N. Bohr".

(vi) Dirac, page 7 of [39]: *"The only object of theoretical physics is to calculate results that can be compared with experiment,* and it is quite unnecessary that any satisfying description of the whole course of the phenomena should be given".

Critics of the Copenhagen interpretation:

(vii) Schrödinger [97]: "the world is given to me only once, not one existing and one perceived. Subject and object are only one. The barrier between them cannot be said to have broken down as a result of recent experience in the physical sciences, for this barrier does not exist". In §2.2, we give more quotations by Schrödinger.

(viii) Einstein, in a letter to Schrödinger in 1928 [98, 99]: "The Heisenberg–Bohr tranquilizing philosophy – or religion? – is so delicately contrived that, for the time being, it provides a gentle pillow for the true believer from which he cannot very easily be aroused."

In 1936 [100]: "The Ψ function does not in any way describe a condition which could be that of a single system; it relates rather to many systems, to an 'ensemble of systems' in the sense of statistical mechanics ... If the function Ψ furnishes only statistical data concerning measurable magnitudes ... the reason lies ... in the fact that the function Ψ does not, in any sense, describe the condition of *one* single system".

(ix) de Broglie [101]: "The interpretation of wave mechanics by Bohr and Heisenberg has many consequences opening new philosophical perspectives. The corpuscle is no longer a well-defined object within the frame of space and time; it is only an ensemble of potentialities to which probabilities are assigned, it is only an entity manifesting itself to us in a fugitive way, sometimes taking an aspect, sometimes another. Professor Bohr, who is in a way the Rembrandt of contemporary physics, since he sometimes shows a clear taste for 'chiaroscuro', said that the corpuscles are 'unsharply defined individuals within finite space-time limits'."

More recent comments:

(x) Bell [55], describing "modern" quantum theory (Copenhagen interpretation) and its relations with cosmology: "it never speaks of events in the system, but only of outcomes of observations upon the system, implying the existence of external equipment[21]" (how, then, do we describe the whole universe, since there can be no external equipment in this case?).

"The problem is this: quantum mechanics is fundamentally about observations. It necessarily divides the world into two parts, a part which is observed and a part which does the observing. The results depend on how this division is made, but no definite prescription for it is given. All we have is a recipe which, because of practical human limitations, is sufficiently unambiguous for practical purposes".

See also his text "Against measurement" where he discusses and criticizes various presentations of the orthodox interpretation [102].

(xi) Mermin [9], summarizing the "fundamental quantum doctrine" (orthodox interpretation): "the outcome of a measurement is brought into being by the act of measurement itself, a joint manifestation of the state of the probed system and the probing apparatus. Precisely how the particular result of an individual measurement is obtained – Heisenberg's transition from the possible to the actual – is inherently unknowable".

(xii) Shimony [10]: "According to the interpretation proposed by Bohr, the change of state is a consequence of the fundamental assumption that the description of any physical phenomenon requires reference to the experimental arrangement".

(xiii) Rosenfeld [103], speaking of the orthodox interpretation: "the human observer, whom we have been at pains to keep out of the picture, seems irresistibly to intrude into it, ...".

(xiv) Gottfried [104] (page 188 of first edition): "The reduction postulate is an independent axiom ... The outcome of these considerations is that quantum mechanics cannot give a complete description of the physical world because there must exist systems (called 'conscious' by Wigner) that are beyond the theory's power of description, i.e. that cannot be incorporated into the part of the world that we treat with the Schrödinger equation".

(xv) Stapp [42] "The interpretation of quantum theory is clouded by the following points: (1) Invalid classical concepts are ascribed fundamental status; (2) The process of measurement is not describable within the framework of the theory; (3) The subject–object distinction is blurred; (4) The observed system is required to be isolated in order to be defined, yet interacting to be observed".

[21] One could add "and of external observers".

Or, as cited by Bell in [55]: "How can a theory which is *fundamentally* a procedure by which gross macroscopic creatures, such as human beings, calculate predicted probabilities of what they will observe under macroscopically specified circumstances ever be claimed to be a complete description of physical reality?" (the completeness of quantum mechanics will be discussed in Chapter 3).

(xvi) Leggett discusses the Copenhagen interpretation [105] in these terms: "... the formalism of QM which goes under the name of the Copenhagen interpretation (though it should probably more correctly be called the Copenhagen non-interpretation, since its whole point is that any attempt to interpret the formalism in intuitive terms is doomed to failure) ... while denying that microscopic objects (electrons, photons, atoms, ...) necessarily have definite properties in the absence of observation, emphatically asserts (or at least implies) that macroscopic objects (counters, cats, etc.) do have such properties at all times, whether they are observed or not. This insistence on the necessity of drawing a sharp line ... between the microscopic world and the macroscopic one of everyday life (including apparatus) is a pervasive theme in the writings of Niels Bohr ...".

In [106], he comments "The real trouble only starts when we take seriously the fact that the measuring apparatus ... is itself a physical system made up of atoms and electrons, and therefore should in principle be describable in quantum-mechanical terms. It should therefore be legitimate to ask what happens if, instead of treating measurement as something quite extraneous to the ordinary behavior of physical systems, we treat it as merely a particular type of physical process and describe it by the linear time-dependent Schrödinger equation". And then, in [107], he comments: "Within the conventional interpretation of quantum mechanics, a system does not possess definite properties until we, as it were, force it to declare them by carrying an appropriate measurement. But is this the only possible interpretation? ... it is perfectly possible that at a deeper level systems do, in fact, have objective properties, whether or not anyone is measuring them ... The apparently random outcomes predicted by the quantum formalism would then simply be due to our ignorance of the details of a deeper level description[22]".

(xvii) Van Kampen [108] has proposed the following interesting caveat: "Whoever endows the state vector with more meaning than needed for computing observable phenomena is responsible for the consequences!" In Appendix A, we discuss how two other references [109, 110] make good use of this warning.

Present situation:

As one can guess from the preceding quotations, even nowadays no general consensus has been reached among physicists concerning the precise meaning of the

---

[22] Chapter 3 gives a much more detailed account of the discussions on the completeness (or incompleteness) of quantum mechanics.

state vector or wave function. They agree on the formalism and how to use the state vector in practice. For all experiments realized so far, a pragmatic choice between the two postulates can be left to the judgment of the physicist; to make practical predictions concerning an experiment, common sense is sufficient (cf. §11.1.1), so that problems related to the foundations of the theory may be put aside. Nevertheless, it would be preferable to have well-defined mathematical laws than just physically reasonable recipes! It is therefore not surprising that, when the question comes to discussing the physical meaning of the mathematical objects and the interpretation of the theory, the debate soon reappears, and may even sometimes become passionate. Moreover, even those who claim to be in perfect agreement with orthodox quantum theory will use, in fact, a variety of nuances (and even sometimes contradictory points of view and inconsistencies) when they are asked to explain in detail their point of view.

To summarize, the orthodox status of the wave function is indeed a subtle mixture between different concepts concerning reality and the knowledge that we have of this reality. Do the fantastic achievements of standard quantum theory mean that it is the ultimate and most precise description of a quantum system that physics will ever provide? Is quantum theory compatible with realism, or does it require positivism? The questions are not settled. Interestingly Bohr himself is generally considered more as a realist[23] than a positivist or an operationalist [21]. In the words of Jammer ([58], page 157): "Bohr, as Von Weizsäcker [111] emphasized, never rejected the notion of reality, he only modified it". If asked about the relations between the wave function and reality, Bohr would probably have said that the wave function is indeed a useful tool, but that the concept of reality can not properly be defined at its level only; it has to include all macroscopic measurement apparatuses that are used to have access to microscopic information (we come back to this point in more detail in §3.3).

At the end, a general question that emerges from all these discussions is what we expect from a good physical theory. Is it sufficient for such a theory to make perfectly correct predictions (no contradiction at all with any result of experiments), without attempting to describe a succession of real events; or should it also describe "what happens" during this experiment, as a succession of physical events? And, in the second case, what type of description is then acceptable? Clearly, as the preceding quotations illustrate, not all physicists agree on the answers to these questions.

---

[23] See for instance in §3.3.2 how he considers that physical reality can be defined without ambiguity. Bohr accepts the notion of reality provided it is properly defined (with a full description of the experiment).

2.6 Unconvincing arguments

We have already emphasized that the invention of the Copenhagen/standard interpretation of quantum mechanics has been, and remains, one of the biggest achievements of all science. One can admire even more, in retrospect, how early its founders conceived it, at a time when experimental data were relatively scarce. Since that time, numerous ingenious experiments have been performed, precisely with the hope of seeing the limits of this interpretation; nevertheless, until now, not a single fact has disproved the theory. It is really a wonder of pure logic that has allowed the early emergence of such a fantastically successful intellectual construction.

This being said, one has to admit that, in some cases, the brilliant authors of this construction may sometimes have gone too far, pushed by their great desire to convince. For instance, authoritative statements have been made concerning the absolute necessity of the orthodox interpretation, which seem exaggerated in retrospect. According to these statements, the orthodox interpretation would give the only possible ultimate description of physical reality; no finer description would ever become possible. From this perspective, the fundamental probabilistic character of microscopic phenomena should be considered as a proven fact, a rule that should be carved into marble and accepted forever by scientists. But, now, we know that this is not necessarily true: yes, one may prefer the orthodox interpretation for various reasons, but this is not the only choice allowed by pure logic. Other interpretations are perfectly possible, and may even be preferred to the standard interpretation (see the discussion of Chapter 11); in particular, determinism in itself is not disproved at all [24]. As discussed for instance in [9], and initially clarified by Bell [5, 6] and Bohm [7, 8] , the "impossibility proofs" put forward by the proponents of the Copenhagen interpretation are logically unsatisfactory for a simple reason: they arbitrarily impose conditions that may be relevant to quantum mechanics (linearity), but not to the theories they aim to dismiss – any theory with additional variables such as the Bohm theory, for instance. Because of the exceptional scientific stature of the authors of the impossibility theorems, it took a long time for the physics community to realize that the theorems were irrelevant; now that this is more widely recognized, the plurality of interpretations is more easily accepted.

---

[24] Provided one accepts nonlocality, see Chapter 4.

3

The Theorem of Einstein, Podolsky, and Rosen

More than 70 years after its publication, the article by Einstein, Podolsky, and Rosen (EPR) [112] is still cited hundreds of times every year in the literature; this is a very exceptional case of longevity for a scientific article! There is some irony in this situation since, for a long time, the majority of physicists did not pay much attention to the EPR reasoning. They considered it as historically interesting, but with no precise relevance to modern quantum mechanics; the argument was even sometimes completely misinterpreted. A striking example is given in the Einstein–Born correspondence [113], where Born, even in comments that he wrote after Einstein's death, clearly shows that he never completely understood the nature of the objections raised by EPR. Born went on thinking that the point of Einstein was a stubborn rejection of indeterminism ("look, Albert, indeterminism is not so bad!"), while actually the major concern of EPR was locality and/or separability (we come back later to these terms, which are related to the notion of space-time). If giants such as Born could be misled in this way, it is no surprise that, later on, many others made similar mistakes!

This is why, in what follows, we will take an approach that may look elementary, but at least has the advantage of putting the emphasis on the logical structure of the arguments and their generality. Doing so, we will closely follow neither the historical development of the ideas nor the formulation of the original article[1], but rather will emphasize the generality of the EPR reasoning. For a more faithful historical report, see chapter 6 of [58], or [117], as well as references contained therein; [118] relates the circumstances under which the EPR article was written,

---

[1] The published version of the EPR article was written by Podolsky. In later comments on the subject, Einstein gave the impression that he thought that the essence of the argument could have been expressed in a simpler way; see for instance a letter from Einstein to Schrödinger [114] ("the writing is not really what I was expecting: what is essential is sort of buried under erudition"); see also [100], [115], and [116], where Einstein explains the argument in his own terms.

and was sent not only to *Physical Review* but also to a newspaper (*New York Times*), which published a report[2] – an initiative of which Einstein disapproved.

3.1 A theorem

One often speaks of the "EPR paradox", although the word "paradox" is not completely appropriate in this case. For Einstein, the basic motivation was not to invent paradoxes or to entertain colleagues inclined to philosophy, it was to build a strong logical reasoning that, starting from well-defined assumptions (roughly speaking: locality and some form of realism), would lead ineluctably to a clear conclusion (quantum mechanics is incomplete, and even: physics is deterministic[3]). To emphasize this logical structure and generality, here we will speak of the "EPR theorem[4]", which formally could be stated as follows:

Theorem: *If the predictions of quantum mechanics are correct (even for systems made of remote correlated particles) and if physical reality can be described in a local (or separable) way, then quantum mechanics is necessarily incomplete: some "elements of reality[5]" exist in Nature that are ignored by this theory.*

The theorem is certainly valid; it has been scrutinized by many scientists who have found no flaw in its derivation. Indeed, the logic that leads from the assumptions to the conclusions is perfectly sound. It would therefore be an error to repeat (a classical mistake!) "the theorem was shown by Bohr to be incorrect", or, even worse, "the theorem is incorrect since experimental results are in contradiction with it[6]". Bohr himself, of course, did not make the error: in his reply to EPR [41], he explained why he thought that the assumptions on which the theorem is based were not relevant to the quantum world, which made it inapplicable to a discussion on quantum mechanics. More precisely, he used the word "ambiguous" to characterize these assumptions, but never claimed that the reasoning is faulty (for more details, see §3.3.2). A theorem that is not applicable in a particular case is not necessarily incorrect: theorems of Euclidean geometry are not wrong, or even useless, because one can also build a consistent nonEuclidean geometry! Concerning possible contradictions with experimental results, we will see that, in a sense, they make the theorem even more interesting, mostly because it can then be used within a "reductio ad absurdum" reasoning.

---

[2] This report ends with comments by E. Condon (Princeton University) that show that he also did not realize the exact point of the EPR article; maybe he did not have the time to read the article in detail before the newspaper was published.

[3] Born's mistake, therefore, was to confuse assumptions and conclusions.

[4] Einstein himself nevertheless sometimes used the word "paradox", for instance in §5 of [100].

[5] These words are carefully defined by the authors of the theorem; see the beginning of §3.3.

[6] The contradiction in question occurs through the Bell theorem (which is therefore sometimes criticized for the same reason), which was introduced as a continuation of the EPR theorem. As we discuss in more detail below, the logical conclusion resulting from this contradiction is that one of the assumptions leading to the theorem should be abandoned, but not that the theorem itself is incorrect.

Figure 3.1 A source S emits two particles, which then propagate in space and reach different regions of space, where Alice and Bob perform measurements with them in their respective laboratories; *a* and *b* are the settings of the two corresponding measurement apparatuses (for instance, orientations of Stern–Gerlach magnets).

Good articles on the EPR argument are abundant; for instance, a classic is the little article by Bell [95]. Another excellent introductory article is, for instance, [56], which contains a complete description of the EPR thought experiment in a particular case (two settings only are used) and provides a general discussion of many aspects of the problem. For a detailed list of references, see for instance [119]. The basic scheme considered is summarized in Figure 3.1: a source S emits two correlated particles, which propagate towards two remote regions of space where they undergo measurements; the types of these measurements are defined by "settings", or "parameters"[7] *a* in the first region and *b* in the second (typically orientations of Stern–Gerlach analyzers), which are the free choice of the experimentalists; in each region, a result is obtained, which can take only two values symbolized by ± 1 in the usual notation. Finally, a crucial assumption is made: for the particular experimental arrangement considered, every time both settings are chosen to be the same value, the results of both measurements are always the same (perfect correlations).

Here, rather than trying to paraphrase the good texts on EPR, we will intentionally take a different presentation, based on an analogy, a sort of a parable. Our purpose is to emphasize an important feature of the argument: the essence of the EPR

---

[7] Depending on the context, we will use the words "settings" and "parameters" indifferently.

reasoning is actually nothing but what is usually called "the scientific method", in the sense discussed by Francis Bacon and Claude Bernard. For this purpose, we will leave pure physics for botany! Indeed, in both disciplines, one needs rigorous scientific procedures in order to prove the existence of relations and causes, which is precisely what we want to do.

3.2 Of peas, pods, and genes

Inferring the properties of microscopic objects from macroscopic observations requires a combination of ingenuity, in order to design meaningful experiments, with a good deal of logic, in order to deduce these microscopic properties from the macroscopic results. Because it is impossible to observe an electron or even a macromolecule with the naked eye, or to take it in one's hand, some abstract reasoning is necessary. The scientist of past centuries who, like Mendel, was trying to determine the genetic properties of plants, had exactly the same problem: he did not have access to any direct observation of the DNA molecules, so that he had to base his reasoning on adequate experiments and on the observation of their macroscopic outcome. In our parable, the scientist grows plants from peas, and later observes the color of flowers (the "result" of the measurement, +1 for red, −1 for blue) as a function of the conditions (temperature for instance) in which the peas are grown – these conditions correspond to the "settings" a or b, also called "experimental parameters", which determine the nature of the measurement. The basic purpose is to infer the intrinsic properties of the peas (the EPR "element of reality") from these observations.

3.2.1 Simple experiments: no conclusion yet

It is clear that many external parameters such as temperature, humidity, amount of light, etc., may influence the growth of vegetables and the color of a flower; in a practical experiment, it seems therefore very difficult to identify and control all the relevant parameters with certainty. Suppose that one observes that the flowers growing in a series of experiments are sometimes blue, sometimes red; the reason behind these variation may therefore be some irreproducibility of the conditions of the experiment, but it can also be something more fundamental. In more abstract terms, a completely random character of the observed result of the experiments may originate either from the fluctuations of uncontrolled external perturbations, or from some intrinsic property that the measured system (the pea) initially possesses. It may even result from the fact that the growth of a flower (or, more generally, life?) is a fundamentally indeterministic process – needless to say, all three reasons can be combined in any complicated way. Transposing the issue to quan-

tum physics leads to the following formulation of the question: are the results of
the experiments random because of the influence of some uncontrolled fluctuating
process acting somewhere (fluctuations in the macroscopic apparatus, fluctuations
of some microscopic property of the measured particles, etc.)? Or are they fundamentally random because they are the result of an irreducible quantum process?

The scientist may repeat the "experiment" a thousand times and even more: as
long as the results are always totally random, there is no way to decide which
interpretation should be selected; it is just a matter of personal taste. Of course,
philosophical arguments may be built to favor or reject one of them, but from a
purely scientific point of view, at this stage, there is no compelling reason to make
one choice or another. Such was the situation of quantum physics before the EPR
argument.

3.2.2 Correlations: causes unveiled

The stroke of genius of EPR was to realize that studying quantum correlations
could lead to a big step forward in the analysis. They exploit the aforementioned
crucial assumption: when the choices of the settings are the same, the observed
results are always identical.

3.2.2.a Same measurement parameters

In our botanical analogy, we assume that the botanist observes correlations between
colors of flowers under appropriate conditions. Peas come together in pods, so that
it is possible to grow peas taken from the same pod and observe their flowers in remote places. When no special care is taken to give equal values to the experimental
parameters (temperature, etc.), or when peas are taken from different pods, nothing
special is observed. But when the two peas come from the same pod and when the
parameters for the growth of each of them are chosen to the same values, then the
colors are systematically the same (both remain random from one pod to the next,
but they always take the same value).

What can we then conclude? Since the peas grow in remote places, there is
no way that they can be influenced by any single uncontrolled fluctuating phenomenon, or that they can somehow influence each other in the determination of
the colors. If we believe that causes always act locally, and that perfect correlations
cannot appear by pure chance (without any cause), we are led to the following conclusion: the only possible explanation of the common color is the existence of some
common property of both peas, which determines the color[8]. It may be difficult to
detect the property in question directly, since it is presumably encoded inside some

---

[8] The fact that the correlations disappear if the parameters are no longer set to the same value shows that the
color is a function of both this common property and of the local parameters of the experiment (settings).

tiny part of a biological molecule, but the property exists and is sufficient to determine the results of the experiments.

This is the essence of the argument, and we could stop at this point. Nevertheless, for completeness, let us make every step of the EPR reasoning even more explicit. The key idea is that the nature and the number of "elements of reality" associated with each pea cannot vary under the influence of some remote experiment, performed on the other pea. Let us first assume that the two experiments are performed at different times: one week, the experimenter grows a pea; the next week another pea from the same pod, in exactly the same conditions of temperature, humidity, etc. We then assume that perfect correlations of the colors are always observed, without any special influence of the delay between the experiments. Just after completion of the first experiment (observation of the first color), but still before the second experiment, the result of that future experiment has a perfectly determined value; therefore, there must already exist one element of reality attached to the second pea that corresponds to this certainty. Clearly, it can not be attached to any other object than the pea, for instance one of the measurement apparatuses; this is because the observation of perfect correlations only arises when making measurements with peas taken from the same pod. Symmetrically, the first pod also had an element of reality attached to it, which ensured that its measurement would always provide a result that coincides with that of the future measurement. We can assume that the elements of reality associated with both peas are coded in some genetic information, and that the values of the codes are the same for all peas coming from the same pod; but other possibilities exist and the precise nature and mechanism involved in the elements of reality do not really matter here. The important point is that, since these elements of reality cannot appear by any action at a distance, they necessarily also existed before any measurement was performed – presumably even before the two peas were separated.

It seems difficult not to agree that the method that led to these conclusions is indeed the scientific method; no tribunal would believe that, in any circumstance, perfect correlations could be observed in remote places by pure chance, without being the consequence of some common characteristics shared by both objects. Such perfect correlations can then only reveal the initial common value of some variable attached to them, which is in turn a consequence of some fluctuating common cause in the past (a random choice of pods in a bag, for instance).

3.2.2.b Different measurements parameters

Now, let us consider any pair of peas, when they are already spatially separated, but before the experimentalist decides what type of measurements they will undergo (values of the parameters, delay or not, etc.). We know that, if the decision turns out to favor measurements with exactly the same parameters, perfect correlations will always be observed; this is true for any local choice of the parameters, provided they are the same at both places. Since elements of reality cannot appear or change their values depending on experiments that are performed in a remote place, the two peas necessarily carry some elements of reality that completely determine the color of the flowers for any experimental condition – not only for the conditions that turn out to be chosen for the experiment. Any theory ignoring these elements of reality is necessarily incomplete.

To take another analogy, let us for instance assume that the peas are replaced with elaborate automata, using the most sophisticated technology available, able to measure the external parameters such as temperature, etc. and to process them with powerful computers[9] to calculate the results ±1 from the parameters. All this computing power is actually useless to simulate the predictions of quantum mechanics for remote experiments, unless the memory of each computer contains initially a common random number. Because each automaton does not have access to the remote parameters, the only way to reproduce quantum mechanics is to feed the memory of both computers with the same random number, and to program both for calculating in the same way all the possible results as a function of the local parameters and the common random number.

3.2.2.c Summary. Schrödinger's parable

To summarize, the reasoning shows that each result of a measurement may be a function of two kinds of variables[10]:

(i) the intrinsic properties of the peas, which they carry along with them;

(ii) the local parameters of the experiment (temperature, humidity, etc.); because correlations disappear when the parameters are different, a given pair that turned out to provide two blue flowers with one choice of common experimental conditions could have provided a pair of red flowers with another choice.

We may also add that:

(iii) the results are well-defined functions of these variables, which means that no fundamentally indeterministic process takes place in the experiments. The only source of randomness is the choice of the initial pair.

(iv) when taken from its pod, a pea cannot "know in advance" to which sort of experiment it will be submitted, since the decision may not yet have been made by the experimenters; when separated, the two peas therefore have to take with them all the information necessary to determine the color of flowers for any kind of experimental conditions.

In other words, each pea carries with it as many elements of reality as necessary

---

[9] We are assuming here that the computers are not quantum computers (if quantum computers can ever be built, which is another question).

[10] In Bell's notation, the A functions depend on λ as well as on a.

to provide "the correct answer" to all possible questions it might be asked in the future. In §13 of [60], Schrödinger offers a comparison with a "schoolboy under examination" who may be asked two different questions by a teacher. During many tests, it appears that "the pupil will correctly answer the first question that is put to him. From that it follows that in every case he knows the answer to both questions", even if afterwards he does not necessarily answers the second correctly. "No school principal will judge otherwise".

3.3 Transposition to physics

We now come back to microscopic physics, as in the original EPR argument.

3.3.1 The EPR argument for two correlated microscopic particles

Historically, EPR introduced their arguments for two correlated particles and measurements of their position and momentum, with continuous results. It is nevertheless convenient to use a version of the argument, often called EPRB, that involves discrete results and spins, initially introduced by Bohm [120].

3.3.1.a Assumptions

We consider two spin 1/2 particles, which are emitted by one source S in a singlet state[11] where their spins are correlated (we choose this example for the sake of simplicity, but the EPR theorem is not limited to spins 1/2 in a singlet state, *cf.* §3.4.1). They then propagate towards two different regions of space, where they undergo the measurement of their spin component along a direction defined by angle a on the left, b on the right (Figure 3.2). We call Alice and Bob the two operators who perform these experiments in two remote laboratories, separated by an arbitrary distance. Alice makes a free choice of the direction a, which defines her "type of measurement", and can only obtain result $+1$ or -1 whatever type of measurement has been chosen; similarly, Bob chooses direction b arbitrarily and obtains one of the results $+1$ or -1. In the EPR thought experiment, it is assumed that the two spins interact only with the measurement apparatuses, with no proper evolution of the singlet state. Standard quantum mechanics then predicts (§4.1.1) that the distance and times at which the spin measurements are performed is completely irrelevant for calculating the probabilities of the various combined results. If for instance a and b are chosen equal (parallel directions of measurements), the prediction is that the results are always opposite, even if the measurements take

---

[11] In this section, we do not need to know the precise definition of a singlet spin state. This definition as well as the quantum probabilities of the results associated with any values of a and b are given in §4.1.1. For equal values of a and b, these probabilities give perfect correlations, which is the only result we need to know to understand the EPR reasoning. See also Chap. 12, equation (12.76).

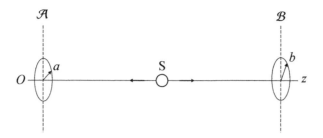

Figure 3.2 Scheme of an EPRB experiment. A source S emits pairs of particles in a singlet spin state. The particles propagate along direction Oz towards two remote regions of space $\mathcal{A}$ and $\mathcal{B}$, where Stern–Gerlach magnets are used to measure the components of their spins along directions in the plane perpendicular to Oz. For the first particle, the direction is defined by angle a, for the second by angle b. Each measurement provides result $+1$ or -1, and correlations between these results are measured when the experiment is repeated many times.

place in very remote places. This remains true whatever common choice of $a = b$ is made, even by two experimenters who operate independently in the two regions of space and make a random choice at the last moment (after the emission of the pair of particles).

The starting point of EPR is to assume that the predictions of quantum mechanics concerning the results of experiments are correct. More specifically, the reasoning assumes that the perfect correlations predicted by this theory are always observed, whatever the distance between the experimental apparatuses may be. In the parable of the peas, the red/blue colors are obviously the analogue of the two results ± 1, the experimental parameters (temperature, etc.) are the analogous to the orientation of these apparatuses. We have assumed that the same colors are always observed for the same pair of peas, provided the experimental conditions are the same for both peas, while for quantum mechanics we just saw that the results are always opposite for parallel settings; to obtain a perfect correspondence with the parable, we have to insert a sign change. For instance we may assume that, in the quantum experiment, Alice attributes a red color to result $+1$ and a blue color to result -1, while Bob chooses the opposite convention[12]. In practice, Alice could complete her measurement apparatus with a device that makes a red lamp flash for

---

[12] Alternatively, one may make no change on the quantum mechanics side and attribute values $+1$ to red flowers and -1 to blue flowers in one place where peas are grown, and choose the opposite convention in the other place.

result $+1$, or a blue lamp flash for result -1; Bob could use a similar device with different connections providing the opposite correspondence between results and colors. This question of sign does not introduce any difficulty; in the context of the EPR argument, there is no fundamental difference between a perfect correlation and a perfect anticorrelation.

Another essential ingredient of the EPR reasoning is the notion of "elements of reality"; EPR first remark that "these elements cannot be determined by a priori philosophical considerations, but must be found by an appeal to results of experiments and measurements". They then propose the following criterion: "if, without in any way disturbing a system, we can predict with certainty the value of a physical quantity, then there exists an element of physical reality corresponding to this physical quantity". In other words, certainty cannot emerge from nothing: an experimental result that is known in advance is necessarily the consequence of some pre-existing physical property. In our botanical analogy, we implicitly made use of this idea in the reasoning of §3.2.2.

A last, but essential, ingredient of the EPR reasoning is the notion of space-time and locality: the elements of reality in question are attached to the region of space where the experiment takes place[13], and they cannot vary suddenly under the influence of events taking place in a very distant region of space. They can even less appear under these conditions[14]. The peas of the parable were in fact not so much the symbol of microscopic objects, electrons or spin 1/2 atoms for instance. Rather, they symbolize regions of space where we just know that "something is propagating inside"; it can be a particle, a field, a biological molecule, or anything else, with absolutely no assumption on its structure or physical description. Actually, in the EPR quotation of the preceding paragraph, one may replace the word "system" by "region of space", without altering the rest of the reasoning. One may summarize the situation by saying that the basic belief of EPR is that regions of space can contain elements of reality attached to them (attaching distinct elements of reality to separate regions of space is sometimes called "separability"– see §3.3.3.c) and that they evolve locally. In the literature, for brevity this is often called "local realism", and this will be the definition of these words throughout this book; the words "local causality" are also often used, in particular because no particular philosophical definition of reality is implied.

---

[13] Einstein writes in [115]: "Things in physics are imbedded in a space-time continuum. These things require an autonomous existence inasmuch as they are in separate 'parts' of space".

[14] In standard quantum mechanics, if one applies the state vector reduction, the unmeasured spin suddenly "jumps" to a state with a perfectly well-defined spin component along the direction of measurement for the other spin. In this sense, quantum mechanics does attribute an element of reality to the unmeasured spin before it undergoes any measurement, in partial agreement with the EPR conclusion. But the disagreement occurs before the first measurement, since then standard quantum mechanics contains no such element of reality whatsoever.

3.3.1.b Conclusions

From these assumptions, by the same reasoning as above, EPR obtain the equivalent of the conclusions of §3.2.2; whatever values are chosen for a, b, the results of the measurements are functions:

(i) of intrinsic individual properties of the spins that they carry with them (the EPR elements of reality);

(ii) of course, also of the orientations a, b of the Stern–Gerlach analyzers.

In addition, they show that:

(iii) the results are given by well-defined functions of these variables, which implies that no indeterministic process is taking place; in other words, a particle with spin carries along with it all the information necessary to provide the result for the measurement, whatever choice is made for the orientation a (for the first particle) or b (for the second). All components of each spin have determined values simultaneously.

(iv) it is possible to envisage future measurements of observables that correspond to two different values b and b' for instance, that is to two different components of the spin that are called "incompatible" in quantum mechanics; the EPR reasoning shows that, in reality, incompatible observables can simultaneously have perfectly well-defined values.

Item (i) may be called the EPR-1 result; it implies that something is missing in quantum mechanics (the description of these intrinsic individual properties before the measurements), which is thus incomplete (EPR require from a complete theory that "every element of physical reality must have a counterpart in the physical theory"). The state vector may be a sufficient description for a statistical ensemble of pairs, but not for one single pair of spins; in this case, it should be completed by some additional information. In other words, inside the ensemble of all pairs, one can distinguish between subensembles with different physical properties.

Item (iii) goes further and establishes the validity of determinism from a locality assumption, combined with correct predictions of quantum mechanics.

Item (iv) can be called the EPR-2 result; it shows that the notion of incompatible observables is not fundamental, but just a consequence of the incomplete character of the theory. It actually provides a reason to reject complementarity (in §3.4.1, we generalize this result to other quantum systems than spins $1/2$ in a singlet state). Curiously, EPR-2 is often presented as the major EPR result, sometimes even with no mention of the others; actually, the rejection of complementarity is almost marginal or, at least, less important for EPR than the proof of incompleteness. In fact, in our reasonings, we will only need EPR-1. Einstein himself did not give much importance to the relation between the reasoning and noncommuting observables[15], and it seems likely that this component was introduced into the article by Podolsky.

Indeed it seems that, for Einstein, the most important result of the EPR article was neither EPR-1 nor EPR-2; it was still another result, which we may call EPR-3. Curiously, this logical element is often overlooked, maybe because it is not mentioned in the abstract of the EPR article. Nevertheless, Einstein considered it as important – for instance it is the main aspect that he emphasizes in his 1936 article [100], one year after the publication of the EPR article[16]. The EPR–3 result is that the description of physical reality given by quantum mechanics is "redundant", since it describes the same physical reality for particle 2 with several different state vectors. This idea is emphasized in an sentence that EPR write even before discussing noncommuting operators: "Thus, it is possible to assign two different wave functions to the same reality (the second system after the interaction with the first)". Indeed, if all pairs of particles are identical as assumed in quantum mechanics, the physical reality attached to particle 2 cannot depend on the type of measurement performed very far away on particle 1 (locality); it will necessarily be the same just after this measurement is performed, whatever physical quantity is measured on particle 1. But quantum mechanics implies that state vector reduction should be applied in a basis of the space of states that does depend on the measurement; therefore, particle 2 will reach several different states, depending on the measurement performed on particle 1. We then end up with too many different state vectors to describe the same physical reality for particle 2; hence a contradiction. It is somewhat paradoxical that the same theory should be both incomplete and redundant! By contrast, if we accept the idea of the EPR elements of reality, the measurement on particle 1 reveals pre-existing properties of this particle, and therefore of the emitted pair, and finally of particle 2 if the two particles are initially correlated; the situation can then be understood within the usual frame of classical correlations.

3.3.2 Bohr's reply

Bohr, in his reply [41, 121], does not criticize the EPR reasoning, but their assumptions, which he considers as inappropriate in quantum physics. For Bohr, the crite-

---

[15] At the end of the already mentioned letter to Schrödinger [114], he writes "as for the fact that the different states of system 2 may be considered as eigenvectors of different operators, I really do not care" ("Das ist mir Wurst" in German).

[16] Einstein's words in this article are (he calls systems A and B the systems we called 1 and 2): "Since there can be only *one* physical condition of B after the interaction, and which can reasonably not be considered as dependent on the particular measurement we perform on the system A separated from B, it may be concluded that the Ψ function is *not* unambiguously coordinated with the physical condition. The coordination of several Ψ functions with the same physical condition of system B shows again that the Ψ function cannot be interpreted as a (complete) description of a physical condition of a unit system".

rion of physical reality proposed by EPR "contains an essential ambiguity when applied to quantum phenomena"; he adds: "their argumentation does not seem to me to adequately meet the actual situation with which we are faced in atomic physics" (in this context, "atomic" is equivalent to "microscopic" or "quantum" in modern language). His text has been studied in detail and discussed by many authors (for a historical review, see for instance Chapter 6 of [58]), but remains delicate to understand in detail. One reason may be that, instead of concentrating his arguments on the precise situation considered by EPR, Bohr emphasizes in general the consistency of the mathematical formalism of quantum mechanics and the "impossibility of controlling the reaction of the object on the measuring instruments". But, precisely, the main point of the EPR argument is to select a situation where these unavoidable perturbations do not exist! Indeed, EPR locality implies that a measurement performed in region $\mathcal{A}$ can create no perturbation on the elements of reality in region $\mathcal{B}$.

Only the second part of Bohr's article really deals with the EPR argument. Bohr then writes the following: "The wording of the above mentioned criterion (the EPR criterion for elements of reality) ... contains an ambiguity as regards the expression 'without in any way disturbing a system'. Of course there is, in a case like that considered (by EPR), no question of a mechanical disturbance of the system under investigation during the last critical stage of the measuring procedure. But even at this stage there is essentially the question of an influence on the very conditions which define the possible types of predictions regarding the future behavior of the system ... the quantum description may be characterized as a rational utilization of all possibilities of unambiguous interpretation of measurements, compatible with the finite and uncontrollable interactions between the objects and the measuring instruments in the field of quantum theory".

Several authors have wondered how exactly these words should be interpreted, what is meant by "mechanical disturbance" and, especially, how to understand the central sentence "there is essentially the question of an influence on the very conditions which define the precise types of predictions regarding the future behavior of the system". What is probably meant is "an influence *of the first measurement performed in $\mathcal{A}$* on the conditions which define the predictions on the future behavior of the system *during the second measurement performed in $\mathcal{B}$*"– or maybe "the future behavior of *the whole system extending in both regions $\mathcal{A}$ and $\mathcal{B}$*". In Bohr's view, physical reality cannot be properly defined without reference to a complete and well-defined experiment. This includes, not only the systems to be measured (the microscopic particles), but also all the measurement apparatuses: "these (experimental) conditions must be considered as an inherent element of any phenomenon to which the term physical reality can be unambiguously applied". As a consequence, a spin measurement performed in $\mathcal{A}$ on one spin, as well as the choice of the direction of the measured component, change the physical reality of the whole physical system, which of course includes the second spin. In other words, Bohr considers that the whole system can be perturbed by a different choice of the type of measurement performed at one end of the experiment; from this point of view, the words "without in any way perturbing the system" are indeed ambiguous. This leads to a rejection of the EPR assumption, according to which the physical reality contained within the space region $\mathcal{B}$ of one spin is independent of measurements performed on the other in an arbitrarily remote region $\mathcal{A}$; therefore, the EPR conclusion is no longer valid.

In the same line, more than ten years later (in 1948), Bohr summarized his point of view by writing [46]: "Recapitulating, the impossibility of subdivising the individual quantum effects and of separating a behavior of the objects from their interaction with the measuring instrument serving to define the conditions under which the phenomena appear implies an ambiguity in assigning conventional attributes to atomic objects which calls for a reconsideration of our attitude towards the problem of physical explanation". What is questioned by Bohr is therefore the very necessity, in the situation considered, of an explanation going into details below a certain level. The quotations given above in §2.5 also show that, for Bohr, a consistent interpretation of the quantum formalism cannot be obtained without the inclusion of the classical concepts associated with each experiment[17].

J.S. Bell summarizes Bohr's reply concisely by writing [95] that, in Bohr's view, "there is no reality below some classical macroscopic level". A more detailed version of this sharp sentence could be "in Bohr's view, below some level, the reality can be defined only in relation with a macroscopic classical environment". Indeed, if one assumes that physical reality can only be defined with respect to macroscopic objects, then EPR's reasoning does not hold: one should not attempt to assign elements of reality to one of the spins only, or to a region of space containing it; this would be incompatible with quantum mechanics and therefore unphysical[18] – even if the region in question is very large and isolated from the rest of the world. Expressed differently, a physical system that is extended over a large region of space is to be considered as a single entity, within which no attempt should be made to dis-

---

[17] Heisenberg did not agree with Einstein either, and characterized his position on quantum mechanics in the following terms (Chapter V of [94]): "When Einstein has criticized quantum theory he has done so from the basis of dogmatic realism. This is a very natural attitude. Every scientist who does research work feels that he is looking for something that is objectively true. His statements are not meant to depend upon the conditions under which they can be verified".
In his Nobel lecture in 1933, he had already written: "The very fact that the formalism of quantum mechanics cannot be interpreted as a visual description of a phenomenon occurring in space and time shows that quantum mechanics is in no way concerned with the objective determination of space-time phenomena".

[18] One could add that the EPR disproval of the notion of incompatible observables (EPR-2) implies that, at least, two different settings are considered for one of the measurement apparatuses; but this should correspond, in Bohr's view, to two different physical realities (corresponding to every different couple a,b), instead of a single one as assumed in the EPR reasoning.

tinguish physical subsystems or any substructure; trying to attach physical reality
to regions of space is then automatically bound to failure. In terms of our Leitmotiv
of §1.1.3, the difference between ordinary space and configuration space, we could
say the following: the two-particle system has a single wave function that propagates in a configuration space with more than three dimensions, and this should be
taken very seriously; no attempt should be made to come back to three dimensions
and implement locality arguments in a smaller space. Bohr's point of view is, of
course, not contradictory with relativity, but it certainly minimizes the impact of
basic notions such as space-time, or events (a measurement process in quantum
mechanics is not local; therefore it is not an event stricto sensu). His point of view
does not fit well with a strict interpretation of relativity.

Many physicists admit that a precise characterization of Bohr's attitude, in terms
for instance of exactly what traditional principles of physics should be given up, is
delicate (see for example the discussion of [10]). What is clear is that Bohr considers that it is vain to attempt to give any physical explanation that goes beyond
orthodox quantum mechanics. In his reply to EPR [41] in *Physical Review*, there
seems to be an influence of previous discussions that he had with Einstein at the
Solvay conferences; this may explain why he just repeats the orthodox point of
view for a single particle submitted to incompatible measurements, even if this is
rather irrelevant in the EPR context. Uncontrollable interactions are emphasized
several times, but locality is not explicitly discussed. Did Bohr fully appreciate
how interesting the discussion becomes for two remote correlated particles, and
the novelty of the EPR argument, which is the starting point of the Bell theorem
for instance[19]? In Pearle's words: "Bohr's rebuttal was essentially that Einstein's
opinion disagreed with his own" [122]. Even Bell confessed that he had strong difficulties understanding Bohr ("I have very little idea what this means ..." – see the
appendix of [95]). But, in any case, the Bohrian point of view remains presently as
strong as ever, while we now know that the EPR assumptions contain self contradictions, as we will see in the next chapter. In this sense, Bohr won the debate.

3.3.3 Locality, relativity, and separability

Locality and separability are two different notions, even if they are often (but not
always) related.

3.3.3.a Various aspects of locality

The notion of locality itself contains several different concepts; various authors
sometimes use the word with different definitions. Most physicists, because of the

---

[19] If Bohr had known the Bell theorem, he could merely have replied to EPR that their logical system was
inconsistent (see §4.1.3)!

strong impact of relativity, tend to immediately associate it with the notion of light-cone and the existence of a maximum possible velocity for the propagation of influences or messages. It is indeed perfectly possible to interpret the EPR reasoning in these terms. Nevertheless the original article is based on more general and basic concept of locality in physics, which predates any kind of relativity (Galileo or Einstein). This concept can be expressed[20] simply as "the influence of distant objects can be ignored if they are sufficiently remote", or more precisely as "the mutual influence of events decreases when their distance is increased, and can therefore made arbitrarily small by varying this distance". This purely spatial notion (no time is involved) is indeed one of the foundations of all experimental sciences: one assumes that the observations made in a laboratory depend on what is contained in the laboratory, but not on arbitrarily remote events (including human choices for experimental settings). In practice, if the observations made in each physics laboratory depended on the parameters chosen in all other locations in the world, it would probably become impossible to do any meaningful experiment! This notion of locality, which appears as a basic component of the scientific method, is sufficient for the EPR reasoning (see §3.3 of [123] for a generalization of the locality concept to systems with stochastic evolution). We will come back later (§4.2.2) to the importance of locality in the context of the Bell theorem.

In the relativistic view of the EPR argument, one can assume for instance that the two measurements are made simultaneously (in the reference frame of the source) at very large distances. Each interaction of one particle with its measurement apparatus then defines an event and a light cone originating from it; because the distance between the measurement apparatuses is large, each of these two events falls oustide the light cone associated with the other measurement. Under these conditions, no influence can propagate from one event to the other. The EPR reasoning then assumes that the elements of reality associated with a physical system at a given point of space-time can change (or appear) only causally, that is under the effect of other events lying inside the (past part of the) light cone of this point. Therefore, the elements of reality of the second spin cannot be affected by the measurement performed on the first spin. In this point of view, one can say that the main purpose of EPR is to restore a description of all physical processes in terms of space-time events influencing each other causally, as opposed to the view of standard mechanics where a measurement process can be a completely delocalized event covering an arbitrary large region of space (Fig. 4.5). EPR do not raise any particular objection against nondeterminism, but they require that the

---

[20] The words used in the EPR article are: "since at the time of measurement the two systems no longer interact, no real change can take place in the second system in consequence of anything that can be done to the first system".

influence of random events should not propagate faster than light. We come back to this discussion in §5.3.

3.3.3.b Relativity

The EPR reasoning is deeply rooted in relativity, which is not surprising for an argument proposed by Einstein. The basic idea is that, if a physical system is isolated and contained in a region R of space, what happens in this region of space can be predicted from all events inside the relativistic past cone of R. What happens inside the region may follow nondeterministic rules[21], but cannot depend on parameters that are arbitrarily adjusted by operators who are far away, outside this past cone. The decisions made by these operators are irrelevant for events taking place inside R (we come back to this subject in more detail in §4.2.2.b). In other words, causes can affect local events, deterministically or not, but these causes have to lie inside the past cone of each event (relativistic causality).

Bohr, on the other hand, took the opposite point of view. He emphasized in the introduction of his famous Como lecture [92] that "This postulate [the quantum postulate] implies a renunciation as regards the causal space-time coordination of atomic processes", and that "We learn from the quantum theory that the appropriateness of our usual causal space-time description depends entirely upon the small value of the quantum of action as compared to the actions involved in ordinary sense perceptions".

3.3.3.c Quantum nonseparability

Instead of invoking the role of all measurement apparatuses to define physical reality, as Bohr does to reject the EPR reasoning, one can introduce the notion of "nonseparability". The idea is that different quantum systems, when they have interacted in the past, no longer have their own physical properties in general; they are both part of a larger system, which is the only one possessing physical properties. One should then not try to separate (conceptually) the whole system into two smaller physical systems and attribute them properties; the whole system is nonseparable[22].

In general, separability is a notion that is conceptually distinct from locality. It is not necessarily related to space: two physical systems could occupy the same

---

[21] In contradiction with what is often read, Einstein himself did not oppose indeterminism per se; what he found shocking was the negation of the structure of space-time and of the notion of relativistic event (a measurement performed on an entangled quantum system is not a space-time event, §4.2.2.b). It is a pity that a sentence that he once wrote in a letter as a joke, "Gott würfelt nicht" (God does not play dice), has now become the only element that so many commentators have selected, while it is somewhat incidental. Even Einstein's close friend, Max Born, never really understood what Einstein was saying, as their correspondence and his later comments show [124] – for John Bell, this misunderstanding between giants was one of the drama of physics.

[22] Peres has proposed a mathematical criterion defining the separability of an ensemble of two physical systems described by a density operator (see §7.2.6).

region of space and remain distinct with their own physical properties (separable is not the same thing as separate). In the EPR reasoning, nevertheless, since the two particles are in remote regions of space, separation is assumed to entail separability. By contrast, quantum nonseparability can be stated as "even when two regions of space are disconnected and very far apart, one cannot always attribute separate physical properties to what they contain".

Quantum nonseparability is rooted in the way the quantum formalism describes systems and subsystems, and clearly related to the notion of entanglement (§7.1): a perfect description of the whole does not contain a perfect description of the parts. We mentioned earlier that Schrödinger considered entanglement as one of the most fundamental properties of quantum mechanics. Entanglement drastically restricts the number of physical properties that can be attributed to the subsystems; this number may even vanish in some cases. In other words, the "best possible description" (with a state vector) is not available to the subsystems; they have an additional level of indeterminacy, which never occurs in classical mechanics.

Invoking intrinsic quantum nonseparability is appealing, since the difficulties associated with the precise definition of a measurement apparatus do not immediately appear. It nevertheless remains delicate, in particular because correlations and entanglement can, according to the Schrödinger equation, propagate towards the macroscopic world (Schrödinger cat paradox) so that this absence of physical properties can reach any scale; it can end up with a situation where only the whole universe has physical properties!

Generally speaking, if the world was completely nonlocal (or completely nonseparable), all physical phenomena being totally interlaced, unravelling them would be beyond reach, and their scientific analysis by the experimental method would become impossible. The scientific progress of mankind have been made possible by the fact that, fortunately, nonlocality and nonseparability remain relatively rare phenomena, appearing mostly in physical situations designed by physicists to observe them.

3.3.4 Experiments

We will introduce in Chap. 4 the Bell inequalities, which are obtained by a direct continuation of the EPR argument. We will also mention a number of experiments testing these inequalities with various physical systems. Experiments illustrating specifically the original EPR argument have also been performed [125] and, more recently, with atomic Bose-Einstein condensates extended in space [126–128].

3.4 Generalizations

We have already mentioned that, in their historical article [112], EPR discuss the measurements of positions and momenta of two spinless entangled particles; the EPR state is chosen so that a measurement of the position of one particle determines the position of the other, while also a measurement of the momentum of one particle determines the momentum of the other. Nevertheless, in §3.3.1, instead of introducing the EPR argument with continuous position and momentum variables, we have chosen to consider the discrete results obtained by measuring components of two spins 1/2 in a singlet state. The reason is that, as we will see in Chapter 5, this leads more naturally to the Bell theorem. But this does not mean that the EPR theorem is limited to this case! In §3.4.1, we show that the reasoning can be generalized to any pair of quantum systems that are described by space of states with the same dimension. The theorem can also be generalized to Bose-Einstein spin condensates, which can be macroscopic; this is discussed in §3.4.2.

3.4.1 Generalized EPR states

The two spins were assumed to be in the singlet state $|\Psi\rangle$ given by:

$$|\Psi\rangle = \frac{1}{\sqrt{2}} [|+,-\rangle - |-,+\rangle] \tag{3.1}$$

where $|\pm,\mp\rangle$ is the common eigenstate of the Oz components of spins 1 and 2 with eigenvalues $\pm\hbar/2$ and $\mp\hbar/2$ respectively. But this singlet state has a property that is crucial for the EPR reasoning: if we choose any direction Ou in space, $|\Psi\rangle$ can also be written (rotation invariance):

$$|\Psi\rangle = \frac{1}{\sqrt{2}} [|+_{\mathbf{u}},-_{\mathbf{u}}\rangle - |-_{\mathbf{u}},+_{\mathbf{u}}\rangle] \tag{3.2}$$

Here $|\pm_{\mathbf{u}},\mp_{\mathbf{u}}\rangle$ are the common eigenstates of the Ou components of spins 1 and 2 with eigenvalues $\pm\hbar/2$ and $\mp\hbar/2$ respectively. Relation (3.2) therefore indicates that $|\Psi\rangle$ keeps the same form when any quantization axis is used. This is crucial for the EPR reasoning, where the measurement of several incompatible observables is essential: whatever arbitrary direction $\mathbf{u}$ is chosen by Alice to perform her measurement, immediately after her measurement, Bob's spin component along the same direction is determined. Assuming local realism/causality, the reasoning then shows that it was already determined before the measurement. All components of Bob's spin are thus determined in the initial state (as well as all components of Alice's spin by symmetry), which is contradictory with standard quantum mechanics (where nonparallel components correspond to incompatible observables).

Hemmick and Shakur discuss a generalization of this result, which they call

"Schrödinger paradox" or "Schrödinger theorem" (chapter 4 of [129]); they show that, if two quantum systems have spaces of states with the same arbitrary dimension N (not necessarily 2), one can construct "generalized EPR states" where all observables of both systems are perfectly correlated. Consider first an observable A and the associated orthonormal basis $\{|\theta_i\rangle\}$ of eigenstates ($i = 1, 2, ..., N$); we assume that the eigenvalues a_i of A are not degenerate. We introduce the following state of the whole system:

$$|\Psi\rangle = \frac{1}{\sqrt{N}} \sum_{i=1}^{N} |1 : \theta_i\rangle |2 : \theta_i\rangle \qquad (3.3)$$

If observable A (1) is measured on the first quantum system and provides a result a_p, after the measurement the projection postulate selects a single term $i = p$ in (3.3). One then immediately sees that, if A (2) is subsequently measured on the second quantum system, the same result a_p is necessarily obtained. We are in an EPR situation of perfect correlations but, for the moment, this result is limited to one observable A only. We now show that any observable of one particle is perfectly correlated with another observable for the other particle.

Consider any observable B and the orthonormal basis of its eigenstates $|\varphi_i\rangle$ ($i = 1, 2, ..., N$). The change of basis from the $\{|\theta_i\rangle\}$ to the $\{|\varphi_i\rangle\}$ is described by an unitary operator U:

$$|\varphi_i\rangle = U |\theta_i\rangle \qquad (3.4)$$

with matrix elements:

$$\langle \theta_i| U |\theta_j\rangle = U_{ij} \qquad (3.5)$$

These elements obey the relation:

$$\sum_{k=1}^{N} U_{ki}^* \, U_{kj} = \delta_{ij} \qquad (3.6)$$

This relation can be shown by writing:

$$\langle \varphi_i |\varphi_j\rangle = \sum_{k} \langle \varphi_i |\theta_k\rangle \langle \theta_k |\varphi_j\rangle$$

$$= \sum_{i} \langle \theta_i| U^\dagger |\theta_k\rangle \langle \theta_k| U |\theta_j\rangle = \sum_{k=1}^{N} U_{ki}^* \, U_{kj} \qquad (3.7)$$

Since $\langle \varphi_i |\varphi_j\rangle = \delta_{ij}$, we then obtain the relations (3.6), which express nothing but that the basis $\{|\varphi_i\rangle\}$ is orthonormal (in other words, that U is unitary).

We now introduce another operator $\overline{U}$, defined by a complex conjugation of the matrix elements of U in the $\{|\theta_i\rangle\}$ basis:

$$\overline{U}_{ij} = \langle \theta_i | \overline{U} | \theta_j \rangle = \langle \theta_i | U | \theta_j \rangle^* = U_{ij}^* \tag{3.8}$$

A complex conjugation of relation (3.6) provides:

$$\sum_{k=1}^{N} \overline{U}_{ki} \, \overline{U}_{kj}^* = \delta_{ij} \tag{3.9}$$

This relation shows that $\overline{U}$ is unitary as well. We can therefore obtain another orthonormal basis $\{|\overline{\varphi}_i\rangle\}$ ($i = 1, 2, ..., N$) by defining the kets:

$$|\overline{\varphi}_i\rangle = \overline{U} |\theta_i\rangle \tag{3.10}$$

We then have:

$$\langle \theta_i | \overline{\varphi}_k \rangle = \langle \theta_i | \overline{U} | \theta_k \rangle = \langle \theta_i | U | \theta_k \rangle^*$$
$$= \langle \theta_i | \varphi_k \rangle^* = \langle \varphi_k | \theta_i \rangle \tag{3.11}$$

Since $\langle \theta_i | \theta_j \rangle = \delta_{ij}$, we can write:

$$|\Psi\rangle = \frac{1}{\sqrt{N}} \sum_{i,j=1}^{N} \langle \theta_i | \theta_j \rangle \; |1 : \theta_i\rangle |2 : \theta_j\rangle$$

$$= \frac{1}{\sqrt{N}} \sum_{i,j,k=1}^{N} \langle \theta_i | \varphi_k \rangle \langle \varphi_k | \theta_j \rangle \; |1 : \theta_i\rangle |2 : \theta_j\rangle$$

$$= \frac{1}{\sqrt{N}} \sum_{i,j,k=1}^{N} \langle \theta_i | \varphi_k \rangle \langle \theta_j | \overline{\varphi}_k \rangle \; |1 : \theta_i\rangle |2 : \theta_j\rangle \tag{3.12}$$

where we have used (3.11) to transform the second line into the third. The summations over i and j then reconstruct the kets $|\varphi_k\rangle$ and $|\overline{\varphi}_k\rangle$ respectively, so that we obtain:

$$|\Psi\rangle = \frac{1}{\sqrt{N}} \sum_{k=1}^{N} |1 : \varphi_k\rangle |2 : \overline{\varphi}_k\rangle \tag{3.13}$$

This shows that $|\Psi\rangle$ remains unchanged when the unitary transformation U is applied to the first particle and the unitary transformation $\overline{U}$ to the second particle.

Therefore $|\Psi\rangle$ can be expanded onto the eigenbasis of any operator $B(1)$ for particle 1, provided the basis for particle 2 is changed according to the unitary operator $\overline{U}$; the $|2 : \overline{\varphi}_k\rangle$ are the eigenstates of another observable $\overline{B}(2)$. The ket $|\Psi\rangle$ then keeps exactly the same form, namely a sum of N products containing orthonormal states for the two particles. As above, if $B(1)$ is measured (we have assumed that this operator has no degenerate eigenvalues), the projection postulate will select only one of the products, which means that the state of the whole systems ends up in a product state that is an eigenstate of $\overline{B}(2)$. In other words, the two measurements give results that are perfectly correlated, so that the measurement of $B(1)$ is equivalent to a measurement of $\overline{B}(2)$, exactly as in the situation with two spins $1/2$ in a singlet state.

The EPR reasoning then shows that, if local realism (or causality) applies, the results of measurements reveal properties of the two quantum systems that must pre-exist before any measurement. Moreover, whatever observable $B(1)$ is chosen, there exists an observable $\overline{B}(2)$ that is perfectly correlated with $B(1)$: measuring $B(1)$ will provide the result of measurement of $\overline{B}(2)$ on particle 2 with certainty. The conclusion is that, if two quantum systems are described by a state such as (3.3), not only some observables correspond to elements of reality that pre-exist any measurement, but all of them!

In the context of the BKS theorem and contextuality (§ 6.4), measurements on particles of spin 1 (which corresponds to $N = 3$) are often considered. What we have seen shows that it is possible to entangle two spin 1 particles in such a way that the perfect knowledge of the value of any observable of one particle can be known with certainty from the measurement of another observable on the other particle; see §4.4.3 of [129] for a detailed calculation of the observables involved, and the relation to the Kochen-Specker theorem. Entangled particles of spin 1 are also considered in the derivation of the Conway-Kochen theorem [130, 131].

3.4.2 The EPR argument for macroscopic systems

Interestingly, the EPR argument also applies in the context of another discovery made by Einstein, namely Bose–Einstein condensation. The original EPR argument involved two microscopical particles, atoms for instance. The essence of Bohr's point of view then hinges on the idea that microscopic systems do not possess physical reality independently of the measurement apparatuses, since physical reality cannot be defined at this level. But it turns out that quantum mechanics also predicts that similar correlations should be observed with systems that are arbitrarily large, and can therefore be macroscopic [132]; this requires that they should be initially in a special quantum state, a "double Fock state". Such a state is not very common, but could be reached for instance through the phenomenon of Bose–Einstein condensation.

Here we just summarize the general idea, since we come back to this subject in §10.4.3. We assume that two large condensates, associated with two Fock states with different spin states, overlap in two different regions of space. Initially, according to quantum mechanics, the relative phase of the two condensates is totally undetermined, so that none of the overlap region contains any transverse spin orientation. But, if spin measurements are performed along transverse directions in one of the regions, a transverse spin polarization spontaneously appears under the effect of quantum measurement in this region. The direction of this spontaneous polarization is completely random. Nevertheless, since this process fixes the relative phase of the two condensates, theory predicts that a spin polarization also appears in the other region of overlap, with a transverse direction that is parallel to that in the measurement region. One then has a remote consequence of the measurement, without any interaction between the two regions, but just under the effect of state vector reduction.

Clearly, the EPR argument then directly applies: how can a spin polarization spontaneously appear in an arbitrarily remote region of space, without any interaction to create it? How can the physical reality associated to a region of space be affected by an arbitrarily distant measurement? But in this case the big difference with the original EPR situation is that the number of particles involved is arbitrarily large, so that the transverse spin polarizations can be macroscopic. If the spins carry magnetic moments, they will create a macroscopic transverse magnetization, which can be detected directly with macroscopic devices such as ordinary compasses. For arbitrarily large macroscopic objects, it seems difficult to invoke the Bohrian argument, and to decide that they have access to physical reality only when they are associated with a well-defined measurement apparatus!

In this case, the EPR element of reality pre-existing the first measurement is the relative phase of the two condensates with spin. The notion is related to spontaneous symmetry breaking and the spontaneous appearance of a phase in a system that undergoes a superfluid transition (Anderson phase [133]). Of course, we do not know what Bohr would have said concerning this macroscopic version of the EPR thought experiment. The relative phase of Bose–Einstein condensates also contains some interesting effects leading to violations of quantum nonlocality [134], but nonlocal effects will be discussed in the next chapter.

4

Bell Theorem

As the EPR argument, the Bell theorem [135] can be seen in several different ways. Initially, Bell invented it as a direct logical continuation of the EPR theorem: his idea was to take the existence of the EPR elements of reality seriously, and to push the reasoning further by introducing them explicitly into the mathematics with the notation λ; he then proceeded to study all possible kinds of correlations that can be obtained from the fluctuations of one or several variables λ, making the condition of locality explicit in the mathematics (locality was already useful in the EPR theorem, but not used in equations). The reasoning develops within determinism (considered as proved by the EPR reasoning) and classical probabilities; it studies in a completely general way all kinds of correlations that can be predicted from the fluctuations of some classical common cause in the past – if one prefers, from some random determination, of the initial state of the system. This leads to the famous inequalities.

This is the way we will first introduce the Bell theorem in §4.1. Another less general view of the theorem is possible, obtained by disconnecting it from the EPR reasoning. It is then seen as an impossibility theorem for additional (or "hidden") variables (see §11.8): if these variables are arbitrarily introduced, and if their evolution is local, the resulting theory cannot reproduce all predictions of standard quantum mechanics, even if a very complicated evolution is postulated. It is unfortunate that this rather limited version of the theorem should be the only one discussed in some texts.

In §4.2, we discuss various generalizations of the Bell theorem, based on different sets of assumptions. In particular, the scope of the theorem is not limited to deterministic theories; for instance, the λ variables may determine the probabilities of the results of future experiments, instead of the results themselves without nullifying the theorem (this is also discussed in more detail in Appendix B).

Finally, in §4.3, we discuss the present situation created by the experimental observation of violations of the Bell inequalities, and provide a presentation of the various "loopholes" that could be invoked to ignore the contradictions between local realism (or local causality, i.e. the EPR assumptions, as we have seen in Chapter 3), and the experimental observations.

Generally speaking, as we will see, the essential condition for the validity of the Bell theorem is locality: all kinds of fluctuations can be assumed, but their effect must affect physics only locally. If we assume that throwing dice in Paris may influence physical events taking place in Tokyo, or even in other galaxies, the proof of the theorem is no longer possible. For nonspecialized general discussions of the Bell theorem, see for instance [56, 95, 136, 137].

In Chapter 5, we will discuss a few other inequalities that also arise from local realism/causality, and we will investigate the relations between the Bell theorem and relativity (no signal can be transmitted at a distance without any delay).

4.1 Bell inequalities

The Bell inequalities are relations satisfied by the average values of products of random variables that are correlated classically – by this we mean that their correlations originate from some fluctuating past event that has influenced their values, as in the preceding chapter for the peas. The inequalities are especially interesting in cases where they are contradictory with quantum mechanics; one of these situations occurs in the spin version of the EPR argument [120], already introduced in §3.3.1, where two spin 1/2 particles undergo measurements in remote regions of space. This is why we begin this section by briefly recalling the predictions of quantum mechanics for such a physical system – the only ingredients we need from quantum mechanics at this stage are the predictions concerning the probabilities of results. Then we will return to the EPR–Bell argument, discuss its contradictions with quantum mechanics, and finally emphasize the generality of the theorem.

4.1.1 Quantum mechanics: two spins in a singlet state

We come back to the experiment shown in Figure 3.2, where two spins 1/2 are emitted by a source S in a singlet spin state and propagate in opposite directions. Their spin state is described by:

$$| \Psi >= \frac{1}{\sqrt{2}} \left[|+, -\rangle - |-, +\rangle \right] \tag{4.1}$$

with the usual notation: the two-spin state $|\pm, \mp\rangle$ contains the first spin in an eigenstate with eigenvalue $\pm\hbar/2$ of its component along Oz^1, and the second in an eigenstate with eigenvalue $\mp\hbar/2$. When they reach distant locations, the spin on the left

---

[1] One can show that the singlet state is invariant under any rotation, which implies that it has the same form (4.1) if any quantization axis is used.

is submitted to a spin measurement with a Stern–Gerlach apparatus oriented along the direction defined by angle a, while the other spin undergoes a similar measurement along the direction defined by b.

If:

$$\theta_{ab} = a - b \tag{4.2}$$

is the angle between the directions defined by a and b, quantum mechanics predicts that the probability for a double detection of results $+1, +1$ (or of $-1, -1$) is:

$$\mathcal{P}_{(+,+)} = \mathcal{P}_{(-,-)} = \frac{1}{2}\sin^2\frac{\theta_{ab}}{2} \tag{4.3}$$

while the probability of two opposite results is:

$$\mathcal{P}_{(+,-)} = \mathcal{P}_{(-,+)} = \frac{1}{2}\cos^2\frac{\theta_{ab}}{2} \tag{4.4}$$

This is all that we need to know, for the moment, of quantum mechanics: probabilities of the results of measurements. An important remark is that, if $\theta_{ab} = 0$ (when the orientations of the measurement apparatuses are parallel), the formulas predict that the two probabilities (4.3) vanish, while the others are equal to 1/2. This means that the condition of perfect correlations required by the EPR reasoning is fulfilled (actually, the results of the experiments are always opposite instead of equal, but it is sufficient to change the arbitrary direction of one of the axes to make them equal – see the discussion of §3.3.1.a).

A state such as (4.1), where the properties of the two subsystems (two spins in this case) are correlated inside the state vector itself, is called an "entangled state" in quantum mechanics. We will come back to the notion of quantum entanglement in Chapter 7 in more detail.

4.1.2 Local realism: proof of the BCHSH inequality

We start from the EPR theorem and, following Bell, we assume that λ represents the "elements of reality" associated with the spins; it should be understood that λ is only a concise notation that may summarize a vector with many components. The number of elements of reality is arbitrary – no limitation is introduced by this notation – and, in fact, one can even include in λ components that play no special role in the problem. The only thing that really matters is that λ should contain all the information concerning the results of possible measurements performed on the spins. Another usual notation is to write these results A and B, and to use small letters a and b for the settings (parameters) of the corresponding apparatuses. Clearly, A and B may depend, not only on λ, but also on the settings a and b; nevertheless, locality requests that b has no influence on the result A (since the distance between the locations of the measurements can be arbitrarily large); conversely, a has no influence on result B. We therefore call $A(a, \lambda)$ and $B(b, \lambda)$ the corresponding functions; their values are either $+1$ or -1.

In what follows, it is sufficient to consider two directions only for each separate measurement; we then use the simpler notation:

$$A \equiv A(a, \lambda) \qquad\qquad A' \equiv A(a', \lambda) \qquad\qquad\qquad (4.5)$$

and:

$$B \equiv B(b, \lambda) \qquad\qquad B' \equiv B(b', \lambda) \qquad\qquad\qquad (4.6)$$

For each pair of particles, λ is fixed, and the four numbers have well-defined values (which can only be ± 1). With Eberhard [138], we notice that the sum of products:

$$M(\lambda) = AB - AB' + A'B + A'B' = A(B - B') + A'(B + B') \qquad (4.7)$$

is always equal to either $+2$ or -2; this is because one of the brackets in the right-hand side of this equation always vanishes, while the other is ± 2. Now, if we take the average $\langle M \rangle$ of $M(\lambda)$ over a large number of emitted pairs (average over λ), we have:

$$\langle M \rangle = \langle AB \rangle_\lambda - \langle AB' \rangle_\lambda + \langle A'B \rangle_\lambda + \langle A'B' \rangle_\lambda \qquad (4.8)$$

where $\langle AB \rangle_\lambda$ denotes the average over λ of the product $AB \equiv A(a, \lambda)B(b, \lambda)$, and similar notations are used for the three other terms in the sum. Since each $M(\lambda)$ is equal to ± 2, we necessarily have:

$$-2 \leq \langle M \rangle \leq +2 \qquad\qquad\qquad (4.9)$$

This is the so-called BCHSH (Bell, Clauser, Horne, Shimony, and Holt) form [139] of the Bell theorem. The inequality is necessarily satisfied[2] by the average values of all possible kinds of pairs of measurements providing random results ± 1, whatever the mechanism creating the correlations may be, as long as locality is obeyed: A does not depend on setting b, and B does not depend on setting a.

Any "local realist" theory must therefore make predictions obeying relation (4.9). Realism is necessary, since we have used the EPR notion of elements of reality to derive the existence of functions A and B. Locality (§3.3.3) is also necessary since it forbids A to depend on b and B to depend on a. In §4.1.4, we give a more detailed discussion of the logical content of the Bell theorem.

In this context, randomness can only arise from the fluctuations of some common cause in the past; but this can also be seen as a very natural general assumption concerning physical processes. Figure 4.1 illustrates this situation, and Figure 4.2

---

[2] In our definition (4.7) of M, the term AB' has a minus sign and the three others a plus sign, but this is arbitrary. By writing $AB + AB' \pm A'B \mp A'B' = A(B + B') \pm A'(B - B')$ and $\pm AB \mp AB' + A'B + A'B' = \pm A(B - B') + A'(B + B')$, we can obtain four inequalities where the minus sign goes to any of the four terms.

shows the corresponding space-time representation of the events; the lines connecting the cause and the effects must remain within the light cone $x = \pm ct$ if relativity
is obeyed (x is the position, t the time, and c is the velocity of light). But the inequality remains valid if, for instance, other fluctuating processes perturb the particles
while they are on their way to the measurement apparatuses, or while they interact
with the measurement apparatuses themselves (Figure 4.3). It then becomes necessary to include into λ new components associated with the corresponding random
processes; this may change the distribution of this multidimensional variable, but
not the fact that the average of a number $M = \pm 2$ necessarily obeys (4.9).

With such a straightforward proof, one can expect the validity of the inequality
to apply in many situations; this is indeed true, as we will discuss in more detail
in §§4.2.1 and 4.2.2. For the moment, we just remark that the result is independent
of the interpretation of the variable λ, which does not have to be defined as an
additional variable or an element of reality. For instance, we can assume that λ is
simply used to label the realization of the experiment: $\lambda = 1$ corresponds to the first
experiment, $\lambda = 2$ to the second, ..., $\lambda = N$ to the last of a series of experiments.
If, for each realization, the four numbers A, B, A', and B' have well-defined values,
all equal to ± 1, the number M is also defined for each realization and is necessarily
equal to -2 or $+2$. Whatever arbitrary values of the four numbers may occur during
any series of N experiments, mathematically, there is no way for the sum of the M
to exceed $2N$ or to be less than $-2N$. Dividing this sum by N then leads to (4.9);
the mere existence of the four numbers for each realization is sufficient to obtain
this result. In other words, the existence of two functions $A(a, \lambda)$ and $B(b, \lambda)$ of the
settings a and b and of the rank of the experiment $\lambda = 1, 2, 3, ..$ is sufficient to
derive the BCHSH inequality.

4.1.3 Contradiction with quantum mechanics

The simplicity and the generality of the proof are such that one could reasonably
think that any sensible physical theory should automatically give predictions that
obey this inequality. The surprise is to realize that quantum mechanics does not.
From (4.3) and (4.4), we can calculate the average value $\langle \Pi(a, b) \rangle$ of the product of
the two results ± 1 obtained by measuring the components of the two spins along
directions a and b with relative angle θ_{ab}:

$$\langle \Pi(a, b) \rangle = \mathcal{P}_{(+,+)} + \mathcal{P}_{(-,-)} - \mathcal{P}_{(+,-)} - \mathcal{P}_{(-,+)} = -\cos \theta_{ab} \qquad (4.10)$$

This gives the quantum equivalent of the average over λ of the product
$A(a, \lambda)B(b, \lambda)$ in a local realist theory. To obtain the quantum equivalent $\langle Q \rangle$ of
the combination of four products in expression (4.7), we calculate the same com-

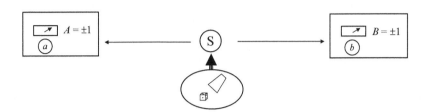

Figure 4.1 A source S emits two particles, which propagate to two remote measurement apparatuses having respective settings a and b; each apparatus provides
a result ±1. The oval containing a die below the source symbolizes a random
fluctuating process that controls the conditions under which the two particles are
emitted, and therefore determines their properties. Correlations are observed between the results obtained with the two remote measurement apparatuses; they are
consequences of the common random properties shared by the particles of each
pair, initially created by the random process at the source.

bination of averages of products of results:

$$\langle Q \rangle = \langle \Pi(a,b) \rangle - \langle \Pi(a,b') \rangle + \langle \Pi(a',b) \rangle + \langle \Pi(a',b') \rangle$$
$$= -\cos\theta_{ab} + \cos\theta_{ab'} - \cos\theta_{a'b} - \cos\theta_{a'b'} \qquad (4.11)$$

We now assume that the four directions are in a same plane, and that each of the
vectors in order $\mathbf{a}$, $\mathbf{b}$, $\mathbf{a}'$, and $\mathbf{b}'$ makes an angle of 45° with the preceding vector
(cf. Figure 4.4); all cosines are then equal to $1/\sqrt{2}$, except $\cos\theta_{ab'}$ which has the
opposite value $-1/\sqrt{2}$. Then $\langle Q \rangle$ reaches value $-2\sqrt{2}$; if the directions of $\mathbf{b}$ and $\mathbf{b}'$
are reversed, one obtains $2\sqrt{2}$. But we have seen that all local realist theories obey
the BCHSH inequality (4.9) with bounds ±2. In both cases, quantum mechanics
therefore predicts a violation of the inequality by a factor $\sqrt{2}$, more than 40% (it
turns out that $\sqrt{2}$ is the maximal possible violation predicted by quantum mechanics – cf. §5.2). As simple as the cosine variation of (4.10) may look, no local realist
theory can obtain it. The EPR–Bell reasoning therefore leads to a quantitative and

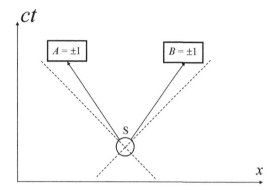

Figure 4.2 Space-time diagram associated with the events shown in Figure 4.1; the horizontal axis represents space in a simplified view (one dimension of space), the vertical axis shows the product ct of time t by the velocity of light c. According to relativity, the two arrows joining the emission of the particles to the measurement events must remain inside the light cone $x = \pm ct$ (dotted lines).

significant contradiction with quantum mechanics, which therefore is not a local realist theory in the EPR sense.

How is this contradiction possible, and how can a reasoning that is so simple be incorrect within quantum mechanics? Different answers are possible:

(i) Bohr died in 1962 and could not react to the Bell theorem, published in 1964. But we know that he rejected the EPR reasoning, as we have seen in Chapter 4; he would probably also have rejected the existence of four pre-existing numbers A, A', B, and B'. Under these conditions, the reasoning of §4.1.2 is not possible, and the BCHSH inequality is not valid. So, Bohr would presumably have considered the theorem as mathematically correct but inappropriate in quantum physics (it cannot be used in the quantum description of the experiment considered).

But, maybe after all, he would have accepted to reason on these numbers as unknown quantities to be determined further, as one often does in algebra; is it then not possible to obtain the inequality? Again, if one takes Bohr's general approach, the answer to the question is still no. As we already discussed in §3.3.2, his view is that only the whole experiment should be considered, without distinguishing in it two separate measurements that would be performed on each particle: a single indivisible two-particle measurement takes place. A fundamentally random pro-

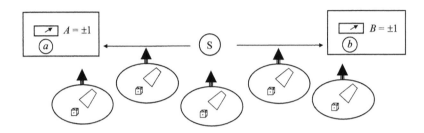

Figure 4.3 Fluctuating uncontrolled causes may influence, not only the emission of particles, but also their propagation and the measurement apparatuses, without changing the validity of the BCHSH inequality.

cess then occurs simultaneously over the whole region of space occupied by the experimental setup; it is delocalized, as schematized in Figure 4.5, to be contrasted with Figures 4.1 and 4.3. The process has no space-time location; it escapes any precise space-time description (we come back in more detail to the discussion of delocalized randomness at the end of §4.2.2.b).

Functions A and B then depend on the two settings, so that they should be written $A(a, b)$ and $B(a, b)$, in an explicitly nonlocal way. Instead of two numbers A and A', we now have four, which are $A = A(a, b)$, $A' = A(a', b)$, as well as $A'' = A(a, b')$ and $A''' = A(a', b')$; similarly, B and B' are replaced by four numbers. Altogether, we have eight numbers instead of four; the proof of the BCHSH inequality is then no longer possible, and the contradiction disappears.

(ii) One may prefer a more local view of the process of measurement and retain the concept of single-particle measurement as meaningful in this context. To avoid the contradiction with the predictions of quantum mechanics, one then considers that it is wrong to attribute well-defined values A, A', B, and B' to each emitted pair, since only two of them can be measured in any realization of the experiment. Therefore, it is not legitimate to speak of these four quantities, or reason on them,

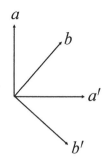

Figure 4.4 The position of the four vectors a, b, a', and b' corresponding to a maximal violation of the BCHSH inequality for two spins 1/2 in a singlet state. These vectors define the components of the spins to measure, along a or a' for the particle on the left, along b or b' for the particle on the right, leading to four distinct experimental setups. The only case where a negative correlation between the results is obtained is with (a, b'), since the angle between the vectors exceeds 90°.

even as unknown quantities (we come back to this point with the discussion of counterfactuality in §4.3.2). As nicely summarized by Peres [140], "unperformed experiments have no result"[3]! Wheeler expresses a similar idea when he writes: "No elementary phenomenon is a phenomenon until it is a registered (observed) phenomenon" (p. 184 of [141]).

(iii) As for Wigner, he emphasizes in [142] that the proof of the Bell inequalities relies on a very simple notion within realism (local or nonlocal): the number of categories into which one can classify all pairs of particles[4] or, equivalently, all realizations of the experiment. Each category is associated with well-defined results of measurements for the various choices of the settings a and b (configurations of the experiment) that are considered; in any long sequence of repeated experiments, each category contributes with some given weight equal to its probability of occurrence (a positive or zero number). For one single type of experiment that can give R different results, the number of categories is R. When one considers P dif-

---

[3] This is distinct from Bohr's view, where unperformed experiments may have results, but where experiments and results can only be expressed in terms of the whole experimental setup, and therefore of both parameters a and b.

[4] In this reference, Wigner actually reasons explicitly in terms of hidden variables; he considers domains for these variables, which correspond to given results for several possible choices of the settings. But, from the EPR point of view, these domains also correspond to categories of pairs of particles, which is why we use this notion of categories.

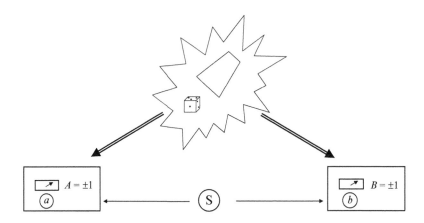

Figure 4.5 Scheme associated with the quantum description of a measurement involving two remote apparatuses. This scheme may be compared to those of Figures 4.1 and 4.3. The big flash at the top of the figure symbolizes a fundamentally random and nonlocal process, by contrast with ovals below the events used in previous figures to represent fluctuating processes resulting from uncontrolled local perturbations. The quantum process is inherently extended in space, so that no space-time description similar to that of Figure 4.2 is possible (randomness "comes from nowhere" in this case).

ferent possible configurations, each with R different possible results, the number of categories becomes R^P.

For the sake of simplicity, here we assume that each setting can take two different values (but, in §5.1, we study Wigner's original case with three different values). If, following Bohr, we consider that the experiment as a whole must be considered, then each configuration is given by the choice of a pair a, b among $P = 4$ possibilities; since each of them can give 4 different pairs of results, the number of categories is 4^4. One can then attribute to each of them appropriate weights and reproduce the predictions of quantum mechanics, and no special contradiction is obtained.

But then Wigner notes that, if one adds the notion of locality, for each pair the result on one side becomes independent of the setting on the other side, keeping only a dependence on the local setting. Each category then becomes the intersection of two subensembles: one associated with the first side of the experiment depending on a only, which is chosen among $2 \times 2 = 4$ possibilities (2 for the possible choices of a, and 2 for the possible results); another associated with the second side of the experiment depending on b only, which is also chosen among 4 possibilities. Altogether, the total number of categories is now only 2^4. To each of these categories, one can ascribe four numbers, A, A', B, and B', which are well-defined and all equal to ± 1, so that their combination in (4.7) has values $M = \pm 2$; the BCHSH inequalities are then immediately obtained. In other words, Wigner points out that the mathematical origin of the Bell inequalities lies precisely in the possibility of distributing all pairs with positive probabilities into a smaller number of categories than from a nonlocal point of view; the difference between these numbers is the origin of the contradiction[5].

4.1.4 Logical content

Figure 4.6 summarizes the two parallel lines of reasoning that lead to contradictory conclusions, one assuming local realism/causality, the other quantum mechanics.

A general way to express the Bell theorem in logical terms is to state that the following system of three assumptions[6] (which could be called the EPR assumptions) is self-contradictory:

(1) Validity of the EPR notion of "elements of reality".

(2) Locality (§3.3.3) .

(3) The predictions of quantum mechanics are always correct.

Indeed, if one starts from the first two assumptions, one can then discover situations where the third assumption is not true, hence a self-contradiction. The Bell theorem then becomes a useful tool to build a "reductio ad absurdum" reasoning: it shows that, among all three assumptions, one (at least) has to be given up. If Einstein's program was to assume local realism and, at the same time, that all predictions of quantum mechanics are correct, this program is impossible. This conclusion is purely logical, independent of any experimental result.

One may notice that the reasoning actually contains a fourth assumption:

(4) the measurement settings a and b are freely chosen by the experimenters[7], and not the physical consequence of some past event.

This is the "free will" hypothesis. It is so general in all experimental scientific disciplines that, often, it is not even mentioned: one naturally assumes that experimenters can freely decide what kind of experiments they will perform, and what parameters they will choose for them. The opposite attitude would be to consider

---

[5] It has also been noticed [143] that the Bell inequalities can be seen as applications of theorems on marginal distributions in the theory of probabilities [144]; the mathematical proof of the inequalities relies on the existence of one common probability space for all the relevant (measured) random variables (this is also related to the notion of counterfactuality).

[6] Here we focus ourselves only on one set of assumptions leading to the theorem, the set we have used to derive it. Other possible sets of assumptions to derive the theorem will be discussed in §4.2.1.

[7] One can even assume that they choose their value after the emission of the two particles by the source while they are still flying towards them.

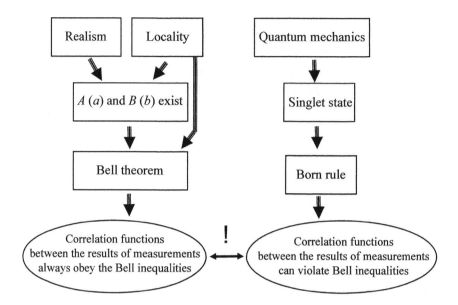

Figure 4.6 Two parallel different reasonings, starting from different assumptions, reach contradictory conclusions: local realism or causality implies that the Bell inequalities should always be obeyed, the predictions of quantum mechanics that they may be violated. Therefore, all three assumptions (EPR realism, locality, the predictions of quantum mechanics are correct for a singlet spin state) cannot be simultaneously valid.

that these decisions are in fact predetermined by some past event and by the propagation of its influences to the distant experimenters (see in §4.3.1.c the discussion of fatalism, or superdeterminism). What is then at stake is the notion of a free variable in a physical theory: a and b are considered as free variables that describe the experimental parameters, which are external to the theory (not solutions of some dynamical equations starting from a common initial condition in the past).

When stated in this way, the Bell theorem appears as general conceptually, but experimentally inaccessible since assumption 3 is obviously too broad to be tested. One may then prefer another form of the logical self-contradiction, where this assumption is replaced by two more specific statements concerning one particular experiment only (for instance, two 1/2 spins in a singlet state, or two photons emitted in an atomic 0-1-0 cascade):

(3′) in this experiment, the prediction of quantum mechanics concerning the total correlations observed when the settings are the same ($a = b$) are correct (this assumption leads to the existence of the EPR elements of reality).

(3″) the predictions concerning the correlations for different settings are also correct.

Removing either (3′) or (3″) from the set of assumptions is also sufficient to remove the self-contradiction. The motivation of the experimental tests of the Bell inequalities was precisely to check if it was not assumption (3′) or (3″) that should be abandoned. Maybe, after all, the Bell theorem is nothing but an efficient pointer towards unexpected situations where the predictions of quantum mechanics are so paradoxical that they are actually wrong? Such was the hope of some theorists, as well as an exciting challenge to experimentalists.

Reference [145] contains a clear discussion of the logical content of the EPR argument and the Bell theorem and their relations to realism, locality, and separability.

4.1.5 Contradiction with experiments

In an experiment, it is of course impossible to obtain at the same time all 4 results corresponding to A, A', B, and B'; for each realization, a given choice has to be made for Alice's and Bob's settings, so that only two of these numbers are registered. Now, if two successive experiments are performed, one with the choice (a,b) for instance, and the other with the choice (a,b') for instance, more results are of course obtained. But, even in realizations for which the observed results A remain the same, nothing guarantees that the values of B and B' are identical (since each of them is measured only once). Generally speaking, there is no way to associate these 4 numbers to a given realization of the experiment – see Peres's and Wheeler's quotations in §4.1.3 (ii). Expression (4.7) is inaccessible in quantum mechanics.

One can nevertheless have access to average values. For this purpose, in a first step one chooses given values for the two settings, one performs an arbitrarily long series of experiments with a given choice, and one registers the results. For each realization, one then calculates the product of the two results, and takes its average value over all realizations; this provides $\langle \Pi(a, b) \rangle$ for instance. Then, repeating the experiment for the other three choices, one obtains other series of results, from which one can derive average values of $\langle \Pi(a, b') \rangle$, $\langle \Pi(a', b) \rangle$, and $\langle \Pi(a', b') \rangle$. Finally, expression (4.11) provides the average $\langle M \rangle$; one can check whether or not the BCHSH inequallity (4.9) is obeyed.

In 1967, an experiment was undertaken by Kocher and Commins [125] with photons emitted during an atomic cascade between three levels of atomic mercury having angular momenta $J = 0 \rightarrow 1 \rightarrow 0$. In this case, quantum mechanics predicts correlations that are similar to those with two spins 1/2 in a singlet state; the directions of photon polarization analyzers play the role of the orientations of the Stern–Gerlach analyzers for spins 1/2 (it is sufficient to divide all measurement angles by 2 to transpose the results from spins to photons). This experiment tested that the predictions of quantum mechanics are correct when the two analyzers are parallel (or perpendicular), corresponding to perfect correlations – in other words, assumption (3′) was tested successfully in this case.

Two years later Clauser, Horne, Shimony, and Holt [139] realized that this kind of experiment could be extended to provide an experimental test of assumption (3″), with measurements of correlations rates at various angles of the polarizers. They proposed a generalization of the Bell theorem to a new form (the BCHSH inequality) that was more suitable to experimental tests than the original Bell inequality. Following this line, in 1972 correlations for different settings (and therefore assumption 3″) were tested by Freedman and Clauser [146] in an atomic cascade $J = 0 \rightarrow 1 \rightarrow 0$ of the calcium atom; this provided a confirmation of the quantum predictions and led to violations of local-realist inequalities by six standard deviations. In 1976, three other experiments were performed: one by Clauser [147] with photons emitted during $J = 1 \rightarrow 1 \rightarrow 0$ cascades in mercury, one by Fry and Thompson [148] with the same cascade, and one with protons by Lamehi-Rachti and Mittig [149]. Again the predictions of quantum mechanics were fully confirmed. Reference [150], in particular its chapter 7, provides a detailed report of the progressive evolution of the ideas, both theoretical and experimental.

In the eighties, the results were made more and more precise and convincing in a series of experiments performed in Aspect's group, using the same calcium transition as [146]. One of these new experiments by Aspect, Grangier, and Roger [151] included a study of the effect of the distance between the detectors and the source on the quantum correlations for various polarizations, in order to check that this distance plays no role as predicted by quantum mechanics. Another used two-channel detections [152] providing actual ±1 signals, while in previous experiments one polarization was just absorbed and gave no signal at all (this experiment led to violations by 15 standard deviations!). The third included a time component in a random choice of the polarizations of the detections [153] – see also [154] for experiments using a two-photon transition between two levels of zero angular momentum in deuterium.

Ever since, the experiments have been constantly improved. One big step forward was the use of parametric down conversion of light to obtain efficient two-photon sources, leading to violations by 22 standard deviations [155], and even measurements of correlations over more than 10 km. [156], more than 100 km.

[157] in 2010, and in 2017 [158][8] more than 1,200 km. Violations with systems of four photons, equivalent to two particles of spin 1 correlated in a singlet state, have been reported [160]. In 2015, three different experiments obtained Bell violations under conditions where the "loopholes" (see §4.3.1) are closed [161–163]. The list of all references is too long to be given here, but we give a few more in §4.3.1 during the discussion of "loopholes"; a general overview of the experimental results can be found in [164].

A summary of the present situation is that, even in these most intricate situations invented and tested by the experimentalists, no one has been able to disprove quantum mechanics. In this sense, we can say that Nature obeys laws that are nonlocal, or nonrealist, or both. It goes without saying that no experiment in physics is perfect, and it is always possible to invent ad hoc scenarios where some physical processes, for the moment totally unknown, "conspire" in order to give us the illusion of correct predictions of quantum mechanics – we come back to this point in §4.3.1 – but the quality and the number of the experimental results does not make this attitude very attractive intellectually.

4.2 Various derivations of the theorem

What is generally meant by the "Bell theorem" is actually not limited to a single inequality applying in one set of specific circumstances, but encompasses a whole set of inequalities that are valid in various situations. We now give a few examples of other possible derivations of the theorem, which start from various sets of assumptions, and broaden the range of validity of the theorem. Then, in §5.1, we will introduce other forms of the theorem: other inequalities that can be obtained if realism as well as locality are assumed, but which are also violated by the predictions of quantum mechanics.

4.2.1 Other sets of assumptions

In §4.1, we gave a derivation of the Bell theorem from one possible set of general assumptions, listed in §4.1.4. The only necessary condition for the derivation of the inequality was the existence of four numbers A, B, A', and B', all equal to ± 1, which have to be well defined (but not necessarily known) for each realization of the experiment; equivalently, one can assume the existence of a joint distribution for the probabilities of the results of the four measurements [165]. But other sets of assumptions can also be used to obtain the theorem. This actually increases the strength of the result: a violation of the Bell inequalities means that, within each

---

[8] Concerning the effects of gravitation, it has even been shown [159] that EPR type correlations can in principle survive up to the largest cosmic scales.

of all these sets, at least one of the assumptions should be rejected, so that the discussion of §4.1.4 can be generalized. To provide a few examples, we now list some possibilities; for the moment, we do not explicitly mention the "free will assumption" (§4.3.1.c) within each set of assumptions, since it is common to all sets.

(i) As mentioned in the introduction of this chapter, the Bell theorem can be seen as a theorem applying specifically to theories with hidden variables. In this case, the λ arise, not from the EPR reasoning and their notion of realism, but from the a priori assumed existence of these variables. The conclusion is then that, if these variables evolve locally, the results of measurements must obey the Bell inequalities. Conversely, a violation of the inequalities means, either that the hidden variables do not exist, or that they evolve nonlocally (or that there is no free will, meaning in this case that a and b are functions of additional components of λ). This point of view is less general than that we have used, but also simpler, which probably explains why it is rather popular. For instance, in one of his celebrated books on quantum mechanics [58], Jammer introduces the Bell theorem within a chapter treating specifically hidden variable theories.

This is nevertheless not the point of view taken by Bell in his historical article [135], where he clearly introduces his reasoning as a continuation of the EPR argument. The title is "On the Einstein Podolsky Rosen paradox", and the first sentences of the introduction are "The paradox of Einstein, Podolsky and Rosen was advanced as an argument that quantum mechanics could not be a complete theory but should be supplemented by additional variables. These additional variables were to restore to the theory causality and locality. In this note, *that idea will be formulated mathematically* (our emphasis) and shown to be incompatible with the statistical predictions of quantum mechanics. It is the requirement of locality ... that creates the essential difficulty ". In other words, the additional variable λ he considers is not introduced arbitrarily, by postulating the existence of a new "hidden variable"; it is just the mathematical object that describes the elements of reality discovered by the EPR reasoning. Bell therefore distinguishes between "additional variables" and "hidden variables", which he then proceeds to discuss in the special case of Bohmian theory: "There have been attempts to show that even without such a separability or locality requirement no 'hidden variable' interpretation of quantum mechanics is possible. These attempts have been examined elsewhere and found wanting. Moreover, a hidden variable interpretation of elementary quantum theory has been explicitly constructed". Bell's point of view is therefore not ambiguous.

(ii) Assuming counterfactuality is another way to derive the Bell theorem: the pre-existence of all the possible results of experiments allows one to derive BCHSH and other inequalities. This is because, as soon as the four quantities A, A', B and B'

(all equal to ± 1) can be defined for each realization of the experiment, the method of §4.1.2 immediately leads to (4.9) and provides a proof of the inequality. Counterfactuality is not necessarily related to the notion of space or to locality[9], so that this view provides another independent logical frame for reasoning.

Peres's quotation in §4.1.3 [140] gives the conclusion to draw from a violation of the inequalities. This is similar to the discussion at the end of §4.1.2: a violation of the BCHSH inequality shows that it is impossible to define two functions $A(a, \lambda)$ and $B(b, \lambda)$ of the settings a and b and of the rank of the experiment providing the values of the experimental results.

(iii) Assuming noncontextuality is still another possibility to obtain the inequalities; this will be discussed in more detail in §6.4. Again, noncontextuality can be seen as a natural consequence of the spatial separation between Alice's and Bob's laboratories, but also as an independent and even more general principle (obeyed for instance in classical physics, local or nonlocal).

4.2.2 Generalizations of the theorem, indeterministic theories

One may in fact obtain the Bell theorem in a whole series of logical contexts by combining various assumptions, such as separability, the existence of causes within past light cones (relativistic causality, as in Figure 4.2), etc. All the derivations obtained in this way are interesting, since they extend the list of sets of assumptions that are incompatible with the predictions of quantum mechanics, and therefore extend the range of the theorem. We now discuss some of them.

4.2.2.a Direct generalizations, role of locality

A first simple generalization of the Bell theorem assumes that the results of an experiment become a function of several fluctuating causes: fluctuations taking place in the source, but also fluctuations in the measurement apparatuses [166], and/or perturbations acting on the particles during their motion towards the apparatuses, etc. Such situations just require the addition of more components of the vector variable λ, one per new fluctuating process. The conclusion of the theorem is not altered when the randomness of the results arises from many different sources.

One could then expect that even fundamentally indeterministic processes could be included in the proof of the theorem. Indeed, the theorem applies to locally indeterministic theories [123, 167] and to situations such as that illustrated by Figure 4.7, where independent random processes occur in two remote regions of space.

---

[9] Counterfactuality can be postulated ab initio, without any particular reference to locality. Conversely, if one assumes locality and realism, the EPR reasoning shows the existence of elements of reality, which play the role of counterfactuals. The EPR theorem nevertheless requires more than locality only, since EPR realism is also assumed.

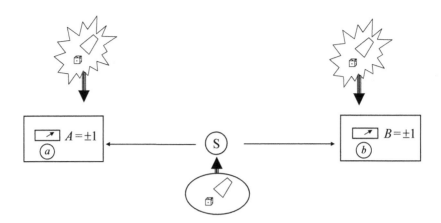

Figure 4.7 Scheme of a fundamentally indeterministic local theory. As in the preceding figures, fundamentally random processes are symbolized by flashes above the events, and the effect of uncontrolled local perturbations are represented by ovals below the events. Actually, for the present discussion, the nature of the stochastic process at the center of the figure, fundamentally indeterministic or not, is irrelevant as long as it remains local; one may replace the central deterministic oval by a stochastic flash above the source acting only on it without changing the conclusions. This scheme may be compared with Figure 4.5 associated with quantum mechanics, where the fundamentally random process is delocalized in space. Such indeterministic, but local, theories lead to predictions that necessarily obey the Bell inequalities.

The derivation of the theorem in this case requires replacing the deterministic functions A and B by probabilities, which is also relatively straightforward [137] (see footnote 10 in [166] and Appendix B). We come back to nondeterministic theories in more detail in §4.2.2.b, in particular to analyze the origin of the factorization of probabilities; see also the discussion of stochastic boxes §§5.4.2.b and 5.4.2.c.

　　More generally, assuming the existence of the A and B functions does not imply a limitation to a particular sort of theory. The role of these functions is just to relate the conditions of production of a pair of particles (or of their propagation) to their behavior when they reach the measurement apparatuses, and to the effects that they produce on them. They are, so to say, solutions of the equations of motion, whatever these equations are; the solutions are just assumed to be well defined but do not need to be specified. They may perfectly include, in a condensed notation, a large variety of physical phenomena: propagation of point particles; propagation of one or several fields from the source to the detectors (see for instance the discussion in §4 of [95]); particles and fields in interaction; or whatever process one may have in mind (even random propagations can be included) – as long as they remain independent of the other setting. The exact mathematical form of the equations of propagation is irrelevant; the essential thing is that the functions exist, even if they are too complicated to be written down explicitly.

Locality, expressed mathematically in terms of a and b, is the crucial ingredient of the derivation of the Bell theorem. What really matters is the dependence with respect to the settings a and b: the function A must depend on a only, while B must depend on b only. For instance, if we wished, we could have assumed that the result A of one measurement is also a function of fluctuating random variables attached to the other apparatus, which introduces a nonlocal process; but this does not create any mathematical problem in the proof, as long as these variables are not affected by setting b. On the other hand, if A becomes a function of a and b (and/or the same for B), it is easy to see that the situation is radically changed: in the reasoning of §4.1.2, we must now associate eight numbers to each pair (since there are two results to specify for each of the four different combinations of settings), instead of four, so that the proof immediately collapses. Appendix C gives another concrete illustration showing that it is mostly locality, not determinism, that is at stake; see also the appendix of [137].

4.2.2.b Spatial propagation of causality, locally explicable correlations

Other derivations of the theorem put more emphasis on the propagation of influences and their consequences on correlations. This idea can be used as a definition of locality: according to Bell (as quoted in [168]), "Locality is the idea that consequences propagate continuously, that they do not leap over distances". It will be also assumed that this propagation cannot occur at a velocity exceeding the velocity of light (propagation always takes place within future light cones, as relativity requires); this creates a closer relation between relativity and the Bell theorem. We now explore the consequences of this notion of relativistic locality on the predictions of theories, deterministic or indeterministic. We will closely follow the analysis given by Bell in his article "La nouvelle cuisine" [169] – see also [170] and Norsen [168].

Consider first the events taking place in a given region of space-time R, as illustrated in Figure 4.8; the horizontal axis represents space (symbolized by one single dimension along axis Ox) and the vertical axis time t (multiplied by the velocity of light c). The causes of these events can be found only in the backward light cone (past cone) of region R and, if an experimenter makes experiments in this region, on the choices she/he makes for the setting a of his experiments ("free will assumption", §4.3.1.c). Now we define a region C of space-time that, as shown in the figure, covers a complete slice of the past light cone of region R. If influences

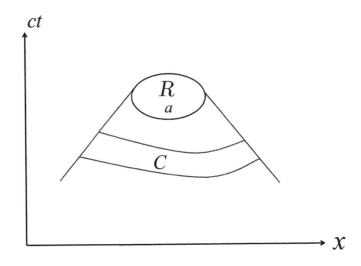

Figure 4.8 In this space-time representation, the horizontal axis Ox represents space (in a simplified one-dimensional view), the vertical axis represents the product ct of time t by c (velocity of light). We consider all events taking place in a region R of space-time; the possible causes of these events can be found only in the backward light cone of R, limited in the figure by slanted lines with slope ± 1. If an experimenter makes an experiment in R and freely chooses his measurement parameter a (setting), the results he obtains can only be consequences of this choice and of causes in this backward light cone. Moreover, if consequences propagate only continuously in space (Bell's definition of locality), all these causes are contained in the events taking place in region C, which "shields" R from causes further in the past (makes them redundant with the content of C).

propagate only continuously, to reach R from far in the past, they have to cross C; therefore, to account for the effect of causes far in the past, it is sufficient to specify all causes that occur within C. In other words, this specification "shields" region R from other causes further in the past, since it makes them redundant with the causes already contained in C (the knowledge of causes further in the past provides only superfluous information).

Variables and beables. At this point, it is useful to specify what sort of causes should be taken into account in this reasoning. Clearly, not every mathematical object appearing in the equations of physics is appropriate to characterize a cause. In classical electromagnetism, for instance, it is known that the Coulomb potential does not propagate in space with any finite velocity, but depends on the instantaneous position of all charges. This is not contradictory with relativity since the instantaneous effects of this potential are compensated by other effects of the vector potential; the "real" electromagnetic fields (electric field **E** and magnetic field **B**) propagate at finite velocity c. So, the equations of physics may contain mathematical functions that are only intermediate entities, from which real physical quantities can be derived; these functions do not necessarily propagate in space with any finite velocity, and just play the role of convenient variables[10]. By contrast with these variables, Bell introduces what he calls "beables", which he defines in [169] as: "the *be*ables of the theory are those entities in it which are, at least tentatively, to be taken seriously, as corresponding to something real". In [170] Bell constrasts beables with observables by writing: "The concept of 'observable' is a rather woolly concept. It is not easy to identify precisely which physical processes are to be given the status of 'observations' and which are to be relegated to be the limbo between one observation and another. So it might be hoped that some increase in precision might be possible by concentration on the *be*ables... because they are there".

Among the beables, some are local [169]: "Local beables are those which are definitively associated with space-time regions"; the electric field **E** and magnetic field **B** (or their integrals over finite regions of space) are such local beables. So, what is meant above by causes contained in a light cone (or in a space-time region C) is the ensemble of all local beables contained in the light cone (or C).

Locally causal stochastic theories. At this point, Bell [169] introduces the notion of a "principle of local causality", which is similar to relativistic causality, and applies to deterministic as well as to stochastic theories: "A theory is said to be locally causal if the probabilities attached to local beables in a space-time region 1 are unaltered by the specification of values of local beables in another space-like separated region 2, when what happens in the backward light cone of 1 is already sufficiently specified". According to this principle, a random event occurring in region R may be influenced by many causes in the past, but all their effects can be summarized in the state of all local beables contained in C, which we note λ for short (λ may have an arbitrary number of components $\lambda_1, \lambda_2, ..., \lambda_N,...$). These beables may be of many sorts, and for instance include those associated with the measurement apparatus, a macroscopic system with many degrees of freedom that can be known only statistically. They may also be influenced by remote events, provided they are contained in the past cone of R. Finally, the probability may also be influenced by local choices made by the experimenter (the setting a of the

---

[10] "Conventions can propagate as fast as may be convenient. But then we must distinguish in our theory what is convention and what is not" [169].

apparatus). The principle of local causality states that the variables associated with all these possible causes are sufficient to determine the probability: once they are specified, it is not necessary to specify anything more, in particular about events taking place in any region lying outside of the past cone of R.

Under these conditions, the probability $\mathcal{P}$ to obtain a result A in the measurement is a function of λ and of the choice of the measurement parameter (setting) a by the experimenter, but nothing else. We note:

$$\mathcal{P}(A \mid a, \lambda) \tag{4.12}$$

the conditional probability of obtaining result A if a has been chosen and if the beables in C take value λ. Locality, as defined previously, forbids that $\mathcal{P}$ should depend on other variables (values of beables attached to regions of space-time that do not belong to the past light cone of R).

We then have the combination of two random processes: the selection of a value of λ, and the fundamentally random processes providing A once λ is determined; we call $\mathcal{P}(A, \lambda \mid a)$ the probability of this double event, for a given choice of the setting a. The probability $\mathcal{P}(A \mid a)$ of obtaining a result A, whatever λ is, is then the sum of probabilities of exclusive events:

$$\mathcal{P}(A \mid a) = \int d\lambda\, \mathcal{P}(A, \lambda \mid a) \tag{4.13}$$

(the sum over $d\lambda$ symbolizes an integral with several dimensions, running over all possible values of λ). Now, the Bayes law of conditional probabilities provides:

$$\mathcal{P}(A, \lambda \mid a) = \mathcal{P}(A \mid a, \lambda)\mathcal{P}(\lambda \mid a) \tag{4.14}$$

where $\mathcal{P}(\lambda \mid a)$ is the probability for the beables in C taking value λ when a has been chosen by the experimenter. But we assume that this value results from a choice with free will ("free will assumption", §4.3.1.c) made in the future of region C, meaning that λ cannot depend on a. We therefore have:

$$\mathcal{P}(\lambda \mid a) = \rho(\lambda) \tag{4.15}$$

where $\rho(\lambda)$ is the distribution of the λ, with:

$$\rho(\lambda) \geq 0 \quad ; \quad \int d\lambda\, \rho(\lambda) = 1 \tag{4.16}$$

The average of the result obtained after many realizations of the experiment with setting a is then:

$$\langle A \rangle = \sum_A A\, \mathcal{P}(A \mid a) \tag{4.17}$$

or, if we insert (4.14) and (4.15) into (4.13):

$$\langle A \rangle = \int d\lambda \, \rho(\lambda) \sum_A A \, \mathcal{P}(A \mid a, \lambda) \tag{4.18}$$

From now on, we assume that A may take only two values ± 1. The sum of the two probabilities $\mathcal{P}(A = \pm 1 \mid a, \lambda)$ is then 1; it is convenient to characterize both of them with a single function $X(A; a, \lambda)$ by the relation:

$$\mathcal{P}(A = \pm 1 \mid a, \lambda) = \frac{1}{2}[1 \pm X(A; a, \lambda)] \tag{4.19}$$

with:

$$-1 \le X(A; a, \lambda) \le +1 \tag{4.20}$$

Equation (4.17) then simplifies into:

$$\langle A \rangle = \int d\lambda \, \rho(\lambda) \, X(A; a, \lambda) \tag{4.21}$$

Measurements in two distant regions of space, consequences of locality. We now apply the preceding analysis to a Bell experiment: in two regions of space-time R_1 and R_2, separated by a large space-like interval (Figure 4.9), two measurements are performed with two quantum systems both originating from a common source. The emission of these quantum systems has taken part far in the past, in a space-time region S belonging to the overlap between the two past light cones of R_1 and R_2. In each of these regions, an experimenter freely chooses a value for her/his measurement parameter (setting), a for the first, b for the second (for instance, the direction of orientation of a Stern–Gerlach analyzer). In the past light cones of R_1 and R_2, we introduce two intermediate space-time regions C_1 and C_2, with C_1 lying outside of the past cone of R_2, and conversely C_2 lying outside of the past cone of R_1. We use the short notation λ_1 for all beables contained in C_1 (including, if necessary, those associated with a statistical description of the measurement apparatus sitting in R_1), λ_2 for all beables contained in C_2.

The same analysis as before applies: the probability of obtaining result A in region R_1 is $\mathcal{P}(A \mid a, \lambda_1)$, which has the same form as (4.12), and is completely independent of events taking place in R_2; similarly, the probability of result B for the measurement performed in region R_2 is $\mathcal{P}(B \mid b, \lambda_2)$, completely independent of events taking place in R_1. The combined probability of these two independent events is therefore the product:

$$\mathcal{P}(A, B \mid a, b, \lambda_1, \lambda_2) = \mathcal{P}(A \mid a, \lambda_1)\mathcal{P}(B \mid b, \lambda_2) \tag{4.22}$$

An important consequence of locality, as defined previously, is that this probability

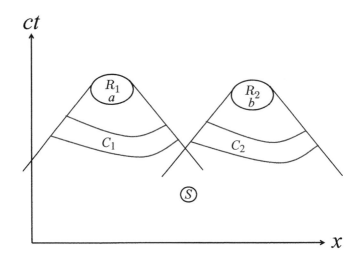

Figure 4.9 Two quantum systems are emitted during a process taking part in a region of space-time S, propagate in space-time from their source, and are then submitted to two measurements in two regions of space-time R_1 and R_2; these two regions are separated by a large space-like interval. In each of these two regions, an experimenter freely chooses a parameter for his measurement, a in R_1, and b in R_2. Two intermediate regions of space-time C_1 and C_2 are introduced, chosen in such a way that C_1 contains all information concerning possible influences acting on local beables in R_1, but lies outside of the past light cone of R_2 (and, conversely, C_2 contains all information concerning influences acting on local beables in R_2, but lies outside of the past cone of R_1). The ensemble of beables contained in C_1 (including those associated with the measurement apparatus) is λ_1, and that contained in C_2 is λ_2.

factorizes for given values of λ_1 and λ_2 (which, of course, does not prevent correlations between the results A and B to occur, when λ_1 and λ_2 fluctuate in a correlated way).

As in (4.15), the free choice of the settings a and b allows us to write:

$$P(\lambda_1, \lambda_2 \mid a, b) = \rho(\lambda_1, \lambda_2) \qquad (4.23)$$

where $\rho(\lambda_1, \lambda_2)$ is the distribution of the values λ_1 and λ_2 of the beables over many realizations of the experiment. We have:

$$\rho(\lambda_1, \lambda_2) \geq 0 \quad ; \quad \int d\lambda_1 \, d\lambda_2 \, \rho(\lambda_1, \lambda_2) = 1 \qquad (4.24)$$

The probability $P(A, B \mid a, b)$ to obtain results A and B when the settings a and b

have been selected is, according to Bayes law:

$$\mathcal{P}(A, B \mid a, b) = \int d\lambda_1 \, d\lambda_2 \, \mathcal{P}(A \mid a, \lambda_1) \mathcal{P}(B \mid b, \lambda_2) \, \rho(\lambda_1, \lambda_2) \qquad (4.25)$$

When the results can take only two values ± 1, it is convenient to introduce two functions $X(A; a, \lambda_1)$ characterizing the probabilities as:

$$\mathcal{P}(A = \pm 1 \mid a, \lambda_1) = \frac{1}{2}[1 \pm X(a, \lambda_1)] \qquad -1 \le X(a, \lambda_1) \le +1$$

$$\mathcal{P}(B = \pm 1 \mid b, \lambda_2) = \frac{1}{2}[1 \pm Y(b, \lambda_2)] \qquad -1 \le Y(a, \lambda_1) \le +1 \qquad (4.26)$$

The average over many realizations of the product of the results then reads:

$$\langle AB \rangle = \int d\lambda_1 \, d\lambda_2 \, \rho(\lambda_1, \lambda_2) \, X(A; a, \lambda_1) \, Y(B; b, \lambda_2) \qquad (4.27)$$

We still obtain a factorization of the a and b dependences of the function inside the integral, which is the key element for the proof of the BCHSH inequality.

Derivation of the BCHSH inequality. We now assume that the experiment in R is made with two different choices of the setting, a and a'; similarly, two different settings b and b' are used for the experiment in R_2. By analogy with (4.5) and (4.6), we use the shorter notation:

$$X(\lambda_1) \equiv X(a, \lambda_1) \quad X'(\lambda_1) \equiv X(a', \lambda_1)$$

$$Y(\lambda_2) \equiv Y(b, \lambda_2) \quad Y'(\lambda_2) \equiv Y(b', \lambda_2) \qquad (4.28)$$

The combination of averages considered in the BCHSH inequality is:

$$\langle M \rangle = \langle XY \rangle - \langle XY' \rangle + \langle X'Y \rangle + \langle X'Y' \rangle$$

$$= \int d\lambda_1 \, d\lambda_2 \, \rho(\lambda_1, \lambda_2) \, F(\lambda_1, \lambda_2) \qquad (4.29)$$

with:

$$F(\lambda_1, \lambda_2) = X(\lambda_1) Y(\lambda_2) - X(\lambda_1) Y'(\lambda_2) + X'(\lambda_1) Y(\lambda_2) + X'(\lambda_1) Y'(\lambda_2)$$

$$= X(\lambda_1)[Y(\lambda_2) - Y'(\lambda_2)] + X'(\lambda_1)[Y(\lambda_2) + Y'(\lambda_2)] \qquad (4.30)$$

Function F, averaged over all possible values of λ_1 and λ_2, is analogous to the function M defined by relation (4.7) of §4.1.2. There is nevertheless a difference: the results $A = \pm 1$ and $B = \pm 1$ appearing in (4.7) are now replaced by differences of probabilities X and Y, which are real numbers never exceeding the limits -1 and $+1$, but not equal to these limits in general. We nevertheless show that, while M is always equal to ± 2, function F obeys the weaker condition:

$$-2 \le F(\lambda_1, \lambda_2) \le +2 \qquad (4.31)$$

To derive this result, let us first assume that each of the four variables X, X', Y, and Y' takes one of its extremal values ± 1. Then, one of the two brackets $[Y(\lambda_2) \pm Y'(\lambda_2)]$ in the second line of (4.30) vanishes, and F is equal to ± 2; one obtains in this way 16 such values associated with all possible combinations of the extremal values of the variables. Now, while keeping all the other variables at their constant values, let us put X to any value between -1 and $+1$; this interpolates F linearly between two values equal to ± 2, which necessarily provides values between -2 and $+2$; one obtains in this way eight such values (for the eight combinations of the possible values of the three remaining variables X', Y and Y'). In a second step, let us move X' to any value between -2 and $+2$; a similar interpolation occurs, and values between -2 and $+2$ are obtained (for all four combinations of the remaining variables Y and Y'). Two more steps in the reasoning are then necessary to give single values to Y and Y'. At the end, one concludes that F is always bound by values -2 and $+2$, whatever intermediate value the four variables take.

Equation (4.29) then defines $\langle M \rangle$ as the average, with a positive and normalized weight function $\rho(\lambda_1, \lambda_2)$, of a function that is bound by -2 and $+2$; we therefore have:

$$-2 \leq \langle M \rangle \leq +2 \tag{4.32}$$

which is nothing but the BCHSH inequality for stochastic processes.

Discussion. Three classes of indeterministic theories can be considered:

- Theories where probabilities of events occurring at different locations are always independent of each other and of any previous event. Such theories cannot account for correlations, and are therefore of limited interest.
- Theories where probabilities of different events occurring at different locations are still independent of each other, but may be functions of other events that have occurred in their past cone (relativistic causality); influences from the past occur and can produce many sorts of correlations, but the fundamental randomness of the theory remains local.
- Theories where probabilities of different events occurring in different locations are all interdependent. When such events occur, a whole indivisible process takes place, which is delocalized in space and includes them all simultaneously.

The derivation we have given of the BCHCH inequality applies to the second class of theories. The basic idea is that fundamentally random processes may occur in Nature, but that they should be space-time events: the probabilities may be affected by influences that propagate from other events in the past light cone, but the fundamental stochasticity remains local.

Conversely, a violation of the Bell inequalities within a stochastic theory means that it belongs to the third category (this is the case for quantum mechanics), where random processes are fundamentally nonlocal: each of them covers an arbitrary large region of space-time; it cannot be decomposed into relativistic events (this impossibility is in the spirit of Bohr's prescription, as discussed in §§5.3.1 and 7.1.1; see also the quotation from his 1933 Nobel lecture, note [17] on page 62). In other words, this indicates the existence of nonlocal randomness, emerging with no particular location[11]; as Gisin expresses it [173]: "These nonlocal correlations seem to emerge, so to say, from outside of space-time!".

Other similar derivations of the inequality for stochastic theories have been proposed by Jarrett, Ballentine, Shimony, and others [174–178]. These authors also make use of the Bayes law, and put the emphasis on notions such as "predictive completeness" [175] or "parameter independence" and "outcome independence" [168, 176], to obtain the factorization (4.25) of the probabilities, and to derive the BCHSH inequality. This class of derivations is summarized in the second section of Appendix B.

4.2.2.c Leggett–Garg inequalities

The role played by the experimental settings in the BCHSH inequalities can be played by time in some cases. For macroscopic systems, Leggett and Garg [179, 180] have shown that BCHSH type inequalities exist for the same physical quantity considered at different times. In the same spririt as the EPR reasoning, they start from two postulates:

(i) Macroscopic realism: a macroscopic system with two (or more) macroscopically distinct states is always in one of these states (it is never in a coherent superposition of macroscopically distinct states).

(ii) Noninvasive measurability at the macroscopic level: it is in principle possible to determine the state of the system with arbitrarily small perturbation of its future evolution.

From these assumptions, a whole family of inequalities involving averages of product of results of measurements performed at different times can then be derived. Consider for instance a macroscopic system having access to two distinct macroscopic states $|\Phi_1\rangle$ and $|\Phi_2\rangle$. At each of four different times t_i ($i = 1, 2, 3,$ and 4) we introduce a number $A_i = \pm 1$, which is equal to $+1$ if the system is in state $|\Phi_1\rangle$, and equal to -1 is the system is in state $|\Phi_2\rangle$. For each given evolution of the system (each realization of the experiment), the four numbers A_i have well-defined values. The reasoning of § 4.1.2 then shows that the combination of numbers:

$$M = A_1 A_2 - A_1 A_4 + A_3 A_2 + A_3 A_4 = A_1 (A_2 - A_4) + A_3 (A_2 + A_4) \qquad (4.33)$$

---

[11] It has been remarked that the existence of quantum correlations with no causal order [171] could be used to control the order of quantum gates and make quantum computation faster [172].

is always equal to ± 2. Now, if the experiment is repeated many times, and if one takes the average value $\langle A_i A_j \rangle$ of each term in this expression, one obtains:

$$-2 \leq \langle A_1 A_2 \rangle - \langle A_1 A_4 \rangle + \langle A_3 A_2 \rangle + \langle A_3 A_4 \rangle \leq +2 \qquad (4.34)$$

This relation is a four-time Leggett–Garg inequality.

A three-time inequality is easily obtained as well. Consider the combination of results:

$$N = A_1 A_2 + A_2 A_3 - A_1 A_3 = A_1 (A_2 - A_3) + A_2 A_3 \qquad (4.35)$$

If $A_3 = A_2$, we have $N = A_2 A_3 = \pm 1$; if $A_3 = -A_2$, we have $N = 2A_1 A_2 - (A_2)^2$, thus $N = 2A_1 A_2 - 1$. In both cases, we check that $N \leq 1$; therefore, if we take the average over many realizations:

$$\langle A_1 A_2 \rangle + \langle A_2 A_3 \rangle - \langle A_1 A_3 \rangle \leq 1 \qquad (4.36)$$

Other inequalities can be obtained by similar methods.

Leggett and Garg [179] have shown that quantum mechanics predicts that some of these inequalities are violated by systems that can indeed be considered as quantum and macroscopic, such as SQUIDs (Superconducting Quantum Interference Devices). The two macroscopically distinct states correspond in this case to situations where a quantum of flux crosses the SQUID in one direction or the opposite. For a review on the inequalities and related experiments, see [181]. Beside SQIDS, the physical systems studied were photons, spins in magnetic resonance experiments, impurities in silicon, etc., which are not macroscopic, and for which the postulate (ii) is not necessarily valid: in general, a microscopic system is perturbed by the measurement (state vector projection). This is why various techniques of "ideal noninvasive measurements", "continuous weak measurement" (§ 9.3), or "negative measurements" (§2.4) have been used. Recent references are for instance [79, 182, 183].

4.2.2.d Other generalizations

Violations of the Bell theorem are not limited to pairs of particles with an entangled state of their spins, that is, their polarizations for photons (Chapter 8). For instance, Grangier et al. [184] have shown that parametrically generated photon pairs mixed with weak coherent beams lead to such violations. Another interesting scheme involving the two photons emitted in cascade by an atom has been proposed by Franson [185], who considers a situation where the frequency of each photon fluctuates (the intermediate atomic state has a very short lifetime) but not their sum (the initial atomic state has a long lifetime). This scheme has been implemented experimentally to obtain significant violations of the Bell inequalities (see for instance [186] and [187]).

More generally, a possible violation of the Bell inequalities is not limited to a few quantum states (singlet for instance), but occurs for all states that are not products [188–191]. With statistical mixtures of states, violations of the BCHSH inequalities are also possible [192], although, conversely, entangled mixtures do not necessarily lead to violations [193]; one then has to use more elaborate sequences of measurements to reveal nonlocality [194, 195]. Two independent sources can be used for observing violations, provided that appropriate interference measurements are performed [196]. The generalization is possible to an arbitrary number of particles, see [197]. Schemes have also been proposed to observe nonlocality even with a single photon [198–200] (but, in the measurement process, more than one photon is involved, for example photons from local oscillators). For a general discussion of the conceptual impact of a violation of the inequalities, we refer to the book collecting Bell's original articles [6].

4.2.3 Status of the theorem; attempts to bypass it

Provided local realism (or causality) is assumed (the assumptions listed in §4.1.4), the Bell theorem is very general; it is therefore difficult to build a reasonable theory that can violate the inequalities, probably more difficult than one might think. All potential authors who believe that they have found a simple explanation for the observed violations should think twice before taking their pen and sending a manuscript to a physics journal. Every year a large number of such texts is submitted, with the purpose of introducing "new" ways to escape the constraints of the Bell theorem, and to "explain" why the experiments have provided results that are in contradiction with the inequalities. For instance, the violations could originate from some new sort of statistics, or from perturbations created by cosmic rays, gas collisions with fluctuating impact parameters, random effects of gravity, etc. Imagination is the only limit to the variety of the processes that can be invoked. Nevertheless, we know from the beginning that all attempts to obtain violations within classical local theories are doomed to failure, however elaborate these theories are.

The situation is somewhat reminiscent of the attempts in past centuries to invent "perpetuum mobile" devices: even if some of these inventions were extremely clever, and if it is sometimes difficult to find the exact reason why they should not work, it remains true that the law of energy conservation allows us to know immediately that they cannot function. In the same way, some of these statistical "Bell beating schemes" may be extremely clever, but we know that the theorem is a very general result in statistics: in all situations that can be accommodated by the mathematics associated with the variables λ and the A and B functions (and there are many!), it is impossible to escape the inequalities. No, nonlocal correlations cannot be explained cheaply; yes, a violation of the inequalities is a very, very, rare situation. In fact, until now, it has never been observed, except of course in experiments designed precisely for this purpose. In other words, if we wanted to build automata including arbitrarily complex mechanical systems and computers, we could never mimic the results predicted by quantum mechanics (at least for remote measurements) without allowing communication between them; it is even possible to calculate the minimum amount of information that must be exchanged to simulate the quantum correlations [201]. This will remain impossible forever – or at least until completely different computers working on purely quantum princi- ples are built[12].

The conclusion of this discussion is that the only way to bypass the Bell theorem is to explicitly give up at least one of the assumptions listed in §4.1.4; in the next chapter (§4.3.1), we come back to one of them, the free will assumption.

4.3 Impact of the Bell theorem, loopholes

In view of the experimental results mentioned in §4.1.5, which were not available when the orthodox interpretation of quantum mechanics was invented, and which are in complete agreement with the predictions of quantum mechanics, some physi- cists conclude triumphantly: "Bohr was right!". Others claim with the same enthu- siasm: "Einstein was right!", and emphasize his precursor role in one more domain of physics. Both these opinions actually make sense, depending on what aspect of the debate one favors. In any case, whether one personally feels closer to the or- thodox quantum camp or to local realism, it remains clear that the line initiated by Einstein and Bell played a decisive role over the last 50 years. They are the first ones who pointed out the crucial importance of the notion of locality in quantum physics; this stimulated much more progress and understanding than the simple restatement of the orthodox position. For instance, even now, the reduction of the state vector is sometimes "explained" by invoking the "unavoidable perturbations that the measurement apparatus brings to the measured system" – see for instance the traditional discussion of the Heisenberg microscope that still often appears in textbooks! But, precisely, the EPR–Bell argument shows us that this is a cheap explanation: in fact, the quantum description of a particle can be modified with- out any mechanical perturbation acting on it, provided the particle in question was previously correlated with another particle. So, a trivial effect such as a classical recoil effect in a photon–electron collision cannot be the real explanation of the

---

[12] In terms of the Mendel parable, an observation of a violation of the Bell inequalities would imply that some- thing inside both peas (maybe a pair of DNA molecules?) remains in a coherent quantum superposition, without decoherence, even if the distance between the peas is large.

deep nature of the state vector reduction; it is actually much more fundamentally quantum and may involve nonlocal effects.

The fact is that, even if quantum mechanics and relativity are not incompatible, they do not fit very well together: the notion of events in relativity, which are supposed to be point-like objects in space-time, and the idea of causality, are still basic notions, but not as universal as one could have thought before the Bell theorem. Indeed, quantum mechanics teaches us to take these notions "with a little grain of salt". Still another aspect is related to the incredible progress that experiments have made in the twentieth century, whether or not stimulated by fundamental quantum mechanics. One gets the impression that this progress is such that it will soon allow us to have access to objects at all levels of scale, ranging from the macroscopic to the microscopic. In Bohr's time, one could argue that the precise definition of the border line between the macroscopic world of measurement apparatuses and microscopic objects was not crucial, or even academic; now, the question may become of real importance. In §3.4.2, we have given one example (macroscopic systems in Fock states) but many other possibilities may emerge and, hopefully, even give rise to experiments soon. All these changes, together, give the impression that the final stage of the theory is not necessarily reached and that conceptual revolutions are still possible, even if for the moment no precise new result has weakened the orthodox interpretation in any way.

Because of the important conceptual impact of the Bell theorem, much attention has been given to the interpretation of the experimental results; many authors have discussed the extent to which they really prove that Nature violates the inequalities and therefore local realism or causality. It is of course true that the interpretation of experiments always involves some assumptions. Actually, even the authors of the experiments themselves have worried about the existence of possible hidden assumptions in the interpretation, and about the possibility to explain their results in terms of unknown local realist theories. This would provide a way to escape from a conflict between the experimental results and local realism; for this reason, the corresponding possibilities are traditionally called "loopholes". Even if the general consensus is that these loopholes have now been closed, one after the other (in particular in a remarkable series of experiments in 2015 described in §4.3.1.e), it remains interesting to examine their nature, since they shed interesting light onto the logical content of the Bell theorem.

4.3.1 Loopholes, conspiracies

There are several ways in which one can deny the existence of any real conflict between the experimental results and local realism/causality. First of all, of course, one can always invoke trivial errors, such as very unlikely statistical fluctuations, to explain why the experiments seem to "mimic" quantum mechanics so well; for instance, some authors have introduced ad hoc fluctuations of the background noise of photomultipliers, which would magically correct the results in a way that would give the impression of exact agreement with quantum mechanics. One could as well even assume that all major experimental results in physics are just erroneous fluctuations! But the number and variety of Bell-type experiments supporting quantum mechanics with excellent accuracy is now large; in view of the results, physicists do not take this explanation seriously.

One could also think of more complicated scenarios: for instance, some local unknown physical variables may couple together in a way that will give the (false) impression of nonlocal results, while the mechanism behind them remains local. In other words, these mysterious variables would "conspire" against the physicists to fool them and lead them to incorrect conclusions. We now discuss some of these scenarios; for reviews on loopholes, see for instance [202], [203], and [204] for the specific case of experiments with photons.

4.3.1.a Pair selection loophole (efficiency loophole)

In the proof of §4.1.2, we have assumed that the emission of pairs is independent of the choice of the settings a and b made by Alice and Bob, for instance because this choice is made after the emission of the particles. The consequence is that the properties of the particle reaching Alice's apparatus are statistically completely independent of b, and conversely that the properties of the particle reaching Bob's laboratory are independent of a. Moreover, we have assumed that every pair of every experiment is always detected; it provides results A and B equal to ± 1, whatever values for a and b are chosen. Then the statistical properties of the detected particles are the same as those of the emitted particles. Under these conditions, a BCHSH inequality containing the averages of the experimental results can be obtained.

Similarly, in the derivation of the BCHSH in §4.2.2.b, we have assumed that the distribution $\rho(\lambda_1, \lambda_2)$, which describes the statistical properties of the detected pairs in the experiment, is independent of a and b; it is easy to see that the proof no longer applies if ρ becomes also a function of a and b. Generally speaking, the derivation of Bell inequalities requires that the statistical properties of each measured particle should be completely independent of the choice of the setting for the apparatus that measures the other particle. The limitation of the relevance of the inequalities to situations where the ensemble of detected pairs is independent of a and b is an important point, which was realized before the first experiments were performed [123, 167, 205]. This is why several authors have pointed out that any experiment leading to a violation of the Bell inequalities, but without any direct experimental proof of this independence, is inconclusive: it can in principle be interpreted within local realism/causality by invoking what is often called the "pair selection loophole" (also called "detection loophole", "efficiency loophole", or "(un)fair sampling loophole", etc.). We now discuss this loophole.

Let us come back to the reasoning of §4.1.2 in more detail. We have assumed that the four numbers A, A', B, and B' are all attached to the same pair. It then makes sense to obtain the ensemble average $\langle M \rangle$ from successive measurements of four average values $\langle AB \rangle$, $\langle AB' \rangle$, etc. But, if M is built from more numbers, such as numbers associated with different pairs, the algebra no longer holds, as we remarked at the end of §4.2.2.a, and the rest of the proof of the inequality collapses[13].

If Alice measures her particles with a Stern–Gerlach apparatus (or a two-channel polarization analyzer for photons), no particle should escape measurement; even in an imperfect experiment, there seems to be no particular reason why the sample of detected particles should depend on the setting a she chooses. Nevertheless, to illustrate the effects of a possible pair selection, let us now assume the opposite: for some reason, Alice's polarization analyzer selects only a subclass (a category) of particles that depends on a; the particles that do not belong to this category are merely ignored by the measurement apparatus (no count). Still keeping a local realist point of view, we can even assume that, as soon as the orientation of the analyzer is changed by a few degrees, a completely different category of particles is selected. Since the properties of the two particles of the pair are initially correlated, for the pairs of particles that are taken into account in the correlation measurement, the properties of the particle reaching Bob's laboratory may then depend on a. It thus becomes possible to ascribe to this (a dependent) category whatever ad hoc physical properties are needed to reproduce any b dependence. One can for instance choose to reproduce in this way the sinusoidal a and b dependence of quantum mechanics, which is known to violate the BCHSH limits. Because each side counts only events that are correlated with a detection on the other side, one actually counts particles that may have properties that depend on the other remote setting; locality can no longer be expressed in simple terms of a and b (see Appendix E for more details). We therefore obtain a case where the inequalities are violated within local realism, just due to a local pair selection process. To summarize the situation:

Experimentally, it is not especially difficult to start with an ensemble of pairs that is independent of the settings, but it is much more difficult to ensure that all pairs are detected. But we have seen that, if some particles escape detection, it becomes

---

[13] Assume for instance that the pairs that are detected on both sides with orientations a and b of the analyzers belong to a first subensemble of pairs, those detected with orientations a and b' to another subensemble. If the choice of the first orientation is a, the result observed locally can then be written either $A_{a,b}$, if the pairs belongs to the first subensemble, or $A_{a,b'}$ if it belongs to the second. The number M is then formed from eight different numbers ± 1 instead of four, and the proof of the Bell limit is not possible.

possible that the statistics of detected particles becomes completely different from that of emitted particles; there is no rigorous reason why the Bell limit should still be a consequence of local realism/causality – see for instance the example studied by Pearle in [205] and Appendix E. Indeed, in most experiments with photons, the detection efficiency is low: only a small fraction of the emitted pairs is collected by the detectors. What is measured is a number of coincidences during a given time, but the number of emitted pairs during the same time is not directly accessible. Of course, in theory, this number depends on known parameters such as the efficiency of the detectors, the collection angles, etc., which are in principle independent of a and b. But, in complete generality and without any theoretical input, it remains true that it could depend on a and b in some unexpected way, opening the possibility of a loophole.

In practice, to interpret most experiments in terms of the BCHSH inequalities, one can assume that no bias is introduced by the pair selection. This is a very plausible assumption: after all, there is no particular reason why the geometrical loss of photons should depend on their polarizations, for instance. But this remains an assumption, and therefore a weak point in the logical proof.

An ideal situation would be provided by a device with a triggering button that could be used by an experimentalist, who could at will launch a pair of particles (with certainty); if the pair in question was always analyzed and detected with 100% efficiency, the loophole would be closed. When discussing thought experiments, Bell introduced in some of his talks the notion of "preliminary detectors" [206], devices that he sketched as cylinders through which any pair of particles would have to propagate before reaching both ends of the experiment (where the a and b dependent measurement apparatuses sit); the preliminary detectors should signal the presence of pairs that, later, would always be detected at both ends, whatever choice of a and b is made. The role of these cylinders was therefore to make the definition of the sample more precise, even if initially the pairs were emitted by the source in all directions. Such systems, which allow a definition of an ensemble that is indeed totally independent of a and b, are sometimes called "event-ready detectors". See also [207] where Bell imagines a combination of veto and go detectors associated with the first detected particles in a ternary emission, precisely for the purpose of better sample definition.

It is also possible to take the opposite point of view and to deliberately choose to bias the sample of detected pairs with a postselection process. One can, then, not only violate Bell inequalities without nonlocality (Appendix E, §E.2), but also exceed the Cirelson bound (§5.2) by a large factor [209–211]. This illustrates the

importance of the fair sampling condition for obtaining clear experimental violations of local realism/causality [14].

Remark: we will see in §5.1.4 that the so-called CH inequality does not require
the assumption that all emitted pairs are detected; nevertheless, it requires that the
distribution of pairs $\rho(\lambda)$ should be independent of the settings a and b, and that
the probability of double detection should factorize for every value of λ.

4.3.1.b Communication loophole

Other loopholes are also possible: even if experiments were done with 100% efficiency, one could also invoke some possibilities for local processes to artificially
reproduce quantum mechanics. One of them is the "communication loophole",
sometimes also called the "conspiracy of the polarizers"[15] (actually, "conspiracy
of the analyzers" would be more appropriate). Assume that, by some unknown
process, each analyzer becomes sensitive to the orientation of the other analyzer;
it can then acquire a response function that depends on the other setting, and the
function A may have a dependence on both a and b. Under these conditions, the
proof of the Bell theorem is no longer possible. The same is true if the choice
of a may somehow influence the propagation of the second particle between the
source and the apparatus with setting b. Unknown physical processes could therefore conspire to reproduce violations of the inequalities without actually violating
local realism/causality, giving the false impression that the latter is not obeyed.

A way to exclude the possibility of these processes is to make use of relativistic
causality. If the distances between the two analyzers and between the analyzers and
the source are sufficiently large, and if the settings a and b are chosen at the very
last moment, then no physical influence can propagate (at the maximum speed of
light) between the different parts of the apparatus; unless of course they violate
relativity, these unknown processes become impossible. The loophole is therefore
closed. A first step in this direction was done by Aspect et al. in 1982 [153]. More
recent experiments [212] with very fast changes of a and b have succeeded in excluding this possibility and completely closing this loophole. Quantum mechanics
still works perfectly well under these more stringent time-dependent conditions.
The 2015 experiments [161–163] also closed this loophole, and will be discussed
in more detail in §4.3.1.e.

---

[14] Schemes others than sample bias may lead to predictions that seem to reproduce those of quantum mechanics
within local realism. See for instance [208] for a discussion of the effects of changing the normalization of
the correlation functions with the mean square deviations of the variables, instead of summing over individual
events.

[15] The word polarizer refers to the experiments performed with photons, where the spin orientation of the particles is measured with polarizing filters; but there is nothing specific on photons in the scenario, which can
easily be transposed to massive spin 1/2 particles.

4.3.1.c Fatalism versus free will

Along a similar line is what is sometimes called the "fatalistic loophole" (or also "superdeterminism"). The idea is to call into question an implicit assumption of the reasoning that leads to the Bell theorem: the completely arbitrary choice of the settings a and b by the experimenters. Usually, a and b are indeed considered as free variables: their values are not the consequence of any preliminary event that took place in the past, but those of free human choice, which may be made just before the measurement. On the other hand, it is true that there is always an overlap between the past cones of two events (here the choice of the settings). It is therefore always possible in theory to assume that they have a common cause; a and b are then no longer free parameters, but uncontrolled variables that can fluctuate (in particular, if this cause itself fluctuates) with all kinds of correlations. In this case, it is easy to see that the proof of the Bell theorem is no longer possible[16]. The "cellular automaton interpretation of quantum mechanics" proposed by G. 't Hooft [213] belongs to this category of theories.

What is then denied is the notion of free will of the experimenters, whose decisions are actually supposed to be always predetermined, even if they are not aware of it; the notion of arbitrary external parameters, which usually define the experimental conditions, no longer makes sense in this scheme. This price being paid, one can in principle build a theory that remains at the same time realist, local, and (super)deterministic, and includes a sort of physical theory of human decision in a way that can introduce violations of the Bell inequalities; along this line, see for instance [214]. This remains, of course, an unusual point of view, since the notion of arbitrary external parameters (free external variables) is generally accepted in physics; in the words of Bell [215]: "A respectable class of theories, including quantum theory as it is practiced, have 'free external' variables in addition to those internal to and conditioned by the theory... They are invoked to represent the experimental conditions. They also provide a point of leverage for free willed experimenters, ...".

In practice, when many values of the settings are chosen randomly in an experiment, they are not decided by a human being, but are automatically created by a random number generator. For instance, in the aforementioned time-dependent experiment [153], a double random generator was used to pilot the values of the settings at both ends of the experiment. Transposing the free will loophole to this situation leads to the question: are the number generators really random? Do they actually provide values that are actually consequences of the fluctuations of some common cause in the past? If so, a and b are functions of some variable λ, and

---

[16] For instance, in the proof that makes use of a probability density $\rho(\lambda)$, if one assumes that a and b become two functions $a(\lambda)$ and $b(\lambda)$, it makes no sense to compare the average values for different fixed values of a and b.

the Bell theorem no longer applies. One could also imagine influences of the random number generator onto the source of particles, which would be emitted in a state that is correlated with the type of measurement they will undergo later; this also would block the proof of the theorem. It is then not totally impossible that λ fluctuates and influences a and b, and/or the source, in a way that reproduces the predictions of quantum mechanics, but from a totally different mechanism that may be local and realist. A method to reduce the plausibility of this possibility is to use very remote measurement apparatuses with independent random number generators, as was initially done in reference [157]; see for instance [216] for the description of experimental work on the development of very fast and pure random number generators.

A fatalistic view of the physical universe leaves little room for experimental scientific disciplines. The scientific method assumes that an experimenter is free to make arbitrary choices concerning the design and settings of his experiments. The theory should then be able to adapt to any values of the variables that he has chosen. By contrast, if the history of the universe is predetermined once and for all, including all experiments that have and will be performed within this universe, it is not possible to "ask questions to Nature" in various ways, since the questions themselves are already consequences of this unique history. It is then no longer clear if the very notion of the scientific experimental method keeps any meaning at all. Needless to say, the fatalist attitude is not very common among scientists; many consider that eliminating the possibility of free observers in this way is too high a price to explain the difficulties of quantum mechanics.

It is nevertheless possible to take an intermediate point of view, where experimenters retain some free will, but not complete. It is shown in [217] that, if the experimenters just give up 14% of their decision independence, within local realism it becomes possible to reproduce the predictions of quantum mechanics for spin measurements with a singlet spin state.

4.3.1.d Experimental progress to close the loopholes

The efforts to close the various loopholes have a long story. As mentioned above, the pair selection loophole had been identified even before the very first experiments [146, 147] were made, and has been constantly studied since, with constant progress. For an early theoretical discussion of the necessary conditions to design a loophole-free experiment in terms of background level and counter efficiencies, see [218].

We have already mentioned that, in most photon experiments, many pairs are simply missed by the detectors. There are several reasons for this situation: the photons are emitted in all directions, while the analyzers collect only a small solid angle and, therefore, only a tiny fraction of the pairs. This was especially true in the initial experiments using photon cascades; in more recent experiments [155], the use of parametric photon conversion processes introduces a strong correlation between the directions of propagation of the photons and a much better collection efficiency, but it still remains low. Moreover, the transmission of the analyzers is less than 1 (it is actually less than 1/2 if ordinary photon polarization filters are used, but experiments have also been performed with birefringent two-channel analyzers [152], which are not limited to 50% efficiency). Finally, the quantum efficiency of particle detectors (for instance photomultipliers for photons) is not 100% either, so that pairs of particles are lost at this stage too. The net result is that many experiments do not provide an independent way to determine the sample of detected pairs, since the detection process itself is obviously a and b dependent; as a consequence, their experimental results become useful only if they are interpreted within a "no-biasing" assumption, or "fair sampling hypothesis", which amounts to considering that the settings of the analyzers do not bias the statistics of events.

This does not mean that these experiments are not conclusive! Their results are as convincing as those of most experiments in physics. Indeed, one should keep in mind that there is no known reason whatsoever why such a sample biasing should take place. The experimentalists are not in the dark concerning the detection efficiencies of their experimental setups. On the contrary, they can calculate them with accuracy, knowing the geometrical characteristics of the apparatuses, the quantum efficiency of the detectors, the rate of excitation of the source, etc. All these parameters can be carefully controlled and allow them to check that the coincidence rates are exactly what one expects from theory. The rates are also compared with those of single-particle detection under the same experimental conditions; experimenters are not restricted to measurements of the relative variations of the coincidence rates as functions of a and b. The same care has been taken to make all possible checks, as in other important experiments in physics; the result of all this work is that everything fits very well with the predictions of quantum mechanics. It would be rather extraordinary if some mysterious physical effect existed that depended so crucially on the detection efficiency; for low efficiencies, it would mimic the results of quantum mechanics and fool the physicists, who would then erroneously think that the results of quantum mechanics are valid; but, for higher efficiencies, it would act completely differently and stop to reproduce quantum mechanics in order to obey the Bell inequalities. This is probably why most physicists have remained so skeptical about this scenario.

For early proposals of really loophole-free experiments[17], see [219, 220] and

---

[17] A perfect correlation between the existence of a detection event on each side (independently of the obtained results) would provide another possible scheme for a loophole free experiment – this, of course, would imply that two channel detectors with a 100% efficiency are used on both ends of the experiment. In itself, the fact that any click at one side is always correlated with a click at the other, independently of the settings a and b, is not sufficient to exclude a setting dependence of the ensemble of detected pairs. But, if one assumes

[221]. The latter reference proposes to use continuous variables that are quantized artificially afterward (the result is $+1$ if the integral of the signal during a pulse is positive, -1 if it is negative). Especially interesting is an idea of "photon subtraction" that does two things at the same time: providing an event–ready detector (in order to close the pair selection loophole) and introducing the necessary quantum state to violate Bell's inequalities. In this case, the predicted violation is only 1%, but there are more elaborate situations where much higher violations are expected [222]. An experiment with $^9\mathrm{Be}^+$ ions was performed with high detection efficiency [223], so that the pair selection loophole was closed; nevertheless, the distance between the two ions was small (a few microns) so that the communication loophole could not simultaneously be closed. It has nevertheless been proposed to use the method of "entanglement swapping" (§7.3.2) where pairs of photons entangle remote ions in different traps [224]; the experiment was performed [225, 226] with $^{171}\mathrm{Yb}^+$ ions at a distance of 1 m. from each other, and led to a clear violation of the Bell inequalities (3 standard deviations). Recent experiments with solid-state Josephson junctions have given results violating the BCHSH inequalities by more than 200 standard deviations, with quantum systems that are macroscopic electrical circuits [227]. Of course, in this case, the problem of capturing pairs of particles does not occur; nevertheless, since the distance between the circuits is small, the communication loophole remains open. Actually, until 2015 no completely loophole-free experiment had been performed.

4.3.1.e 2015–2016 experiments

In 2015, several experiments succeeded in closing all loopholes simultaneously. One of them relied on the BCHSH inequality, the two others on CH-Eberhard type inequalities. Generally speaking, the BCHSH inequality is more sensitive to the detection loophole than the Eberhard or CH inequality. This is because the BCHSH inequality compares detection averages to pure numbers, ± 2. As we have seen above, averages are experimentally obtained by dividing number of coincidences (contents of coincidence counters) by a total number of pairs, which is not directly measured. If the detection efficiency is low, the total number of emitted pairs will exceed that of detections by a large factor, leading to small averages, so that the middle term of the BCHSH inequality will be a small number, with an absolute value much smaller than 2. The inequality can then never be violated, for trivial reasons that have nothing to do with local realism or causality.

---

locality at this stage also, a simple reasoning shows that a perfect detection correlation is sufficient to ensure the independence: if a selection occurs for the first particle, it is determined by the detection of the second particle with setting b, independently of the value chosed for a; it is then independent of a. Symetrically, the detection of the second particle is independent of b. Therefore the detection of the pair is independent of a and b.

In other words, locality arguments may be used, not only for the results of the apparatuses (the functions A and B), but also in order to specify the ensemble of observed pairs (the distribution function ρ).

For the BCHSH inequality, an accurate way to measure the number of events is therefore required, for instance an event-ready device as described at the end of §4.3.1.a. By contrast, the Eberhard inequalities contain only correlation probabilities on both sides of the inequality, and no pure number; they are therefore much less sensitive to the precise knowledge of the number of events. As for the CH inequality, it compares correlations rates on one side to single particle detection rates on the other. This could create problems if the former were much larger than the latter. Nevertheless, in experiments where photons are emitted by parametric down conversion, the directions of emission are strongly correlated, so that the single particle detection and double particle detection rates are comparable; obtaining violations is therefore perfectly possible.

(i) BCHSH inequality with spins.

In the 2015 experiment described in [161], the elementary measured quantum systems were not photons, but the electron spins of two NV centers (nitrogen-vacancy defects localized in the crystal lattice) in diamond. Since the centers are located inside two solid crystals, each contained in a cryostat, they are static; the problem of capturing them does not occur as it would for photons. The entanglement between the spins of the NV centers were created by a method [228] that is similar to entanglement swapping (§7.3.2), but does not require to start from pairs of entangled photons. It is based on the elaborate scheme proposed by Barrett and Kok [229], which involves the interference of fluorescence photons and two successive projections by measurements. The two NV centers are initially optically pumped in order to define their spin state, then put in a coherent superposition of the two spin states by a RF pulse, and then excited by laser pulses in a selective way (one spin state is excited, the orthogonal state is not). The photons that are spontaneously emitted are directed onto a beam splitter, so that they interfere. The observation of a photon at a given output channel of the beam splitter then guarantees the preparation of the expected entangled state of the two spins, and therefore provides an "event-ready" signal. The subsequent measurement on each spin is performed by selective absorption of a detection laser, possibly after a controlled radio-frequency pulse to change the basis of measurement. This provides an efficient measurement of the spins and avoids any detection loophole. A large spatial separation of more than 1 km. ensures that the locality and no-communication conditions are fulfilled. The end result is a violation of the BCHSH inequality by more than two standard deviations.

(ii) CH and Eberhard inequalities with photons.

Another brilliant success was simultaneously achieved by two other groups [162, 163], this time with pairs of photons emittted by parametric down conversion. The tested inequalities were slight variants of the CH and Eberhard inequalities (CH-E inequalities). Random generators of the type described in [216] were used, in order to close the free will loophole as well as possible. The distance between the two measurements processes was of more than 50 m. in one case, of more than 18 m. in the other [163]. In both cases, a particular care was taken to ensure that the sequence of emission and detection is so fast that the communication loophole is completely closed. In both cases as well, the observed violation was very clear, more than 11 standard deviations in [162] for instance.

(iii) BCHSH with atoms and "event-ready test".

In 2016, another group performed an experiment that closes the detection and communication loopholes simultaneously [230]. As for the experiment with NV centers [161], the two measured entangled systems are material particles: Rb atoms in optical traps, at a distance of 400 m. The method for creating the initial entanglement is nevertheless simpler. Initially, the two atoms are excited with lasers, and each emits a photon, with a polarization that is entangled with the Zeeman sublevel reached by the atom. The two photons then propagate to a laboratory where they are made to interfere; counters are put at the end of the interferometer and register the clicks in outputs; this provides the event-ready signal. Entanglement swapping (§7.3.2) then occurs, so that both atoms in their ground state become entangled. This completes the preparation stage of the two quantum particles. Measurements of their quantum internal state are performed by state selective ionization with additional blue lasers with adjustable polarizations. The resulting Rb^+ ions are accelerated and sent to electron multipliers. The final result is a violation of the BCHSH inequalities by about five standard deviations or more.

Generally speaking, in all these experiments, the only loopholes that remain possible are those that can never, even in principle, be removed (such as the possibility of an event far away in the past influencing both detectors). It therefore seems reasonable to conclude that 2015 is the year when all loopholes were closed, except of course those that can never be closed!

4.3.1.f Credibility of loopholes

No perfect experiment has ever been performed in physics, and it is not surprising that it should be forever impossible to close all loopholes in the interpretation of any experiment. The clearest example is the fatalist loophole, which will never be completely eliminated: it is always possible to assume that all measurements settings of any experiments are just consequences of some event that has taken place in the past, or even during the big bang. As we mentioned before, believing that this is the case and that "everything was written long ago and cannot be changed by any human decision" amounts to rejecting the possibility of any experimental scientific method. Clearly, if loopholes still exist, they have to take a strong ad hoc character. The explanations in question do not rest on any specific theory: no one has precise ideas about the nature of the physical processes involved in a possible conspiracy aiming at a perfect simulation of quantum mechanics within a completely local realist frame. The only thing one can do (see Appendix E) is to build ad hoc models to reproduce quantum mechanics. But by what kind of mysterious process would experiments mimic quantum mechanics so perfectly and produce apparent violations of local realism/causality, in a sort of conspiracy against the analysis of physicists? Bell himself was probably the one who should have most liked to see that his inequalities could indeed be used as a logical tool to find the limits of quantum mechanics; nevertheless, he found these explanations too unaesthetic to be really plausible.

4.3.2 Is quantum mechanics itself nonlocal? Counterfactuality

As we will discuss in more detail in §5.3, it is not possible to express all rules of quantum mechanics in terms of space-time events; the explanation of Bell experiments always requires at some stage to include an ingredient that does not belong to space-time, even if no frontal contradiction with relativity is involved (superluminal signaling remains impossible). For a theory to be really local, it is not sufficient that locality is obeyed sometimes, or even most of the time; it must be local all the time. Does this mean that quantum mechanics is nonlocal?

Whether or not quantum mechanics in itself is nonlocal has been the subject of debate for years. The consequence is that one can find in the literature various attitudes concerning the exact relation between quantum mechanics and locality and realism. Some authors consider that the nonlocal character of quantum mechanics is a fact. For instance, in [169], Bell writes "Quantum mechanics is non locally causal", and adds "quantum mechanics cannot be embedded in a local causal theory". As for Popescu and Rohrlich (see §5.4.4), they even propose to consider nonlocality as one of the axioms of quantum mechanics [231]. For others, quantum nonlocality is just an artifact created by the introduction into quantum mechanics of notions that are foreign to it (typically the EPR elements of reality). They add that state vector reduction is not an essential component of quantum mechanics (see for instance §§1.2.2.b and 11.1.2.b) and that this theory never introduces any violation of relativistic causality. Their conclusion is then, in general, that quantum mechanics is contradictory with realism, but not with locality.

The proponents of a local nonrealist theory often remark that the correlations between the results observed by Alice and Bob in distant galaxies are unobservable directly; to be measured, they require that Alice and Bob should exchange messages, or travel to some common place, and then compare their results. Since neither they nor their messages can travel faster than the velocity of light, the observation of the correlations is indeed subject to the minimum time prescribed by relativity. Indeed, a perfectly relativistic space-time diagram accounting for the whole experiment can be drawn, from the emission of the entangled particles to the observation of the correlations. The proponents of nonlocality then reply that a basic component of the scientific method is to trust the observation made by other scientists. Since Alice and Bob report that they observed the results long before they met at the end of the experiment, the correlation already existed when they made their measurements, and was therefore necessarily nonlocal. But the first group then remarks that this reasoning implies a definition of the reality in the past, a delicate concept that is not at all indispensable if realism is abandoned, etc. Lively discussions to decide whether or not quantum mechanics in itself is inherently nonlocal have taken place and are still active [232–234]; see also [24], [123, 235], the general discussion of realism by Leggett [236], or a more recent contribution in [237]. Delicate problems of logic are involved and we will not discuss the question in more detail here.

What is easier to grasp for the majority of physicists is the notion of "counterfactual statement" or, more generally, "counterfactuality" [238]. A counterfactual reasoning considers the results of possible experiments that can be envisaged for the future as well-defined quantities, and as valid mathematical functions to use in equations, even if they are still unknown – of course, in algebra one writes unknown quantities in equations all the time. This is very natural: as remarked by d'Espagnat [239, 240] and by Griffiths [241], "counterfactuals seem a necessary part of any realist version of quantum theory in which properties of microscopic systems are not created by the measurements". One can also see the EPR criterion of reality as a statement of the existence of counterfactuals.

But it also remains true that, in practice, it is never possible to realize more than one of the four experiments that are necessary to obtain a violation of the BCHSH inequalities: for a given pair, one has to choose a single orientation of the analyzers for the measurement, so that all other orientations will remain forever in the domain of speculations. For instance, in the reasoning of §4.1.2, at least two of the numbers A, A', B, and B' are counterfactuals. One could then conclude that counterfactuality is the notion to reject from quantum mechanics. We have already quoted a sentence by Peres [140], who wonderfully summarizes the situation in orthodox theory: "unperformed experiments have no results"; as Bell once regretfully remarked [215]: "it is a great inconvenience that the real world is given to us once only"!

But, after all, one can also accept counterfactuality and take a point of view that is still perfectly consistent with quantum mechanics, provided one then also accepts a more explicit appearance of nonlocality. The Bell theorem is not an impossibility theorem, either for counterfactuality, or for hidden variables – a sort of new version of the von Neumann theorem. Accepting explicit nonlocality is after all natural: why require that theories with counterfactuality/additional variables should be explicitly local at all stages, while this is not required from standard quantum mechanics? Indeed, in this theory, neither the state vector itself, nor the state vector reduction postulate – or the calculation of correlation of experimental results in the correlation point of view (§11.1.2) – corresponds to mathematically local calculations. We will see in the discussion of §5.3.2, but also in §5.4.2.b, that quantum mechanics does not lead to local stochastic boxes. In other words, even if one can discuss at a fundamental level whether or not quantum mechanics is local or not, it is perfectly clear that its formalism is not; it would therefore be just unfair to request a local formalism from a nonorthodox theory – especially when the theory in question is built in order to reproduce all results of quantum mechanics! As an illustration of this point, we quote Goldstein [18]: "in recent years it has been common to find physicists ... failing to appreciate that what Bell demonstrated with his theorem was not the impossibility of Bohmian mechanics, but rather a more radical implication – namely nonlocality – that is intrinsic to quantum theory itself".

5

Other Inequalities, Cirelson's Limit, Signaling

This chapter discusses more issues related to the Bell theorem; it is a direct continuation of the preceding chapter. We begin in §5.1 by a discussion of other inequalities that are similar to the Bell theorem; they are also consequences of local realism/causality but violated by quantum mechanics. Then, in §5.2, we show that the $\sqrt{2}$ violation factor of the BCHSH inequality that was obtained in chapter 4 is not specific of spins $1/2$ or of any particular types of measurements; it is universal, and directly arises from the very structure of the space of states of pairs of two-level systems (Cirelson theorem). In §5.3, we study the relation between the violations of Bell inequalities and relativity, and very briefly discuss field theory. Since relativity implies that no faster than light signal transmission is possible, in §5.4 we introduce the "nonsignaling condition" that a relativistic theory has to obey, and more generally discuss logical and quantum boxes (in particular Popescu-Rohrlich boxes).

5.1 Other inequalities

Many forms of inequality[1] have been proposed as consequences of local realism (or of the other logical sets of assumptions discussed in §4.2.1), while in contradiction with the predictions of quantum mechanics. Here we will give a few examples, starting with the initial 1964 Bell inequality. The purpose of several of them is mostly to illustrate the general ideas, and to make them more intuitive by referring to situations taking place in common life (Mermin, Bell's game). Others have a more practical motivation (Clauser-Horne, Eberhard) and have been used in real experiments to test quantum mechanics.

Following a well established tradition, we will call Alice and Bob the partners

---

[1] It has been noted [143] that the Bell inequalities can be seen as applications of theorems on marginal distributions in the theory of probabilities [144]; the mathematical proof of the inequalities relies on the existence of one common probability space for the relevant random variables (this is related to the notion of counterfactuality).

who, each in one of two remote laboratories, perform experiments on the particle they receive for each emitted pair by the source.

5.1.1 Bell 1964

In his initial 1964 article [135], Bell did not introduce the BCHSH inequality (4.7)-(4.9), but another inequality that is similar but mathematically distinct. His derivation was also different from that given above, but here we show that his inequality can be obtained as a special case of the BCHSH inequality. Assume we replace (4.5) and (4.6) by:

$$A = A(a, \lambda) \quad ; \quad B = B(b, \lambda) \quad ; \quad A' = A(-b, \lambda) \quad ; \quad B' = (c, \lambda) \qquad (5.1)$$

where a, b, and c are three angles defining three different settings (orientations of the Stern-Gerlach analyzers), instead of four as previoulsy; the notation $-u$ is used for the direction opposite to u, obtained by turning the analyzer by half a turn. Since this operation interchanges results $+1$ and -1, we have, for any u:

$$A(-u, \lambda) = -A(u, \lambda) \qquad (5.2)$$

Now, we assume that, as quantum mechanics predicts, the results obtained by Alice and Bob are always perfectly correlated when they use opposite directions for their measurements (which is equivalent to assuming that they are perfectly anticorrelated for parallel directions of measurements). We therefore have:

$$A'B = A(-b, \lambda)B(b, \lambda) = 1 \qquad (5.3)$$

so that (4.9) becomes:

$$-2 \le \left\langle A(a, \lambda)B(b, \lambda) - A(a, \lambda)B(c, \lambda) - A(b, \lambda)B(c, \lambda) \right\rangle + 1 \ \le +2 \qquad (5.4)$$

If we note $E(a, b)$ the average of the product of results obtained by Alice and Bob:

$$E(a, b) = \langle A(a, \lambda)B(b, \lambda) \rangle \qquad (5.5)$$

the upper bound condition in (5.4) then becomes:

$$E(a, b) - E(a, c) \le 1 + E(b, c) \qquad (5.6)$$

Now, if we replace a by $-a$ and perform the same calculation, we obtain:

$$-E(a, b) + E(a, c) \le 1 + E(b, c) \qquad (5.7)$$

or, grouping (5.6) and (5.7):

$$|E(a, b) - E(a, c)| \le 1 + E(b, c) \qquad (5.8)$$

which is the 1964 Bell inequality. We remark that its proof requires the perfect correlations contained in (5.3), which is not the case in the BCHSH inequalities.

Quantum mechanics predicts violations of this inequality: if we choose $\theta_{ab} = 45°$ and $\theta_{ac} = 135°$, using (4.10) we obtain $\sqrt{2}$ for the left-hand side of (5.8) while the right-hand side is equal to 1, corresponding to a violation by a factor $\sqrt{2}$.

5.1.2 Wigner and d'Espagnat inequalities

Other local-realist inequalities contain directly probabilities, instead of average values. To derive these inequalities, as in §5.1.1 (and as is the case for two spins $1/2$ in a singlet quantum state), we assume that, when Alice and Bob choose parallel directions of measurements, their results are always perfectly anticorrelated.

For each realization of the experiment (each emission of a pair of particles), the EPR local-realist reasoning shows that Alice's result $A(a)$ and Bob's result $B(b)$ are well defined, for any choice of the measurement parameters a and b. At the end of §4.1.3, we discussed Wigner's point of view, which ascribes a category to each realization; we assumed that two different values could be chosen for each of the parameters a and b. Here, as in the original article [142], we assume that Alice may choose among three different values a, a', and a''; similarly, Bob may choose among the same three values for his measurement parameter.

In a first step, we will focus our attention only on Alice's choice and result. For each realization, we define three numbers, all equal to ± 1 :

$$A \equiv A(a) \qquad A' \equiv A(a') \qquad A'' \equiv A(a'') \tag{5.9}$$

We consider a large number N of realizations of the experiment, and we denote $p_3(A, A', A'')$ the proportion of these realizations falling into the category[2] defined by three given results A, A' and A''. Similarly, we note $p_2(A, A')$ the proportion of realizations for which only the two variables A and A' are specified, but not A''; in the same way, we introduce the proportions $p_2(A, A'')$ and $p_2(A', A'')$. We then have:

$$p_2(A = +1, A' = -1) = p_3(+1, -1, +1) + p_3(+1, -1, -1) \tag{5.10}$$

The proportion $p_2(A = +1, A'' = -1)$ is given by:

$$p_2(A = +1, A'' = -1) = p_3(+1, +1, -1) + p_3(+1, -1, -1) \tag{5.11}$$

and finally the proportion $p_2(A' = -1, A'' = +1)$ is:

$$p_2(A' = -1, A'' = +1) = p_3(+1, -1, +1) + p_3(-1, -1, +1) \tag{5.12}$$

Now, consider the sum of the two right-hand sides of (5.11) and (5.12); it contains

---

[2] There are 8 different possible groups of results A, A' and A'', and therefore 8 such categories.

the two terms in the right hand side of (5.10), plus other terms. Since the proportions are ratios between two numbers that are positive, these other terms are also positive or zero. We then obtain:

$$p_2(A = +1, \ A' = -1) \le p_2(A = +1, \ A'' = -1) + p_2(A' = -1, \ A'' = +1) \quad (5.13)$$

an inequality which puts limits on the possible proportions for different values of the measurement parameter.

This inequality contains numbers that cannot be measured directly: Alice can not use simultaneously two different measurement parameters a and a'. But the assumed property of perfect anticorrelation allows us to express the same inequality in terms of results of measurements performed by Alice and Bob. If we set:

$$B \equiv B(a) \qquad B' \equiv B(a') \qquad B'' \equiv B(a'') \quad (5.14)$$

we have $B = -A$, $B' = -A'$ and $B'' = -A''$. We then obtain[3]:

$$p_2(A = +1, \ B' = +1) \le p_2(A = +1, \ B'' = +1) + p_2(A' = -1, \ B'' = -1) \quad (5.15)$$

where each term is now experimentally measurable.

In quantum mechanics, this inequality may be violated. This is because equation (4.3) provides:

$$p_2(A = +1, \ A' = -1) = p_2(A = +1, \ B' = +1) = \frac{1}{2} \sin^2 \frac{\theta_{aa'}}{2} \quad (5.16)$$

where $\theta_{aa'}$ is the angle between directions a and a'. Inequality (5.13) then becomes:

$$\sin^2 \frac{\theta_{aa'}}{2} \le \sin^2 \frac{\theta_{aa''}}{2} + \sin^2 \frac{\theta_{a'a''}}{2} \quad (5.17)$$

In the particular case where a, a'', and a' are equally distributed in a plane and in this order, we have $\theta_{aa''} = \theta_{a''a'} = \theta$ and $\theta_{aa'} = 2\theta$, and the preceding inequality becomes (after simplyfying by $2 \sin^2 \theta/2$):

$$2 \cos^2 \frac{\theta}{2} \le 1 \quad (5.18)$$

which is violated[4] if $-\pi/2 < \theta < +\pi/2$. We therefore have another case where the predictions of quantum mechanics contradict those of local realism/causality.

The method of counting events associated with both sides of inequality (5.15) illustrates its local realist physical content, in particular since it specifies which events make the difference between the two sides of the inequality. It is also well suited for illustrations of the occurrence of such inequalities in ordinary life. In "A la recherche du réel" ([25], p. 27), d'Espagnat formulates the inequality as "The

---

[3] Needless to say, similar inequalities exist for all probabilities $p_2(A = \pm 1, \ B' = \pm 1)$.
[4] Function $\sin^2 \theta_{aa''}/2 + \sin^2 \theta_{a'a''}/2 - \sin^2 \theta_{aa'}/2$ has a minimum $-1/4$ when $\theta_{aa''} = \theta_{a''a'} = \pi/3$ and $\theta_{aa'} = 2\pi/3$.

number of young women is less or equal to the number of woman smokers plus the number of young nonsmokers". In his essay "Bertlmann socks and the nature of reality" [95], Bell uses analogies with consumer test for collections of new socks. Nevertheless, experimentally, an inequality containing absolute probabilities is not necessarily the best choice for experimental tests of quantum mechanics against local realism/causality.

5.1.3 Mermin inequality

Another inequality, proposed by Mermin [242], can be obtained with the same experimental configuration, but different angles than those considered by Wigner. In §3.2.2.c of Chapter 3, we mentioned Schrödinger's comparison to illustrate the local realist reasoning: he represents the particles by schoolboys passing exams, the measurement apparatuses by teachers asking them questions, and the results of measurements by the answers of the schoolboys, assumed to be binary answers (+1 for yes, −1 for no). Extending this comparison, we assume that two school-boys, Albert and Bernard, are each asked a question by two different teachers in separate rooms; each teacher chooses his question at random among three possi-ble questions, with a probability 1/3 each (these three questions correspond to the three possible values of the settings a or b). The plan of the schoolboys is to give answers to the question that reproduce the predictions of quantum mechanics as well as possible; they know in advance the nature of the three questions, but not which precise question they will have to answer; moreover, when they meet one teacher, they do not know what question has been selected by the other (since the rooms are separate). However, before the examination, they are free to elaborate together any common strategy and to decide what answer each of them will give to each question. The total number of possible strategies is $2^3 \times 2^3 = 64$.

As previousiy, in order to reproduce the quantum predictions, every time the questions are the same the schoolboys have to make opposite answers. Obtaining this result is not very difficult: they decide in advance which answer Albert will make to every possible question, and agree that Bernard will choose the opposite answer to the same question. This brings back the number of possible strategies to 8; each of them can be noted (A, A', A''), where the three numbers between the brackets are equal to ±1 and give Albert's answers to the three possible questions. Among these 8 strategies, two are $(+1, +1, +1)$ and $(−1, −1, −1)$: the answers of the first schoolboy are always identical, whatever question is asked; for the 6 other strategies $(+1, +1, −1)$, $(+1, −1, −1)$, etc., two answers have the same sign and the third is opposite.

We now assume that the experiment is repeated a large number of times (the schoolboys may change their strategy each time if they wish); at the end of a series of realizations, one retains only the subensemble where the two questions have been different. For each such realization, two cases may occur:

(i) Either the strategy selected by the schoolboys is one among the two where the three answers of Albert are independent of the question; since those of Bernard are always the opposite of those of Albert, the two answers of the two schoolboys (then results) are then necessarily opposite, whatever questions are asked on both sides.

(ii) Or one of the 6 other strategies has been chosen; since each teacher chooses his question randomly among three possibilities (with equal probabilities 1/3), and since the two questions are different, there is one chance out of three that the questions asked by the two teachers will correspond to the precise pair of questions for which Albert has planned to give the same answers, and two chances out of three that the two selected questions correspond to cases where Albert has planned to give opposite answers. In the former case, because of the selected strategy, the two teachers get opposite answers, corresponding to a probability 1/3; in the latter case, the two teachers get the same answer with probability 2/3.

Therefore, whatever strategies are chosen, the probability to get opposite answers in one realization of the experiment is the probability associated with case (i) plus 1/3, which cannot be smaller than 1/3. In terms of classical probabilities, this yields:

$$P(+1,-1) + P(-1,+1) \geqslant \frac{1}{3} \qquad (5.19)$$

Let us now return to the quantum problem with two spins in a singlet state, and assume that the three a, a', and a'' are coplanar directions, all making 120° with each other (Figure 5.1), directions b, b', and b'' being the same as above. Relations (4.4) show that the probability to obtain different results is:

$$P_{(+,-)} + P_{(-,+)} = \frac{1}{2}\left(\frac{1}{4} + \frac{1}{4}\right) = \frac{1}{4} \qquad (5.20)$$

But this probability is smaller than 1/3, the minimum possible value according to (5.19); we therefore obtain one more case where the predictions of quantum mechanics are incompatible with those of local realism or causality.

5.1.4 Clauser–Horne inequality

Clauser and Horne [167] have derived a form of inequalities, often called CH inequalities, that is especially useful for experiments with photons. This form can be derived without any assumption that the detected particles fairly represent the ensemble of emitted particles; this allows one to close the "pair selection loophole"(see §4.3.1.a). Moreover, the CH inequalities provide a natural generalization

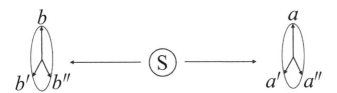

Figure 5.1 Scheme of a quantum experiment providing a large violation of inequality (5.19). The source S in the center emits two spins in a singlet state propagating to two apparatuses. Measurements of the spin components along three directions a, a', and a'' can be performed with one particle, and along three possible directions b, b', and b'' with the other.

of Bell inequalities to stochastic theories (§4.2.2.b). In many experiments with photons, polarization analyzers are inserted in front of photomultipliers detecting the particles. The rotation of these analyzers changes the transmitted polarization, and determines the value of the experimental settings, a for Alice, and b for Bob. If the photon is transmitted by the analyzer, it can then be detected as a count in the photomultiplier, which is registered in the electronic equipment; if the photon is absorbed by the analyzer, nothing is registered. In other words, instead of two kinds of results that can be obtained in an idealized experiments, here only one kind of result can be obtained. For each combination of the settings a and b, one can then measure a coincidence rate for the (almost) simultaneous registration of a count at each of the two photomultipliers.

To study this situation, Clauser and Horne (see also Freedman's thesis [146]) consider what they call "objective local theories", a very general class of theories, which can be either deterministic or stochastic. In these theories, for each realization of the experiment, the state of the source is defined by a variable λ of any kind, which does not need to be specified in more detail; when Alice chooses value a for her measurement setting, and when Bob chooses b, the probability $\mathcal{P}_{a,b}(\lambda)$ that both Alice and Bob obtains a count in their respective photomultipliers is the product:

$$\mathcal{P}_{a,b}(\lambda) = \mathcal{P}_a(\lambda) \times \mathcal{P}_b(\lambda) \qquad (5.21)$$

The factorization results from the assumption that, for each state λ of the source, independent random events determine the two observed detections: the probability $\mathcal{P}_a(\lambda)$ of registering a photon in Alice's laboratory is independent of the choice of the setting b made by Bob, and of the fact that he detects a photon or not; conversely, $\mathcal{P}_b(\lambda)$ is independent of a and of any event taking place in Alice's laboratory. When the distance between the two laboratories is very large, the factorization results from a locality assumption (a more detailed discussion of probability factorization is given in §4.2.2.b). Of course, when the experiment is repeated a large number of times, λ may fluctuate, so that an average over this variable becomes necessary to account for these fluctuations; this can introduce strong correlations between the two results. If $\rho(\lambda)$ is the probability distribution of the states λ of the source, the averaged joint detection probability $D(a, b)$ is the integral:

$$D(a, b) = \int d\lambda \, \rho(\lambda) \, \mathcal{P}_a(\lambda) \times \mathcal{P}_b(\lambda) \tag{5.22}$$

5.1.4.a An inequality with 6 numbers

Now consider six real numbers x_1, x_2, y_1, y_2, X and Y obeying the inequalities:

$$0 \le x_1 \le X \quad ; \quad 0 \le x_2 \le X$$
$$0 \le y_1 \le Y \quad ; \quad 0 \le y_2 \le Y \tag{5.23}$$

and introduce the function U as:

$$U = x_1 y_1 + x_2 y_2 + x_2 y_1 - x_1 y_2 - Y x_2 - X y_1 \tag{5.24}$$

We then have:

$$-XY \le U \le 0 \tag{5.25}$$

To prove these inequalities, we first assume that $x_1 \ge x_2$. Since the coefficient $(x_2 - x_1)$ of y_2 in U is then negative (or zero), U is always larger than (or equal to) the value obtained when y_2 takes its maximal value $y_2 = Y$; we therefore have:

$$U \ge x_1 y_1 + x_2 y_1 + (x_2 - x_1) Y - Y x_2 - X y_1 = x_1 (y_1 - Y) + y_1 (x_2 - X) \tag{5.26}$$

In the right-hand side of this inequality, the coefficient $(y_1 - Y)$ of x_1 is also negative (or zero); this term is therefore larger than (or equal to) its value when x_1 takes its maximal value $x_1 = X$, so that:

$$U \ge X (y_1 - Y) + y_1 (x_2 - X) = x_2 y_1 - XY \tag{5.27}$$

Finally, since y_1 is positive (or zero), the right hand side of this inequality is minimal when $x_2 = 0$ (which is compatible with our assumption $x_1 \ge x_2$), so that:

$$U \ge -XY \tag{5.28}$$

Moreover, still when $x_1 \geq x_2$ so that the coefficient $(x_2 - x_1)$ of y_2 is negative (or zero), U is always smaller than (or equal to) the value obtained when y_2 takes its minimal value $y_2 = 0$:

$$U \leq (x_1 + x_2) y_1 - Y x_2 - X y_1 \qquad (5.29)$$

In the right hand side of this inequality, the coefficient $(y_1 - Y)$ of x_2 is negative (or zero), so that this term is smaller than the value obtained when $x_2 = 0$; this leads to:

$$U \leq (x_1 - X) y_1 \qquad (5.30)$$

Since y_1 is positive, the right-hand side is smaller than the value obtained when x_1 takes its maximal value $x_1 = X$, so that:

$$U_{\max} \leq 0 \qquad (5.31)$$

Finally, if $x_2 > x_1$ the reasoning is almost the same, but the values of the variables and extrema are interchanged. In this case, U is larger than the value obtained for $y_2 = 0$, already written in (5.29); the right hand side of this inequality is larger than (or equal to) the value obtained when $x_2 = X$, which is $x_1 y_1 - XY$; this expression is larger than (or equal to) $-XY$, which leads again to (5.28). In addition, U is smaller than its value obtained when $y_2 = Y$, which is written in the right hand side of (5.26). Since $(y_1 - Y)$ is negative, this term is smaller than the value obtained when $x_1 = 0$, which is $y_1 (x_2 - X)$. This term is negative since it is the product of a positive factor by a negative factor, so that we obtain again (5.31). This completes the proof of inequalities (5.25).

5.1.4.b Application to a Bell experiment

The preceding inequality can be applied to a Bell experiment such as that shown schematically in Figure 4.1. We assume that only one kind of measurement result can be obtained in each laboratory: a count in a photomultiplier. For each realization of the experiment, we assume that Alice chooses one of the two values a or a' for the orientation of her polarization analyzer; similarly, Bob chooses either b or b' of the orientation of his analyzer. We then set:

$$\begin{aligned} \mathcal{P}_a(\lambda) = x_1 \quad &; \quad \mathcal{P}_{a'}(\lambda) = x_2 \\ \mathcal{P}_b(\lambda) = y_1 \quad &; \quad \mathcal{P}_{b'}(\lambda) = y_2 \end{aligned} \qquad (5.32)$$

(i) A first form of the CH inequalities

Since probabilities are positive (or zero) numbers smaller than (or equal to) unity, the inequalities (5.23) are then obeyed if we choose $X = Y = 1$. We then obtain:

$$-1 \leq \mathcal{P}_{a,b}(\lambda) + \mathcal{P}_{a',b'}(\lambda) + \mathcal{P}_{a',b}(\lambda) - \mathcal{P}_{a,b'}(\lambda) - \mathcal{P}_{a'}(\lambda) - \mathcal{P}_b(\lambda) \leq 0 \qquad (5.33)$$

We assume that the experiment is repeated many times, and call $D_2(a, b)$ the fraction of realizations where counts are obtained by Alice and Bob in coincidence; $D_2(a, b)$ is nothing but the average of $\mathcal{P}_{a,b}(\lambda)$ over λ. We call $D_1(a)$ the fraction of experiments where Alice obtains one count, whether or not Bob observes one; similarly, $D_1(b)$ is the fraction of experiments where Bob obtains one count, whether or not Alice gets one. Taking the average of (5.33) then leads to the inequality:

$$-1 \leq D_2(a, b) + D_2(a', b') + D_2(a', b) - D_2(a, b') - D_1(a') - D_1(b) \leq 0 \quad (5.34)$$

Equivalently, the second inequality can be written as:

$$D_2(a, b) + D_2(a', b') + D_2(a', b) - D_2(a, b') \leq D_1(a') + D_1(b) \quad (5.35)$$

We now use quantum mechanics to calculate the term that corresponds to the term in the middle of (5.34). In the results of §4.1.1, we assume that the + result corresponds to a count in the detector, and the − result to the absence of detection. Using relations (4.3) and (4.4), we obtain the following correspondence:

$$D_2(a, b) \Leftrightarrow \mathcal{P}_{(+,+)}(a, b) = \frac{1}{2} \sin^2 \frac{\theta_{ab}}{2} = \frac{1}{4}[1 - \cos \theta_{ab}]$$

$$D_1(a) \Leftrightarrow \mathcal{P}_{(+,+)}(a, b) + \mathcal{P}_{(+,-)}(a, b) = \frac{1}{2} \quad (5.36)$$

The term in the middle of (5.34) therefore corresponds to the quantum average value:

$$\langle R \rangle = -\frac{1}{4}[\cos \theta_{ab} + \cos \theta_{a'b'} + \cos \theta_{a'b} - \cos \theta_{ab'} + 2] \quad (5.37)$$

which is nothing but:

$$\langle R \rangle = \frac{\langle Q \rangle}{4} - \frac{1}{2} \quad (5.38)$$

where $\langle Q \rangle$ is written in (4.11). The discussion of §4.1.3 therefore shows that $\langle R \rangle$ ranges from $-\left(1 + \sqrt{2}\right)/2$ to $+\left(\sqrt{2} - 1\right)/2 > 0$, so that both inequalities (5.34) are indeed violated.

In this calculation, we have assumed that all emitted particles are measured by Alice and Bob. If this is not the case, and if the pair detection rate is much lower than the single particle detection rate, this inequality cannot be violated. For instance, in experiments where the source involves a radiative atomic cascade, there is no correlation between the directions of the two emitted photons: most photons are missed by the detectors. The rate of observed coincidences is then much lower than the rate of single photon counts, so that the D_2 terms can be neglected when compared to the terms in D_1. Moreover, the single photon detection terms are smaller than $1/2$. As a consequence, the inequality is trivially obeyed.

Inequality (5.35) is very useful in all experiments where the rate of coincidences is comparable to the rate of simple counts. This is the case when photon pairs are obtained by parametric down conversion of light in a nonlinear crystal, providing sources where the directions of emission of the two photons are strongly correlated. This method led to experiments where very clear violations of Bell inequalities were observed, in conditions in which the "pair selection loophole" (§4.3.1.a) was closed [162, 163, 243, 244] (some of these experiments used inequalities that are slightly different from the CH inequalities, but mathematically equivalent).

(ii) CH inequalities with the no-enhancement assumption

To obtain an inequality that can be used in experiments where the pair detection rate is significantly lower that the single particle detection rate, Clauser and Horne have added one more assumption: some experiments are performed, either with one of the polarization analyzers removed from the front of the photomultiplier, or both; in such situations, the setting variable a and/or b associated with the suppressed analyzer obviously disappears from the probability. Following the notation of Clauser and Horne, we call $\mathcal{P}_{a=\infty}(\lambda)$ the detection probability in Alice's laboratory when her analyzer has been removed, $\mathcal{P}_{b=\infty}(\lambda)$ the corresponding probability for Bob. We can then set:

$$\mathcal{P}_{a=\infty}(\lambda) = X \quad ; \quad \mathcal{P}_{b=\infty}(\lambda) = Y \qquad (5.39)$$

and introduce a "no-enhancement assumption": for every emission characterized by λ, the probability of a count with an analyzer in front of the photomultiplier is less than or equal to the probability in the absence of the analyzer. Since the polarization analyzer is a purely passive device, which can only absorb some polarizations and transmit others, physically this is a very plausible assumption. For any values of a, b, and of λ, we then have:

$$\mathcal{P}_a(\lambda) \leq \mathcal{P}_{a=\infty}(\lambda) \quad ; \quad \mathcal{P}_b(\lambda) \leq \mathcal{P}_{b=\infty}(\lambda) \qquad (5.40)$$

Conditions (5.23) are then fulfilled.

Then inequality (5.25) becomes:

$$-\mathcal{P}_{a=\infty}(\lambda)\,\mathcal{P}_{b=\infty}(\lambda)$$
$$\leq \mathcal{P}_{a,b}(\lambda) + \mathcal{P}_{a',b'}(\lambda) + \mathcal{P}_{a',b}(\lambda) - \mathcal{P}_{a,b'}(\lambda) - \mathcal{P}_{b=\infty}(\lambda)\,\mathcal{P}_{a'}(\lambda) - \mathcal{P}_{a=\infty}(\lambda)\,\mathcal{P}_b(\lambda)$$
$$\leq 0$$

$$(5.41)$$

We now take an average over λ, as in (5.22), and call $D(\infty, b)$ the coincidence rate when Alice removes her analyzer, $D(a, \infty)$ the coincidence rate when Bob removes his analyzer, and finally $D(\infty, \infty)$ the coincidence rate when both remove them. We then obtain:

$$-D(\infty,\infty) \le D(a,b) + D(a',b') + D(a',b) - D(a,b') - D(a',\infty) - D(\infty,b) \le 0$$
$$(5.42)$$

This inequality contains only coincidence rates, instead of a combination of single and double particle detection rates. The same calculation used earlier shows that quantum mechanics predicts violations of this inequality.

Experiments are often performed with equal angles between a and b, b and a', and a' and b', as in Figure 4.4. When the coincidence rate $D(\phi)$ depends only on the angle ϕ between the two orientations of the analyzers, relation (5.42) takes the simpler form:

$$-D(\infty,\infty) \le 3D(\phi) - D(3\phi) - D(a',\infty) - D(\infty,b) \le 0 \qquad (5.43)$$

This type of inequality was used in several experiments, for instance those of References [146, 151, 153].

5.1.5 Eberhard inequalities

The BCHSH inequalities apply to situations where the measurement apparatuses always provide results $+1$ or -1, for each measurement performed on each pair emitted by the source. Nevertheless, in practice, whether the particles are photons or not, the measurement devices never detect all of them. A first reason is that they do only capture those emitted within a limited solid angle from the source; some pairs give rise to only one, or zero, detection. Another reason is that, even if the particle hits the detector, the detection efficiency is less than 100%. The Eberhard inequality [218] takes such no-detection events into account; by convention, we attribute a measurement result 0 to the absence of detection.

We consider an experiment where, as previously, a source emits particles propagating towards two remote laboratories, those of Alice and Bob, who both perform measurements on the particles they receive. Alice chooses one of the two measurements settings a or a', Bob one of the two measurements settings b or b', so that four experimental configurations are possible. The difference with the usual case, nevertheless, is that three measurement results (instead of two) are observable by Alice and Bob, the three results $+1$, -1, and 0 (no detection). For each emission of a pair, one of these three results is registered by each of the apparatuses of the two operators.

5.1.5.a Emissions and detections

In what follows, we distinguish between two types of events taking place during each realization of the experiment:

(i) Emission of pairs of particles; these events are ranked $n = 1, 2, ..., N$.

(ii) Double detection event; each is characterized by one of the 4 possible configurations of the measurement apparatuses and by one of the $3 \times 3 = 9$ possible results. It is therefore associated with one of the 36 boxes marked with an asterisk in the following table:

		a			a'		
		+1	0	−1	+1	0	−1
b	+1	*	*	*	*	*	*
	0	*	*	*	*	*	*
	−1	*	*	*	*	*	*
b'	+1	*	*	*	*	*	*
	0	*	*	*	*	*	*
	−1	*	*	*	*	*	*

When the experiment is realized a large number of times, one can measure the relative frequency with which each box is obtained, and derive a probability.

We now examine how to characterize emission events within a realist local theory and follow the EPR reasoning; our purpose is to sort them by different categories, as in Wigner's argument [142], already mentioned in §4.1.3, (iii). The EPR reasoning shows that, during the emission of the particle in Alice's direction, this particle takes with it properties that determine one among the 3 possible results if Alice chooses setting a, as well as one among the 3 possible results if she chooses a'. Therefore, 9 different categories of particles can reach Alice's laboratory. The same is obviously true for Bob. For the pair of particles emitted by the source, 81 different categories are possible[5]. We do not need to make any assumption concerning the probability of appearance of every category, but we assume that the source is stable and that these probabilities do not vary in time: if the experiment is repeated a large number of times, the proportion of each category tends to a well-defined constant.

The preceding table has been introduced to specify the category of a double detection event; we now show how it can also be used to indicate the category of a pair emission event. More information is needed: the knowledge of the two results for one given choice of the two measurement settings has to be completed with the information on the results that would be obtained with the three other experimental configurations (obviously, the choice of one box among the 36 in the

---

[5] For a nonlocal theory, the whole experimental setup should be taken into account, with 4 possible configurations; for each of these configurations, 9 results are possible. The number of categories for the emitted pairs would then be 9^4.

table is not sufficient to identify such a category among the 81 that are accessible). A category of emission events is actually characterized by 4 boxes in the table. Moreover, since locality implies that the result obtained by Alice does not depend on the setting chosen by Bob, and conversely, these three boxes are necessarily the 4 corners of a rectangle in the table, as for instance the 4 boxes labeled X in the following diagram:

		a			a'		
		+1	0	−1	+1	0	−1
	+1	*	X	*	X	*	*
b	0	*	*	*	*	*	*
	−1	*	*	*	*	*	*
	+1	*	*	*	*	*	*
b'	0	*	X	*	X	*	*
	−1	*	*	*	*	*	*

5.1.5.b Obtaining inequalities

We now suppose that the experiment is realized a large number of times for each of the 4 experimental configurations. If these results are compatible with those of a local relativistic theory, in principle one can sort them among the 81 emission categories. We have seen, nevertheless, that a detection event can result from several categories; the knowledge of the results is not sufficient to attribute a given detection event to one of them, even if we know that this is possible in a subjacent way. We will now investigate the possibility to determine, through an inequality, if such subjacent categories may exist within a given list of results.

Consider one of the 36 boxes in the table, for instance that associated with two measurements with settings a and b leading both to results +1; this box is shown with a T in the following table:

		a			a'		
		+1	0	−1	+1	0	−1
	+1	T	*	*	X	x'	x"
b	0	*	*	*	*	*	*
	−1	*	*	*	*	*	*
	+1	y	*	*	*	*	*
b'	0	y'	*	*	*	*	*
	−1	y"	*	*	*	*	*

The corresponding probability of occurrence $\mathcal{P}_{a,b}(+1,+1)$ is the sum of the probabilities of all emission categories leading to this result. It is therefore the sum of the probabilities of all categories having the upper-right corner of their representative rectangle at one of the boxes labelled x, x' and x" in the table, the lower-left corner being one of the 3 boxes labelled y, y' and y" in the table (the lower right corner then falls into one of the 9 blank boxes in the lower-right boxes of the table); there are 9 such categories.

From this sample, we now withdraw the emission events that include one of the boxes x' and x". The probability $\overline{\mathcal{P}}$ associated with the remaining categories obeys the inequality:

$$\overline{\mathcal{P}} \geq \mathcal{P}_{a,b}(+1,+1) - \mathcal{P}_{a',b}(0,+1) - \mathcal{P}_{a',b}(-1,+1) \qquad (5.44)$$

This is because probability $\mathcal{P}_{a',b}(0,+1)$ includes, not only emission categories leading to joint results $a' \Longrightarrow 0, b \Rightarrow +1$ and $a \Rightarrow +1$, but also those for which measurement a leads to results 0 and -1; the difference in (5.44) therefore subtracts an excess of probability, so that the right-hand side is smaller than $\overline{\mathcal{P}}$. Similarly, probability $\mathcal{P}_{a',b}(-1,+1)$ suppresses too large a number of emission categories, since it does not select the result for a measurement a.

In a second step, we now subtract from the new sample all emission events associated with one of the boxes y or y'. The remaining probability $\tilde{\mathcal{P}}$ is then associated with a well-defined category of emission events, that associated with the 4 boxes T, X, Y and Z at the corners of a square in the following table:

		a			a'		
		+1	0	-1	+1	0	-1
b	+1	T	*	*	X	*	*
	0	*	*	*	*	*	*
	-1	*	*	*	*	*	*
b'	+1	Y	*	*	Z	*	*
	0	*	*	*	*	*	*
	-1	*	*	*	*	*	*

This probability $\tilde{\mathcal{P}}$ is positive, and the same reasoning as before shows that it obeys the relation:

$$\tilde{\mathcal{P}} \geq \overline{\mathcal{P}} - \mathcal{P}_{a,b'}(+1,0) - \mathcal{P}_{a,b'}(+1,-1)$$
$$\geq \mathcal{P}_{a,b}(+1,+1) - \mathcal{P}_{a',b}(0,+1) - \mathcal{P}_{a',b}(-1,+1) - \mathcal{P}_{a,b'}(+1,0) - \mathcal{P}_{a,b'}(+1,-1)$$
$$(5.45)$$

But $\tilde{\mathcal{P}}$ cannot exceed the probability associated with any of the corners T, X, Y or Z in the table; if for instance we choose Z, we obtain:

$$\mathcal{P}_{a',b'}(+1,;+1) \geq \tilde{\mathcal{P}} \tag{5.46}$$

The conjunction of this relation and of (5.45) leads to the Eberhard form of a Bell inequality:

$$\mathcal{P}_{a',b'}(+1,;+1)$$
$$\geq \mathcal{P}_{a,b}(+1,+1) - \mathcal{P}_{a',b}(0,+1) - \mathcal{P}_{a',b}(-1,+1) - \mathcal{P}_{a,b'}(+1,0) - \mathcal{P}_{a,b'}(+1,-1) \tag{5.47}$$

It expresses that the probability associated with the lower-right corner of the square in the table exceeds the difference between the probability associated with the opposite diagonal corner and the 4 probabilities associated with measurements on the same line or same column.

5.1.5.c Generalization

In the preceding calculation, we started from the upper-left corner of the square, but we could have started from any other corner, for instance the opposite diagonal corner, as in the diagram:

		a +1	0	−1	*a′* +1	0	−1
	+1	Z	*	*	Y	*	*
b	0	*	*	*	y'	*	*
	−1	*	*	*	y"	*	*
	+1	X	x'	x"	T	*	*
b′	0	*	*	*	*	*	*
	−1	*	*	*	*	*	*

where the subtracted detection events are again shown with letters x', x" and y',y". The corresponding inequality is then:

$$\mathcal{P}_{a,b}(+1,+1)$$
$$\geq \mathcal{P}_{a',b'}(+1,+1) - \mathcal{P}_{a,b'}(0,+1) - \mathcal{P}_{a,b'}(-1,+1) - \mathcal{P}_{a',b}(+1,0) - \mathcal{P}_{a',b}(+1,-1) \tag{5.48}$$

which is obtained from (5.47) by merely interchanging a and a', as well as b and b'.

We may actually locate a corner in any of the 9 boxes in the 3×3 square in the upper left, and the opposite diagonal corner in any of the 9 boxes of the 3×3 square in the lower right, and obtain new inequalities. For instance, the diagram:

		a			a'		
		+1	0	−1	+1	0	−1
	+1	*	*	*	*	*	*
b	0	*	*	*	*	*	*
	−1	*	*	T	x'	x"	X
	+1	*	*	y'	*	*	*
b'	0	*	*	y"	*	*	*
	−1	*	*	Y	*	*	Z

leads to:

$$\mathcal{P}_{a',b'}(-1,;-1)$$
$$\geq \mathcal{P}_{a,b}(-1,-1) - \mathcal{P}_{a',b}(0,-1) - \mathcal{P}_{a',b}(+1,-1) - \mathcal{P}_{a,b'}(-1,0) - \mathcal{P}_{a,b'}(-1,+1) \tag{5.49}$$

This time, results $+1$ and -1 are interchanged with respect to inequality (5.47).

In fact, any of the 81 rectangles associated with the emission categories gives rise to inequalities but, of course, all of them are not independent. We will focus our interest onto (5.47) and (5.49).

5.1.5.d Relation with the BCHSH inequality

It is interesting to relate the Eberhard inequalities to the BCHSH inequality of relation (4.9), which contains average values of products such as:

$$\langle AB \rangle = \mathcal{P}_{a,b}(+1,+1) + \mathcal{P}_{a,b}(-1,-1) - \mathcal{P}_{a,b}(+1,-1) - \mathcal{P}_{a,b}(-1,+1) \tag{5.50}$$

Because the sum of the 4 probabilities is 1, we also have:

$$\langle AB \rangle = 2\left[\mathcal{P}_{a,b}(+1,+1) + \mathcal{P}_{a,b}(-1,-1)\right] - 1$$
$$= 1 - 2\left[\mathcal{P}_{a,b}(+1,-1) + \mathcal{P}_{a,b}(-1,+1)\right] \tag{5.51}$$

The 3 other mean values $\langle A'B \rangle$, $\langle AB' \rangle$ and $\langle A'B' \rangle$ are obtained in the same way; for $\langle A'B \rangle$, it is for instance sufficient to replace in the right-hand side of (5.51) a by a'.

Assume now for a moment that, as is usually assumed for deriving the BCHSH inequality, results 0 are never obtained; all probabilities $\mathcal{P}$ vanish as soon as one of the results is zero, so that only 4 probabilities remain in each of the right hand sides of (5.47) and (5.49). If we take the sum of these two inequalities and use (5.51), we obtain:

$$\frac{1}{2}\left[\langle A'B' \rangle + 1\right] \geq \frac{1}{2}\left[\langle AB \rangle + 1\right] - \frac{1}{2}\left[-\langle A'B \rangle + 1\right] - \frac{1}{2}\left[-\langle AB' \rangle + 1\right] \tag{5.52}$$

or, if we multiply by 2:

$$\langle A'B' \rangle - \langle AB \rangle - \langle A'B \rangle - \langle AB' \rangle \geq -2 \tag{5.53}$$

This is equivalent to:

$$\langle AB \rangle + \langle A'B \rangle + \langle AB' \rangle - \langle A'B' \rangle \leq 2 \tag{5.54}$$

We therefore recover one of the BCHSH inequalities (*cf.* note [2]).

5.1.5.e Relation with the CH inequality

The Clauser–Horne and the Eberhard inequalities are closely related. Why this is the case can be seen by uniting the results -1 and 0 in the Eberhard inequality into one single category of results. Reference [204] then shows that a simpler inequality is obtained:

$$\mathcal{P}_{a,b}(+1,+1) - \mathcal{P}_{a,b'}(+1,0) - \mathcal{P}_{a',b}(0,+1) - \mathcal{P}_{a',b'}(+1,+1) \leq 0 \tag{5.55}$$

which is called the CH-E inequality.

5.1.5.f Quantum violations

When result 0 is never obtained, we know that the BCHSH may be violated by a factor $\sqrt{2}$, an important factor. Since relations (5.47) and (5.49) lead to the BCHSH inequality, this shows that at least one of them is not obeyed; its left hand side is sufficiently smaller than its right-hand side to create a large violation of the BCHSH inequality. Now assume that we include the possibility of results 0 (no-detection of particles); if the corresponding probabilities remain low, continuity indicates that the left hand-side remains lower than the right-hand side, so that the BCHSH violation remains valid. This qualitative reasoning indicates why, as long as 0 results are not too frequent, quantum mechanics must lead to violations of the Eberhard inequalities.

In Reference [218], Eberhard provides a more precise calculation of the maximal violations predicted by quantum mechanics. His method uses the calculation of the eigenvalues of a 4×4 matrix, and involves a numerical part. This leads to the determination of the optimal experimental conditions, when no-detection events are taken into account as well as background noise in the detectors. These conditions specify an optimal state vector for the spins of the two entangled particles, as well as values for the measurement settings; both differ from the optimal parameters for the BCHSH inequality. Concerning the state vector, an interesting result is that its optimal value differs from the "maximally entangled state" (such as a singlet state for two spins), and even comes arbitrarily close to an uncorrelated product state if the detection efficiency is low (but the violation simultaneously tends to zero). With the BCHSH inequality, a minimum detection efficiency of 82.8% is necessary for establishing a violation of local realism without any possible loophole (§4.3.1); meanwhile, a 66.7% efficiency turns out to be sufficient with the Eberhard inequalities and the corresponding optimized experimental conditions.

Experiments using these inequalities have been performed in 2012 [244] and 2015 [162].

5.1.6 Bell's game

In order to write the results of experiments in terms of a binary variable (equal to ± 1), we introduce the numbers $G_\pm$ defined by:

$$G_\pm(A, B) = \frac{1 \pm AB}{2} \tag{5.56}$$

and define S as:

$$S = G_+(A, B) + G_+(A, B') + G_+(A', B) + G_-(A', B') = 2 + \frac{\overline{M}}{2} \tag{5.57}$$

where:

$$\overline{M} = AB + AB' + A'B - A'B' \tag{5.58}$$

This number is very similar to that defined as M in (4.7), the only difference being the position of the minus sign; but this difference disappears if the definitions of b and b' are exchanged. Therefore, M and $\overline{M}$ have the same properties and, in terms of S, the BCHSH inequality $-2 \le \overline{M} \le +2$ becomes $1 \le S \le 3$. Therefore, within local realism/causality the average value $\langle S \rangle$ of S obeys:

$$1 \le \langle S \rangle \le 3 \tag{5.59}$$

By contrast, the results of §4.1.3 (with the interchange of b and b') show that quantum mechanics predict violations of this inequality: $\langle \overline{M} \rangle$ can reach the value $2\sqrt{2}$ and therefore $\langle S \rangle$ reach the value $2 + \sqrt{2} = 3.141$, which is more than 3.

    This calculation can be used to introduce a game played by Alice and Bob, which Gisin calls "Bell's game" [173]. Alice and Bob sit in different rooms, with no possible communication between them; nevertheless, before the game starts, together they can decide a strategy they wish to try and win the game (this is a collaborative game). As soon as the experiment begins, every 10 seconds (for instance) a light flashes in their room, randomly either red or green. In each room, the color is determined independently by a local random number generator (or by the free will of one local organizer of the game per room); neither Alice nor Bob can influence this choice and they are completely uncorrelated in both rooms. As soon as Alice and Bob see the color, each of them decides to press either a button "yes", or a button "no". The rule of the game is as follows: for every event where at least one of the lights is green, the score of the players increases by 1 if their answers are the same, otherwise it does not change; for every event where both lights are red, their score increases by 1 if their answers are different, otherwise it does not change.

A computer registers all events, including the colors of the flashes and the corresponding subsequent choices of Alice and Bob. Taking first into account all events with two green flashes, the computer calculates the average number of times where the score has increased by one unit (it divides the number of successes by the total number of green-green events); then it does the same computation for the three other kinds of events (green-red, red-green, red-red). Finally it adds the 4 averages to obtain the final score S. If the sum is more than 3, Alice and Bob win the game, otherwise they lose.

If Alice and Bob press the buttons "yes" or "no" at random, the chances of success are $1/2$, and the sum S necessarily takes a value that is close to 2; they lose the game. Can they use a less trivial strategy and win? Actually, if they do not use quantum physics, they have no way to win the game[6]. They can elaborate a common strategy, deciding in advance what Alice will do during each event in both cases, green or red light, and similarly for Bob; but the difficulty is that, to optimize the gain, Alice should make opposite choices when the light is red or green in Bob's room, which she can only try to guess. Each of the players can even take with her/him a table with, say, $1,000$ lines corresponding to the first $1,000$ events and containing the choices ("yes" or "no") to make in both cases (green or red). But this simply amounts to defining a function $A(a, \lambda)$ for Alice, $B(b, \lambda)$ for Bob; the values of these functions can then be inserted into (5.57). Now, this expression corresponds exactly to the calculation of the final score S provided a and b are associated with green, and a' and b' with red: the three first terms provide $+1$ in case of success when at least one light flashes green, and the fourth term has the same result when both lights are red. So, within classical physics, Alice and Bob can only lose the game. Therefore, if Alice and Bob follow any predetermined strategy, even if it is very elaborate, S can never exceed 3; they always lose.

Nevertheless they can win with the help of quantum physics: if, for every event, each of them has access to one spin from a common singlet pair, they can use the color information to decide the direction of analysis of their local Stern-Gerlach analyzer, and then push the button "yes" if result $+1$ is observed, "no" if result -1 is observed. The previous quantum calculation ensures that they will then obtain $2 + \sqrt{2}$ on average. They will then win!

Other versions of the game are possible. We chose a version that is directly in the line of Schrödinger's parable with schoolboys (§3.2.2.c), who do not choose the random question they are asked (the setting of the measurement apparatus), but the answer (the result of the experiment). Here, Alice and Bob do not choose the color of the flashes, which are determined by a local random number generator, but the answer , "yes" or "no". In another equivalent version [173], the opposite is

---

[6] Except of course very rarely, if by pure chance they benefit of some big random fluctuation; this becomes extremely unlikely when the number of events increases

assumed: Alice and Bob choose the settings a and b randomly, and each of them applies this choice as an "input" to some physical device (a computer for instance); the role of this device is to provide a result that is a function of the local setting. This result is $A(a)$ for Alice and $B(b)$ for Bob. In fact, these functions may change from one event to the next; they then become $A(a, \lambda)$ and $B(b, \lambda)$, where λ is the rank of the event (as discussed at the end of §4.1.2). This version of the game leads to the same inequality as before; a violation of the inequality indicates that no functions $A(a, \lambda)$ and $B(b, \lambda)$ can describe the results provided by the two devices, for each input that Alice and Bob can decide to enter into them.

5.2 Cirelson's theorem

In §4.1.2, we have introduced the BCHSH inequality, a combination $\langle M \rangle$ of four averages that can never exceed 2 within local realism. In §4.1.3 we have seen that, in quantum mechanics and with two spins in a singlet state, the corresponding combination of averages $\langle Q \rangle$ can reach $2\sqrt{2}$. This is of course already a large violation of the BCHSH inequality, but one could nevertheless hope to find other quantum systems or other measurements that lead to even larger violations. After all, each of the four average values $\langle AB \rangle$, $\langle A'B \rangle$, etc. appearing in $\langle Q \rangle$ is bounded between -1 and $+1$; if these averages were independent variables, mathematically the maximal value accessible to $\langle Q \rangle$ would be 4, larger than $2\sqrt{2}$. One could then wonder if better quantum situations could be found where the violation is larger than what is obtained with two spins in a singlet state.

Within quantum mechanics, Cirelson's theorem [245, 246] shows that, with a system made of two subsystems, it is actually impossible to go beyond this value of $2\sqrt{2}$, whatever series of measures is made on the two subsystems and whatever the initial state of the whole system is. The limitation arises from the very structure spaces of states, which is the tensor product of two spaces having each two dimensions.

5.2.1 Measurements on two-level subsystems

Consider a physical system made of two quantum subsystems on which separate measurements can be performed, each leading to two possible results, $+1$ and -1. Each of the subsystems has a space of states that is spanned by two levels which, without loss of generality, we can consider as the two eigenstates of the Oz components of a (pseudo) spin 1/2. We denote $\sigma_x(1)$, $\sigma_y(1)$, and $\sigma_z(1)$ as the three components for the first subsystem – cf. equation (12.59) of Chapter 12 – which we group into a vectorial operator $\sigma(1)$. A similar notation $\sigma_x(2)$, $\sigma_y(2)$, $\sigma_z(2)$, and $\sigma(2)$ is taken for the second subsystem. Any measurement performed on the first subsystem corresponds to an Hermitian operator $O(1)$ acting in its space of states, which can be expressed as a linear combination of the three components of $\sigma(1)$ and of the unit operator $\widehat{I}(1)$ in the form:

$$O(1) = \alpha \, \widehat{I}(1) + \boldsymbol{a} \cdot \boldsymbol{\sigma}(1) \tag{5.60}$$

where α and the three components of $\boldsymbol{a}$ are real scalar parameters. Nevertheless, since the results of the measurement can only be ± 1, the two eigenvalues must have the same values, so that $\alpha = 0$ and $|\boldsymbol{a}| = 1$. We then denote $\sigma_a(1)$ and $\sigma_{a'}(1)$ as the spin operators associated with two kinds of measurement performed on the first subsystem with the settings a and a':

$$\sigma_a(1) = \boldsymbol{a} \cdot \boldsymbol{\sigma}(1) \qquad \sigma_{a'}(1) = \boldsymbol{a'} \cdot \boldsymbol{\sigma}(1) \tag{5.61}$$

and similarly, for the second subsystem:

$$\sigma_b(2) = \boldsymbol{b} \cdot \boldsymbol{\sigma}(2) \qquad \sigma_{b'}(2) = \boldsymbol{b'} \cdot \boldsymbol{\sigma}(2) \tag{5.62}$$

(the norms of vectors a, a', b, and b' are all 1). Since the square of any Pauli matrix (§12.1.7) gives the identity operator $\widehat{I}$, we have:

$$[\sigma_a(1)]^2 = [\sigma_{a'}(1)]^2 = \widehat{I}(1) \quad ; \quad [\sigma_b(2)]^2 = [\sigma_{b'}(2)]^2 = \widehat{I}(2) \tag{5.63}$$

The quantum average $\langle Q \rangle$ that generalizes (4.11) is given by $\langle \Psi | Q | \Psi \rangle$, where $|\Psi\rangle$ is any normalized state of the two-particle system, and where Q is the operator given by:

$$Q = [\sigma_a(1)] \; [\sigma_b(2)] - [\sigma_a(1)] \; [\sigma_{b'}(2)] + [\sigma_{a'}(1)] \; [\sigma_b(2)] + [\sigma_{a'}(1)] \; [\sigma_{b'}(2)] \tag{5.64}$$

Our purpose is to show that the modulus of the average value of this operator cannot exceed $2\sqrt{2}$, whatever choice is made for $|\Psi\rangle$ and the four vectors a, b, a', and b'.

5.2.2 Maximal quantum violation

We first calculate the square of this operator, which contains three kinds of terms: the squares of each of the four operators appearing in (5.64), the crossed terms where one of the operators between brackets occurs twice, and finally the crossed terms where all σ operators are different. Since the squares of the Pauli matrices are $\widehat{I}$, the contribution of the first kind of terms is:

$$4 \times \widehat{I} \tag{5.65}$$

As for the terms involving one operator σ twice, they give the contribution:

$$\begin{aligned}
&\left[-[\sigma_a(1)]^2 + [\sigma_{a'}(1)]^2 \right] \left[\sigma_b(2)\sigma_{b'}(2) + \sigma_{b'}(2)\sigma_b(2) \right] + \\
&\left[[\sigma_b(2)]^2 - [\sigma_{b'}(2)]^2 \right] \left[\sigma_a(1)\sigma_{a'}(1) + \sigma_{a'}(1)\sigma_a(1) \right]
\end{aligned} \tag{5.66}$$

which vanishes since the squares of all Pauli matrices are equal to the unit matrix (§12.1.7). The crossed terms containing four different σ operators give:

$$\sigma_a(1)\sigma_{a'}(1)\sigma_b(2)\sigma_{b'}(2) + \sigma_{a'}(1)\sigma_a(1)\sigma_{b'}(2)\sigma_b(2)$$
$$-\sigma_a(1)\sigma_{a'}(1)\sigma_{b'}(2)\sigma_b(2) - \sigma_{a'}(1)\sigma_a(1)\sigma_b(2)\sigma_{b'}(2)$$
(5.67)

which is nothing but the product of two commutators:

$$[\sigma_a(1), \sigma_{a'}(1)]\,[\sigma_b(2), \sigma_{b'}(2)]$$
(5.68)

Collecting all these results, we obtain the equality [246]:

$$Q^2 = 4 \times \widehat{I} + [\sigma_a(1), \sigma_{a'}(1)]\,[\sigma_b(2), \sigma_{b'}(2)]$$
(5.69)

The commutation relation of the Pauli matrices:

$$[(a \cdot \sigma), (a' \cdot \sigma)] = 2i\,(a \times a') \cdot \sigma$$

can then be used to arrive at:

$$Q^2 = 4 \times \widehat{I} - 4\,[(a \times a') \cdot \sigma(1)]\,[(b \times b') \cdot \sigma(2)]$$
(5.70)

The operator $(a \times a') \cdot \sigma(1)$ has eigenvalues $\pm |a \times a'|$, which in general have a modulus smaller than 1 since the length of the vector $a \times a'$ cannot exceed 1; similarly, operator $(b \times b') \cdot \sigma(2)$ has eigenvalues with modulus equal or smaller than 1. Therefore, the modulus of the average value of the product of operators between brackets in the right hand side of (5.70) cannot exceed 1, so that:

$$\langle Q^2 \rangle = \langle \Psi | Q^2 | \Psi \rangle \le 4 + 4 \times 1 = 8$$
(5.71)

But the square of the average $\langle Q \rangle^2$ of a Hermitian operator is always smaller[7] than the average value of its square $\langle Q^2 \rangle$. We therefore always have:

$$-2\sqrt{2} \le \langle Q \rangle \le 2\sqrt{2}$$
(5.72)

This inequality is Cirelson's theorem.

Remarks:

(i) The proof of this inequality is useful to predict conditions under which the bound $\pm 2\sqrt{2}$ can be reached, that is when (5.71) and (5.72) become equalities. A first condition is that the operators between brackets in (5.70) must have eigenvalues of unit modulus; this implies that the vectors $a \times a'$ and $b \times b'$ should have unit length, which requires that a and a' should be perpendicular, as well as b and b'. Another condition is that the state $|\Psi\rangle$ should be an eigenvector with eigenvalue -1 of the product of the component of the first spin along axis $a \times a'$ by the component of the second spin along axis $b \times b'$; in the coplanar configuration of the four

---

[7] This Schwarz inequality is obtained by writing that the average value of $[Q - \langle Q \rangle]^2$ is positive; it appears in the definition of the square of the mean square value ΔQ.

vectors, this corresponds to spin components along the same axis, which we may call Oz. Indeed, a singlet state is such an eigenstate of $\sigma_z(1)\sigma_z(2)$ – but this is also the case of a triplet state with zero component along Oz, for which one can check that the same violation is obtained as with the singlet state.

(ii) In §6.2, we study a case where, by considering three subsystems instead of two, one can find cases where quantum mechanics predicts that $|\langle Q \rangle|$ exceeds $2\sqrt{2}$ and reaches the mathematical limit 4.

5.3 Relativity, locality, field theory

As we have seen in §3.3.3.a, relativity can be invoked in the EPR reasoning, even if a more general notion of locality is sufficient. The same is true for the Bell theorem, which can also be derived as a consequence of the relativistic impossibility of physical influences propagating faster than light, combined with the notion of EPR realism; this is what Bell himself did in lectures at conferences [170]. As we have emphasized in §3.3.3.a, one can see the EPR reasoning, and its continuation the Bell theorem, as attempts to reintroduce into quantum mechanics a complete description of physical phenomena in terms of relativistic space-time events, related by causal influences.

To illustrate this, let us try to describe a Bell experiments in terms of space-time events, the most basic notion in relativity. Can we explain what happens in space-time when Alice and Bob make their separate measurements and observe results at a macroscopic scale? Two possibilities may be considered:

– either we consider that each result is the consequence of something that pre-existed before the measurement, through some unspecified process, following the line of EPR. But, from the Bell theorem, we know that something should evolve in a nonlocal way to reproduce all predictions of quantum mechanics. We are then necessarily led to a nonrelativistic description.

- or we consider that Alice's result emerges from nothing during some fundamentally nondeterministic process (and the same for Bob's result). The difficulty, then, is to explain the emergence of highly correlated macroscopic results at an arbitrary large distance, without any causes. Postulating this spontaneous emergence amounts to postulating the existence of a sort of "double space-time event" that may take place simultaneously in two remote galaxies. It is certainly not a single space-time event, but a random process that is delocalized in space-time, as schematized in Figure 4.5.

This second possibility fits well with Bohr's views, when he points out that the only correct way to describe the measurement process is to consider the experimental apparatus (and measured system) as a whole. So, in this case also, we end up with a description of the experiment that does not fit well with relativity (§5.3.1

below). Actually, no one has succeeded to propose a formalism that remains explicitly relativistic from the beginning to the end and reproduces all predictions of quantum mechanics; as we discuss in more detail in §5.3.2, a nonrelativistic ingredient is always necessary at some point.

At the end of his essay "Speakable and unspeakable in quantum mechanics" (chapter 18 of [6]), Bell writes: "We have an apparent incompatibility, at the deepest level, between the two fundamental pillars of contemporary theory" (quantum mechanics and relativity). Of course, "apparent incompatibility" does not mean "contradiction": the theory resting on these two pillars is not self-contradictory as long as a violation of the Bell theorem does not imply superluminal signaling (we come back to this important property in more detail in §5.4). Shimony expresses this idea by writing: [247] "In this sense there may be a peaceful coexistence between quantum mechanics and relativity".

5.3.1 Bohr and space-time

Already at the beginning of quantum mechanics, Bohr had already strongly emphasized that the quantum postulate, symbolized by a nonzero value of the Planck quantum of action, implies a "failure of our ordinary space-time description". In the proceeding of the famous Como lecture [92], he expresses this general idea about ten times, each time with different words. For instance, he writes: "We learn from the quantum theory that the appropriateness of our usual causal space-time description depends entirely on the small value of the quantum of action, as compared to the actions involved in ordinary sense perceptions"; in other words, such a description is no longer possible for physical processes involving very small values of the action. Later, he adds: "The difficulties with which a causal space-time description is confronted in the quantum theory, and which have been the subject of repeated discussions, are now placed into the foreground by the recent developments of the symbolic methods". Bohr sees this as one more illustration of complementarity: "This circumstance may be regarded as a simple symbolical expression for the complementary nature of the space-time description and the claims of causality". But, in relativity, one usually uses at the same time a space-time description and causality without considering them as exclusive!

As mentioned previously, in this view a measurement process is not necessarily a space-time event, in the relativistic sense. The two measurements performed by Alice and Bob should be considered as a single quantum process that extends in space-time over an arbitrarily large distance; it certainly does not correspond to a single point in four-dimension space. It is, if one likes, a nonseparable pair composed of two single events fused into one, with no possible relativistic or causal connection between them. The Born rule directly provides the probability of this double event. Alternatively, one can state that a measurement process should be considered as occupying a sort of extended "bubble" in space-time, which should not be decomposed into smaller elements and events (as EPR tried to do). Bohr expresses a similar idea when he writes: "This entails, however, that in the interpretation of observations, a fundamental renunciation regarding the space-time description is unavoidable".

5.3.2 *Does field theory solve the problem?*

It is sometimes argued that quantum field theory solves these difficulties, because it suppresses wave functions from the formalism, and makes use of field operators that obey perfectly causal evolution equations. The commutators of field operators vanish outside of light cones, as relativistic causality requires. Moreover, the dynamics of quantum fields can be derived from a Lagrangian in which all symmetries of the Poincaré group have been included, so that its relativistic invariance is automatically ensured. The interactions between various fields are described by perfectly local Hamiltonians. Doesn't this provide a perfectly locally causal formalism that immediately eliminates any difficulty?

Actually, it does not. The reason is that the dynamics of quantum operators is not the only ingredient of quantum field theory. Operators are mathematical objects that cannot be used alone: in order to provide physical predictions, they must act on a vector belonging to the space where they are defined. Indeed, to obtain the expressions of physical probabilities, one must introduce a state vector (or a density operator), which is an inherently nonlocal object; it may contain in a nonseparable way all the information on two entangled systems, even if these systems are very distant from each other. This excludes any decomposition of physical phenomena in terms of space-time events, exactly as Bohr said. Generally speaking, any method that attributes numbers (quantum averages) to any product of operators (observables), even when they relate to very remote regions of space, contains an irreducible nonlocal component. As a consequence, while it is true that most calculations within field theory remain perfectly causal and relativistic, at some point a nonlocal ingredient becomes unavoidable.

Let us consider for instance a Bell experiment involving two entangled photons. The field operators propagate from the source towards the two measurement apparatuses in a perfectly causal way. Nevertheless, one has to calculate probabilities of measurement results. This requires to calculate averages of operators in a quantum state that is entangled and delocalized over space, and therefore affects both probabilities at the same time, without any relativistically causal constraint. Locality is then no longer obeyed, and this is the reason why the Bell inequalities are violated in relativistic quantum field theory, exactly as strongly as in more elementary quantum calculations. The propagation of fields is indeed perfectly local, but the calculation of averages and probabilities is not.

In fact, no completely relativistic description of a Bell experiment is possible, independently of the formalism used. The calculation of probabilities must include some nonlocal ingredient somewhere, making the measurement a process that is "outside of space-time". The predictions of quantum mechanics are incompatible with any theory that remains completely and explicitly relativistically causal from the beginning to the end.

5.4 No instantaneous signaling

The theory of relativity requires that it should be fundamentally impossible to transmit signals containing information between two distant points faster than the speed of light (relativistic causality); suppressing this absolute limit would lead to serious internal inconsistencies in the theory. But one could then wonder if a violation of the BCHSH inequalities does not imply the possibility of a communication at an arbitrary velocity, since the distance between the two operators Alice and Bob has no influence on the appearance of correlations between the results of distant measurements. Indeed, from a local realist point of view, we have seen that the only possibility to reproduce the predictions of quantum mechanics is to include an instantaneous influence of the value of a local experimental setting on the result at the other end. Moreover, even within standard quantum mechanics, the instant determination of the state of the second particle by a measurement performed on the first seems to indicate an influence at an arbitrary distance. We now check that quantum mechanics does not lead to such a possibility of instantaneous signal transmission.

What methods of transmission could be envisaged? The first idea that comes to mind is to imagine a system having some analogy with the Morse telegraph, where the results $+1$ and -1 are used by Alice to code a message to be sent to Bob (for the sake of simplicity, here we assume that each of the measurements can only provide two results, but a generalization is possible). But, obviously, such a system would not work since the results of measurements are completely random, so that Alice has no control to code any message. A telegraphic system can work only if what is used to code a message is chosen by the experimenters, for instance, the experimental settings instead of the results. Is it then possible for Bob to determine the value a or a' of the setting that Alice has chosen, knowing that Bob can only make local observations in his laboratory, and that he is completely free to choose whatever value for his measurement setting b? This signaling technique would not imply any particular delay for the transmission of information, creating a direct conflict with relativity (in which the minimal delay is proportional to the distance between Alice and Bob).

We now discuss under which conditions a conflict with relativity can be avoided. For a deterministic theory, we will see that this amounts to a condition of "setting independence". For a stochastic theory, this also introduces so called "nonsignaling" conditions; if more stringent conditions of locality are assumed, this introduces a condition of "outcome independence". See [174] and [175] for a discussion in terms of "strong locality" and "predictive completeness" (or "parameter independence" and of "outcome independence" in [248]).

5.4.1 Nonsignaling (NS) conditions

From a general point of view, and without restricting the discussion to quantum mechanics, what are the general mathematical conditions ensuring that a theory is "nonsignaling" (NS conditions), meaning that it does not allow Alice to transmit instantaneous signals to Bob (and conversely)?

For a deterministic theory, these conditions are very simple: the result $A = \pm 1$ of each measurement performed by Alice must depend only on the measurement setting a, while the result $B = \pm 1$ of each measurement performed by Bob must depend only on b. This is the condition of "setting independence".

For a stochastic theory, and for each well-defined experimental setup (as meant by Bohr) corresponding to given measurement settings a and b, four probabilities $\mathcal{P}(A, B|a, b)$ are associated with the four events $A = \pm 1$ and $B = \pm 1$, with a sum 1:

$$\sum_{A,B} \mathcal{P}(A, B|a, b) = 1 \quad \text{for every couple } a, b \tag{5.73}$$

In what follows, we will assume that the first setting can take only two values a and a', while the second can take only two values b and b', so that four experimental setups are possible for the whole experiment. We then have 16 probabilities obeying 4 sum relations similar to (5.73); the most general probabilistic model therefore depends on 12 free parameters.

When the experiment is repeated, since Bob does not have access to Alice's results, the only thing that he can measure is the occurrence frequency of his own results. This corresponds to the probabilities obtained by a summation over A of the preceding probabilities (sum of probabilities associated with exclusive events):

$$\sum_{A} \mathcal{P}(A, B|a, b) \tag{5.74}$$

The NS condition amounts to assuming that this probability is independent of a; we therefore obtain the condition:

$$\sum_{A} \mathcal{P}(A, B|a, b) = \sum_{A} \mathcal{P}(A, B|a', b) \quad \text{for any value of } b \tag{5.75}$$

One could write this relation for any value of B but, when only two results can occur, the sum rule (5.73) implies that the two conditions for different values of B are
not independent (their sum provides the trivial equality $1 = 1$); we can therefore
retain only one condition, that obtained for one value of B. Since we have assumed
two different values of b, the NS condition relative to the Alice-to-Bob communication channel then implies two linear relations (5.75) between the probabilities.
For the reverse communication channel, one gets two other relations:

$$\sum_B \mathcal{P}(A, B|a, b) = \sum_B \mathcal{P}(A, B|a, b') \quad \text{for any value of } a \qquad (5.76)$$

Altogether, we therefore have four NS conditions.

These conditions are indeed satisfied in quantum mechanics, when Alice and
Bob perform local experiments in two remote laboratories, without any possible
influence of one of the subsystems on the other. For the singlet state studied in
§4.1.1, we immediately obtain $\mathcal{P}_{(+,\pm)} + \mathcal{P}_{(-,\pm)} = 1/2$ and $\mathcal{P}_{(\pm,+)} + \mathcal{P}_{(\pm,-)} = 1/2$,
which show that relations (5.75), and (5.76) are obeyed. The general proof is given
in Appendix F, whatever entangled state is chosen for the quantum system, and
whatever measurements are performed by Alice and Bob. This property of quantum
mechanics avoids a frontal conflict with relativistic causality.

5.4.2 Logical boxes

Following Popescu and Rohrlich [231] (see also [249]), we define logical "boxes"
as devices with which Alice and Bob can arbitrarily choose a binary value of "input
variables", $a = \pm 1$ for Alice and $b = \pm 1$ for Bob; the box returns to them binary
"output variables" $A = \pm 1$ and $B = \pm 1$ (left part of Figure 5.2). We will distinguish
between deterministic boxes, where A and B are given functions of a and b, and
stochastic boxes defined by probabilities for the output variables that depend on a
and b.

5.4.2.a Deterministic boxes

When the input variables can take only two values, 4 distinct experimental setups
are possible for the whole system; when the output variables can take only two
values, each setup can provide 4 different pairs of results. The number of distinct
deterministic boxes is therefore $4^4 = 256$. But, if the NS condition is satisfied,
output A must be a function of a only, so that 4 different functions $A(a)$ are possible
for Alice (since she uses two values of her variable a), 4 also for Bob; the total
number of deterministic boxes is then only 16.

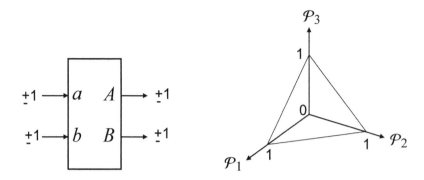

Figure 5.2 The left part of the figure gives a schematic representation of a logical box which, from the values ± 1 of the input variables a and b, provides the values of the output variables $A = \pm 1$ and $B = \pm 1$. The box is deterministic if A and B are given functions of a and b, stochastic if A and B are determined by a and b dependent probabilities.

The right part of the figure illustrates how a stochastic logical box is characterized by a point belonging to a polytope in probability space $\mathcal{P}_i(a, b)$ ($j = 1,...,4$). For any given value of the input variables a and b, the sum of probabilities is equal to one, so that it is sufficient to plot the three first probabilities on the three axes; the point associated with the logical box belongs to the inside (or surface) of a tetrahedron with unit sides. Since four values are possible for the (a, b) couple, the polytope is the product of four tetrahedra. The deterministic boxes correspond to points at the corners.

5.4.2.b Stochastic boxes

The number of stochastic boxes is not finite, since they depend on continuous parameters. A stochastic box associates to each experimental setup 4 probabilities $\mathcal{P}_i$ ($j = 1,...,4$) with a unit sum. If we choose three axes $\mathcal{P}_1, \mathcal{P}_2, \mathcal{P}_3$, each box is associated with a point in the corresponding three-dimensional space, while $\mathcal{P}_4$ is given by $1 - \mathcal{P}_1 - \mathcal{P}_2 - \mathcal{P}_3$. Since all probabilities $\mathcal{P}_i$ remain nonnegative, the corresponding point therefore lies inside (or at the surface) of a tetrahedron having a corner at the origin and the three other corners on the axes at a unit distance from the origin (right part of Figure 5.2). Each of the corners corresponds to a deterministic box. Since the box characterizes in fact the output values A and B for 4 different values of the pairs of input variables a and b, it is actually defined by 4 independent points inside (or at the surface of) 4 thetrahedra. We can group these 4 points into one single point G in a space with $4 \times 3 = 12$ dimensions. The space that is accessible to G is a polytope, with boundaries corresponding to one vanishing probability, and corners associated with deterministic situations. A general discussion of such polytopes is given in [249].

For a NS box, the 4 points are no longer independent, since they must satisfy the 4 relations (5.75) and (5.76); the stochastic NS boxes therefore depend on 8 parameters, and the new polytope accessible to the parameters is smaller than before; in particular, the corners are limited to those associated with deterministic NS boxes.

5.4.2.c Local stochastic boxes

A subcategory of the stochastic NS boxes is given by the local stochastic boxes, which have the property of "outcome independence"[8]. Among them the simplest are those with probabilities that factorize:

$$\mathcal{P}(A, B|a, b) = p(A|a) \times p(B|b) \tag{5.77}$$

where $p(A|a)$ and $p(B|b)$ are two local probabilities between 0 and 1 obeying two separate normalization conditions:

$$\sum_A p(A|a) = 1 \qquad \sum_B q(B|b) = 1 \tag{5.78}$$

The corresponding boxes depend on $2 \times 2 = 4$ continuous parameters only.

But, in the spirit of the Bell theorem, it is possible to generalize them by assuming the presence of fluctuating causes characterized by a random variable λ and a normalized distribution $\rho(\lambda)$:

$$\int d\lambda \, \rho(\lambda) = 1 \tag{5.79}$$

One then assumes that, for a given value of λ (which may be a multidimensional parameter), the conditions of the box are sufficiently well defined to obtain a factorization from a locality argument:

$$\mathcal{P}_\lambda(A, B|a, b; \lambda) = p_\lambda(A|a; \lambda) \times p_\lambda(B|b; \lambda) \tag{5.80}$$

Now the local probabilities p_λ and q_λ depend on the statistical parameter λ, while satisfying (5.78) for each value of λ. When λ fluctuates, the probabilities become:

$$\mathcal{P}(A, B|a, b) = \int d\lambda \, \rho(\lambda) \, p_\lambda(A|a) \times p_\lambda(B|b) \tag{5.81}$$

In general, this probability does not factorize, as opposed to (5.77) and (5.80).

---

[8] As we saw in §4.2.2.b, the Bayes theorem provides the general relation $\mathcal{P}(A, B|a, b) = p(A|a, b) \times p(B|A, a, b)$, where $p(B|A, a, b)$ is the conditional probability to obtain B if result A has been obtained. Assuming "setting independence" amounts to replacing $p(A|a, b)$ by $p(A|a)$ as well as $p(B|A, a, b)$ by $p(B|A, b)$. Assuming "outcome independence" amounts to writing that this conditional probability $p(B|A, b)$ is independent of the outcome A, and therefore leads to (5.77). The same assumptions can be made when one has to take into account the fluctuating variable λ, and then lead to (5.80).

These boxes are classical, and can be seen as a natural continuation of local realism/causality as introduced by EPR and Bell; they obey the Bell theorem. It is easy to check that these boxes are NS since:

$$\sum_A \mathcal{P}(A, B|a, b) = \int d\lambda \, \rho(\lambda) \, p_\lambda(B|b) \qquad (5.82)$$

which is indeed independent of a (the proof is similar for the summation over B).

Quantum mechanics can be used to build stochastic NS boxes, since it obeys relations (5.75) and (5.76). Since quantum mechanics can lead to violation of the Bell inequalities, in the corresponding situations these boxes do not belong to the category of local stochastic boxes with outcome independence.

5.4.3 Popescu–Rohrlich boxes

Any quantum scheme with experiments depending on two experimental settings a and b and leading to binary results can be used to calculate probabilities and then define a logical box, called a quantum box. We mentioned above that conditions (5.75) and (5.76) are always satisfied; all quantum boxes are therefore NS. One can then ask the reverse question: can any logical NS box be a quantum box? Is it possible, by starting from an appropriate quantum state for two subsystems and performing adequate measurements on them, to reconstruct any NS box? We will see that the answer to this question is no: the category of theories leading to logical boxes that remain compatible with relativity is actually larger than quantum mechanics.

Introducing the "Popescu–Rohrlich box" [231] is a way to prove this result by producing an example, a box that is sometimes called a "PR box". The idea is simple: in expression (4.11), we obtained a quantum violation by considering values of the input variables (settings) for which the three cosines with the same sign have value $1/\sqrt{2}$, while that with opposite sign has value $-1/\sqrt{2}$, so that the sum reaches $2\sqrt{2}$; we can try to increase this sum even more by building a model where the correlation rates are enhanced to their maximal value 1. For the 3 pairs of values (a, b), (a', b), and (a', b') of the input variables, we then assume that the output variables are perfectly correlated, and therefore always equal; moreover, to preserve the NS conditions at best, we minimize the transmission of information by attributing equal probabilities to the two pairs of possible results $(-1, -1)$ and $(+1, +1)$:

$$\mathcal{P}(-1, -1) = \mathcal{P}(+1, +1) = 1/2 \qquad (5.83)$$

so that the probability of a different value vanishes:

$$\mathcal{P}(-1, +1) = \mathcal{P}(+1, -1) = 0 \qquad (5.84)$$

For the couple (a, b'), the situation is the opposite, and the output variables are never equal:

$$\mathcal{P}(-1, -1) = \mathcal{P}(+1, +1) = 0$$
$$\mathcal{P}(-1, 1) = \mathcal{P}(+1, -1) = 1/2 \tag{5.85}$$

We now check that this box is NS. In both the right- and left- hand sides of (5.75), only one term contributes in the summation over A if the probabilities (5.83), (5.84), and (5.85) are assumed. This is because the same value of B is never associated with two different values of A, for any experimental setting; the two sides have therefore value $1/2$ and the equality holds. The same is true of (5.76). The Popescu–Rohrlich box thus does not permit instantaneous transmission of signals; it is compatible with relativity[9].

Let us now calculate the value obtained for the combination of products of results appearing in (4.7):

$$\langle M \rangle = \langle AB \rangle (a, b) - \langle AB \rangle (a, b') + \langle AB \rangle (a', b) + \langle AB \rangle (a', b') \tag{5.86}$$

The calculation is very simple since the probabilities we have chosen imply that the product AB is always $+1$ for the three pairs of inputs (a, b), (a', b), and (a', b'), while relation (5.85) implies that this product is always -1 for (a, b'). Therefore:

$$\langle M \rangle = 4 \tag{5.87}$$

which shows that the mathematical limit for $\langle M \rangle$ is indeed saturated.

The very fact that the Cirelson bound (§5.2) is exceeded immediately signals an incompatibility with quantum mechanics: no quantum setup with two subsystems undergoing measurements, no initial quantum state is able to reproduce a Popescu–Rohrlich box, even if such a box forbids instantaneous transmission of signals. This box[10] therefore provides an example of "superquantum" correlations.

It is also possible to simulate such a box by using postselection of detected events, which amounts to introducing a bias in the sampling of detected pairs (§4.3.1.a). With a sufficient selection one can, not only exceed the Cirelson limit $2\sqrt{2}$, but virtually reach the mathematical limit of 4 [209–211].

---

[9] A Popescu–Rohrlich box also obeys the no-cloning theorem (§8.2.1).

[10] Physically, it may seem more plausible to consider continuous input variables, instead of binary variables. This is because they correspond to experimental settings (such as angles of measurement for instance) which, in most cases, are continuous. Assuming relations (5.83) to (5.85) for all values of the input variables would then lead to discontinuities. But, in quantum mechanics, the maximal violation $2\sqrt{2}$ is obtained only for some values of the variables. Similarly, here, one can assume that these probability relations are satisfied only for some values of the parameters, and continuously interpolate between them in a second step; the violations are then limited to some range of the variables, but this does not change the proof of existence of superquantum correlations.

5.4.4 How to characterize quantum mechanics?

Popescu and Rohrlich propose new axioms to introduce quantum mechanics [231]. Instead of considering indeterminism as a fundamental postulate, as one usually does in quantum mechanics, they suggest choosing two different axioms: the theory is nonlocal (meaning that it sometimes predicts violations of the BCHSH inequalities), but nevertheless remains compatible with relativistic causality. Indeterminism is then a consequence of these postulates, as a "reductio ad absurdum" reasoning shows. Indeed, if the theory was deterministic, it would provide results A and B as functions of the experimental parameters a and b, and the compatibility with relativistic causality would immediately imply that A depends on a only, and B on b only; the proof of §4.1.2 would then apply, forbidding any violation of BCHSH inequalities. In other words, within determinism, a BCHSH violation would immediately create another violation of relativistic causality; it is quantum indeterminism that avoids this conflict.

Nevertheless, the example of the Popescu–Rohrlich box shows that these axioms do not define the theory in a unique way. What is defined is a broader ensemble of theories, among which is quantum mechanics with the Cirelson bound $2\sqrt{2}$, while other theories can reach the mathematical limit 4. A general discussion of the properties of nonsignaling theories is given in [250]: intrinsic randomness, impossibility of perfect cloning, monogamy, etc. (we come back to the meaning of these terms in §§7.2.5 and 8.2.1). One can nevertheless take a point of view whereby the BCHSH inequalities and the degree of their violation are considered more as indicators of the degree of correlations permitted by a theory than as a proof of nonlocality. The question is then: why are the correlations predicted by quantum mechanics not maximal among those that remain compatible with relativistic causality, but more restricted? What is the additional physical principle obeyed by quantum mechanics that selects the value $2\sqrt{2}$?

A partial answer to this question is given in [251] by remarking that stronger correlations would result in a world in which communication complexity is not trivial[11]. It has been proposed in [252] that the principle in question is a generalization of the principle of noninstantaneous communication: for any quantum system shared by Alice and Bob, if she sends to him m bits through a classical channel, whatever local operations and measurements Bob makes (these operations may depend on the received bits) he cannot obtain information that exceeds m.

In this scheme, Alice receives a chain of N binary numbers a_i ($i = 0, 1, N -$
1); her objective is to transmit as much information as possible concerning this chain to Bob, knowing that she can only send m bits to him through a classical

---

[11] Alice and Bob wish to compute some Boolean function $F(a, b)$ of variable a, chosen by Alice and known to her only, and of variable b chosen by Bob (and known to him only). The communication complexity of F is said to be trivial if the operation can be performed by the transmission of a single classical bit of communication.

channel ($m < N$). She chooses some of the numbers of her chain as values for the measurement parameters she performs (inputs for her side of the box), and obtains outputs (results); the message she sends to Bob can be built by combining her inputs and outputs in an arbitrary way. Bob, when he receives the message, is free to use its bits to perform any operation with his own logical system, in order to obtain the maximum possible number of values of the a_i.

(i) We first assume that Alice and Bob share a logical system made of Popescu–Rohrlich boxes. One can then show [252] that such a device allows Bob to exactly determine the value of any series of m different a_i, and to choose which ones – but never to determine more than m values. All bits in Alice's database are accessible to Bob, but the number of bits he can acquire is strictly limited to m.

(ii) Now assume that the system initially shared by Alice and Bob is a quantum system in any state. It is shown in [252] that there exists a relation between the rate of violation of the BCHSH inequality and the maximum amount of information obtained by Bob on the values of the series of a_i. It defines the information causality principle as: "the information gain that Bob can reach about a (previously unknown to him) data set of Alice, by using all his local resources and m classical bits communicated by Alice, is at most m bits". It then shows that the information causality principle is violated exactly when the Cirelson limit $2\sqrt{2}$ is exceeded.

The information causality principle is, so to say, a principle of nonamplification of classical information. For $m = 0$, one recovers the NS condition of relativistic causality: no information at all can pass from Alice to Bob through simple local measurements that they can perform, each in his/her own laboratory. For larger values of m, the new principle posits that, for any local measurements performed by the two partners, the amount of transmitted information is not increased; these measurements are, in a way, useless. This provides an explanation of the Cirelson bound. It then becomes natural to consider that the information causality principle belongs to those defining quantum mechanics, or even to conjecture that it may be used to perfectly define the theory.

The authors of [253] propose a different approach to the characterization of quantum nonlocal correlations. Assuming that quantum mechanics holds locally, they show that the nonsignaling condition then implies that all possible correlations between the two distant parties should be those of quantum mechanics as well. In other words, if any experiment provided nonlocal correlations beyond those of quantum mechanics, relativity would imply that quantum mechanics could not be valid locally either. The Reference [254] generalizes the characterization of quantum correlations to those observed between N partners, instead of two. It discusses a nonlocal game where each partner is given an input variable x_i and where, by using quantum correlations provided by a common quantum system in an entangled state $|\Psi\rangle$, each partner tries to guess the x_j variable of his neighbor. It turns out that, in this case, quantum correlations do not perform better than classical correlations. This result suggests that quantum mechanics might be characterized by a multi partner no-improvement criterion, a generalization of the nonsignaling principle.

6

More Theorems

The Bell theorem can take the form of several inequalities, as we have seen in §5.1 of Chapter 5. Moreover, since it was discovered, the theorem has stimulated the discovery of several other mathematical contradictions between the predictions of quantum mechanics and those of local realism/causality . We review a few of them in this chapter: GHZ contradictions (§6.1) and their generalizations, Cabello's inequality (§6.2), and Hardy's impossibilities (§6.3). In §3.4.1, we show that the EPR reasoning is not limited to pairs of spin 1/2 particles: whenever two quantum systems have spaces of states with the same dimension, one can construct an entangled state where any observable of one system is perfectly correlated with an observable of the other, so that the EPR argument applies to these observables. In §6.4, we discuss the important notion of contextuality and introduce the BKS theorem (Bell–Kochen–Specker). Finally, in §6.5 we review a few theorems concerning the reality of the quantum state.

6.1 Quantum properties of GHZ states

6.1.1 GHZ contradictions

For many years, everyone thought that Bell had basically exhausted the subject by considering all really interesting situations, and that two-spin systems provided the most spectacular quantum violations of local realism. It therefore came as a surprise to many when in 1989 Greenberger, Horne, and Zeilinger (GHZ) showed that systems containing more than two correlated particles may actually exhibit even more dramatic violations of local realism [255, 256]. They involve a sign contradiction (100% violation) for perfect correlations, while the BCHSH inequalities are violated by about 40% (Cirelson bound) and deal with situations where the results of measurements are not completely correlated. In this section, we discuss three-particle systems, but generalizations to N particles are possible (§6.1.3).

6.1.1.a Derivation

GHZ contradictions may occur in various systems, not necessarily involving spins. Initially, they were introduced in the context of entanglement swapping (§7.3.2) for four particles [255] or entanglement of three spinless particles [256]. Here, following Mermin [257], we will consider a system of three 1/2 spins, since this simple example is sufficient to discuss the essence of the ideas. We assume that the three spins are described by the quantum state:

$$|\Psi\rangle = \frac{1}{\sqrt{2}}\left[|+,+,+\rangle + \eta|-,-,-\rangle\right] \tag{6.1}$$

where the $|\pm\rangle$ states are the eigenstates of the components of the spins along the Oz axis of an orthonormal frame $Oxyz$; in the three-particle kets, the first sign refers to the state of the first spin, the second to the state of the second spin, and, similarly, for the third spin; the number η is either $+1$ or -1:

$$\eta = \pm 1 \tag{6.2}$$

We now calculate the quantum probabilities of measurements of the spins $\sigma_{1,2,3}$ of the three particles, either along direction Ox, or along a perpendicular direction Oy (Figure 6.1) . We first consider a measurement of the product $\sigma_{1y} \times \sigma_{2y} \times \sigma_{3x}$; a straightforward calculation (made explicit in §6.1.3 in a more general case[1]) then shows that $|\Psi>$ is an eigenvector of this product with eigenvalue $-\eta$, so that $-\eta$ is the only possible result with state (6.1). Therefore the corresponding probability is:

$$\mathcal{P}(\sigma_{1y} \times \sigma_{2y} \times \sigma_{3x} \Longrightarrow -\eta) = 1 \tag{6.3}$$

while the probability $\mathcal{P}(\sigma_{1y} \times \sigma_{2y} \times \sigma_{3x} \Longrightarrow +\eta)$ of the other possible result vanishes. Similarly, we find that $|\Psi>$ is an eigenvector of the two products $\sigma_{1x} \times \sigma_{2y} \times \sigma_{3y}$ and $\sigma_{1y} \times \sigma_{2x} \times \sigma_{3y}$, with eigenvalues $-\eta$, so that the corresponding probabilities are:

$$\begin{aligned} \mathcal{P}(\sigma_{1x} \times \sigma_{2y} \times \sigma_{3y} \Longrightarrow -\eta) &= 1 \\ \mathcal{P}(\sigma_{1y} \times \sigma_{2x} \times \sigma_{3y} \Longrightarrow -\eta) &= 1 \end{aligned} \tag{6.4}$$

All three products therefore take the value $-\eta$; the results are known with certainty before the measurement[2]. Now, if we consider the product of three spin components along Ox, it is also easy to check (§6.1.3) that the same state vector is also

---

[1] With the notation of that section, here we have $\eta = e^{i\xi}$, $\varphi_1 = \varphi_2 = \pi/2$, and $\varphi_3 = 0$, so that $\zeta = e^{i(\xi-\varphi_1-\varphi_2-\varphi_3)} = -\eta$, which provides the eigenvalue. Similarly, for a measurement of the product $\sigma_{1x} \times \sigma_{2x} \times \sigma_{3x}$, we have $\zeta = e^{i\xi} = \eta$, and the eigenvalue is $+\eta$.

[2] The product is fixed, but each of the individual components still fluctuates between results $+1$ and -1.

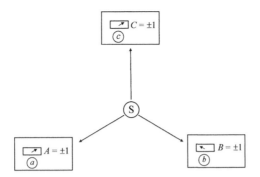

Figure 6.1 Scheme of a GHZ experiment, where three spins in state (6.1) undergo measurements in three different regions of space, with three measurement apparatuses. Each of them includes a button allowing the local experimenter to choose one among two possible values of the setting, which correspond to the measurement of a spin component along direction either Ox or Oy. Whatever choices are made, the three apparatuses provide results $A = \pm 1$, $B = \pm 1$ and $C = \pm 1$.

an eigenstate of the product operator $\sigma_{1x} \times \sigma_{2x} \times \sigma_{3x}$, but now with eigenvalue $+\eta$, so that:

$$\mathcal{P}(\sigma_{1x} \times \sigma_{2x} \times \sigma_{3x} \Longrightarrow +\eta) = 1 \tag{6.5}$$

This time the result takes the value $+\eta$ with certainty.

Let us now investigate the predictions of an EPR local realist point of view in this kind of situation. Since the quantum calculation is so straightforward when the initial state is an eigenstate of all observables considered (all the results are perfectly certain), one could expect that nothing special will be found. But, actually, we will see that a complete contradiction emerges from this analysis! The local realist reasoning is a direct generalization of that given in §4.1.2. First, the perfect correlations imply that the result of measuring the component along Ox (or Oy) of the spin of any particle can be inferred from the results of measurements made on other particles, at arbitrarily large distances. Therefore, the local realist EPR reasoning shows the existence of elements of reality attached to these two components, which we note $A_{x,y} = \pm 1$ for the first spin. In the same way, for the second spin we introduce the elements of reality $B_{x,y} = \pm 1$, and $C_{x,y} = \pm 1$ for the third. According to the EPR reasoning, for each realization of the experiment (for instance each emission of a trio of particles by a source), these six numbers have well-defined values, even if they may be unknown before any measurement. They are merely the results that will be obtained in possible future measurements. For instance, a measurement performed on the first spin will necessarily provide the result A_x if the direction of measurement is chosen along Ox, or A_y if it is chosen along Oy, independently of the type of measurement performed on the two other spins.

To reproduce the three equalities written in (6.3) and (6.4), we need to have:

$$
\begin{aligned}
A_y B_y C_x &= -\eta \\
A_x B_y C_y &= -\eta \\
A_y B_x C_y &= -\eta
\end{aligned}
\tag{6.6}
$$

Now locality implies that the same values of A, B, and C can be used again for the experiment where the three Ox components are measured: the result is merely the product $A_x B_x C_x$. But, since the squares A_y^2 etc. are always equal to $+1$, we can obtain this result by multiplying all three lines of equation (6.6), which provides:

$$
A_x B_x C_x = -\eta
\tag{6.7}
$$

But equality (6.5) predicts that the measurement of $\sigma_{1x} \times \sigma_{2x} \times \sigma_{3x}$ will always give the result $+\eta$, which has the opposite sign! The contradiction between the predictions of local realism and those of quantum mechanics cannot be more pronounced.

6.1.1.b Discussion

The GHZ contradiction looks even more dramatic than for the Bell inequalities, since the quantum and local realist predictions do not differ by some significant fraction (about 40%), but are completely opposite. In a thought experiment, all fluctuations are eliminated since all of the results (the products of the three components) are perfectly known before measurement: the 100% contradiction is obtained with 100% certainty! Apart from this, what differentiates a GHZ situation from a usual Bell experiment with two spins? Different points of view are possible.

(i) One point of view assumes that the three spins are measured individually in each realization of the experiment. The three spins may be in different regions of space; when the spatial variables are included, (6.1) symbolizes a ket that can be written more explicitly as:

$$
|\Psi\rangle = \frac{1}{\sqrt{2}} |1 : \varphi_a\rangle |2 : \varphi_b\rangle |3 : \varphi_c\rangle \otimes \left[|1 : +; 2 : +; 3 :\rangle + \eta \, |1 : -; 2 : -; 3 : -\rangle \right]
\tag{6.8}
$$

where $|\varphi_{a,b,c}\rangle$ are three orbital states with nonoverlapping wave functions. These functions can be entirely localized in separate boxes, where the spin measurements are performed, so that no particle is missed and each of them is addressed separately. The way to proceed is that, after choosing a component, Ox or Oy for each spin, one performs the three corresponding measurements, obtains three re-sults $A_{x,y}$, $B_{x,y}$, and $C_{x,y}$, and then calculates their product. Averaging over many realizations of the experiment provides the average value $\langle A_{x,y}B_{x,y}C_{x,y}\rangle$. With this procedure, one first measures $\langle A_yB_yC_x\rangle$, $\langle A_xB_yC_y\rangle$, and $\langle A_yB_xC_y\rangle$ to check the per-fect correlations predicted by quantum mechanics, from which the EPR reasoning leads to the existence of the six separate elements of reality. Then one measures $\langle A_xB_xC_x\rangle$ and, if quantum mechanics still gives correct predictions, obtains the op-posite sign; the conclusion is then that local realism is violated. Equivalently, one can conclude that the value obtained by measuring, say, σ_{1x} depends on whether the spin component measured for the other particles is along Ox or Oy, even if the corresponding operators commute with σ_{1x}. In other words, one arrives at the notion of "quantum contextuality", which we will discuss in more detail in §6.4.

(ii) A different point of view is to consider that some experimental procedure has been found to measure directly the products of operators, without obtaining infor-mation on the three separate factors inside each product. All four products of three operators then commute with each other[3], which introduces an important concep-tual difference with violations of Bell type inequalities, where the noncommutation is essential. Here, at least in principle, it is not impossible to measure all of them with a single setup; Bohr could not have invoked the incompatibility of experimen-tal arrangements. Under these conditions, where is the contradiction between the local realist reasoning and quantum mechanics? While, within local realism, a mea-surement of the product of three operators is equivalent to a separate measurement of each of them, in quantum mechanics this is not the case. Within quantum me-chanics, it is possible (in principle) to conceive an apparatus that measures all four products of spin components, but impossible to design a single experimental setup to have access to all six factors $A_{x,y}$, $B_{x,y}$, and $C_{x,y}$ (since, for instance, A_x and A_y correspond to the results of two incompatible measurements). This provides access to four results, but it is not possible to find six values of the individual components that reconstruct these four results by multiplication. What quantum mechanics then violates is the "product rule" discussed in §6.4.3.

An ideal GHZ experiment would therefore involve only measurements of com-muting products, more precisely the simultaneous measurements of four products (without measuring separately each factor in the products). The beauty of this the-oretical possibility is that loopholes such as the "biased sample loophole" (§4.3.1)

---

[3] Any pair of these products contains the same component of one spin, but different components of the two others. Since Pauli matrices anticommute (§12.1.7), interchanging the order of two of them introduces one minus sign. Since two such changes are made, two minus signs cancel each other, the products of three operators commute.

would then automatically be closed[4]. On the other hand, measuring a product without measuring each factor may raise a real experimental challenge.

Even with separate measurements, experimental tests of the GHZ equality still require that three particles should be put in state (6.1), surely a nontrivial task. Nevertheless, with the help of elaborate techniques of quantum electronics, GHZ entanglement has been observed with three photons [258], leading to a successful experimental test of the GHZ nonlocality [259]. Similar results have also been obtained with four entangled photons [260]. The techniques of NMR (nuclear magnetic resonance) have also been used to test the GHZ equality, at least with microscopic distances [261] – see also [262, 263]. All these experiments have confirmed the predictions of quantum mechanics.

6.1.2 GHZ–Mermin inequalities

It is also possible to derive local-realist inequalities that are violated by a quantum system in state (6.1). This makes the contradiction more similar to the usual Bell inequalities; using an inequality is also convenient for experiments, where perfect (anti)correlations are never observed (experiments are briefly discussed at the end of this section). We have seen in § 6.1.1.a that this state is a common eigenvector with eigenvalue $-\eta$ of the three products $\sigma_{1y}\sigma_{2y}\sigma_{3x}$, $\sigma_{1x}\sigma_{2y}\sigma_{3y}$, and $\sigma_{1y}\sigma_{2x}\sigma_{3y}$. It is also an eigenvector of the product $\sigma_{1x}\sigma_{2x}\sigma_{3x}$ with eigenvalue η, and therefore an eigenstate of the operator:

$$M = \sigma_{1x}\sigma_{2x}\sigma_{3x} - \sigma_{1y}\sigma_{2y}\sigma_{3x} - \sigma_{1x}\sigma_{2y}\sigma_{3y} - \sigma_{1y}\sigma_{2x}\sigma_{3y} \qquad (6.9)$$

with eigenvalue 4η. Therefore, its quantum average in state (6.1) is:

$$\langle M \rangle = 4\eta \qquad (6.10)$$

with:

$$\eta = \pm 1 \qquad (6.11)$$

Within local realism, the expression corresponding to M is:

$$\begin{aligned} M &= A_x B_x C_x - A_y B_y C_x - A_x B_y C_y - A_y B_x C_y \\ &= A_x \left(B_x C_x - B_y C_y \right) - A_y \left(B_y C_x + B_x C_y \right) \end{aligned} \qquad (6.12)$$

where all numbers A, B, C appearing in the right-hand side are equal to ± 1; each of them is equal to its inverse. Assume first that the product $B_x B_y C_x C_y$ is equal to

---

[4] One could not assume that different experimental setups select different groups of particles, since only one is used.

+1. Then:

$$B_x C_x = \frac{1}{B_y C_y} = B_y C_y \tag{6.13}$$

so that $\mathcal{M}$ reduces to $-A_y \left(B_y C_x + B_x C_y \right)$ and:

$$-2 \le \mathcal{M} \le +2 \tag{6.14}$$

Now, if the product $B_x B_y C_x C_y$ is equal to -1, we have:

$$B_y C_x = -\frac{1}{B_x C_y} = -B_x C_y \tag{6.15}$$

so that $\mathcal{M}$ reduces to $A_x \left(B_x C_x - B_y C_y \right)$ and inequality (6.14) is still valid.

The comparison between this inequality and relation (6.10) shows that quantum mechanics can violate the local-realist inequality by a factor 2, which is more than the factor $\sqrt{2}$ of violation of the BCHSH inequalities. Mermin has generalized the inequality to an arbitrary number N of particles [264], and shown that a N-particle GHZ state can lead to a violation by a factor that grows exponentially with N.

As mentioned previously, imperfections are unavoidable in experiments: the detectors are never perfectly efficient, they have intrinsic noise, the data are never obtained with infinite accuracy, etc. Inequalities are then a convenient way to quantify the violations of local realism in terms of experimental error bars. Other inequalities have been proposed to deal with these problems [265, 266]. Experiments to test these inequalities have been performed [267, 268]. The use of particles originating from three independent sources with quantum interference effects that disprove local realism has been discussed in [269], and later generalized to an arbitrary large number of particles [197]. The transposition to continuous position and momentum variables is discussed in [270]. Two-particle interferometers can also lead to GHZ contradictions for spinless particles [271].

6.1.3 Generalizing GHZ (all or nothing states)

We now generalize the GHZ state (6.1) by introducing an N-particle state $|\Psi\rangle$ defined as:

$$|\Psi\rangle = \frac{1}{\sqrt{2}} \left[|+, +, ..., +\rangle + e^{i\xi} |-, -, ..., -\rangle \right] \tag{6.16}$$

where ξ is a real parameter defining the relative phase of the two components; $|\Psi\rangle$ is a coherent superposition of two N-particle states where all, or none, of the spins are in the individual state[5] $|+\rangle$, hence the name "all-or-nothing state" that we use

---

[5] An all-or-nothing state does not necessarily require that all spins be up in the first component of the state vector, down in the second; what actually matters is that every spin changes to an orthogonal spin state when

here[6] – but "many-particle GHZ states" would be equally appropriate. One also sometimes finds the name "maximally entangled states", or[7] "NOON states", in the literature.

The state we consider here is very special, since it is a superposition of two N-particle components where all particles are in different, orthogonal, individual states. This state should not be confused with a different state where many particles have each been prepared in a coherent superposition. For instance, one can prepare individually N spins 1/2 by sending them through a spin filter (polarization-dependent beam splitter, Stern–Gerlach analyzer, etc.), which prepares a coherent superposition $\alpha|+\rangle + \beta|-\rangle$ for each spin. The effect of the filter on the group of N particles is to put them into a state that is a product of single-particle coherent states, namely:

$$|\Psi'\rangle = \Big[\alpha|1:+\rangle + \beta|1:-\rangle\Big] \otimes \Big[\alpha|2:+\rangle + \beta|2:-\rangle\Big] \otimes$$
$$\otimes ... \otimes \Big[\alpha|N:+\rangle + \beta|N:-\rangle\Big] \tag{6.17}$$

This state contains components of the state vector where some of the spins are up, some down, in various proportions. In (6.17), each particle is in the same spin state, and the situation is somewhat analogous to a Bose–Einstein condensate where all particles are in the same coherent state – for instance, a condensate located on both sides of a potential barrier coupled by tunnel effect, as in the Josephson effect. By contrast, in (6.16) the coherence is essentially an N-body coherence without any lower order coherence, so that the entanglement is more subtle.

Consider now the component of spin j along a direction in plane xOy making angle φ_j with Ox; it corresponds to the operator:

$$\sigma_j(\varphi_j) = \frac{1}{2}\Big[e^{-i\varphi_j}\,\sigma_+(j) + e^{i\varphi_j}\,\sigma_-(j)\Big] \tag{6.18}$$

where $\sigma_\pm$ denotes the operator $\sigma_x \pm i\sigma_y$ as usual, with the following action:

$$\sigma_\pm|\pm\rangle = 0 \quad \text{and} \quad \sigma_\pm|\mp\rangle = 2|\pm\rangle \tag{6.19}$$

We now introduce the N-particle operator corresponding to the product of various components of the spins:

$$Q(\varphi_1, \varphi_2, ..., \varphi_N) = \sigma_1(\varphi_1)\,\sigma_2(\varphi_2)...\sigma_N(\varphi_N) \tag{6.20}$$

which is equal to:

one goes from one N-particle component of $|\Psi\rangle$ to the other (in other words, the quantization axis may vary from spin to spin)

[6] Strictly speaking, we should use the name "all and nothing state", since both possibilities are simultaneously present in $|\Psi\rangle$.

[7] These states can be written as a superposition of kets with occupation numbers $N, 0$ and $0, N$, hence the name.

$$Q(\varphi_1, \varphi_2, ..., \varphi_N) = \prod_{j=1}^{N} \frac{1}{2} \left[e^{-i\varphi_j} \sigma_+(j) + e^{i\varphi_j} \sigma_-(j) \right] \qquad (6.21)$$

The action of this operator onto state (6.16) is easily obtained by using (6.19). If one takes the first component of $| \Psi >$, the only way to obtain nonzero by application of operator (6.21) is to select the term in $e^{i\varphi_j} \sigma_-(j)$ in each bracket of the product over j; therefore:

$$Q(\varphi_1, \varphi_2, ..., \varphi_N) |+, +, ..., +\rangle = e^{i(\varphi_1 + \varphi_2 + ... + \varphi_N)} |-, -, ..., -\rangle \qquad (6.22)$$

In the same way:

$$Q(\varphi_1, \varphi_2, ..., \varphi_N) |-, -, ..., -\rangle = e^{-i(\varphi_1 + \varphi_2 + ... + \varphi_N)} |+, +, ..., +\rangle \qquad (6.23)$$

Therefore, if we set:

$$\zeta = e^{i(\xi - \varphi_1 - \varphi_2 - ... - \varphi_N)} \qquad (6.24)$$

we obtain:

$$Q(\varphi_1, \varphi_2, ..., \varphi_N) |\Psi\rangle = \frac{1}{\sqrt{2}} \left[\zeta |+, +, ..., +\rangle + \frac{1}{\zeta} e^{i\xi} |-, -, ..., -\rangle \right] \qquad (6.25)$$

We then see that, if the condition:

$$\zeta = \frac{1}{\zeta} = \pm 1 \qquad (6.26)$$

is fulfilled, the action of Q on $|\Psi\rangle$ reconstructs exactly $\zeta |\Psi\rangle$, which means that $|\Psi\rangle$ is an eigenstate of Q with eigenvalue ζ. In other words, provided the sum of all angles φ_j is equal to ξ (plus some integer multiple of π), the product of all operators $\sigma_j(\varphi_j)$ corresponds to a measurement that gives a result $\zeta = \pm 1$ with certainty.

We can calculate the quantum average of Q from (6.25) and (6.24); for any value of the angles, we obtain:

$$\langle Q \rangle = \cos (\xi - \varphi_1 - \varphi_2 - ... - \varphi_N) \qquad (6.27)$$

For instance, if all angles φ_j are equal to the same value φ, this formula predicts that the average value of Q oscillates rapidly as a function of φ if N is large. Whatever the value of N is, it turns out that it is completely impossible to reproduce the oscillations contained in (6.27) within local realism [272]. In the case $N = 2$, this of course reduces to the usual Bell theorem. But, as soon as N becomes 3 or takes a larger value, the contradiction becomes even more dramatic. In [272], it is assumed that a local probabilistic theory reproduces (6.27) only for some sets of particular value of the angles φ (those for which the result is certain). One can then show that the theory in question necessarily predicts that $\langle Q \rangle$ is independent of all angles φ:

the average then keeps a perfectly constant value $+1$! Indeed, the very existence of the oscillation predicted by (6.27) is a purely quantum nonlocal effect.

As we have already remarked, the effect is essentially a coherent N body effect: if a single spin is missed in the measurement, in (6.21) the number of spin operators is no longer sufficient to transform the ket $|+, +, ..., +\rangle$ into $|-, -, ..., -\rangle$ as in (6.22). The result then becomes completely independent of the relative phase $e^{i\xi}$ of the two components; no quantum coherence effect then takes place. Actually, it is easy to check that the average value of any product of $N - 1$, $N - 2$, etc. components of the spins vanishes; the quantum interference effect leading to (6.27) takes place only if all N spins are measured.

This is not the only remarkable property of all-or-nothing states. For instance, it can be shown that, for large N, they lead to exponential violations of the bounds put by local realist theories [264]. In the context of a possible use of quantum correlations to reduce the quantum noise in spectroscopy [273], it has been pointed out [274] that these states (called in this context "maximally correlated states") have remarkable properties in terms of spectroscopic measurements: the frequency uncertainty of measurements decreases as $1/N$ for a given measurement time, therefore much faster than the usual $1/\sqrt{N}$ factor obtained with measurements on independent particles[8]. A discussion of this method, as compared to others making use of Bose–Einstein condensates, is given in [275]. Such quantum states may also be the source of "quantum lithography", using a nonlinear optics process (many photon absorption process) where the usual diffraction limit $\lambda/2$ obtained in classical physics is divided by N [276–278]. The quantum correlation of these states may turn out to be, one day, the source of improved accuracy in various experiments.

We have already mentioned in §6.1 that entanglement with $N = 3$ was reported in [258, 279] and used to test the GHZ equality [259]. Proposals for methods to generalize to larger values of N with ions in a trap were put forward by Mølmer *et al.* [280]; the idea exploits the motion dependence of resonance frequencies for a system of several ions in the same trap, as well as some partially destructive interference effects. The scheme was successfully put into practice in an experiment [281] where "all-or-nothing states" were created for $N = 2$ as well as $N = 4$ ions in a trap. GHZ states are not the only possibility to obtain violations of local realism with three particles; one can also use three particles originating from independent sources, and send them into an appropriate interference device [269].

---

[8] One can give a simple physical interpretation of this sensitivity improvement. Suppose that the two individual spin states $|\pm\rangle$ represent single particles propagating in the two arms of a Mach–Zehnder interferometer. The ket (6.16) then describes a N particle state where all particles propagate together, either in one arm of the interferometer, or in the other. In other words, what propagates in the interferometric device is a cluster of N particles having N times the mass and N times the energy of a single particle. Since the de Broglie wavelength of a quantum object with mass Nm and energy Ne is $\hbar/N \sqrt{em}$, it is N times shorter than for a single particle, which improves the sensitivity of the interferometer by the same factor.

6.2 Cabello's inequality

We now discuss a scheme introduced by Cabello [282], which leads to violations of BCHSH type inequalities exceeding the Cirelson bound (§5.2) – in fact, they reach the maximal value 4 mathematically compatible with the definition of the sum of averages appearing in these inequalities. To obtain this result, we consider a system made of three subsystems, and we cast the GHZ results into a form in which they appear as components of a BCHSH inequality; the idea is, so to say, to transform three-particle GHZ correlations into two-particle BCHSH correlations.

Again, we consider three spins in state (6.1) with $\eta = +1$ and measurements of the components of the three spins along Ox or Oy. More explicitly, we write the initial quantum state as:

$$|\Psi\rangle = \frac{1}{\sqrt{2}} |\varphi_a(1)\rangle |\varphi_b(2)\rangle |\varphi_c(3)\rangle \otimes \left[|1:+,2:+,3:+\rangle + |1:-,2:-,3:-\rangle \right]$$
(6.28)

where $|\varphi_{a,b,c}\rangle$ are three orbital states located in three different regions of space A, B, and C, where the separate spin measurements take place (Alice, Bob, and Carol).

Assume for instance that we measure the three Ox components of all spins. We have shown that the product of the results is always $+1$, so that two cases are possible. If two results -1 are obtained, we call i and j the corresponding regions and spins (i being attributed to the first region in the alphabetical order A, B, C), and k the region where result $+1$ has been obtained. If the three results are equal[9] to $+1$, $i = A$, and we choose $j = B$, and $k = C$. We always have:

$$\langle \sigma_x(i)\sigma_x(j)\rangle = 1$$
(6.29)

which is indeed a two-spin relation.

6.2.1 Local realist point of view

To each of the three regions of space we attach three numbers X, Y, and Z, each equal to ± 1. The EPR local realist theorem and the existence of complete GHZ correlations between remote spins show the existence of nine numbers $X_{a,b,c}$, $Y_{a,b,c}$ and $Z_{a,b,c}$ for regions A, B, and C; these nine numbers give the result that a measurement of the spin component along Ox, Oy, or Oz in each region will provide. The transposition of the results of quantum mechanics implies that, for each realization of the experiment (each triplet of particles), the product $X_a X_b X_c$ is equal to 1; either two of the X are equal to -1, or none.

---

[9] Expanding $|\Psi\rangle$ onto the eigenvectors of the Ox components of the three spins shows that this happens one time out of four.

This allows us to define three indices i, j, and k for all triplets of spins (not only those for which all Ox components of the spins are measured) as follows:

- if two of the X are equal to -1, they get indices i and j (among these two regions, i is associated with the region that is first in alphabetical order), and k is for the third region; X_k is then always equal to $+1$.

- if all the X are equal to $+1$, then $i = a$, $j = b$, and $k = c$.

For each realization of the experiment, this defines a perfect correspondence between the regions and the three indices i, j, and k, but of course this correspondence may change from one experiment to the next. Nevertheless, because local realism has introduced the counterfactuality (§4.3.2) on which the numbering process is based, this correspondence is always possible whatever measurements are actually performed, as opposed to what happens in quantum mechanics (since the correspondence is only defined for one kind of measurements).

In the proof of the BCHSH inequality for two spins in §4.1.2, we have considered in (4.7) a quantity of the form:

$$X\left(Y + mY'\right) + nX'\left(Y - mY'\right) \tag{6.30}$$

with:

$$m = \pm 1 \quad ; \quad n = \pm 1 \tag{6.31}$$

Expression (6.30) is always equal to ± 2 when the four numbers involved are ± 1. Here, we introduce the following combination of the nine numbers:

$$X_i\left(X_j + mY_j\right) + nY_i\left(X_j - mY_j\right) \tag{6.32}$$

which is also always -2 or $+2$; this is because either $\left(X_j + mY_j\right)$ or $\left(X_j - mY_j\right)$ vanishes, and all numbers have modulus equal to 1. Since $Y_k = \pm 1$, we may choose $m = n = -Y_k$, so that this expression becomes:

$$X_i\left(X_j - Y_kY_j\right) - Y_iY_k\left(X_j + Y_kY_j\right) \tag{6.33}$$

which provides:

$$X_iX_j - X_iY_kY_j - Y_iX_jY_k - Y_iY_j = \pm 2 \tag{6.34}$$

But since, by definition of index k, we have $X_k = 1$, equation (6.34) gives:

$$X_iX_jX_k - X_iY_jY_k - Y_iX_jY_k - Y_iY_jX_k = \pm 2 \tag{6.35}$$

Therefore, within local realism, the product of three measurements along Ox minus the three products of two Oy components with one Ox components is always equal to -2 or $+2$. The average value over many realizations is therefore always comprised between -2 and $+2$:

$$-2 \leq \langle X_i X_j X_k \rangle - \langle X_i Y_j Y_k \rangle - \langle Y_i X_j Y_k \rangle - \langle Y_i Y_j X_k \rangle \leq +2 \qquad (6.36)$$

The average value of the product of the three different X, minus the three different combinations of one X and two Y, is always bounded by ± 2.

6.2.2 Contradiction with quantum mechanics

In quantum mechanics, the average of the product of the three different Ox spin components, minus the three different combinations of one Ox and two Oy components, is:

$$\begin{aligned} Q &= \langle \sigma_x(1)\sigma_x(2)\sigma_x(3) \rangle - \langle \sigma_x(1)\sigma_y(2)\sigma_y(3) \rangle \\ &\quad - \langle \sigma_y(1)\sigma_x(2)\sigma_y(3) \rangle - \langle \sigma_y(1)\sigma_y(2)\sigma_x(3) \rangle \end{aligned} \qquad (6.37)$$

But we have seen above (§6.1) that the first term is $+1$, while all the others are -1. Therefore quantum mechanics predicts $+4$ for this quantity, in strong violation of (6.36) – by a factor 2.

The conclusion is that, by combining elements of the BCHSH reasoning with others of the GHZ reasoning, it is possible with three particles to obtain violation of generalized BCHSH inequalities that saturate the absolute mathematical limit. Using a third particle as a "marker", so to say, allows one to select the appropriate BCHSH inequality for the two other particles and to exceed the Cirelson bound. The reference [283] proposes another scheme, involving pre- and postselection, which also provides the maximum value 4 to the BCHSH term.

6.3 Hardy's impossibilities

Another scheme illustrating contradictions between local realism and the predictions of quantum mechanics was introduced by Hardy [72]. As the initial Bell theorem, it involves two correlated particles, but it is conceptually different: Hardy's theorem leads to conclusions concerning the very possibility (or impossibility) of occurrence for some type of events – instead of mathematical constraints on correlation rates. A general discussion of this interesting contradiction is given in [284]. As in §4.1.2, we assume that the first particle may undergo two kinds of measurements, characterized by two values a and a' of the first setting.

Within local realism, we call A and A' the corresponding results. Similar measurements can be performed on the second particle, and we call B and B' the results. Let us now consider three types of situations (Figure 6.2):

(i) If the settings are a, b, the result $A = 1$, $B = 1$ is sometimes obtained.

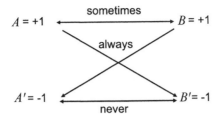

Figure 6.2 Logical scheme associated with the Hardy impossibilities. If results $A = B = 1$ are sometimes obtained, and if B' (resp. A') is always equal to -1 when A (resp. B) is equal to 1, then event $A' = B' = -1$ must sometimes occur. But one can construct a quantum state for which the first two conditions are obeyed while such events never occur, which creates a contradiction with local realism.

(ii) When "crossed measurements" are performed (either a, b' or a', b), we have certainties: if $A = 1$, the value of B' is always -1; if $B = 1$, the value of A' is always -1. One can also express this as an exclusion: "double ones" never occur in crossed measurements, neither $A = 1$, $B' = 1$ nor $A' = 1$, $B = 1$ are ever obtained.

(iii) When both settings are primed, we assume that "double minus one" is impossible: $A' = -1$, $B' = -1$ is never observed.

It turns out that these three assumptions are in fact incompatible. To see why, let us consider the logical scheme of Figure 6.2, where the upper part corresponds to the possibility opened by statement (i); statement (ii) then implies that, if $A = 1$, one necessarily has $B' = -1$, which explains the first diagonal in the figure; the second diagonal follows by symmetry. Then we see that all events corresponding to the results $A = B = 1$ also necessarily correspond to $A' = B' = -1$, so that a contradiction with statement (iii) appears. Indeed, the "sometimes" of (i) is contradictory with the "never" of the exclusion of proposition (iii).

But it turns out that quantum mechanics does allow a simultaneous realization of all three propositions! To see how, let us consider a two-spin state vector of the form:

$$|\Psi\rangle = \alpha\,|+,\ -\rangle + \beta\,|-,\ +\rangle + \gamma\,|+,\ +\rangle \qquad (6.38)$$

where the kets $|\pm,\ \pm\rangle$ refer to the common eigenstates of the Pauli operator $\sigma_z(1)$ for spin 1 and the spin operator $\sigma_z(2)$ for spin 2, with a given quantization axis Oz. We assume that $A' = \sigma_z(1)$ and $B' = \sigma_z(2)$, so that:

$$|\pm, \pm\rangle = |A' = \pm, B' = \pm 1\rangle \qquad (6.39)$$

The absence of any $|\Psi\rangle$ component on $|A' = -1, B' = -1\rangle$ ensures that proposition

(iii) is automatically true. As for the measurements without prime, we assume that they are both performed on the spins along a direction in the plane xOz that makes an angle 2θ with Oz; the single-spin eigenstate with eigenvalue $+1$ is then:

$$\cos\theta\,|+\rangle + \sin\theta\,|-\rangle \tag{6.40}$$

Proposition (ii) (diagonals in Figure 6.2) amounts to excluding the two-spin state:

$$\cos\theta\,|+,+\rangle + \sin\theta\,|-,+\rangle \tag{6.41}$$

as well as state:

$$\cos\theta\,|+,+\rangle + \sin\theta\,|+,-\rangle \tag{6.42}$$

The two exclusion conditions are obtained by cancelling the scalar product of $|\Psi\rangle$ and these two vectors. This is equivalent to the condition:

$$\alpha\sin\theta + \gamma\cos\theta = \beta\sin\theta + \gamma\cos\theta = 0 \tag{6.43}$$

or:

$$\alpha = \beta = -\gamma\cot\theta \tag{6.44}$$

Within an arbitrary coefficient, we can then write $|\Psi\rangle$ in the form:

$$|\Psi\rangle = -\cos\theta\left(|+,-\rangle + |-,+\rangle\right) + \sin\theta\,|+,+\rangle \tag{6.45}$$

The scalar product of this ket with that where the two spins are in the state (6.40) is:

$$-\sin\theta\cos^2\theta \tag{6.46}$$

The final step is to divide the square of this result by the square of the norm of ket (6.45) in order to obtain the probability of the process considered in (iii); this is a straightforward calculation (Appendix D), but here we just need to point out that the probability is not zero; the precise value of its maximum with respect to θ is found in Appendix D to be about 9%. This proves that the pair of results considered in proposition (i) can sometimes be obtained together with (ii) and (iii): indeed, in 9% of the cases, the predictions of quantum mechanics are in complete contradiction with those of a local realist reasoning.

An interesting aspect of the preceding propositions is that they can be generalized to an arbitrary number of measurements [285]; it turns out that this permits a significant increase in the percentage of "impossible events" (impossible within local realism) predicted by quantum mechanics – from 9% to almost 50%! The generalization involves a chain (Figure 6.3), which keeps the two first lines (i) and (ii) unchanged, and iterates the second in a recurrent way by assuming that:

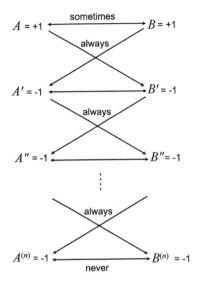

Figure 6.3 Logical scheme for iterating the Hardy impossibilities

(iii) For measurements of the type (a', b'') or (a'', b'), one never gets opposite results[10];

(iv) Similarly, for measurements of the type (a'', b''') or (a''', b''), one never gets opposite results,

etc...

(n) But, for measurement of the type (a^n, b^n), one never gets -1 and -1.

The incompatibility proof is very similar to the preceding; it is summarized in Figure 6.3.

In both cases, the way to resolve the contradiction is the same as for the Bell theorem: in quantum mechanics, for a given pair of spins, it is not correct to reason on all four quantities A, A', B, and B', even considered as quantities that are unknown and that could be determined in a future experiment. This is simply because, with a given pair, it is obviously impossible to design an experiment that will measure all of them: the measurements are incompatible. For a discussion of nonlocal effects with other states, see [286].

GHZ contradictions, or Hardy impossibilities, may appear as a stronger contradiction with quantum mechanics than just a violation of statistical BCHSH in-

---

[10] In fact, the reasoning just requires that the pair -1, $+1$ is never obtained, and does not require any statement about $+1$, -1.

equalities. But a closer examination shows that, in fact, both are logically related and that all violations discussed so far involve inequalities at a deeper level [287].

6.4 Bell–Kochen–Specker theorem, contextuality

Observing violations of the various Bell inequalities requires, as we have seen, the measurement of correlations between two entangled quantum systems (described by a nonproduct state). One could then get the impression that correlations and entangled states between two systems are necessary conditions to obtain paradoxical quantum results (impossible to explain within classical realism). It turns out that this is not the case: even with a single quantum system, without any entanglement, quantum mechanics predicts results that are in contradiction with naive classical realism as well. Locality is no longer an issue here, but rather the notion of contextuality. What we will see is that quantum mechanics is incompatible with the combination of:

– realism: physical systems possess physical properties that can be defined independently of their measurement.

– noncontextuality: these properties can be measured directly. This sentence means that the obtained values (the results of measurement) are independent of the apparatus, and in particular independent of the fact that other compatible measurements are performed simultaneously in the same experiment, or not.

In other words, measurements can reveal pre-existing properties of quantum physical systems in a way that does not depend on the combination of different properties that are simultaneously measured. The general idea was introduced independently and almost simultaneously by Bell (§V of [5]) as well as by Kochen and Specker [288]. While the details of the arguments given by these authors are different, the physical conclusions are the same, which explains why one often uses the generic name "BKS theorem" for such results. The assumption of realism may take different forms, EPR realism for instance. But we have seen that it is also possible to directly assume the existence of additional variables. In this case, an additional variable attached to a physical system is said to be "noncontextual" if, when the physical quantity it describes is measured, the result that is obtained is independent of the other physical quantities that are measured in the same experiment (in quantum mechanics, they necessarily correspond to commuting observables). If, on the other hand, the additional variable determines a measurement result that depends of the other observables that the experimenter may decide to measure at the same time, the additional variable is called "contextual". The notion of distance and of locality are no longer relevant in this context; for instance, the theorem applies to a single system with no extension in space. We now discuss a few examples.

6.4.1 A spin 1 particle

Following Reference [288] (§4), we consider a spin 1 particle in quantum mechanics, with the three eigenstates $|+1\rangle$, $|0\rangle$, and $|-1\rangle$ of the component of the spin along Oz; these states provide a basis of its space of states. The three components S_x, S_y, and S_z do not commute (they obey the usual commutation relation for the angular momentum). For instance, the matrices associated with S_x and S_z are, for a spin 1:

$$S_z = \hbar \begin{pmatrix} 1 & 0 & 0 \\ 0 & 0 & 0 \\ 0 & 0 & -1 \end{pmatrix} \qquad S_x = \frac{\hbar}{\sqrt{2}} \begin{pmatrix} 0 & 1 & 0 \\ 1 & 0 & 1 \\ 0 & 1 & 0 \end{pmatrix} \qquad (6.47)$$

While these matrices do not commute, it is easy to check that their squares do; more generally, the squares of S_x, S_y, and S_z commute (this is a specific property of angular momentum 1). Moreover, the sum of these squares is a constant (an operator proportional to the identity operator, sometimes called a "c-number") since:

$$S_x^2 + S_y^2 + S_z^2 = 2\hbar^2 \qquad (6.48)$$

It is therefore not against any fundamental principle of quantum mechanics to imagine a triple simultaneous measurement of the observables S_x^2, S_y^2, and S_z^2; we know that the sum of the three results will always be 2 (from now on we drop the factor $\hbar^2$, which plays no role in the discussion). Needless to say, the choice of the three orthogonal directions is completely arbitrary, and the compatibility is ensured for any choice of this triad, but not more than one: the measurements for different choices of the triad remain incompatible (S_z^2 commutes with the squares of the components of the spin along any axis that is perpendicular to Oz, but not otherwise).

In passing, we note that the measurement of the square S_x^2 of one component cannot merely be seen as a measurement of S_x followed by a squaring calculation made afterwards by the experimentalist! Not obtaining some information (the sign) is not equivalent to ignoring the information after it has been obtained by a measurement (we come to this point in more detail, in terms of interferences and decoherence, at the end of §11.1.2.c). There is indeed less information in S_x^2 than in S_x itself, since the former has only two eigenvalues (1 and 0), while the latter has three (since result -1 is also possible). What is needed to measure S_x^2 is, for instance, a modified Stern–Gerlach system where the components of the wave function corresponding to results ± 1 are not separated, or where they are separated but subsequently grouped together in a way that makes them impossible to distinguish. Generally speaking, in quantum mechanics, measuring the square of an operator is certainly not the same physical process as measuring the operator itself!

Now, suppose that we try to attach to each individual spin an EPR element of reality (additional variable) that corresponds to the result of measurement of S_x^2; by symmetry, we will do the same for the two other components, so that each spin now gets three additional variables λ to which we may attribute values that determine the possible results: 1 or 0. The results are described by functions of these variables, which we note $A_{x,y,z}$:

$$A_x = 0 \text{ or } 1; \; A_y = 0 \text{ or } 1; \; A_z = 0 \text{ or } 1 \qquad (6.49)$$

At first sight, this seems to provide a total number of eight possibilities; but, if we want to preserve relation (6.48), we have to select among these eight possibilities only those three for which two A are 1, and one is 0. For this particular spin, we then attribute colors to the three orthogonal directions Ox, Oy, and Oz: the two directions that get an $A = 1$ are painted in red, the last in blue [289].

The same operation can obviously be made for all possible choices of the triplet of directions $Oxyz$. A question which then naturally arises is: for an arbitrary direction Ox, can one attribute a given color (a given value for A_x) that remains independent of the context in which it was defined? Indeed, we did not define the value as a property of an Ox direction only, but in the context of two other directions Oy and Oz; the possibility of a context independent coloring is therefore not obvious. Can we for instance fix Oz and rotate Ox and Oy around it, and still keep the same color for Oz? We are now facing an amusing little problem of geometry that we might call "ternary coloring of all space directions". Kochen and Specker have shown that this is actually impossible; for a complete proof, see the original articles[11]; a simpler version can be found in §IV of the review [9] given by Mermin.

The conclusion is that any theory where measurements reveal a property that the system already had before measurement must be contextual to reproduce the predictions of quantum mechanics: the results of several compatible measurements performed at the same time must depend on the nature of all these measurements, otherwise contradictions appear. An important feature of this theorem is that it does not require the system to be in a precise quantum state, but applies to any state (we come back to state-independence in §6.4.4). It deals with a single quantum system and does not involve any quantum entanglement, as opposed to the Bell theorem.

6.4.2 Pentagram inequality

Spin-1 particles provide other contradictions between quantum mechanics and non-contextual realism. We now discuss a version proposed by Klyachko et al. [291]. According to relation (6.47), the matrix associated with the component S_θ of **S**

---

[11] The original proof by Kochen and Specker involves projectors over 117 different directions in real space; since then proofs involving a smaller number have been given, for instance in [290] where 18 directions are used.

along the direction of the xOz plane making angle θ with Oz is:

$$S_\theta = S_z \cos\theta + S_x \sin\theta = \begin{pmatrix} \cos\theta & \frac{1}{\sqrt{2}}\sin\theta & 0 \\ \frac{1}{\sqrt{2}}\sin\theta & 0 & \frac{1}{\sqrt{2}}\sin\theta \\ 0 & \frac{1}{\sqrt{2}}\sin\theta & -\cos\theta \end{pmatrix} \qquad (6.50)$$

We denote $|1\rangle$, $|0\rangle$, and $|-1\rangle$ the eigenvectors of S_z with eigenvalues $+1, 0$, and -1 respectively, and $|0_\theta\rangle$ the eigenvector of S_θ with zero eigenvalue. This eigenvector is equal to:

$$|0_\theta\rangle = \frac{\sin\theta}{\sqrt{2}}|1\rangle - \cos\theta|0\rangle - \frac{\sin\theta}{\sqrt{2}}|-1\rangle \qquad (6.51)$$

since it is easy to check that the action of operator (6.50) gives zero; this ket is called "neutrally polarized state" in [291]. The scalar product $\langle 0|0_\theta\rangle = -\cos\theta$ vanishes when $\theta = \pi/2$. Since any two orthogonal directions can be chosen as the Ox and Oz axis of a reference frame, this shows that the scalar product of two "neutrally polarized states" $|0_{\theta,\varphi}\rangle$ and $|0_{\theta',\varphi'}\rangle$ is zero whenever the two directions defined by the polar and azimuthal angles θ, φ and θ', φ' are orthogonal:

$$\langle 0_{\theta,\varphi}|0_{\theta',\varphi'}\rangle = 0 \quad \text{for orthogonal directions} \qquad (6.52)$$

6.4.2.a Quantum system

For any unit vector $\mathbf{u}$, we define the observable $A_\mathbf{u}$ as:

$$A_\mathbf{u} = 2\,[\mathbf{u}\cdot\mathbf{S}]^2 - 1 \qquad (6.53)$$

This operator has the two eigenvalues $+1$ (twice degenerate) and -1. It can also be written:

$$A_\mathbf{u} = 1 - 2\,|0_\mathbf{u}\rangle\langle 0_\mathbf{u}| \qquad (6.54)$$

where $|0_\mathbf{u}\rangle$ is the normalized eigenvector of $S_\mathbf{u} = \mathbf{u}\cdot S$ of zero eigenvalue:

$$S_\mathbf{u}|0_\mathbf{u}\rangle = 0 \qquad (6.55)$$

Two operators $A_\mathbf{u}$ and $A_{\mathbf{u}'}$ commute if $\mathbf{u}$ and $\mathbf{u}'$ are orthogonal, since we have seen in §6.4.1 that the squares of the components of S then commute.

We now consider (figure 6.4) a regular pentagram with vertices M_i (where i ranges from 1 to 5) and center P, and a point O on its axis (perpendicular in P to the plane of the pentagram). The five directions of vectors OM_i define five unit vectors $\mathbf{u}_i$. The angle between these vectors depends on the position of O, which we can adjust so that the angle between consecutive vectors is $90°$; the scalar product $\mathbf{u}_i \cdot \mathbf{u}_{i+1}$ then vanishes for any i (we introduce a cyclic permutation by considering that $i+1 = 1$ when $i = 5$). But we have seen in §6.4.1 that the squares of the components

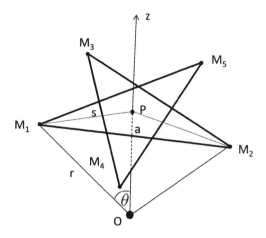

Figure 6.4 Five points M_1, M_2, ..,M_5 are the vertices of a regular pentagram with center P. Point O, on the axis Oz of the pentagram, is such that the angle between OM_i and OM_{i+1} is $\pi/2$. The length a is that of the segment M_iM_{i+1} (for any i), s that of segment PM_i, r that of segment OM_i, and θ is the angle between Oz and OM_i.

of an angular $J = 1$ momentum along perpendicular directions commute. The directions $\mathbf{u}_i$ then define five operators $A_{\mathbf{u}_i}$ obeying the commutation relations:

$$[A_{\mathbf{u}_i}, A_{\mathbf{u}_{i+1}}] = 0 \qquad \forall i \tag{6.56}$$

In principle, it is possible to measure two of these observables simultaneously, but not more, since there is no reason why $A_{\mathbf{u}_i}$ should commute with $A_{\mathbf{u}_{i+2}}$ for instance.

According to (6.54) and (6.52), the product P_i of two consecutive $A_{\mathbf{u}_i}$ is equal to:

$$P_i = A_{\mathbf{u}_i}A_{\mathbf{u}_{i+1}} = 1 - 2\left(|0_{\mathbf{u}_i}\rangle\langle 0_{\mathbf{u}_i}| + |0_{\mathbf{u}_{i+1}}\rangle\langle 0_{\mathbf{u}_{i+1}}|\right) = A_{\mathbf{u}_i} + A_{\mathbf{u}_{i+1}} - 1 \tag{6.57}$$

Therefore the sum T of all P_i is:

$$T = \sum_{i=1}^{5} P_i = \sum_{i=1}^{5} A_{\mathbf{u}_i}A_{\mathbf{u}_{i+1}} = 2\sum_{i=1}^{5} A_{\mathbf{u}_i} - 5 \tag{6.58}$$

Consider a spin 1 system in the state $|0_{\mathbf{u}=\mathbf{e}_z}\rangle$, where $\mathbf{e}_z$ is the unitary vector along the axis Oz of the pentagram (figure 6.4). The average value of $A_{\mathbf{u}}$ in this state is:

$$\langle A_{\mathbf{u}}\rangle = \langle 0_{\mathbf{u}=\mathbf{e}_z}|[1 - 2|0_{\mathbf{u}}\rangle\langle 0_{\mathbf{u}}|]|0_{\mathbf{u}=\mathbf{e}_z}\rangle = 1 - 2\left|\langle 0_{\mathbf{u}=\mathbf{e}_z}|0_{\mathbf{u}}\rangle\right|^2 = 1 - 2\cos^2\theta \tag{6.59}$$

where θ is the angle between the directions $\mathbf{u}$ and $\mathbf{e}_z$ – for obtaining the third equality, we have used (6.51). A simple geometrical calculation (which we defer to §6.4.2.d) shows that the cosine of the angle θ between any of the $\mathbf{u}_i$ and $\mathbf{e}_z$ is given by:

$$\cos^2 \theta = \frac{1}{\sqrt{5}} \qquad (6.60)$$

The average value of operator T is therefore equal to:

$$\langle T \rangle = \left\langle 0_{\mathbf{u}=\mathbf{e}_z} \middle| S \middle| 0_{\mathbf{u}=\mathbf{e}_z} \right\rangle = 2(5 - 2\sqrt{5}) - 5 \simeq -3.94.. \qquad (6.61)$$

It is in principle not impossible to measure T directly (operator T is Hermitian), but this would probably correspond to a very complicated measurement device. We therefore prefer to assume that five different types of measurements are performed in succession: $A_{\mathbf{u}_1}A_{\mathbf{u}_2}$, $A_{\mathbf{u}_2}A_{\mathbf{u}_3}$, ...,$A_{\mathbf{u}_5}A_{\mathbf{u}_1}$. They cannot be made simultaneously since these operators are not compatible: for instance, the product $A_{\mathbf{u}_1}A_{\mathbf{u}_2}$ does not commute with $A_{\mathbf{u}_2}A_{\mathbf{u}_3}$ since $A_{\mathbf{u}_1}$ and $A_{\mathbf{u}_3}$ do not commute (directions $\mathbf{u}_1$ and $\mathbf{u}_3$ are not orthogonal). We can then assume that, for each run of the experiment, A_i and A_{i+1} are measured, and that the product of the two results is calculated (alternatively, one can assume that a single measurement apparatus is used to directly measure the product $P_i = A_{\mathbf{u}_i}A_{\mathbf{u}_{i+1}}$). Keeping the same value of i, the experiment is performed many times; the average of this product over many realizations provides $\langle P_i \rangle$. The whole procedure is then repeated for other values of i and, finally, the sum of these averages for all values of i leads to $\langle T \rangle$.

6.4.2.b Classical system

Consider now a classical system for which five different quantities A_i ($i = 1, ...5$) have the values ± 1. We form the combination:

$$T = \sum_{i=1}^{5} A_i A_{i+1} = A_1 A_2 + A_2 A_3 + A_3 A_4 + A_4 A_5 + A_5 A_1 \qquad (6.62)$$

The product of the five products $A_i A_{i+1}$ is always equal to $+1$, since it contains the product of all $(A_i)^2$; therefore, only an odd number of products $A_i A_{i+1}$ can be equal to -1, and at least one of them is equal to $+1$. The minimum value of T is therefore -3:

$$T \geq -3 \qquad (6.63)$$

This equation is called the "pentagram inequality". It is clearly incompatible with the quantum result (6.61).

6.4.2.c Discussion, role of contextuality

Where does this contradiction come from? In classical physics, the value measured for A_i is the same, whether or not another A_j is measured; if another operator is indeed measured, it does not depend on which A_j is selected (noncontextuality). In quantum mechanics, the situation is different. The conclusions depend on the type of measurement considered:

(i) Suppose that each P_i is measured by calculating the product of the results of two successive compatible measurements, that of $A_{\mathbf{u}_i}$ and that $A_{\mathbf{u}_{i+1}}$. Then nothing garantees that the result ± 1 obtained for the measurement of $A_{\mathbf{u}_{i+1}}$ has to be the same when this observable is measured after $A_{\mathbf{u}_i}$ or before $A_{\mathbf{u}_{i+2}}$. As a consequence, the number A_{i+1} appearing in (6.62) is not well-defined, so that the proof of (6.63) is no longer possible. What matters here to evade the conclusion of this inequality is contextuality: the result of a measurement of a given observable can depend of the other measurements that are associated with it ("quantum contextuality"). The conclusion is the same, whether the two measurements are successive or simultaneous.

(ii) Suppose now that measurement devices are available to directly measure each of the products $A_{\mathbf{u}_i} A_{\mathbf{u}_{i+1}}$, one by one. In this case, knowing only the product, there is no unique way to attribute values ± 1 to each factor in the product and, again, expression (6.62) contains undefined quantities A_i. Inequality (6.63) cannot be derived.

A violation of the pentagram inequality therefore illustrates that it is impossible, for each realization of a quantum experiment, to assign pre-existent values to all results of experiments, independently of the measurement apparatuses. If these values exist, they must result from a combined effect of the system and of the apparatus; in this sense, one can say that they are created by the act of measurement itself. We finally note that, as opposed to that of §6.4.1, this result is state dependent: it requires the quantum system to be in a specific state, namely state $\left|0_{\mathbf{u}=\mathbf{e}_z}\right\rangle$.

6.4.2.d Geometrical calculation

We complete the proof by deriving relation (6.60). In figure 6.4, all points M_i are at the same distance of O, which is the center of a sphere of radius r containing all vertices of the regular pentagram. We call a the length of segment $M_1 M_2$. Since we assume that the triangle $M_1 O M_2$ has a square angle in O, we have:

$$a = \sqrt{2}r \tag{6.64}$$

We call P the center of the pentagram and Oz its perpendicular axis. All lengths $M_i P$ are equal to the same distance r, and triangle $M_1 P M_2$ is isosceles, with angle

$4\pi/5$ in P. The two other angles are therefore $\pi/10$, and:

$$a = 2s \cos\left(\frac{\pi}{10}\right) \tag{6.65}$$

Now, triangle M_1PO has a square angle in P, so that:

$$\sin\theta = \frac{s}{r} = \frac{1}{\sqrt{2}\cos(\pi/10)} \tag{6.66}$$

We now calculate:

$$\cos^2\theta = 1 - \sin^2\theta = \frac{2\cos^2(\pi/10) - 1}{2\cos^2(\pi/10)} = \frac{\cos(\pi/5)}{1 + \cos(\pi/5)} \tag{6.67}$$

Finally, since:

$$\cos\left(\frac{\pi}{5}\right) = \frac{1 + \sqrt{5}}{4} \tag{6.68}$$

we obtain:

$$\cos^2\theta = \frac{1 + \sqrt{5}}{5 + \sqrt{5}} = \frac{1}{\sqrt{5}} \tag{6.69}$$

6.4.3 Two spin 1/2 particles, product rule

In a similar line, Peres [292] has shown that the results of quantum mechanics are incompatible with the two following propositions:

(i) the results of the measurement of an operator A depend solely on A and on the system being measured (noncontextuality)

(ii) if operators A and B commute, the result of a measurement of their product AB is the product of the results of separate measurements of A and B (product rule).

Peres shows the existence of the incompatibility by considering two spin 1/2 particles in a singlet state. Mermin has generalized the result [9, 293] to any state vector of the two spins, and we follow his proof to show that the two assumptions lead to a sign contradiction with the predictions of quantum mechanics[12]. We consider two spin 1/2 particles and the following table of nine quantum variables (we use the same notation as in §6.1):

$$\begin{array}{ccc} \sigma_{1x} & \sigma_{2x} & \sigma_{1x}\sigma_{2x} \\ \sigma_{2y} & \sigma_{1y} & \sigma_{1y}\sigma_{2y} \\ \sigma_{1x}\sigma_{2y} & \sigma_{1y}\sigma_{2x} & \sigma_{1z}\sigma_{2z} \end{array} \tag{6.70}$$

All operators have eigenvalues ±1. The three operators belonging to the same line commute, as well as the three operators belonging to the same column (the

---

[12] We note in passing that the reasoning is very close to that of §6.1, which illustrates again the similarity between the GHZ theorem and this form of the BKS theorem.

products of two anticommuting σ are commuting operators, since changing their order introduces two -1 signs with cancelling effects). Moreover, the products of all three operators is always $+1$, except the last column[13] for which it is -1. Here, instead of an infinite number of triplets of directions in space, we have three groups of three operators, but the same question as before arises: can we attribute a value ± 1 to each of the nine elements of matrix (6.70) in a way that is consistent with the results of quantum mechanics? For this consistency to be satisfied, all lines and columns should contain either three $+1$ values or one $+1$ and two -1, except the last column that must contain one or three -1.

This little matrix problem is much simpler[14] than the geometrical coloring problem mentioned in §6.4.1. One can calculate the product of all matrix elements, either as the product of the products inside each line or as the product of the products inside columns. The product of all lines has to be $(+1)^3 = +1$, while the product of all columns is $(+1)^2 (-1)$, which is -1; a sign contradiction arises. It is therefore impossible to find nine numbers satisfying all conditions.

For another illustration of this sort of impossibility, see also §VI of [9] which deals with three 1/2 spins instead of two.

6.4.4 Contextuality versus local realism

From the previous discussions, we can conclude that the predictions of quantum mechanics are incompatible with a noncontextual view on the EPR elements of reality/additional variables. Now, is this result stronger, or weaker, than the Bell theorem, based on locality instead of contextuality?

The Bell theorem, seen as a continuation of the EPR argument, does not assume additional variables: their existence is derived from the local realist EPR argument applied to perfect correlations. In the BKS theorem, the existence of additional variables is just assumed, as well as some specific properties of these variables (noncontextuality). In this sense, the Bell theorem is more general. Moreover, as Bell noted [5], "the result of an observation may reasonably depend not only on the state of the system (including hidden/additional variables) but also on the complete disposition of the apparatus". Since the measured system S and the measurement apparatus M interact during the measurement process, there is no reason why the result of measurement should depend only on the initial properties of S, and not at all on those of M. Indeed, different measurement apparatuses introduce different dynamics, and it is hard to see why given values of the additional variables of S

---

[13] This can easily be checked from the well-known properties of the Pauli matrices; the minus sign for the third column comes from the product of the two i, arising from the relation $\sigma_x \sigma_y = i\sigma_z$; on the other hand, in the third line one gets $i \times (-i) = 1$ because of the change of order of the operators.

[14] The complication of the geometrical problem of the original BKS theorem is entirely avoided by going to a space of states of dimension 4 instead of dimension 3.

should necessarily lead to the same indication of the pointers for a given observable, independently of the others that are simultaneously measured. Mathematically, in a theory with additional variables λ, the function A giving the result of a measurement depends very naturally on all settings of the apparatus interacting with S. For instance, in the case studied in §6.4.1, the function should be written as $A(a, b, \lambda)$, where a and b define the two directions along which the squared components of the angular momentum are measured; no difficulty then occurs to reproduce the quantum predictions.

Moreover, one can build a theory where additional variables are also attributed to the apparatuses, and where both kinds of additional variables collaborate in order to determine the observed a and b dependent results. An example of such a collaborative effect is given for instance in [294]; it illustrates how the dBB theory (§11.8.1) has no problem with the BKS theorem, and can easily reproduce the results of quantum mechanics. Generally speaking, the BKS theorem does not take into account the possible effect of hidden variables associated with the measurement apparatus, which could play a role in the observed result. By contrast, the Bell local realist theorem remains valid in this case, provided of course the variables in question are local. Violations of the Bell theorem by quantum mechanics are therefore generally considered as more significant quantum manifestations than violations of the BKS theorem.

On the other hand, if one limits oneself from the beginning to hidden variable theories (the Bell theorem is then disconnected from the EPR argument), and assumes that these variables are contained in the measured system only (not in the measurement apparatus), the situation is different. Local hidden variables then appear as a special type of noncontextual hidden variables [295] (the Bell conditions of setting independence are a special case of noncontextuality). Seen in this way, the Bell theorem appears as less general than the BKS relations.

6.4.5 *Various versions; experiments; CSM interpretation*

Both the Bell and the BKS theorems can be expressed in forms that are either state-dependent, or state-independent and therefore more general. Examples of state-dependent forms were given earlier when we assumed that the system should be put in a given quantum state, such as (4.1) or (6.1); by contrast, the reasonings used in §§ 6.4.1 and 6.4.3 are state-independent. For a general discussion of the status of the various "impossibility theorems" with emphasis on the BKS theorems, see [9, 293].

We have seen in Chapter 4 that local realism can give rise to several forms of the Bell theorem, leading to different equalities or inequalities. In the same way, noncontextual realism can be expressed in various mathematical ways. An all-or-nothing state-independent test involving two spins 1/2 was proposed in [296]; extending the idea, [297] has proposed an all-or-nothing test that can be performed with single particles. This resulted, a few years later, in an experiment done with single photons [298], which provided results in agreement with quantum mechanics, eliminating noncontextual hidden variables. Tests of the pentagram inequality (6.63) have been made with single photons propagating along three modes [299]; a violation of the inequality by more than 120 standard deviations was obtained, eliminating again noncontextuality. Experiments have also been performed with neutrons [300], trapped ions [301], and nuclear spins in a solid studied by nuclear magnetic resonance [302], providing other examples of good agreement with the predictions of quantum mechanics.

The notion of contextuality has been used by Auffèves and Grangier to propose a realist interpretation of quantum mechanics, based on the notion of contextual objectivity [303], and called CSM; see also §11.6. In this formulation, realism is defined in a way that is reminiscent of the EPR criterion of reality, but emphasizes the role of the whole experimental setup (holism) rather than local properties of subsystems or regions of space – in other words, in a way that is in line with Bohr's point of view. Under these conditions, clearly, neither the EPR incompleteness argument nor the Bell theorem apply, and contradictions with the standard formulation of quantum mechanics are avoided.

6.5 Reality of the quantum state, ψ-ontology theorems

Does the quantum state represent reality, or our knowledge of reality? The question has been asked repeatedly since the inception of quantum mechanics. It is related to our discussion of the status of the state vector (§1.2.3), and also appeared in several other sections of this book (for instance, in the quotations of §2.5). We now discuss a few theorems that are directly related to the status of the state vector, but not particularly to locality or contextuality, as opposed to the theorems in the preceding sections of this chapter.

6.5.1 Ontic or epistemic ψ ?

In 2008, Harrigan and Spekkens published an article discussing the relation between Einstein's incompleteness argument (§ 3.3.1) and the "epistemic view of quantum states" [304]. Many authors introduce the words "ontic" and "epistemic" in this context, even if they are more common in philosophy than in physics. An analogy with classical physics is useful to understand their meaning: in classical physics, the position and momenta of all particles of a physical system provide a direct description of its physical properties; this is an "ontic" description. But if the system is described only in a probabilistic way by a statistical distribution in phase space (as in the Liouville theorem) reflecting a partial knowledge of its state, this description is called "epistemic". Similarly, in quantum mechanics, one can distinguish between "ψ-ontic" and "ψ-epistemic" views of the state vector[15]: in the first (ontic) view, ψ directly represents the reality of the physical system, or at least some elements of this reality; in the second view, ψ has an epistemological role, and represents only our knowledge (statistical information) about this reality. Asked in these terms, the question becomes: should we think of ψ as an ontic state or an epistemic state, or (after all, why not?) something completely different from both? Following Leifer [305], we propose a list of possible answers to the question:

(i) ψ is ontic, and there is no other ontic degree of freedom (ψ is "complete").

(ii) ψ is ontic, but may be completed with additional degrees of freedom (additional variables).

(iii) ψ is epistemic, and there is no deeper underlying physical reality that is accessible to physical theory.

(iv) ψ is epistemic, but there exists some underlying ontic physical state that can be characterized by physical variables; then quantum mechanics is the statistical theory of these ontic states (in analogy with classical statistical mechanics).

Discussions about the ontological meaning of the wave function are of course as old as the beginning of quantum mechanics; they have not resulted in a general consensus. See for instance [306], where Aharonov, Anandan and Vaidman propose to use "protective measurements" to give an ontological meaning to the wave function; see nevertheless the reply by Unruh [307], who considers that what is actually investigated is the ontological significance of certain operators in the theory, while the wave function plays its usual epistemological role.

6.5.2 *PBR theorem*

In 2012, in an article entitled "On the reality of the quantum state" [308], Pusey, Barrett and Rudolph (PBR) introduced a theorem that is now called the PBR theorem. The purpose of their article is to discuss whether a quantum state corresponds directly to reality, or must be interpreted statistically since it represents only information (§§ 11.1.3 and 11.2). Of course, the theorem requires some assumptions, which we list in §6.5.2.a. Its conclusion is that any model in which a quantum state ψ represents mere information about an underlying physical state must make predictions that are incompatible with those of quantum theory: if quantum theory makes correct predictions (and of course if the assumptions of the theorem are correct) the statistical/informational view of ψ of option (iv) should be rejected.

---

[15] We simplify the notation of the quantum state $|\psi\rangle$ into ψ, as is done in most discussions of the reality of the state vector.

6.5.2.a Assumptions of the theorem

PBR first assume that any physical system S has physical properties at any time, and that these properties can be described by an ensemble of variables P (the authors use the notation λ, but we have already used this letter for additional variables). The purpose of the analysis is then to clarify the possible relations between ψ and P. If ψ directly represents reality, ψ may be a subensemble of P, or even coincide with P. By contrast, in the informational view of the state vector, ψ and P are of a different nature; ψ gives only some statistical information on P. A given ψ must then correspond to several different P: otherwise ψ would characterize P perfectly well, and therefore be equivalent to the direct determination of the physical properties of the system. Conversely, a given P should be obtained from several different ψ (two at least): if every P was obtained from a single ψ, then again ψ could be identified with at least part of the physical properties of S (a subensemble of P). The consequence is that ψ must have an overlap with some other quantum sate ψ', the two quantum states sharing common physical properties P in some domain. Because a quantum state is usually associated with a preparation procedure of S (§1.2.3.b), this means that several different experimental procedures can be used to obtain the same physical properties P of the quantum system.

To the existence of the physical properties P, the authors add the following assumptions:

• The predictions of quantum mechanics are correct (same assumption as EPR and Bell).

• To any orthogonal basis in the space of states of S corresponds a possible measurement performed on S with an appropriate apparatus M (this is an usual assumption in standard quantum mechanics).

• The state vector ψ contains an information related to the preparation process (this is also standard quantum mechanics, §1.2.3.b).

• In a measurement of S with an apparatus M, the possible results are random, and their probabilities are determined by the ensemble of the physical properties P of S (determinism is not assumed); of course, they may also depend on the physical properties P_M of M.

• Independently prepared systems S_1 and S_2 have independent physical properties P_1 and P_2; the properties of the whole system $S = S_1 + S_2$ are determined by the reunion of their individual properties.

These assumptions are, of course, perfectly natural within the context of the PBR reasoning.

6.5.2.b Proof

In the statistical view of ψ, consider two different quantum states ψ and ψ' and assume that they have an overlap containing an ensemble of common P; if the system is prepared in state ψ, there exist a certain probability $p \neq 0$ that the system will have physical properties that could also have been obtained by a preparation in state ψ' (and conversely). PBR then consider a single system S with a two-dimension space of states, spanned by the two states $|+\rangle$ and $|-\rangle$, and the two nonorthogonal quantum states:

$$|\psi_a\rangle = |+\rangle$$

$$|\psi_b\rangle = \frac{1}{\sqrt{2}}[|+\rangle + |-\rangle] \tag{6.71}$$

They then assume that, for each run of the experiment, two such systems S and S' are prepared in a completely independent way. With probability p^2, the two systems have physical properties P and P' falling in the common range of properties of $|\psi_1\rangle$ and $|\psi_2\rangle$. In other words, for these particular realizations of the experiment, the ensemble of the two systems can be described indifferently by the four state vectors:

$$|\Psi_{aa}\rangle = |\psi_a\rangle \otimes |\psi_a\rangle$$

$$|\Psi_{ab}\rangle = |\psi_a\rangle \otimes |\psi_b\rangle = \frac{1}{\sqrt{2}}[|+,+\rangle + |+,-\rangle]$$

$$|\Psi_{ba}\rangle = |\psi_b\rangle \otimes |\psi_a\rangle = \frac{1}{\sqrt{2}}[|+,+\rangle + |-,+\rangle]$$

$$|\Psi_{bb}\rangle = |\psi_b\rangle \otimes |\psi_b\rangle = \frac{1}{2}[|+,+\rangle + |+,-\rangle + |-,+\rangle + |-,-\rangle] \tag{6.72}$$

(We use a notation where the first ket in the tensor product specifies the quantum state of S_1, the second ket that of S_2).

Two systems are then sent to the same measurement apparatus (Figure 6.5) and submitted to an entangled measurement (similar to that discussed in § 7.3.2) associated with the following measurement eigenstates:

$$|M_1\rangle = \frac{1}{\sqrt{2}}[|+,-\rangle + |-,+\rangle]$$

$$|M_2\rangle = \frac{1}{2}[|+,+\rangle \ - \ |+,-\rangle \ + \ |-,+\rangle \ + \ |-,-\rangle]$$

$$|M_3\rangle = \frac{1}{2}[|+,+\rangle \ + \ |+,-\rangle \ - \ |-,+\rangle \ + \ |-,-\rangle]$$

$$|M_4\rangle = \frac{1}{\sqrt{2}}[|+,+\rangle - |-,-\rangle] \tag{6.73}$$

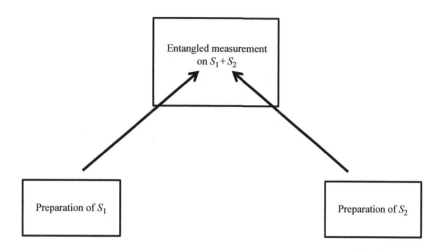

Figure 6.5 Scheme of the experiment discussed in the PBR reasoning. Two systems S_1 and S_2 are prepared independently, and send to the same measurement apparatus, where an entangled measurement operation is performed.

It can easily be checked that these four states provide an orthonormal basis. But each of them is also orthogonal to one of the states (6.72):

$$\langle M_1 | \Psi_{aa} \rangle = 0$$
$$\langle M_2 | \Psi_{ab} \rangle = 0$$
$$\langle M_3 | \Psi_{ba} \rangle = 0$$
$$\langle M_4 | \Psi_{bb} \rangle = 0 \qquad (6.74)$$

The Born rule then specifies that, if $|\Psi_{aa}\rangle$ is chosen to describe the pair, the first result can never be obtained; if $|\Psi_{ab}\rangle$ is used instead, the second result can never be obtained; if $|\Psi_{ba}\rangle$ is chosen, the third can never be obtained; finally, if $|\Psi_{bb}\rangle$ is used, the fourth result can never be obtained. Our assumptions have led to an absurd situation where no result can be obtained for the entangled measurement! Clearly, with probability p^2, a strong contradiction with the predictions of quantum mechanics is obtained. We note in passing that the exact value of p does not really matter, as long as it is not zero. The initial assumption according to which the two quantum states (6.71) may sometimes describe the same ensemble of physical properties turns out to be contradictory with the previously listed PBR assumptions.

Until now, our discussion has been restricted to two particular states, $|\psi_a\rangle$ and

$|\psi_b\rangle$; this gives a general idea of the essence of the theorem, but of course a generalization to arbitrary quantum states is necessary. This is done in Reference [308] by introducing more than two systems. A similar contradiction can indeed be obtained with N uncorrelated quantum systems. One then has 2^N equivalent quantum states that are compatible with the same physical properties of the ensemble of systems. An appropriate entangled measurement is then designed so that each of these quantum states correspond to a zero probability for one result; this measurement involves a circuit with N quantum gates (§8.4.2) acting on the individual systems considered as qubits, an entangling gate acting on the N particle state, and again N individual gates (Hadamard gates). This completes the proof of the theorem.

6.5.3 Similar theorems

The publication of the PBR theorem was quickly followed by several other contributions in the same line, several of them reaching similar conclusions from a different set of assumptions. We briefly mention a few of them in this section.

6.5.3.a Various assumptions and generalizations

The possible one-to-one correspondence between the wave function of a physical system and its elements of reality has been discussed by Colbeck and Renner [309]. These authors propose to replace the assumption concerning multiple preparations of the system by another assumption concerning the free choice of the measurement settings by the experimenter. This creates an interesting connection with the assumptions of the Bell theorem (§4.3.1.c). Combing the fact that ψ is contained in P and the nonextendibility of quantum mechanics, Colbeck and Renner conclude that ψ and P are indeed in one-to-one correspondence.

The general implications of the PBR theorem have been discussed in more detail by Schlosshauer and Fine in [310]. They introduce the notion of response function $\mathcal{A}(R, P)$ giving the probability that, if the system has properties P, a measurement of A will provides results belonging to set or results R. If $\mathcal{A}$ depends on the state vector, they call the model ψ-dependent; otherwise it is ψ-independent. They then show that PBR theorem concerns only ψ-independent models, and analyze the required assumptions for composite systems. One of the conclusions is that the PBR theorem can be seen as illustrating a difficulty for theories with additional variables in forming composite of identically prepared systems. This general point of view provides an interesting connection with the BKS theorem (§6.4).

The authors of Ref. [311] have also emphasized the essential character of the assumption of "preparation independence" if one wants to exclude the possibility of several quantum states being consistent with an single physical state P. Dropping the preparation independence assumption allows them to construct ψ-epistemic models that are perfectly consistent with all predictions of quantum mechanics.

6.5.3.b Hardy's interferometer

In [312], Hardy discusses the reality of quantum states by making assumptions that are also different from those of PBR; considering many copies of the physical system is not required. The key new assumption is "ontic indifference": a change of the experimental arrangement in a way that does not affect the state vector does not affect the physical variables P either. Hardy illustrates the ideas with an experiment involving a Mach-Zehnder interferometer with two detectors D_1 and D_2 at the output (Figure 6.6). In relation (6.71), state $|\psi_a\rangle$ corresponds to a particle that propagates in arm (a) of the interferometer, while the second state $|\psi_b\rangle$ describes a particle that crosses the interferometer in a coherent superposition of two states propagating in the two arms (a) and (b). The interferometer is adjusted so that all particles in state $|\psi_b\rangle$ will reach detector D_1; nevertheless, a phase plate can be inserted into arm (b) to drive all these particles into detector D_2 instead.

Ontic indifference ensures that the physical properties P_a associated with $|\psi_a\rangle$ propagating in arm (a) are not changed when the a phase plate is inserted in arm (b). The particles in state $|\psi_a\rangle$ can reach either detector, and have physical properties described by the variables P_a. The ensemble Λ_a of all possible P_a is therefore the sum (union):

$$\Lambda_a = \Lambda_a^{D_1} \cup \Lambda_a^{D_2} \tag{6.75}$$

of the set of properties $\Lambda_a^{D_1}$ of particles that will reach D_1 and of the set $\Lambda_a^{D_2}$ of particles that will reach D_2. These two subensembles are compared to the set Λ_b of possible physical properties of a particle in state $|\psi_b\rangle$ before it crosses the phase plate (Λ_b is therefore independent of the possible presence of the plate). $\Lambda_a^{D_2}$ has an empty intersection with the properties of Λ_b since, when no phase plate is inserted, the two sets of properties describe particles that will reach different detectors; similarly, $\Lambda_a^{D_1}$ has an empty intersection with Λ_b, for the same reason when the phase plate is inserted. Therefore, according to (6.75):

$$\Lambda_a \cup \Lambda_b = 0 \tag{6.76}$$

which shows that the two quantum states (6.71) cannot share common physical properties. As in the initial proof of the PBR theorem, a generalization is possible to any distinct quantum states. Hardy concludes that, if the assumptions are correct, the quantum state is a "real thing", because it is written into the underlying variables that describe reality.

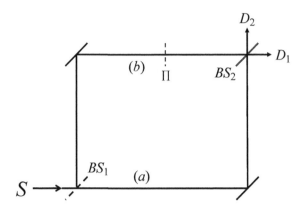

Figure 6.6 Scheme of the experiment considered by Hardy in Ref. [312]. A Mach-Zehnder interferometer receives particles from a source S; the beam splitters BS_1 and BS_2 split and recombine two interfering paths (a) and (b); D_1 and D_2 are the two detectors at the output of BS_2. Two elements of the interferometer are removable: the entrance beam splitter BS_1, and an phase plate Π in path (b). If BS_1 is removed, the particle follows only path (a), no interference effect takes place, and the quantum state is $|\psi_a\rangle$; the phase plate plays no role. If BS_1 is inserted, the particle is described by a quantum superposition $|\psi_b\rangle$ that includes both paths; the path difference between (a) and (b) is adjusted so that all particles go to detector D_1 in the absence of the phase plate, and to D_2 when the phase plate is inserted. The purpose of the reasoning is to show that $|\psi_a\rangle$ and $|\psi_b\rangle$ cannot share common physical properties.

6.5.3.c Reviews and experiments

An extended and general review of the subject can be found in [305]. Other rejections of ψ-epistemic models can be found in References [313] and [314]. A more recent discussion by Colbeck and Renner [315] assumes that P gives a complete description of the physical state of the system, and examines the conditions under which ψ is uniquely determined by P. As in the previous work of these authors [309], they assume a (sufficient) condition of "free choice", related to the notion of "experimental parameters", which are similar to Bell's "external variables" already discussed in §4.3.1.c (precise definitions of free choice and complete description are proposed in [315]). The conclusion of this analysis is that, if the corresponding conditions are satisfied, ψ is as objective as P.

An experimental test with trapped ions is described in [316], and with photons in [317] and [318]; all these results confirm the predictions of quantum mechanics.

7

Quantum Entanglement

In this chapter, we study the properties of quantum entanglement, and more generally the way correlations can appear in quantum mechanics. Quantum entanglement is an important notion that we have already discussed, for instance in the context of the von Neumann chain or of the Schrödinger cat, but here we give more details on its properties.

In classical physics, the notion of correlation is well known. It hinges on the calculation of probabilities and on linear averages over a number of possibilities. A distribution gives the probability of having the first system in a some given state and the second in another state. If this distribution is not a product, the two systems are correlated. If it is a product, they are uncorrelated; measuring the properties of one system does not bring any information on the other. This is in particular the case if the state of each of the two systems is perfectly defined (which also defines the state of the whole system perfectly well). The notion of correlation between subsystems therefore stems from the multiplicity of possible states of the whole system: if its state is perfectly defined, no correlations occur. Fluctuations of the state are necessary to give its full meaning to the classical notion of correlation.

In quantum mechanics, the situation is different: as we have seen (in particular in Chapter 4), even a physical system that is perfectly defined by a given state vector already contains fluctuations. This leads naturally to another notion of correlation, independent of any fluctuation of the state. For instance, the components of two $1/2$ spins fluctuate with strong correlations in a singlet state (§4.1.1), which is a pure state. This is because the principle of linear superposition of quantum mechanics allows one to introduce superpositions of different states directly inside the state vector; this is very different from superpositions of probabilities, which are quadratic functions of this state vector. In other words, in quantum mechanics one may introduce correlations directly at the level of probability amplitudes, one level "below" linear combinations of probabilities; this may create cross terms in the amplitudes resulting in quantum interference effects. Since it is also possible in quantum mechanics to assume that the state of the whole system is imperfectly known and defined only by probabilities, the two levels may coexist, so that the quantum notion of correlation is much richer than in classical physics.

We come back in more detail to these points in §7.1, and to methods to characterize entanglement in §7.2. In §7.3, we discuss how quantum entanglement may be created in experiments, as well as an opposite process, decoherence, which tends to destroy entanglement[1]. The dynamics of a subsystem of a large quantum system and the corresponding master equation are studied in §7.4.

7.1 A purely quantum property

In quantum mechanics, the relation between parts and the whole is very special, and certainly nonintuitive. We have already mentioned that Schrödinger was the first to use the words "quantum entanglement" in 1935, when he wrote (page 555 of [319]): "When two systems, of which we know the states by their respective representatives, enter into temporary physical interaction due to known forces between them, and when after a time of mutual influence the systems separate again, then they can no longer be described in the same way as before, viz. by endowing each of them with a representative of its own. I would not call that *one* but rather *the* characteristic trait of quantum mechanics, the one that enforces its entire departure from classical lines of thought. By the interaction the two representatives [the quantum states] have become entangled ... Another way of expressing the peculiar situation is: the best possible knowledge of a whole does not necessarily include the best possible knowledge of all its parts, even though they may be entirely separate and therefore virtually capable of being 'best possibly known', that is, of possessing, each of them, a representative of its own. The lack of knowledge is by no means due to the interaction being insufficiently known – at least not in the way that it could possibly be known more completely – it is due to the interaction itself."

7.1.1 The part and the whole

In classical mechanics, the description of the part is simply contained in the description of the whole[2]: when a system is made of two subsystems, 1 and 2, a full

---

[1] More precisely, it tends to make entanglement propagate further and further into the environment, so that in practice it rapidly becomes impossible to observe; for all practical purposes, one can consider that entanglement has disappeared.

[2] Heisenberg published in 1969 a book with the title *The Part and the Whole*, relating how quantum mechanics emerged from discussions between him and other physicists. Nevertheless this title does not seem to be specifically related to entanglement, but to a more abstract concept: "There is a fundamental error in separating the parts from the whole, the mistake of atomizing what should not be atomized. Unity and complementarity constitute reality" (sentence attributed to Heisenberg [320]).

description of the whole system 1+2 immediately provides a full description of each subsystem. Actually, the description of the dynamical state of the whole system is nothing but the sum of the descriptions of the subsystems: if one specifies the values of the positions and the momenta of all particles (as well as the values in space of the fields), all dynamical variables associated with each subsystem are automatically known. For instance, a perfect description of the solar system contains a perfect description of all the positions and velocities of the planets and their satellites; a perfect description of the state of the electromagnetic field contains a perfect description of the electric field, as well as of the magnetic field, in all space.

Strangely enough, and as emphasized by Schrödinger, this is no longer true in quantum mechanics. If the whole system is described with the most accurate way accessible in quantum mechanics, namely a state vector, the subsystems may be described less accurately; one can assign them just probabilities of occupying quantum states – in other words, as we will see in more detail, they are not in a pure state but in a statistical mixture, described mathematically by a density operator. Therefore, parts may be known only statistically, while the whole is perfectly known (with probability unity).

To see how this can happen in simple terms, suppose that a system is made of a first subsystem 1 with quantum states $|\varphi\rangle$, and of a second subsystem 2 with quantum states $|\chi\rangle$. If the whole system is described by a simple product quantum state[3]:

$$|\Psi\rangle = |1 : \varphi\rangle |2 : \chi\rangle \tag{7.1}$$

the two subsystems are uncorrelated; the first is described by a quantum state $|\varphi\rangle$, while the second is described by $|\chi\rangle$. All three systems (whole and two subsystems) are then associated with state vectors, the most accurate description that standard quantum mechanics can assign to any physical system.

But assume now that the whole system is in state:

$$|\Psi\rangle = \alpha |1 : \varphi_a\rangle |2 : \chi_a\rangle + \beta |1 : \varphi_b\rangle |2 : \chi_b\rangle \tag{7.2}$$

where $|1 : \varphi_{a,b}\rangle$ are two orthonormal states for the first system, and $|2 : \chi_{a,b}\rangle$ two other orthonormal states associated with the second; α and β are two (nonvanishing) complex numbers satisfying $|\alpha|^2 + |\beta|^2 = 1$. The whole system is still in a perfectly well-defined quantum state, but the first subsystem has a probability $|\alpha|^2$

---

[3] The notion of tensor product is discussed in Chapter 12, see Equation (12.68). We use three equivalent notations for the tensor product:

$$|1 : \varphi\rangle \otimes |2 : \chi\rangle \equiv |1 : \varphi\rangle |2 : \chi\rangle \equiv |1 : \varphi; 2 : \chi\rangle$$

Depending on the context of each calculation, one or the other may be more convenient and correspond better to the general use.

of being in state $|\varphi_a\rangle$ and a probability $|\beta|^2$ of being in state $|\varphi_b\rangle$, so that it is described by a statistical mixture instead of a pure state[4] – we come back to this point later in terms of density operators. The subsystems are then described less accurately than the whole system, a situation without any classical equivalent. The words "quantum entanglement" are used to describe such situations.

The formalism of the density operator (§12.1.6) and the use of partial traces (§12.2.3) is useful to express this property in a more quantitative way. The density operator ρ associated with state $|\Psi\rangle$ is:

$$\rho = |\Psi\rangle\langle\Psi| \tag{7.3}$$

which is merely the projector over state $|\Psi\rangle$, obeying $\rho^2 = \rho$ since:

$$\rho^2 = |\Psi\rangle\langle\Psi|\Psi\rangle\langle\Psi| = |\Psi\rangle\langle\Psi| = \rho \tag{7.4}$$

One then says that ρ describes a "pure state", a case characterized by the equality:

$$Tr\{\rho^2\} = 1 \tag{7.5}$$

(the trace of ρ itself is always 1 by definition). The subsystems are described by partial traces of ρ; for instance, the first subsystem is described by the operator:

$$\rho_1 = Tr_2\{\rho\} \tag{7.6}$$

with the following matrix[5] in the basis $|\varphi_a\rangle$, $|\varphi_b\rangle$:

$$(\rho_1) = \begin{pmatrix} |\alpha|^2 & 0 \\ 0 & |\beta|^2 \end{pmatrix} \tag{7.7}$$

This diagonal expression shows that system 1 is described by a statistical mixture of states $|\varphi_a\rangle$ and $|\varphi_b\rangle$ with probabilities $|\alpha|^2$ and $|\beta|^2$.

We also have:

$$(\rho_1)^2 = \begin{pmatrix} |\alpha|^4 & 0 \\ 0 & |\beta|^4 \end{pmatrix} \tag{7.8}$$

---

[4] Similarly, the second subsystem has a probability $|\alpha|^2$ of being in state $|\chi_a\rangle$ and a probability $|\beta|^2$ of being in state $|\chi_b\rangle$, and is therefore also associated with a statistical mixture.

[5] According to equation (12.84) of Chapter 12 defining a partial trace, the matrix elements of ρ_1 are given by:

$$\langle\varphi_i|\rho_1|\varphi_j\rangle = \sum_{k=a,b,...} \langle 1:\varphi_i;2:\chi_k|\Psi\rangle\langle\Psi|1:\varphi_j;2:\chi_k\rangle$$

(where i,j and k label one of the basis vectors, a or b). From expression (7.2) of $|\Psi\rangle$, we then obtain:

$$\langle\varphi_i|\rho_1|\varphi_j\rangle = (\alpha\delta_{i,a})(\alpha^*\delta_{j,a}) + (\beta\delta_{i,b})(\beta^*\delta_{j,b})$$

which is matrix (7.7).

and:

$$Tr\left\{(\rho_1)^2\right\} = |\alpha|^4 + |\beta|^4 = \left[|\alpha|^2 + |\beta|^2\right]^2 - 2\,|\alpha|^2\,|\beta|^2$$
$$= 1 - 2\,|\alpha|^2\,|\beta|^2 \leq 1 \qquad (7.9)$$

If one of the coefficients α or β vanishes, we are back to the case where the sub-systems are uncorrelated as in (7.1); one then obtains $(\rho_1)^2 = \rho_1$ and the trace of $(\rho_1)^2$ is equal to 1, which indicates that the first subsystem is in a pure state (as well as the second subsystem). But, when none of the coefficients vanishes, the subsystems are correlated, we have $(\rho_1)^2 \neq \rho_1$, and the trace of $(\rho_1)^2$ is smaller than 1; this indicates that the subsystem is described with less accuracy than if it was in a pure state, by a statistical mixture.

Of course, one can add more than two terms in the sum of (7.2), which tends to make the knowledge on each subsystem less and less precise, while the whole system still remains perfectly defined. For instance, if $|\Psi\rangle$ contains a linear su-perposition of three terms, all containing orthogonal individual states, the density operator ρ_1 describing system 1 becomes a 3×3 diagonal operator and introduces three possible states for this system (eigenstates of ρ_1). Moreover, in general and as remarked by Schrödinger [319], the determination of these possible states is not necessarily unique; one then cannot even list the states that are accessible to the subsystem without ambiguity[6]. One can even reach a situation where the description of each subsystem is completely random, so that no information at all is available on their states, all kets in their space of states being equally probable. In such cases of extreme entanglement, the whole system has some physical properties, those associated with its state, but nothing can be said about any specific property of the subsystems. A simple example of extreme quantum entanglement is given by the singlet state of two spins 1/2 – cf. equation (4.1) – or its direct generalization:

$$|\Psi\rangle = \frac{1}{\sqrt{2}}\left[|+,-\rangle + e^{i\xi}\,|-,+\rangle\right] \qquad (7.10)$$

(where $e^{i\xi}$ is an arbitrary phase factor). With this state, the whole system is perfectly defined, while no information at all is available on the state of each individual spin, which has equal probabilities to be in state $|+\rangle$ or $|-\rangle$.

The purely quantum character of entanglement is clear. In classical mechanics, if the properties of two physical systems are initially defined with the best possible accuracy, after mutual interactions each system is still defined as accurately as possible. This is not the case in quantum mechanics: even if one starts from systems with the best possible definition compatible with their quantum nature, at the end

---

[6] This happens each time two probabilities (eigenvalues of the reduced density operator) are equal.

of the process one can reach situations where they are defined with less accuracy. This is a signature of the quantum character of the entanglement created by the interaction. In terms of entropy of the two systems (§7.2.3), if initially they both have a zero entropy, the same is true at the end of the process, while in quantum mechanics this is not the case.

7.1.2 Two possible origins of correlations

Entanglement and correlation are obviously closely related notions. As we have already mentioned in the introduction to this chapter, correlations may arise in quantum mechanics in two ways:

(i) By a process that is the direct transposition of classical correlations: one attributes random correlated quantum state vectors to each of the two subsystems. The statistical average is then performed linearly with respect to probabilities, and one then says that the whole system is described by a statistical mixture of product states (or by a density operator). This leads to classical correlations, which necessarily obey the Bell inequalities (§4.1 and Appendix C).

(ii) By a purely quantum process involving quantum entanglement: the whole system is described by a state vector that is neither random nor a product. The linear superposition then occurs at the level of the state vector itself, which is a coherent sum of several components, with a relative phase that plays an important role. The correlations then have a purely quantum origin, and may violate the Bell inequalities.

The former process takes place at the level of events, properties of physical systems, results of measurement, etc. as in classical physics. The latter takes place "at a lower level" involving directly state vectors and probability amplitudes (instead of probabilities themselves). It can also result in a broader range of correlations, as illustrated for instance by the Bell theorem.

Both possibilities may also be combined: one can assign to the whole system itself a statistical mixture of states that are not necessarily products. The formalism of the density operator (§12.1.6) makes it possible to include both into the same operator, which is very convenient for compact calculations; on the other hand, it obscures the origin of correlations[7], classical or quantum, which may also sometimes be inconvenient. One sometimes calls "proper mixtures" the situations

---

[7] The formalism of the density operator, or matrix, is elegant and compact, but precisely because it is compact it sometimes partially hides the physical origin of the mathematical terms. The density matrix allows one to treat in the same way classical probabilities (proper mixtures), arising from nonfundamental uncertainties and imperfect knowledge of a physical system, and purely quantum probabilities (improper mixtures) which are more fundamental and have nothing to do with imperfect knowledge. But mathematical analogies should not obscure conceptual difficulties!

associated with usual quantum statistics, and "improper mixtures" those associated with purely quantum correlations.

7.2 Characterizing entanglement

We now study how quantum entanglement can be characterized more quantitatively, with the help of tools such as the Schmidt decomposition or the statistical entropies. We begin with simple considerations concerning the ambiguity of entanglement.

7.2.1 Ambiguous entanglement

For a given $|\Psi\rangle$, the expansion as a sum of products of states is not unique, which implies that the detailed nature of the entanglement itself may be ambiguous. For instance, the expression (7.2) of $|\Psi\rangle$ can be changed by expanding the kets $|1 : \varphi_{a,b}\rangle$ and $|2 : \chi_{a,b}\rangle$ onto arbitrary bases $\{|u_i\rangle\}$ and $\{|v_l\rangle\}$ spanning the respective space states of systems 1 and 2. One then obtains expansions of $|\Psi\rangle$ over the kets $|u_i, v_l\rangle$ that differs from the initial expansion; if the dimensions of the spaces of states are P and Q, these new expansions contain in general PQ terms instead of 2.

Among all expansions, one could then select only those containing the smallest number of terms; nevertheless, even with this condition, the expansion is not always unique. As a simple counterexample, let us consider the following entangled state:

$$|\Psi\rangle = \frac{1}{\sqrt{2}} \left[|1 : \varphi_a\rangle |2 : \chi_a\rangle + |1 : \varphi_b\rangle |2 : \chi_b\rangle \right] \tag{7.11}$$

Looking at this expression of the vector, one could have the impression that it directly indicates that it is the states $|1 : \varphi_{a,b}\rangle$ that are entangled with the states $|2 : \chi_{a,b}\rangle$, in other words, that the basis of individual states in which the entanglement is naturally expressed is determined. But we can introduce the new basis of states for system 1:

$$|\xi_{\pm}\rangle = \frac{1}{\sqrt{2}} \left[|\varphi_a\rangle \pm e^{i\xi} |\varphi_b\rangle \right] \tag{7.12}$$

(if $|\varphi_a\rangle = |+\rangle$ and $|\varphi_b\rangle = |-\rangle$ are the two eigenstates of the Oz component of a $1/2$ spin, the $|\xi_{\pm}\rangle$ are the eigenstates of the spin component along axis $O\xi$ in plane xOy making angle ξ with Ox). We then notice that $|\Psi\rangle$ can be written as well:

$$|\Psi\rangle = \frac{1}{\sqrt{2}} \left\{ |1 : \xi_+\rangle \otimes |2 : \Xi_+\rangle + |1 : \xi_-\rangle \otimes |2 : \Xi_-\rangle \right\} \tag{7.13}$$

with:

$$|\Xi_\pm\rangle = \frac{1}{\sqrt{2}} \left[|\chi_a\rangle \pm e^{-i\xi} |\chi_b\rangle \right] \tag{7.14}$$

Equation (7.13) has exactly the same form as (7.11); it is sufficient to replace the two kets $|\phi_{a,b}\rangle$ with the two other kets $|\xi_\pm\rangle$ as well as the $|\chi_{a,b}\rangle$ by the $|\Xi_\pm\rangle$. Looking at expression (7.13) of the entangled state vector, one now gets the impression that the natural basis to characterize the entanglement is that of the $|\xi_\pm\rangle$ (for a spin 1/2, it would be the spin component along direction $O\xi$ that is entangled with the other system, instead of the Oz component) and the $|\Xi_\pm\rangle$; in this sense, entanglement is ambiguous[8]. This remark can easily be generalized to individual systems that have access to more than two different quantum states.

The conclusion is that the basis of individual states over which an entangled state can simply be expanded is not always uniquely defined. This situation is very different from what happens with classical correlations, which involve correlations between well-defined classical states of individual systems.

7.2.2 Schmidt decomposition of a pure state

Deciding whether or not a pure state describing two quantum systems contains entanglement is not obvious in general. For instance, starting from a tensor product of states, if one expands each term of the product on an arbitrary basis, the simple product becomes the sum of a large number of products; it is then difficult to see that it can be factorized, and therefore contains no entanglement. Tools that allow one to detect the degree of entanglement contained in a pure state in a systematic way are therefore useful.

We generalize the discussion of §7.1.1 in order to introduce particular decompositions of a pure state with a minimum number of components. The quantum system containing subsystems 1 and 2 is described by a normalized state vector $|\Psi\rangle$, or equivalently by the density operator ρ given by the projector onto $|\Psi\rangle$; relation (7.5) is obeyed, as expected for a pure state. Each of the subsystems is described by density operators obtained by partial traces:

$$\rho_1 = Tr_2\{\rho\} \quad ; \quad \rho_2 = Tr_1\{\rho\} \tag{7.15}$$

These two operators are Hermitian, non negative, and with unit trace; their matrices can therefore be diagonalized, with eigenvalues that are real, lying between 0 and 1. We call $|u_i\rangle$ the eigenvectors of ρ_1 (index i takes P different values, where P is the dimension of the space of states of subsystem 1) with eigenvalues q_i, all positive or zero (but not necessarily all distinct). Similarly, the kets $|v_l\rangle$ are the eigenvectors of

---

[8] In §7.2.2, we show that this ambiguity arises from the particular form (7.11); it does not occur for any value of α and β in the general state (7.2).

ρ_2 (where l takes Q different values, Q being the dimension of the space of states of the second subsystem) and r_l the corresponding eigenvalues. The two partial density operators can then be written as:

$$\rho_1 = \sum_{i=1}^{P} q_i \, |u_i\rangle \langle u_i| \qquad \rho_2 = \sum_{l=1}^{Q} r_l \, |v_l\rangle \langle v_l| \tag{7.16}$$

with $0 \le q_i, r_l \le 1$.

7.2.2.a Obtaining the decomposition

We can now expand $|\Psi\rangle$ over the basis of tensor products $\{|1 : u_i\rangle \otimes |2 : v_l\rangle\}$, for which we use the simplified notation $\{|u_i, v_l\rangle\}$; we therefore write:

$$|\Psi\rangle = \sum_{i,l} x_{i,l} \, |u_i, v_l\rangle \tag{7.17}$$

where the $x_{i,l}$ are the components of $|\Psi\rangle$ in this basis. If we introduce the ket $|\overline{w}_i\rangle$, which belongs to the space of states of subsystem 2, by:

$$|\overline{w}_i\rangle = \sum_{l} x_{i,l} \, |v_l\rangle \tag{7.18}$$

expression (7.17) becomes simpler, with a single summation:

$$|\Psi\rangle = \sum_{i} |u_i, \overline{w}_i\rangle \tag{7.19}$$

According to the definition of a partial trace – see (12.84) – the matrix elements of ρ_1 are:

$$\langle u_i| \rho_1 |u_j\rangle = \sum_{m} \langle u_i, v_m |\Psi\rangle \langle \Psi| u_j, v_m\rangle \tag{7.20}$$

where relation (7.19) provides:

$$|\Psi\rangle \langle \Psi| = \sum_{i',j'} |u_{i'}, \overline{w}_{i'}\rangle \langle u_{j'}, \overline{w}_{j'}| \tag{7.21}$$

When we substitute this result into (7.20), the only nonzero terms correspond to $i' = i$ et $j' = j$, so that:

$$\langle u_i| \rho_1 |u_j\rangle = \sum_{m} \langle v_m| \overline{w}_i\rangle \langle \overline{w}_j| v_m\rangle = \langle \overline{w}_j| \overline{w}_i\rangle \tag{7.22}$$

and we obtain:

$$\rho_1 = \sum_{i,j} |u_i\rangle \langle u_i| \rho_A |u_j\rangle \langle u_j| = \sum_{i,j} |u_i\rangle \langle u_j| \times \langle \overline{w}_j| \overline{w}_i\rangle \tag{7.23}$$

But, by construction of the basis $\{|u_i\rangle\}$ we have used in the calculation, ρ_1 is diagonal and given by expression (7.16); comparing with (7.23) shows that we necessarily have:

$$\langle \overline{w}_j | \overline{w}_i \rangle = \delta_{i,j} \times q_i \tag{7.24}$$

For all values of index i that correspond to nonzero eigenvalues q_i, this relation implies that we can define an ensemble of orthonormal kets $|w_i\rangle$ belonging to the space of states of subsystem 2 by:

$$|w_i\rangle = \frac{1}{\sqrt{q_i}} |\overline{w}_i\rangle \tag{7.25}$$

For all values of index i associated to zero eigenvalues q_i, the same relation shows that the kets $|\overline{w}_i\rangle$ must vanish.

Therefore, expression (7.19) becomes:

$$|\Psi\rangle = \sum_i \sqrt{q_i} |u_i, w_i\rangle \tag{7.26}$$

where the $|u_i\rangle$ are a set of orthonormal vectors in the space of states of the first system, and where the $|w_i\rangle$ are another set of orthonormal vectors in the second space of states. This expression is the Schmidt decomposition of a pure entangled state; it is also called the "bi orthonormal decomposition", and plays an important role in the modal interpretation of quantum mechanics (§11.9).

7.2.2.b Discussion

If we now come back to the partial traces ρ_1 and ρ_2, a calculation from (7.26) provides two symmetric expressions:

$$\rho_1 = \sum_i q_i |u_i\rangle \langle u_i| \tag{7.27}$$

(already known) and:

$$\rho_2 = \sum_i q_i |w_i\rangle \langle w_i| \tag{7.28}$$

This restores the symmetry between the two systems: the $|u_i\rangle$ were defined as eigenvectors of ρ_1, but we see that the $|w_i\rangle$ are also eigenvectors of ρ_2; moreover, the two partial density operators always have the same eigenvalues[9], with a sum equal to 1, since the two operators have unit trace. In the special case where they are all zero except one, each of the subsystems is in a pure state. But, in general, several eigenvalues are non zero, and we then immediately see that $(\rho_1)^2$ is not equal to ρ_1; a similar property is true for ρ_2. We then find again a case where the two subsystems are described by statistical mixtures while the whole system is in a pure state.

---

[9] These properties are specific to pure states for the whole system; they are not necessarily true if it is described by a statistical mixture.

The number of nonzero eigenvalues q_i, in other words the number of effective terms in (7.26), is called the "Schmidt rank" R of $|\Psi\rangle$. If $R = 1$, the state of the whole system is not entangled, and the two subsystems are in pure states. If $R = 2$, the situation is that discussed in the example of §7.1.1; if $R = 3$, the entanglement is more complicated, etc. The entanglement is shared by the subsystems, 1 or 2, in a symmetric way; for instance, one of the subsystems cannot be in a pure state while the other is in a statistical mixture. The dimension of the space of state of subsystem 2 with which 1 is entangled gives an upper bound to the number of independent kets $|w_i\rangle$, and therefore to the rank R; actually, R cannot exceed the dimension of any of the spaces of states of the subsystems: a high rank entanglement between them therefore requires that both should have spaces of states with sufficient dimensions.

When all eigenvalues q_i of ρ_A (and of ρ_B) are different, the expansions (7.16) and (7.28) of ρ_B over the projectors on its eigenvectors necessarily coincide ; the series of the $|w_i\rangle$ coincide with that of the $|v_l\rangle$. The eigenvectors of the partial density operators directly give the Schmidt decomposition, which is therefore unique in this case. It is also the decomposition of $|\Psi\rangle$ over products of orthonormal bases that contains the minimum number of terms: this is because the summation of (7.19) never contains more terms than the summation of (7.17).

When some eigenvalues q_i are degenerate, several bi-orthonormal decompositions become possible. For instance, for a singlet state, the two partial density matrices have two eigenvalues equal to $1/2$. This singlet can be expanded in the same way onto products of eigenstates of components of the spins along an arbitrary direction. It therefore has an infinity of Schmidt decompositions. In § 7.2.1, we have seen another case where the eigenvalues are degenerate so that "ambiguous entanglement" takes place.

7.2.3 Statistical entropies

We can associate a statistical entropy (§12.1.6.d) to each density operator ρ. We now compare the entropy of ρ to those associated with the two partial density operators ρ_1 and ρ_2.

In the case studied in the preceding section, the whole physical system is in a pure state, its density operator is the projector (7.3) over a single state $|\Psi\rangle$, with a vanishing associated entropy $S = 0$. Nevertheless, the two subsystems are not in general in pure states (except if $|\Psi\rangle$ is a tensor product, without any entanglement), so that:

$$S_1 = -k_B Tr\{\rho_1 \ln \rho_1\} \geq 0$$
$$S_2 = -k_B Tr\{\rho_2 \ln \rho_2\} \geq 0$$

(7.29)

We then have:

$$S_1 + S_2 \geq S \qquad\qquad (7.30)$$

The equality corresponds to the particular case where $|\Psi\rangle$ is a product, and where the Schmidt rank is equal to 1.

In a more general situation, the whole system is described by a density operator ρ that does not necessarily correspond to a pure state, so that its entropy S does not vanish either. One can nevertheless show that this entropy S always remains smaller than, or equal to, the sum of the two entropies of the subsystems [10]; in other words, relation (7.30) remains valid in this case. The equality is obtained only when ρ is a product:

$$\rho = \rho_1 \otimes \rho_2 \qquad\qquad (7.31)$$

which corresponds to a case where both subsystems are described by statistical mixtures, but remain completely uncorrelated. The difference $S_1 + S_2 - S$ therefore gives a quantitative estimation of the loss of accuracy when going from the quantum description of the whole system to the separate quantum descriptions of the two subsystems.

7.2.4 Measures of entanglement

If the whole system $S_1 + S_2$ is in a pure state $|\Psi\rangle$, the Schmidt decomposition shows that:

$$S_1 = S_2 = \sum_i q_i \ln q_i \geq 0 \qquad\qquad (7.32)$$

It is then natural to choose S_1 as the definition of the degree of entanglement of the two subsystems. When the whole system is in a pure state, the two subsystems are said to be maximally entangled when the partial entropies S_1 and S_2 reach their maximal value. If the dimensions of the spaces of states of the two subsystems are both equal to the same value D, this happens when the two partial density matrices are diagonal with all their diagonal elements equal to $1/D$. This is for instance what happens for two spins $1/2$ in a singlet state. The opposite case occurs when the whole system is in a product state; both partial entropies then vanish, and so does the degree of entanglement.

But one may also wish to define a precise measure of the degree of entanglement of two quantum subsystems when the whole system is not in a pure state. To be physically acceptable, this measure must obey several criteria. For instance, it must obviously vanish if the whole system is in a state that is a tensor product

---

[10] One sometimes speaks of "subadditivity of the entropy" to express the fact that the entropy of the whole system is less than the sum of the entropies of the two subsystems.

where the two subsystems are not correlated. It is also necessary that the degree of entanglement should remain invariant when Alice and Bob, each acting on two remote subsystems, perform local operations on them [321]. For instance, maximally entangled states, such as those we have studied in §6.1.3, keep the same form if different unitary transformations are applied to the spins (footnote [5] of that section). Several such measures have been proposed, generally related to von Neumann entropies; for a review, see for instance [322]. When the number of entangled systems is three or more, the definition of their degree of entanglement becomes more difficult; for the moment, there is no generally accepted definition of the degree of entanglement of a multipartite system.

One could also expect that a system giving rise to maximal violations of the BCHSH inequalities should reach the maximal degree of entanglement. Actually, it turns out that the notions of entanglement and of nonlocality are not directly related, except in a few special cases (two spins 1/2, for instance). Curiously, maximally entangled states generally produce fewer nonlocal effects than nonmaximally entangled states [218, 323]. nonlocality and entanglement are therefore really different concepts.

7.2.5 Monogamy

A state such as (7.10) is a state where the two spins are strongly correlated, while a GHZ state such as (6.1) can be regarded as the equivalent for three spins. One might naively think that the GHZ state conserves the entanglement between the two first spins available in (7.10), while they have also become entangled with the third. In fact, this is wrong: while going from (7.10) to (6.1), it is true that one entangles the third spin with the group of the two others, but at the same time one totally destroys the entanglement within the initial group of two spins. Actually, we have already noticed this property, when we remarked in §6.1.3 that these states exhibit strong correlations between the spins only if all of them are measured; if the measurements relate to two spins instead of three, no correlation between them appears whatsoever.

Under these conditions, what can we do to add an additional spin without destroying the correlation between the two initial spins? We can assume the following form of the three-spin state:

$$| \Psi > = \frac{1}{\sqrt{2}} \Big[|1 : +; 2 : -\rangle + e^{i\xi} |1 : -; 2 : +\rangle \Big] \otimes |3 : \theta\rangle$$

$$= \frac{1}{\sqrt{2}} \Big[|1 : +; 2 : -; 3 : \theta\rangle + e^{i\xi} |1 : -; 2 : +; 3 : \theta\rangle \Big] \qquad (7.33)$$

(where $|\theta\rangle$ is any normalized state for the third spin), with a factorized state for the additional spin; this choice retains the same entanglement between spins 1 and 2 as (7.10)[11]. But then the third spin is totally uncorrelated with the two former!

A compromise between the two preceding attempts is:

$$\frac{1}{\sqrt{2}} \left[|1:+;2:-;3:\theta_1\rangle + e^{i\xi} |1:-;2:+;3:\theta_2\rangle \right] \qquad (7.34)$$

If $|\theta_1\rangle = |\theta_2\rangle$, one recovers (7.33), and the additional spin is not entangled with the two first; if $|\theta_1\rangle$ and $|\theta_2\rangle$ are orthogonal, one recovers a GHZ state where none of the three pairs has any entanglement, this property being restricted to all three particles. When $|\theta_1\rangle$ and $|\theta_2\rangle$ are neither parallel nor orthogonal, we have an intermediate situation: the more parallel they are, the more entangled the two first spins remain (we will see in the next section that the coherent terms contain the scalar product $\langle\theta_1 | \theta_2\rangle$), but the third then has little entanglement. Conversely, the more orthogonal they are, the less entangled the two initial spins are: they lose their correlations in favor of a three-spin correlation. The third spin actually plays for the two others a role that is similar to the role of the environment in decoherence (§7.3.3.a): indeed, the environment destroys the coherence of the initial state more efficiently when it correlates to its components with states that are almost orthogonal.

This is a general property: if two systems are maximally entangled, a principle of mutual exclusion makes it impossible to be entangled with a third system. Mathematically, the property can be expressed through the Coffman–Kundu–Wootters inequality[12] [324, 325]. It has no classical equivalent, since classically nothing forbids the correlation of a third system with two others without destroying the initial correlation. One often calls this property "entanglement monogamy". It is for instance possible to show that, if two quantum subsystems A and B are mutually entangled as well as to a third system C, and if two of them are sufficiently correlated to create violations of the BCHSH inequalities with measurements on these subsystems, then the inequalities are necessarily satisfied for all measurements related to the two other pairs of subsystems [326, 327].

7.2.6 *Separability criterion for density operators*

We have seen in §3.3.3.c that the EPR argument and the Bell theorem are strongly related to the notion of quantum nonseparability; they are also related to the notion of quantum entanglement, since violations of the Bell inequalities require the pres-

---

[11] In the calculation of the effect of the coherent term (in $e^{i\xi}$) with the method used in §6.1.3, it is now sufficient to flip two spins, the third remaining always in state $|\theta\rangle$.

[12] Reference [324] makes use of a measure of entanglement called "concurrence". It shows that the sum of the squares of the concurrence between A and B and of that between A and C cannot exceed the concurrence between A and the BC pair.

ence of quantum entanglement. In terms of density operators, how is it possible to recognize quantum nonseparability?

Consider a quantum system made of two subsystems 1 and 2 and described by a density operator ρ. Since ρ belongs to a space that is the tensor product of the space of operators acting on 1 by the space of operators acting on 2, it can always be expanded as a sum of products of operators acting on both systems. If ρ can be expanded with real positive coefficients w_n in terms of a series of density operators ρ_1^n and ρ_2^n relative to each subsystem according to:

$$\rho = \sum_n w_n \, \rho_1^n \otimes \rho_2^n \qquad (7.35)$$

one says that ρ is separable[13] [193, 328]. When the traces of all density operators are normalized to 1, one obtains:

$$1 = \sum_n w_n \quad \text{and therefore} \quad 0 \le w_n \le 1 \qquad (7.36)$$

which shows that the w_n can be interpreted as probabilities[14]; the event where first subsystem is described by ρ_1^n and the second by ρ_2^n has probability w_n. If the expansion (7.35) necessarily contains coefficients w_n that are not real positive numbers, one says that the density operator ρ is "nonseparable" and contains quantum entanglement[15].

If one performs separate measurements on subsystems 1 and 2, and if the whole system has a separable density operator, the results always obey the Bell inequalities [193] (nevertheless, the opposite is not necessarily true: nonseparable density operators do not necessarily give rise to violations of the inequalities). In such a situation, even if each of the two subsystems has strong quantum properties, the way correlations between them are introduced remains purely classical. A separable system cannot have quantum entanglement, even if each subsystem exhibits extreme quantum properties. The Peres–Horodecki criterion [328, 329] indicates that a necessary condition for a density matrix to correspond to a separable density operator ρ is that the matrix obtained by partial transposition (where only the indices relative to one of the subsystems are transposed, not for the other) should have nonnegative eigenvalues. The appearance of negative eigenvalues may then signal quantum entanglement, with a better sensitivity than Bell violations.

---

[13] In his initial article, Werner uses the words "classically correlated" [193], but the word "separable" chosen by Peres [328] is more frequently used nowadays.

[14] If one replaces the discrete variable n by a continuous variable λ, one obtains the case studied in Appendix C.

[15] For two spins 1/2, one can easily show that the density operator associated with a singlet state, or its direct generalization (7.10), necessarily involves negative expansion coefficients. Two spins in a singlet state are therefore not separable, and possess quantum entanglement.

7.3 Creating and losing entanglement

Historically, at Schrödinger's time, quantum entanglement between remote physical systems was seen as a rather rare phenomenon, playing a role mostly in thought experiments. Nowadays, many experimental methods have been invented to obtain entanglement. Actually, quantum entanglement has now become an essential part of quantum information, quantum cryptography, teleportation, etc. (Chapter 8). All experiments already mentioned in §4.1.5 involve pairs of entangled particles, often photons with entangled polarization variables.

7.3.1 Entanglement created by local interactions

As Schrödinger had initially suggested (see quotation in §7.1), one way to obtain entanglement between physical subsystems is to make use of local interactions between particles. An atom emitting two photons in succession may provide such a scheme, which in fact has been used in many experiments. Initially, the atom is in an excited state, then emits a first photon, and reaches an intermediate state that depends on the polarization of the emitted photon; at this stage, the atom plus photon system is described by an entangled state, with coherent components on several polarization states associated with intermediate states of the atom. Each of these components then gives rise to the emission of a second photon with different polarizations, while the atom itself reaches a ground state that is independent of the polarizations of the emitted photons. This corresponds to the case $|\theta_1\rangle = |\theta_2\rangle$ in (7.34), where the state of one of the three particles (here, the final state of the atom) factorizes; the atom leaves the quantum entanglement party, which allows the two photons to enter maximal entanglement (according to the monogamy rule).

An often cited example is that of the atomic cascade $J = 0 \rightarrow 1 \rightarrow 0$ of the calcium atom. It provides, by successive spontaneous emissions, two photons in a state:

$$| \Psi >= \frac{1}{\sqrt{2}} \left[|1:H;2:H\rangle + |1:V;2:V\rangle \right] \tag{7.37}$$

where $|H\rangle$ and $|V\rangle$ are two states of polarization (horizontal and vertical) – for photons[16] these states are analogous to states $|+\rangle$ and $|-\rangle$ for a spin $1/2$. We have also seen that the techniques of parametric down-conversion in nonlinear optics provide similarly entangled photons, often in more favorable experimental conditions than an atomic cascade.

Quantum entanglement is not limited to photons, but can also be obtained with

---

[16] These polarizations are mutually perpendicular as well as perpendicular to their direction of propagation. For the sake of simplicity, we limit our discussion to the polarization variables of the photons, which are the essential variables in the discussion of the BCHSH inequalities. Nevertheless, photons have other variables, frequency (energy) and direction of propagation, which we consider as fixed here.

particles with rest mass. For instance, the experiment described in [149] studies the correlation between the spins of two protons after a low-energy collision between a beam of protons and an hydrogen target; the reference [220] proposes using the dissociation of Mercury dimers (isotope 199 has a nuclear spin 1/2) to obtain atoms with correlated spin variables. Experiments have also been done where two atoms are entangled because they exchange a single photon confined in a cavity with extremely high-quality factor [330]. We have already discussed in §4.3.1.e the Bell experiments performed with the spins of NV centers in diamond.

Trapped ions provide other possibilities to obtain quantum entanglement, in a case where the particles are localized and may be observed for a long time. The reference [331] describes how such an entanglement has been obtained with Beryllium ions sitting in a radio-frequency Paul trap, according to the method proposed by Cirac and Zoller [332]. For a review of experiments with entangled ions, see [333]. Entanglement may also be observed in solid-state physics with superconducting currents [334] involving a very large number of electrons (macroscopic systems). We now discuss still another method, entanglement swapping, which produces entangled particles without any interaction between them, under only the effect of the process of quantum measurement acting on other particles.

7.3.2 Entanglement swapping

The "entanglement swapping" method can entangle two particles created by two independent remote sources (the particles then have no common past) under the effect of the process of quantum measurement [335, 336]. Assume that we have two independent sources S_{12} and S_{34}, each emitting a pair of entangled photons, 1 and 2 for the former, 3 and 4 for the latter (Figure 7.1). The state describing both pairs is a tensor product of terms that are similar to (7.37):

$$| \Psi >= \frac{1}{2} [|1:H;2:H\rangle + |1:V;2:V\rangle] \otimes [|3:H;4:H\rangle + |3:V;4:V\rangle] \quad (7.38)$$

If we introduce the 4 Bell states relative to particles i, j defined as:

$$\begin{aligned} | \Phi^B_{i,j} >_{(\pm)} &= \frac{1}{\sqrt{2}} [|i:H;j:H\rangle \pm |i:V;j:V\rangle] \\ | \Theta^B_{i,j} >_{(\pm)} &= \frac{1}{\sqrt{2}} [|i:H;j:V\rangle \pm |i:V;j:H\rangle] \end{aligned} \quad (7.39)$$

we obtain an orthonormal basis in the space of states associated with particles i and j. Since:

$$| \Phi^B_{1,4} >_{(+)} \otimes | \Phi^B_{2,3} >_{(+)} + | \Phi^B_{1,4} >_{(-)} \otimes | \Phi^B_{2,3} >_{(-)} = [|HHHH\rangle + |VVVV\rangle]$$

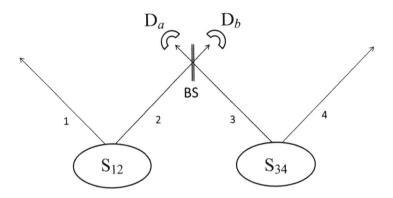

Figure 7.1 Schematic representation of entanglement swapping. Two independent sources S_{12} and S_{34} emit each a pair of entangled particles, numbered 1 and
2 for the former, 3 and 4 for the latter. A half-reflecting beam splitter BS is inserted on the trajectory of particles 2 and 3, and two detectors D_a and D_b measure
the number of particles in the two output channels. This measurement can project
the state vector and put particles 1 and 4 into a completely entangled state, even
if they have never interacted with each other.

(to simplify the notation, we implicitly assume that, in the right-hand side, the
particles are always in the order 1, 2, 3 and 4) and:

$$| \Theta^B_{1,4} >_{(+)} \otimes | \Theta^B_{2,3} >_{(+)} + | \Theta^B_{1,4} >_{(-)} \otimes | \Theta^B_{2,3} >_{(-)} = [|HHVV\rangle + |VVHH\rangle]$$

$$(7.40)$$

we can write state (7.38) in the form:

$$| \Psi >= \tfrac{1}{2} \Big[| \Phi^B_{1,4} >_{(+)} \otimes | \Phi^B_{2,3} >_{(+)} + | \Phi^B_{1,4} >_{(-)} \otimes | \Phi^B_{2,3} >_{(-)} +$$
$$+ | \Theta^B_{1,4} >_{(+)} \otimes | \Theta^B_{2,3} >_{(+)} + | \Theta^B_{1,4} >_{(-)} \otimes | \Theta^B_{2,3} >_{(-)} \Big]$$

$$(7.41)$$

Suppose now that the experiment shown schematically in Figure 7.1 is performed: particles 2 and 3 undergo an interference experiment with eigenvectors
that are the four Bell states for these two particles – this can be obtained by sending the two particles onto a semireflecting beam splitter and observing at which
detectors, D_a or D_b, they are found in the output channels[17]. The projection onto

---

[17] Among the four Bell states, the only one for which each output channel contains one particle is $| \Theta^B_{23} >_{(-)}$;

one of the four Bell states for these particles projects the state of the two others onto the same Bell state. The two nonobserved particles are then put into a completely entangled state. What is remarkable in the process is that, initially, the 1, 2 pair is internally entangled, but not with the 3, 4 pair, which also has only internal entanglement. By performing an appropriate measurement on one particle from each pair, one projects the two remaining particles into a completely entangled state, even if these two particles have never interacted.

It is interesting to compare the correlations provided by entanglement swapping to classical correlations. Suppose two independent sources emit pairs of correlated objects, numbered 1 and 2 for the first source, 3 and 4 for the second, as in figure 7.1. In each run of the experiment, every source emits two objects with some common property (as for instance the same color, or opposite angular momenta, etc.), but the two sources are completely uncorrelated (objects emitted in the same run by the two different sources have no correlation between their colours, or angular momenta, etc.). Nevertheless, if we select events where particles 2 and 3 have some given correlation (same colour, different colours, parallel or antiparallel angular momenta, etc.), it is clear that particles 1 and 4 will also be correlated in the same way, even if they never interacted in the past. This is just a consequence of a selection operated in a classical probability distribution, and can be called "classical correlation swapping". But, even with this selection, if a Bell experiment is performed with objects 1 and 4, the observed correlations will always be such that the Bell inequalities are obeyed. By contrast, the entanglement swapping method can lead to large violations of the inequalities. Entanglement swapping is therefore a more powerful method to create correlations than classical correlation swapping. For instance, we have seen in §4.3.1.e several applications of the method to perform Bell experiments.

We have already mentioned the possibility to entangle a larger number of particles [279] with similar schemes. Other protocols have been implemented to create quantum entanglement, and to put six ions in a NOON state [337], and up to eight into a W state (coherent superposition of states where a single excitation is localized on any of the ions, with the same probability amplitude) [338]. In quantum optics, the techniques of parametric down conversion in nonlinear crystals have been used in experiments providing entanglement in 2, 4, or 6 photons [339, 340]. In §4.3.1.d, we have discussed how entanglement swapping has been used to create quantum correlations between remote ions with the help of photons (in Figure 7.1, particles 1 and 4 are then ions, particles 2 and 3 are photons), leading to Bell inequality violations.

the three other states correspond to situations where the two particles always exit through the same channel (Hong–Ou–Mandel effect). If the two detectors D_a and D_b each register a particle, the measurements project the remaining particles 1 and 4 into the completely entangled state $| \Theta_{14}^B >_{(-)}$.

Various experimental methods have also been developed to create entanglement in neutral atoms [341, 342], in particular a recent method called "carving entangled states", which consists in starting from a product state for two atoms, and removing components from this state in order to make it strongly entangled [343, 344].

7.3.3 Decoherence

We now discuss more precisely a phenomenon that we have already introduced, decoherence, which acts to reduce the lifetime of coherent superpositions. The process becomes extremely fast when the number of correlated particles increases.

We have introduced decoherence in §2.1 as the initial part of the phenomenon associated with the von Neumann infinite regress: coherent superpositions tend to constantly propagate toward the environment, involving more and more complex correlations with it. During decoherence, entanglement does not actually disappear. It remains present, but goes further and further into the environment; as a consequence, it becomes rapidly completely impossible to detect in practice.

7.3.3.a Mechanism

To see more precisely how this happens, let us for instance consider a state:

$$|\Psi\rangle = \left[\alpha\,|\varphi_a\rangle + \beta\,|\varphi_b\rangle\right] \otimes |\mathbf{k}_0\rangle \qquad (7.42)$$

which is the product of two states: one describes an atom in a coherent superposition of two orthogonal states $|\varphi_a\rangle$ and $|\varphi_b\rangle$, localized in two different regions of space; the other describes another particle, a photon for instance, initially in state $|\mathbf{k}_0\rangle$ (we assume that all these states are normalized).

Initially, the atom is described by a state with quantum properties that depend on the relative phase of α and β, which are therefore coherent; decoherence is a process where these coherent properties will disappear. This happens when the photon interacts with the atom and is scattered into a quantum state that depends on the location where scattering took place: if the scattering atom is in state $|\varphi_a\rangle$, the photon reaches state $|\mathbf{k}_+\rangle$; if it is in state $|\varphi_b\rangle$, it reaches state[18] $|\mathbf{k}_-\rangle$ (states $|\mathbf{k}_\pm\rangle$ are normalized). After scattering, the atom has not changed its position, and returns to the same internal ground state (assuming that momentum transfer is negligible), so that the state vector describing the whole system after interaction becomes:

$$|\Psi'\rangle = \alpha\,|\varphi_a\rangle \otimes |\mathbf{k}_+\rangle + \beta\,|\varphi_b\rangle \otimes |\mathbf{k}_-\rangle \qquad (7.43)$$

---

[18] We could also have assumed that the photon is focused so that it can interact only with atoms in one of the states, but is not scattered by the other state; the conclusions of our discussion would not have been changed.

Assume now that we are interested only in the atom, not the photon; the reason might be, for instance, that detecting the scattered photon is impossible or very difficult (e.g., it may be a far-infrared photon). It is then useful to calculate the partial trace (§12.2.3) over this photon, in order to obtain the density operator that describes the atom only. A calculation similar to that leading to (7.23) shows that this partial trace can be written, on the basis of the two states $|\varphi_a\rangle$ and $|\varphi_b\rangle$:

$$\rho = \begin{pmatrix} |\alpha|^2 & \alpha\beta^* \langle \mathbf{k}_- | \mathbf{k}_+ \rangle \\ \alpha^*\beta \langle \mathbf{k}_+ | \mathbf{k}_- \rangle & |\beta|^2 \end{pmatrix} \tag{7.44}$$

If the scalar product $\langle \mathbf{k}_- | \mathbf{k}_+ \rangle$ was equal to 1, the density matrix of the atoms would not be affected at all by the scattering of the single photon. But this would be assuming that the photon is scattered exactly into the same state, independently of the spatial location in $|\varphi_a\rangle$ or $|\varphi_b\rangle$ of the scatterer; in other words, that it carries no information about the location of the atom! This cannot be the case if the distance between the two locations is much larger than the photon wavelength. Actually, it is much more realistic to assume that this scalar product is close to zero, which means that the off-diagonal element of (7.44) is also almost zero. We then conclude that the scattering of even a single photon destroys the coherence between atomic states, as soon as they are located in different places.

The loss of coherence becomes even worse when more and more photons (assumed to be all in the same initial state $|\mathbf{k}_0\rangle$) are scattered, since one should then replace (7.43) by state:

$$|\Psi''\rangle = \alpha |\varphi_a\rangle \otimes |\mathbf{k}_+\rangle |\mathbf{k}'_+\rangle |\mathbf{k}''_+\rangle \dots + \beta |\varphi_b\rangle \otimes |\mathbf{k}_-\rangle |\mathbf{k}'_-\rangle |\mathbf{k}''_-\rangle \dots \tag{7.45}$$

with obvious notation (the states with n primes correspond to the $(n-1)$th scattered photon); the same calculation as above then provides the following value for the partial trace ρ:

$$\begin{pmatrix} |\alpha|^2 & \alpha\beta^* \langle \mathbf{k}_- | \mathbf{k}_+ \rangle \langle \mathbf{k}'_- | \mathbf{k}'_+ \rangle \dots \\ \alpha^*\beta \langle \mathbf{k}_+ | \mathbf{k}_- \rangle \langle \mathbf{k}'_+ | \mathbf{k}'_- \rangle \dots & |\beta|^2 \end{pmatrix} \tag{7.46}$$

Since we now have, in the off-diagonal elements, the product of many scalar products $\langle \mathbf{k}_- | \mathbf{k}_+ \rangle$, it is clear that these elements have even smaller modulus than when a single photon is scattered. Actually, as soon as the two scattered states of the photon are not strictly identical, the off-diagonal elements tend exponentially to zero when the number of scattering events increases.

This is a completely general property: all objects have a strong tendency to leave a trace in the environment because they develop correlations with any elementary particle which passes by, so that they lose their coherence in the process. To illustrate this property, one can for instance in (7.42) replace the coherent state of a

single atom by a GHZ type state (coherent state of a large number of atoms) and
write the initial state as:

$$|\Psi\rangle = \left[\alpha \, |\varphi_a, \varphi_a, ..., \varphi_a\rangle + \beta \, |\varphi_b, \varphi_b, ..., \varphi_b\rangle \right] \otimes |\mathbf{k}_0, \mathbf{k}_0, ...\rangle \qquad (7.47)$$

The larger the number of scattering atoms initially in states $|\varphi_a\rangle$ and $|\varphi_b\rangle$, the sooner
many photons will be scattered, and the faster the off-diagonal elements will tend
to zero. Nevertheless, the scattering of a single photon into two orthogonal states is
already sufficient to completely destroy the coherence. Consequently, the large size
of a physical system in a coherent superposition makes the superposition extremely
fragile. The coherence actually propagates into a coherence involving the environment and more and more complex correlations with it (a scattered photon may in
turn correlate with other particles); soon the coherence becomes practically impossible to detect. The phenomenon is unavoidable, unless of course the scattering
properties of both states symbolized by $|\varphi_a\rangle$ and $|\varphi_b\rangle$ are exactly the same, which
excludes any significant spatial separation between the states. This illustrates how
fragile coherent superpositions of macroscopic objects are, as soon as they involve
states that can be seen as distinct[19].

7.3.3.b Revisiting the Schrödinger cat and macroscopic uniqueness

We are now in a position where we can come back in more detail to some questions
that we already discussed in §2.2, and which are related to decoherence and/or the
Schrödinger cat. Of course, a cat is not a two-level quantum system, as the atom
described by the density matrix (7.46); a huge number of quantum states are associated with a cat that is alive, and the same is true of a dead cat. If $\mathcal{N}$ is the number
of these states, each matrix element of (7.46) should be replaced by a square block
of $\mathcal{N}$ lines and $\mathcal{N}$ columns containing matrix elements. This being changed, the
rest of the reasoning remains the same as before. In particular, a cat that is alive or
dead will not scatter all photons in the same way, otherwise we could not even see
the difference! Clearly, decoherence will take place almost immediately in this case
also. Does this simple remark provide an obvious solution to the famous paradox?
More generally, can we regard decoherence as an "explanation" of the postulate
of state vector reduction: when the superposition of the initial system becomes
incoherent, are we not in the presence of a statistical mixture that resembles the
description of a classical object with well-defined (but unknown) properties? A
countless number of authors constantly rediscover this idea, and claim that they

---

[19] In §9.1.2 we study the effects of the decoherence induced by the environment of a measurement apparatus,
and introduce the notions of "pointer states".

have solved the difficulties related to quantum measurement[20]. Is this a reasonable explanation?

The answer to this question was already given in §§2.2.2 and 2.2.3: this explanation is unsatisfactory because the purpose of the postulate of state vector reduction is not to explain decoherence, already contained in the Schrödinger equation, but the uniqueness of the result of the measurement, which is beyond this equation. In fact, the effect of the state vector reduction may sometimes be just the contrary: it puts back the measured subsystem into a pure state, the perfect opposite of a statistical mixture, so that the real question is then to understand how the (re)emergence of a pure state should be possible [57], not of a statistical mixture. Indeed, in common life, as well as in laboratories, one never observes superposition of results; Nature seems to operate in such a way that a single result always emerges from a single experiment. How can something that was indeterminate become determinate, and where does this disturbance come from? This will never be explained by the Schrödinger equation, since all that it can do is to endlessly extend its ramifications into the environment, without ever selecting one of them.

Another way to answer the question is to emphasize its logical structure. The central point of the paradox is the necessity for some kind of limit of the validity of the linear Schrödinger equation, since a linear equation can never predict the emergence of a single result in an experiment. The difficulty is where and how to create this border. Logically, it is clear that this problem will never be solved by invoking any process that is already contained inside the linear Schrödinger equation[21], such as decoherence or any other similar process that remains linear. No one doubts that a typical measurement process will involve decoherence at some preliminary stage, but the real questions are what happens after, and how does a single result emerge.

Indeed, once in a preliminary stage the off-diagonal elements of the density matrix have disappeared under the effect of decoherence, the question is to understand what happens to the diagonal terms. To obtain a quantum dynamics that is compatible with the fact that a single result is obtained in a measurement (or, more generally, to predict macroscopic uniqueness), one would need a process that could collapse the diagonal elements into a single one (for a two-level atom; for the cat, all diagonal elements should collapse into elements in a single block, either these describing a cat that is alive, or those describing a dead cat). How is this possible? This is the difficult part of the measurement problem: explain why, at the end

---

[20] Bell deplored the constant resurgence of this idea. In [102] for instance, after emphasizing the deep difference between a state where the diagonal terms of the density operator after decoherence coexist, and a state where these terms appear as exclusive alternatives ('and' is different from 'or', cf. footnote [11] of §2.2.3), he writes: "The idea that the elimination of coherence, in one way or another, implies the replacement of 'and' by 'or' is a very common one among the solvers of the measurement problem. It always puzzled me ".

[21] Staying in the middle of a country, one never reaches its borders...

of the measurement process, the diagonal elements can coalesce. Sure, these elements have all the mathematical properties that can be required for probabilities: they are positive numbers and their sum is unity (*cf.* the discussion in §2.2.3); they could actually be called "pre-probabilities" [345]. But these preprobabilities occur simultaneously in each realization of the experiment, while true probabilities characterize exclusive events (see Bell's quotation in note [20]). Turning preprobabilities into real probabilities is an important step that requires, for instance, the introduction of von Neumann's reduction postulate.

Pressed to this point of the discussion, some physicists reply that one can always assume that, somehow and at some later stage, the quantum solution of the Schrödinger equation has spontaneously resolved into one of its branches only. This is certainly true, but this would amount to first throwing a problem out through the door, and then letting it come back through the window! In fact, this amounts to saying that the standard dynamic equation cannot always be valid, which was precisely the point of the cat paradox. A more consistent attitude would be to adhere to the Everett interpretation of quantum mechanics (§11.12), considered as a natural complement of decoherence. Indeed, this provides a consistent interpretation of quantum mechanics, where the emergence of a single result does not have to be explained, since it is assumed never to take place (the Schrödinger equation then has no limit of validity). But, of course, if one takes this point of view, one has to deal with all the intrinsic difficulties of the Everett interpretation, which we will discuss later (§11.12). A general discussion of the relations between decoherence and the measurement problem, and its role in the various interpretations of quantum mechanics, is given in [346].

Concerning terminology, we have already mentioned in §2.2 that, during the last few years, it has become rather frequent to use the words "Schrödinger cat" (SC) to characterize coherent states such as (6.16), even sometimes for small values of N (sometimes for a single ion!). This is a redefinition of the words, since the essential property of the original cat was precisely to have a macroscopic number of degrees of freedom, as opposed to the few radioactive atoms or ions. Let us then assume that someone succeeded in preparing an all-or-nothing state with a very large value of N – would that be a much better realization of the Schrödinger cat as meant by its inventor? To some extent, yes, since the cat can be seen as a symbol of a system of many particles that go to an orthogonal single-particle quantum state, when one goes from one component of the state vector to the other. Indeed, it is likely that many of the atoms of a cat take part in different chemical bonds if the cat is alive or dead, which puts them in a different quantum state. But it seems rather hard to invent a reason why every atom, every degree of freedom, should necessarily be in two orthogonal states in the original story, while this is the essential property of "all-or-nothing states". In a sense, they do more than realizing a standard

Schrödinger cat; the two concepts are related but remain somewhat different, even for large values of N.

7.3.3.c Quantum reservoir engineering

The result of the coupling of a quantum system with its environment is not always negative, but can also be very useful. It can actually be used to "engineer" the master equation (§ 7.4) that gives the time evolution of the density operator of the system, and to drive it from a statistical mixture into interesting quantum states; the method is called "quantum reservoir engineering" [347–349]. A well-known example is the method of optical pumping [350, 351], which can transfer an ensemble of atoms in a thermal statistical mixture of all Zeeman sublevels of their ground level into a pure state in which they occupy a single sublevel, or a coherent superpositions of these sublevels. In this case, the role of the environment (the "reservoir") is played by all the modes of the electromagnetic radiation, which are coupled to the atoms by the process of spontaneous emission. Many other examples of reservoir engineering have been proposed since; see for instance [352–355] and [356] for an example of experimental realization with trapped ions. Therefore, dissipative coupling to a reservoir does not necessarily lead to states without quantum coherence; in fact, controlled decoherence may be used to prepare interesting quantum states.

7.3.3.d Theory and experiments

The problem of the evolution of a microscopic quantum system coupled to a macroscopic quantum environment covers many different physical situations, which are sometimes grouped under the terms of "quantum dissipation". This subject has a long history, which we can only skim through here. In 1963, Feynman and Vernon [357] studied the coupling between a test system and an external environment made of combinations of harmonic oscillators, as well as an environment that can be treated approximately in this way. They showed how the evolution of the test system can be obtained in terms of its own variables only, while the influence of the external environment is included in "influence functionals" of the variables of the test system. In 1983, Caldeira and Leggett [358], motivated by the study of the quantum tunneling effect of the electric current in a SQUID (Superconducting Quantum Interference Device), elaborated a theory where the environment is also assumed to have a linear response to the perturbation created by the test system. This was followed by a general study of the dynamics of a two level system coupled to a dissipative environment made of a bath of oscillators [359]. Nevertheless, many environments cannot be modeled as such oscillator baths. A general review of the topic, along with detailed analysis of "spin bath" environments (including a two-level system coupled to a bath of two-level systems – the so-called "central spin" model) can be found in the article by Prokof'ev and Stamp [360]. The references [361] and [362] provide a historical introduction to decoherence in a more philosophically oriented point of view. More references in the context of quantum measurement and the determination of the "pointer basis" will be given in §9.1.2.

Experiments on decoherence have been done with a variety of quantum systems. For instance, superpositions of photon coherent quantum states with different phases have been studied in Reference [363], providing information about the time constants associated with the decoherence process in this case. Another interesting quantum system is provided by a coherent superposition of opposite persistent currents in a SQUID [364]; further studies have shown that the decoherence process can be mastered and studied precisely [365]. Decoherence has also be experimentally studied in crystals of quantum molecular magnets [366]; in this case, theory predicts the existence of three main contributions to decoherence: phonons, nuclear spins, and intermolecular interactions, which is a good illustration of the large variety of possible decoherence processes.

In a series of lectures given in 1961 at Caltech [367], Feynman pointed out that gravitation can also produce effects that are equivalent to decoherence. This is because gravitational waves can introduce distortions of space-time that change the paths in an interferometer, and therefore modify the measured differential phase shift. If the fluctuations of this shift are sufficiently large during the time of measurement, the interference effect is blurred, which amounts to the effect of a quantum decoherence. For example, the presence of a constant stochastic background of gravitational waves determines the maximum size of objects that can be used in matter-wave interferometry [368]. It turns out that the scattering of stochastic gravitational waves, rather than the scattering of photons, dominates the decoherence of the motion of planets [369], due to the large effective temperature of the gravitational wave background (even if this effect produces a negligible damping). In general relativity, gravitational fields can also rotate the polarization of light along its direction of propagation; in principle, the fluctuations of gravitational waves can therefore change quantum correlations in a photon EPR experiment, but this is an extremely small effect [159].

7.3.4 Purification, distillation

Distributing entangled states among partners may play an important role in several processes, such as quantum cryptography or teleportation, which we will study in §8.2. But, to realize this distribution, it is not sufficient to have a source that emits pairs of particles in a strongly entangled state. The reason is that the particles must propagate toward Alice and Bob, who, in many applications, may be in remote places; this propagation may then change the state of the pair, transforming it either into a different pure state or a statistical mixture, and reduce the entanglement. The problem that then arises is to find a method to restore the initial entanglement, and to obtain particles that return to a state of complete entanglement. For this purpose, the only possible actions are local operations: each partner, in his/her laboratory, can modify the property of his/her particles by applying appropriate evolutions (for instance, they can apply a magnetic field to locally rotate one spin, or a phase plate to change the polarization of a photon), or perform local measurements with results that can be communicated to the other partner by classical channels.

It has been remarked that this operation is indeed compatible with the rules of quantum mechanics [370, 371], provided one agrees to reduce the number of useful pairs available to Alice and Bob. The loss of quality of entanglement is, so to say, converted into a reduction of the efficiency of the source, but then one is sure that the remaining pairs are indeed in the requested quantum state. This process is possible only if the initial loss of quality of the entanglement is not too strong. It conserves the entropy of entanglement, inasmuch as the local Hamiltonian operations performed by Alice and Bob do not change the von Neumann entropies of their respective subsystems. In fact, the efficiency of the process (number of pairs after the entanglement concentration operation divided by the total number of received pairs) is merely given by this entropy. The method can actually be used to produce any state of the two particles, provided the two partners can exchange information through a classical channel and act accordingly. It has a deep relation with the protocols for quantum correction of errors, an essential component of quantum computation (§8.4) [372].

Experimentally, this method has been successfully tested [373], nevertheless with a success rate that is smaller than the maximal rate in theory; the involved protocols are difficult to implement, so that making some compromise is necessary.

7.4 Quantum dynamics of a subsystem

We now study the evolution of a test system S interacting with another (possibly much larger) quantum system B. Under general conditions, we will see that one can obtain a "master equation" giving the time evolution of the partial density operator ρ_S of S. This is in particular the case when the coupling between S and B is sufficiently weak, and for instance when B is a large system playing the role of a thermal reservoir. The master equation gives more quantitative information on the process of decoherence, generally called "transverse relaxation" in this context. It also contains other types of evolution such as "longitudinal relaxation" or "effective Hamiltonian" due to the coupling with B. In order to introduce a master equation in a general and convenient way, it is useful to first introduce the notion of Kraus operators.

7.4.1 Kraus operators

The space of states of S is an N_S-dimensional space $\mathcal{E}_S$ spanned by an orthonormal basis $\{|u_n\rangle\}$, with $n = 1, 2, ..., N_S$. Initially, S is in a state:

$$|\varphi_0\rangle = \sum_{n=1}^{N_S} c_n |u_n\rangle \tag{7.48}$$

while B is in a state $|\Phi_0\rangle \in \mathcal{E}_B$, where $\mathcal{E}_B$ is the space of states of B spanned by the N_B kets of an orthonormal basis $\{|\Theta_q\rangle\}$. We assume that the initial state $|\Psi\rangle$ of the whole system at $t = 0$ is a product:

$$|\Psi\rangle = |\varphi_0\rangle \otimes |\Phi_0\rangle \tag{7.49}$$

Between time $t = 0$ and $t = \tau$, the two systems interact and become entangled; we call $|\Psi'\rangle$ their state at time τ after interaction. We wish to study state $|\Psi'\rangle$ and the corresponding reduced density operator ρ'_S of system S.

7.4.1.a A first calculation

We begin with a simple calculation in order to define N_B operators M_q acting in the space of states of S, called Kraus operators. In the next section, we refine this definition and show how the number of Kraus operators can be limited in a way that depends only on the dimension N_S of $\mathcal{E}_S$, instead of N_B.

Assume that the initial state of S is one of the $|u_n\rangle$:

$$|\Psi\rangle = |u_n\rangle \otimes |\Phi_0\rangle \tag{7.50}$$

The corresponding final state $|\Psi'\rangle$ can then be expanded over the basis of the tensor products $|u_{n'}\rangle \otimes |\Theta_q\rangle$ as:

$$|\Psi'\rangle_n = \sum_{n'=1}^{N_S} \sum_{q=0}^{N_B-1} x_{n',q}^{(n)} |u_{n'}\rangle \otimes |\Theta_q\rangle \tag{7.51}$$

(For reasons that will become clear in §7.4.3, it is convenient to let the values of index q range from 0 to $N_B - 1$). We then have:

$$|\Psi'\rangle_n = \sum_{q=0}^{N_B-1} |\bar{u}_q^{(n)}\rangle \otimes |\Theta_q\rangle \tag{7.52}$$

with:

$$\left|\bar{u}_q^{(n)}\right\rangle = \sum_{n'=1}^{N_S} x_{n',q}^{(n)} \left|u_{n'}\right\rangle \qquad (7.53)$$

The states $\left|\bar{u}_q^{(n)}\right\rangle$ of S appearing in this expansion are neither necessarily orthogonal nor normalized. We can introduce the N_B linear operators M_q acting in $\mathcal{E}_S$ defined by:

$$M_q \left|u_n\right\rangle = \left|\bar{u}_q^{(n)}\right\rangle \qquad (7.54)$$

for any value of n. Equation (7.52) then becomes:

$$\left|\Psi'\right\rangle_n = \sum_{q=0}^{N_B-1} M_q \left|u_n\right\rangle \otimes \left|\Theta_q\right\rangle \qquad (7.55)$$

Assume now that the initial state of S is any state (7.48) in $\mathcal{E}_S$. By linearity, the state after interaction can be written as:

$$\begin{aligned}
\left|\Psi'\right\rangle &= \sum_{n=1}^{N_S} c_n \sum_{q=0}^{N_B-1} M_q \left|u_n\right\rangle \otimes \left|\Theta_q\right\rangle \\
&= \sum_{q=0}^{N_B-1} M_q \left|\varphi_0\right\rangle \otimes \left|\Theta_q\right\rangle
\end{aligned} \qquad (7.56)$$

Since the unitary interaction between S and B does not change the norm of the state vector, this state is still normalized to 1. But by construction the $\left|\Theta_q\right\rangle$ are orthonormal; we therefore have:

$$\sum_{q=0}^{N_B-1} \left\langle\varphi_0\right| M_q^\dagger M_q \left|\varphi_0\right\rangle = 1 \qquad (7.57)$$

for any $\left|\varphi_0\right\rangle$. This can only be true[22] if:

$$\sum_{q=0}^{N_B-1} M_q^\dagger M_q = 1 \qquad (7.58)$$

7.4.1.b An upper limit of the number of Kraus operators

The summations over q in the preceding expressions contain N_B terms; this situation may become intractable when the dimension N_B of the space of states of the quantum system b is very large (for instance if B is a macroscopic thermal reservoir). We now show that these summations can actually be limited to a maximum

[22] Operator $\sum_{q=0}^{N_B-1} M_q^\dagger M_q$ is Hermitian, and can therefore be diagonalized. Relation (7.57) then shows that all its eigenvalues are necessarily equal to 1, which means that the operator itself is equal to the identity operator acting in $\mathcal{E}_S$.

of $(N_S)^2$ terms, that is the square of the dimension of the space of state of S. The general idea is that the Kraus summation does not need to run over a basis spanning the whole space of states $\mathcal{E}_B$ of B; it is sufficient to span the subspace $\mathcal{E}'_B$ that is accessible from $|\Phi_0\rangle$ by interaction with S.

In the expression (7.51) of $|\Psi'\rangle_n$, instead of first performing the summation over index n', let us begin by a summation over index q and write:

$$\left|\overline{\Theta}_{n'}^{(n)}\right\rangle = \sum_{q=0}^{N_B-1} x_{n',q}^{(n)} |\Theta_q\rangle \tag{7.59}$$

We then obtain:

$$|\Psi'\rangle_n = \sum_{n'=1}^{N_S} |u_{n'}\rangle \otimes \left|\overline{\Theta}_{n'}^{(n)}\right\rangle \tag{7.60}$$

If now the initial state of S is not $|u_n\rangle$ but the general superposition (7.48), the state after interaction reads:

$$|\Psi'\rangle = \sum_{n=1}^{N_S} c_n \sum_{n'=1}^{N_S} |u_{n'}\rangle \otimes \left|\overline{\Theta}_{n'}^{(n)}\right\rangle \tag{7.61}$$

The $(N_S)^2$ kets $\left|\overline{\Theta}_{n'}^{(n)}\right\rangle$ span a subspace $\mathcal{E}'_B$ of $\mathcal{E}_B$ with dimension $N'_B \leq (N_S)^2$. In this subspace, we choose an orthonormal basis of N'_B kets $|\Xi_q\rangle$. By expanding all the $\left|\overline{\Theta}_{n'}^{(n)}\right\rangle$ onto this basis, we obtain:

$$|\Psi'\rangle = \sum_{n=1}^{N_S} c_n \sum_{n'=1}^{N_S} \sum_{q=0}^{N'_B-1} \left\langle \Xi_q \left|\overline{\Theta}_{n'}^{(n)}\right\rangle \right| |u_{n'}\rangle \otimes |\Xi_q\rangle \tag{7.62}$$

Now, if we introduce the N'_B operators M_q by:

$$M_q |u_n\rangle = \sum_{n'=1}^{N_S} \left\langle \Xi_q \left|\overline{\Theta}_{n'}^{(n)}\right\rangle \right| |u_{n'}\rangle \tag{7.63}$$

we obtain:

$$|\Psi'\rangle = \sum_{n=1}^{N_S} c_n \sum_{q=0}^{N'_B-1} M_q |u_n\rangle \otimes |\Xi_q\rangle$$

$$= \sum_{q=0}^{N'_B-1} M_q |\varphi_0\rangle \otimes |\Xi_q\rangle \tag{7.64}$$

This expression has the same form as (7.56), but the summation is now restricted to a number of terms N'_B that is smaller or equal to $(N_S)^2$. This range of summation may be much smaller if the dimension of the space of states $\mathcal{E}_B$ is very large. The unitarity relation (7.58) still holds, for the same reasons as before. Note also that the Kraus operators are not uniquely defined: there is a large flexibility for the choice of the basis $|\Xi_q\rangle$ in the subspace of dimension N'_B, which may result in different M_q operators.

7.4.2 Density operator, Kraus sum

The density operator of system S after the interaction is given by a trace over the states of B:

$$\rho'_S = Tr_B \left\{ |\Psi'\rangle\langle\Psi'| \right\} \tag{7.65}$$

where $|\Psi'\rangle$ is given by (7.64). Since the $|\Xi_q\rangle$ are orthonormal, we have:

$$\rho'_S = \sum_{q=0}^{N'_B-1} M_q |\varphi_0\rangle\langle\varphi_0| M_q^\dagger \tag{7.66}$$

Therefore, the density operator of S after interaction can be expressed as a function of its value $\rho_0 = |\varphi_0\rangle\langle\varphi_0|$ before entanglement as:

$$\rho'_S = \sum_{q=0}^{N'_B-1} M_q \rho_0 M_q^\dagger \tag{7.67}$$

where the Kraus operators obey the normalization condition (7.58) and where $N'_B \leq (N_S)^2$. The right-hand side of (7.67) is called a "Kraus sum"; it provides the general expression of the partial density operator of S at the end of the process.

The Kraus operators therefore characterize, not only the way the state of the whole system has evolved from an initial product state (7.49), but also the evolution of the density operator ρ_S of the test system S. If, initially, the state of system S is defined by a density operator that is a statistical mixture and not a pure state, a simple linear superposition shows that relation (7.67) is still valid with the same operators M_q. If, initially, system B is also defined by a statistical mixture, a linear superposition can still be used. Nevertheless, since the definition of the Kraus operators depends on the initial state $|\Phi_0\rangle$, adequate averages should be introduced with the square roots of the probabilities associated with the initial states of B. Physically, it is not surprising that the Kraus operators M_q, which give the evolution of the partial density operator ρ_S, should depend on the state of the physical system B to which S is coupled; for instance, if B is a thermal bath, the evolution of ρ_S depends on the temperature of this bath.

7.4.3 Master equation, Lindblad form

We now assume that the interaction between the two systems is weak and, for the sake of simplicity, we ignore their proper evolution between times $t = 0$ and $t = \tau$. The evolution operator $U(0, \tau)$ is then not very different from 1, and $|\Psi'\rangle$ not very different from $|\Psi\rangle$. If we choose a basis $\{|\Theta_q\rangle\}$ having $|\Phi_0\rangle$ as its first vector, we have:

$$|\Psi'\rangle \simeq |\Psi\rangle = |\varphi_0\rangle \otimes |\Theta_0\rangle \tag{7.68}$$

and the contribution of value $q = 0$ is then dominant in (7.56), with $M_0 \simeq 1$. We can write M_0 as:

$$M_0 = 1 + J + iK \tag{7.69}$$

with the following definitions of J and K:

$$J = \frac{M_0 + M_0^\dagger}{2} - 1 \qquad K = i\frac{M_0^\dagger - M_0}{2} \tag{7.70}$$

They are both Hermitian operators, with small values if the interaction is weak. To first order in this interaction, we have:

$$M_0^\dagger M_0 = 1 + 2J + \dots \tag{7.71}$$

and the contribution of the term $q = 0$ in the right-hand side of (7.67) is:

$$\rho_0 + [J, \rho_0]_+ + i[K, \rho_0] + \dots \tag{7.72}$$

where $[C, D]_+$ is the anticommutator $CD+DC$ and $[C, D]$ the commutator $CD-DC$. If we insert (7.71) into (7.58), we obtain:

$$1 + 2J + \sum_{q=1}^{N'-1} M_q^\dagger M_q = 1 \tag{7.73}$$

or:

$$J = -\frac{1}{2} \sum_{q=1}^{N'-1} M_q^\dagger M_q \tag{7.74}$$

To first order, we can then write (7.67) in the form:

$$\rho'_S - \rho_0 = i[K, \rho_0] + \sum_{q=1}^{N'-1} \left(M_q \rho_0 M_q^\dagger - \frac{1}{2} M_q^\dagger M_q \rho_0 - \frac{1}{2} \rho_0 M_q^\dagger M_q \right) \tag{7.75}$$

where $\rho_0 = |\varphi_0\rangle \langle\varphi_0|$ is the initial density operator of S.

Now we assume that, during time dt, system S interacts in succession with ndt identical systems B, all identical and initially in the same state $|\Phi_0\rangle$ (n has the dimension of an inverse time). We then obtain the variation $d\rho_S$ of ρ_S during this time in the form:

$$\frac{d\rho_S}{dt} = \frac{-i}{\hbar}\left[H_{eff}, \rho_S\right] + \sum_{q=1}^{N'-1}\left(L_q \rho_S L_q^\dagger - \frac{1}{2}L_q^\dagger L_q \rho_S - \frac{1}{2}\rho_S L_q^\dagger L_q\right) \qquad (7.76)$$

where H_{eff} is the "effective Hamiltonian" acting on S:

$$H_{eff} = -\hbar n K \qquad (7.77)$$

and L_q the "Lindblad operator":

$$L_q = \sqrt{n}M_q \qquad (7.78)$$

Differential equation (7.76) is called a "master equation" for the evolution of the partial density operator ρ_S. The right-hand side term has a general form called a "Lindblad form". The number of Lindblad operators can exceed neither the square of the dimension of the space of states of the test system, nor the dimension of the space of states of the coupled system B.

This type of master equation occurs in many physical situations. Its derivation can be generalized in several ways:

(i) If S is initially described by a density operator ρ_0 corresponding to a statistical mixture, instead of a pure state, this operator can be expanded as:

$$\rho_0 = \sum_m p_m \left|\varphi_0^m\right\rangle\left\langle\varphi_0^m\right| \qquad (7.79)$$

The above reasoning applies to each component $\left|\varphi_0^m\right\rangle\left\langle\varphi_0^m\right|$ of this operator and leads to (7.75) with ρ_0 replaced by $\left|\varphi_0^m\right\rangle\left\langle\varphi_0^m\right|$. Linearity then implies that equations (7.75) and (7.76) are still valid.

(ii) If B is also described by a density operator corresponding to a statistical mixture and if the initial density operator of the whole system is the product of this operator by ρ_0, the situation is different since the M_q operators depend on the initial state $|\Phi_0\rangle$ of B. Nevertheless, Lindblad forms can still be obtained in this case (see for instance §4.3 of [374]).

(iii) We have assumed that system S interacts in succession and for a short time with many other systems B. A more frequent situation is that S constantly (but weakly) interacts with a single system B, for instance a heat bath. In such a case, one can show that similar pilot equations can be obtained, provided the correlation times associated with B are sufficiently short. Physically, it is plausible that a big system with short correlation times should be equivalent to a series of independently prepared systems. We refer for instance to chapter IV of [375] for more details.

(iv) The proper time evolution of systems S and B can be included in the equations.

8

Applications of Quantum Entanglement

Quantum entanglement does not only provide a field of fundamental studies; it can also be harnessed as a powerful tool for applications. In this chapter, after introducing two general theorems that are useful in the context of this discussion (§8.1), we propose a few examples: quantum cryptography (§8.2), teleportation (§8.3), and quantum computing and simulation (§8.4). For the moment, only quantum cryptography has given rise to real applications, and has been used in practical (and even commercial) applications; its purpose is the sharing of cryptographic keys between several partners by using a protocol where privacy is guaranteed by fundamental laws of physics. As for quantum computation, it is based on the general manipulation of quantum information. It may be a more futuristic field of research in terms of applications, but it is certainly a domain of intense activity throughout the world.

Strictly speaking, none of these subjects in itself brings a really new view on the interpretation of quantum mechanics. Nevertheless, in addition to their strong intrinsic interests, they provide a very direct and particularly interesting application of its basic principles. This is the reason why we study them in this chapter even if this may be, in a sense, a slight digression. We will therefore only summarize the main ideas; the interested reader is invited to read the proposed references.

8.1 Two theorems

The two theorems that follow are similar; one deals with the creation and duplication of quantum states, the other with their determination.

8.1.1 No-cloning

The duplication of a quantum state, often called "quantum cloning", is an operation where one starts from one particle in an arbitrary quantum state and reaches a situation where two particles are in the same state $|\varphi\rangle$. However, it turns out that, within the laws of quantum mechanics, this operation is fundamentally impossible, a property often called the "no-cloning theorem" [376, 377]. The proof involves the linearity and the unitarity of the evolution of the state vector (the norm and the scalar product between states are conserved during the evolution).

We assume that the complete system contains:

– the "source", initially in an arbitrary state $|\varphi\rangle$

– the "target", a physical system having the same space of states as the source (or an isomorphic space of states). Initially, the target is in a normalized state $|\xi_0\rangle$ and should be transferred into the same state $|\varphi\rangle$ as the source.

– an environment, with any space of states, and which is initially in the normalized state $|\Phi_0\rangle$.

We wish to study what kind of evolution can transform the initial state:

$$\left|\Psi_i\right\rangle = |\varphi\rangle \otimes |\xi_0\rangle \otimes |\Phi_0\rangle \tag{8.1}$$

into a final state $\left|\Psi_f\right\rangle$ given by:

$$\left|\Psi_f\right\rangle = |\varphi\rangle \otimes |\varphi\rangle \otimes \left|\Phi_f(\varphi)\right\rangle \tag{8.2}$$

We now consider two different values $|\varphi_1\rangle$ and $|\varphi_2\rangle$ of $|\varphi\rangle$, associated with initial states $\left|\Psi_i^{(1)}\right\rangle$ and $\left|\Psi_i^{(2)}\right\rangle$ for the whole system; according to (8.1), the scalar product of these two states is:

$$\left\langle\Psi_i^{(1)}\middle|\Psi_i^{(2)}\right\rangle = \langle\varphi_1|\varphi_2\rangle \tag{8.3}$$

After evolution, the product scalar should become, according to (8.2):

$$\left\langle\Psi_f^{(1)}\middle|\Psi_f^{(2)}\right\rangle = [\langle\varphi_1|\varphi_2\rangle]^2 \left\langle\Phi_f(\varphi_1)\middle|\Phi_f(\varphi_2)\right\rangle \tag{8.4}$$

This can be equal to (8.3) only in two cases:

– either:

$$\langle\varphi_1|\varphi_2\rangle = 0 \tag{8.5}$$

– or:

$$\langle\varphi_1|\varphi_2\rangle \times \left\langle\Phi_f(\varphi_1)\middle|\Phi_f(\varphi_2)\right\rangle = 1 \tag{8.6}$$

If this second possibility occurs, since all vectors are normalized, each of the scalar products in this expression has a modulus that is less than, or equal to, 1 (Schwarz inequality); the equality necessarily implies that the vectors themselves are equal (within an irrelevant phase factor). Relation (8.6) therefore implies at the same time that $|\varphi_1\rangle = |\varphi_2\rangle$ and $\left|\Phi_f(\varphi_1)\right\rangle = \left|\Phi_f(\varphi_2)\right\rangle$, while we have assumed previously that states $|\varphi_1\rangle$ and $|\varphi_2\rangle$ are different. We then reach a contradiction, and the only remaining possibility is the first: the scalar product $\langle\varphi_1\,|\varphi_2\rangle$ should vanish. Unitarity therefore requires that, if state $|\varphi_1\rangle$ can be cloned, the only other states $|\varphi_2\rangle$ that can be cloned are orthogonal; it is impossible to clone arbitrary linear combinations of source states.

Suppose that we release one of the assumptions, namely that state $|\varphi\rangle$ is invariant in the whole process; can we then make cloning possible? We now assume that, instead of putting two systems into the initial state $|\varphi\rangle$ of the source, the process puts them both into another state $|\overline{\varphi}\rangle$ that is a function of $|\varphi\rangle$ given by:

$$|\overline{\varphi}\rangle = \overline{U}\,|\varphi\rangle \tag{8.7}$$

where $\overline{U}$ is a unitary operator. Equation (8.2) is then replaced by:

$$|\Psi_f\rangle = |\overline{\varphi}\rangle \otimes |\overline{\varphi}\rangle \otimes |\Phi_f(\varphi)\rangle \tag{8.8}$$

After all, this would also be a useful sort of cloning, since from the knowledge of $|\overline{\varphi}\rangle$ one could infer that of $|\varphi\rangle$. But this process also is forbidden by the rules of quantum mechanics. To see why, it is sufficient to apply the same reasoning as before to the unitary operator obtained by multiplying the evolution operator by the product $\overline{U}^{-1}$(source) $\times \overline{U}^{-1}$(target); one then obtains the same equations and reaches the same contradictions. Releasing the invariance of the state of the source therefore does not help.

If it is impossible to clone states exactly, is it possible at least to do it in some approximate way? We now show that this is impossible as well. Without writing the strict equality (8.2) for the final state, let us now assume a weaker condition:

$$|\Psi_f\rangle = |\overline{\varphi}\rangle \otimes |\overline{\varphi}\rangle \otimes |\Phi_f(\varphi)\rangle \tag{8.9}$$

where $|\overline{\varphi}\rangle$ and $|\widetilde{\varphi}\rangle$ are good approximations of the initial target state $|\varphi\rangle$. The conservation rule of the scalar product then gives:

$$\langle\varphi_1\,|\varphi_2\rangle = \langle\overline{\varphi}_1\,|\overline{\varphi}_2\rangle\,\langle\widetilde{\varphi}_1\,|\widetilde{\varphi}_2\rangle\,\langle\Phi_f(\varphi_1)\,|\Phi_f(\varphi_2)\rangle \tag{8.10}$$

Since $\langle\overline{\varphi}_1\,|\overline{\varphi}_2\rangle \simeq \langle\varphi_1\,|\varphi_2\rangle$, we necessarily have
 – either:

$$\langle\varphi_1\,|\varphi_2\rangle \simeq 0 \tag{8.11}$$

 – or:

$$\langle\overline{\varphi}_1\,|\overline{\varphi}_2\rangle \times \langle\Phi_f(\varphi_1)\,|\Phi_f(\varphi_2)\rangle \simeq 1 \tag{8.12}$$

which implies that $\langle\overline{\varphi}_1\,|\overline{\varphi}_2\rangle \simeq 1$ and, by symmetry, that $\langle\widetilde{\varphi}_1\,|\widetilde{\varphi}_2\rangle \simeq 1$. As a consequence, the states that can be approximately cloned are necessarily either almost orthogonal, or almost identical; again, arbitrary linear combinations cannot be cloned.

8.1.2 *No single-shot state determination*

A similar theorem is the following: given a single quantum system in a state $|\varphi\rangle$, it is impossible to determine $|\varphi\rangle$ exactly with any sequence of measurements. The reason is that, whatever first measurement is performed on the system, the information provided by the knowledge of the result is only that the state vector $|\varphi\rangle$ is not orthogonal to the corresponding eigenstate of measurement; a probability cannot be obtained from a single result. Moreover, no information whatsoever is obtained about the relative phases of the components of $|\varphi\rangle$ on these eigenvectors. Further measurements would therefore be necessary to determine $|\varphi\rangle$. But the state vector of the system has been modified by the first measurement (state vector reduction), so that additional measurements have access only to this modified state, which makes the accurate determination of the initial state $|\varphi\rangle$ impossible.

This theorem is necessary to ensure consistency with the no-cloning theorem: if it were possible to determine the quantum state with arbitrary accuracy, one could then construct a filter (Stern–Gerlach system for spins) to put an arbitrary number of particles into the same state $|\varphi\rangle$, in violation of the no-cloning theorem.

The no-determination theorem is valid only when a single realization of the quantum system is given: if many copies of the system in state $|\varphi\rangle$ are available, then it becomes possible to determine this state better and better, with an accuracy that is an increasing function of the number of copies. Different methods have been proposed, in particular "quantum tomography", which relies on the successive measurements of incompatible observables with several realizations of the same quantum state, and a mathematical reconstruction of the most probable initial quantum state [378–380]; it has been used in many experiments. See also the discussion of chapter 15 of [381] and the method of so-called "measurements of quantum weak values" (§9.3.1) [382–384], which has been used to determine the "wave function" of single photons with the appropriate use of small birefringent phase shifter, a Fourier filter, and polarization measurements [385].

A related interesting theoretical result is the following: if several copies of the initial system are available and described by the same density operator ρ (pure state or statistical mixture), it turns out that a single experimental setup is sufficient to determine the quantum state [386]. A condition is that a larger auxiliary quantum system is allowed to interact in a controlled way with the initial system; at the end of the process, one measures a factorized observable acting on the two systems. The interaction with the auxiliary system (sometimes called "ancilla") transforms the information related to noncommuting observables in the initial system into commutative information for the total quantum system.

8.1.3 Consequences on signaling

If state cloning was possible in an EPR experiment, Bob could apply it to the particle he receives after Alice has made her measurement along direction *a*. He could then put many particles in the same quantum state as the received particle, and then determine their common state, as we have seen in §8.1.2. But the direction of polarization of all these copies of the same particle would reveal the direction Alice has chosen to make her measurement, so that Bob could learn Alice's choice. This method could therefore be used to transmit information from Alice to Bob, without any minimal delay proportional to the distance, and therefore in contradiction with relativity (no-signaling condition discussed in §5.4). The impossibility of cloning is therefore essential to ensure compatibility between quantum mechanics and relativity. Appendix F gives a more detailed discussion of why signal transmission is impossible, even for instance if the experiment is repeated many times. See [387] for a historical discussion of this theorem, and [388] for a study of multiple cloning.

8.2 Quantum cryptography

Quantum cryptography is not, as one could believe from these words, a method of cryptography to code or decode messages by quantum methods. The purpose of the method is only the transmission of a secret key between two partners through quantum measurements, the subsequent use of this key remaining classical. This explains why the more accurate name "quantum key distribution" is often used for it.

8.2.1 Sharing cryptographic keys

The method makes use of the specific properties of quantum systems to ensure the transmission of a cryptographic key between two partners, without any risk of a third person having access to the key by monitoring the transmission line. The basic idea is to build a scheme that leads to a perfectly secure way to transmit a cryptographic key – such a key is a random sequence of 0 and 1, which is used to code, and then decode, a secret message. In a first step, the two remote correspondents Alice and Bob share this key, and then use it to code[1] and decode all

[1] The simplest method is to write the message with a binary coding of its characters, and then to perform a binary sum of each bit of the message with the corresponding bit of the key (the bit with the same rank); the result, which looks totally random and is therefore completely unreadable, is sent by Alice to Bob. He can then perform the same binary addition again and recover the initial message.

messages they exchange. If the key is perfectly random, and if each of its elements is used only once (which means that the key should be at least as long as the messages themselves), it becomes totally impossible for anyone to decode any message without knowing the key, even if the message is sent publicly.

But the risk is that, during the initial process of key communication, a spy (traditionally named Eve, after "eavesdropping") is able to have access to the communication line used by Alice and Bob in order to learn the key; from this moment, she will be able to decode all messages sent with this key. The exchange of keys is therefore a crucial step in cryptography. The usual strategy is to make the best possible use of the traditional methods of confidentiality – storing the key in a safe, secure transportation, etc. – but it is always difficult to assess the safety of such methods where human factors play such an important role.

By contrast, the quantum key distribution relies on fundamental laws of physics, which are impossible to break: as clever and inventive spies may be, they will not be able to violate the laws of quantum mechanics! The basic idea is that Alice and Bob will create their common key by making quantum measurements, for instance on particles in an EPR correlated state; in this way they can generate series of random numbers that can be subsequently used as a secret communication key. What happens then if Eve tries to intercept the photons during the process of key creation? For example, she could couple some elaborate optical device to the optical fiber where the photons propagate between Alice and Bob, and then make measurements. If Eve just absorbs the photons she measures, this immediately changes the correlation properties observed by Alice and Bob, and the spying process can be detected. Of course, this does not necessarily stop the spying, but at least Alice and Bob know which data have been perturbed and can use only the others in the future as a perfectly safe cryptographic key.

8.2.2 Examples of protocols for key exchange

Quantum cryptography is now a big field of research. There is a large variety of schemes and protocols for quantum cryptography, some making use of series of single-particle events [389] as in the BB84 protocol, others involving several entangled particles in an EPR quantum state [390, 391] . For a general introduction, see [392]; for a review with more details, see [393]. Some of the schemes have been implemented in practice, and provide the distribution of keys over distances exceeding 100 kilometers. Here, we will restrict ourselves to a few examples.

8.2.2.a BB84 protocol

We assume that Alice sends to Bob single photons, one by one, either in state $|H\rangle$ with horizontal polarization to signal a bit 0 of the key, or in state $|V\rangle$ with vertical polarization to signal a bit 1 (Figure 8.1). Bob, when he measures the polarization of the photons he receives, can reconstruct the key, but there is no guarantee that Eve has not been able to interfere and obtain the value of the bit. This is because, when Eve does the same polarization measurement as Bob in her laboratory along the transmission line, even if she has to absorb the photon, she can also resend another photon with the polarization corresponding to the result of her measurement. Bob, when measuring the polarization of the photons he then receives from Eve, obtains exactly the same result as if Eve had not made any measurement; neither he nor Alice can know that they have been spied upon.

The BB84 (for Bennett and Brasssard, 1984 [389]) protocol provides a way to suppress this risk. The protocol involves two different bases for the polarizations, that of the two states $|H\rangle$ and $|V\rangle$ (horizontal and vertical polarizations) and a second basis with two other states:

$$
\begin{aligned}
|A\rangle &= \tfrac{1}{\sqrt{2}}\,[|H\rangle + |V\rangle] \\
|B\rangle &= \tfrac{1}{\sqrt{2}}\,[|H\rangle - |V\rangle]
\end{aligned}
\tag{8.13}
$$

These two states correspond to the two linear polarizations at $\pm 45°$ with respect to polarizations $|H\rangle$ and $|V\rangle$. By convention, $|H\rangle$ and $|A\rangle$ are associated with a bit of the key equal to 0, while $|V\rangle$ and $|B\rangle$ are associated with a bit equal to 1 (this is just a possibility; any other convention associating 0 with one state of each basis, and 1 with the two others, would work equally well).

A crucial point is that, to send each bit of the key, Alice must choose in a completely random way to use either basis $\{|H\rangle, |V\rangle\}$ or basis $\{|A\rangle, |B\rangle\}$. On his side, Bob also chooses his direction of measurement completely randomly, so that there is a 50% chance that his basis is different from Alice's. In this case, relations (8.13) then show that he can obtain his two results with the same probability $1/2$, independently of the bit sent by Alice, so that he receives no information. But if, by chance, he chooses the same basis, the information is contained in his measurement. The protocol therefore has two steps: first, Alice sends particles to Bob and, second, and only later, they communicate with each other publicly (through a classical channel, which does not have to be secret) to exchange information about what basis they used. Of course, they do not communicate any information on their results, otherwise privacy would be lost! This allows Bob to retain only the bits corresponding to cases where the two bases are the same (perfectly correlated results), and to reject all the others. The secret key is then made of the retained results only, with a number of bits that is approximately half the number of particles sent by Alice. This method may look complicated, but we will see that it ensures that any attempt by Eve to get access to the data becomes detectable by Alice and Bob.

What can Eve do to have access to the key sent by Alice? While the particles

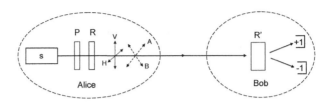

Figure 8.1 Scheme of the BB-84 protocol for secure exchange of a cryptographic key. Alice sends photons one by one to Bob and, for each of them, randomly chooses a polarization among four possible values: H (horizontal), V (vertical), A (first diagonal at 45°), and B (second diagonal at −45°). For this purpose, she uses two optical devices P and R that produce these four possible polarizations. By convention, H and A correspond to a bit 0, V and B to a bit 1. Bob, on his side, uses an optical system R′ to measure the polarizations of the photons he receives in succession; for each measurement, he chooses randomly the basis of polarizations (H,V) or (A,B), and obtains a result ±1. How the shared secret key is extracted from these results is discussed in the text and summarized in Figure 8.2.

are travelling from Alice to Bob, she may attempt to capture each of them and to perform a measurement of their polarizations; to remain unnoticed, each time she measures a particle, she should resend towards Bob another particle with the polarization corresponding to the result she has obtained. But, before Alice and Bob communicate about what basis they choose, no information is available on this basis, so that Eve does not know if she must analyze the particle with basis $\{|H\rangle, |V\rangle\}$ or basis $\{|A\rangle, |B\rangle\}$. We have seen in §8.1 that she cannot determine the quantum state of a single particle, and moreover that the no-cloning theorem prevents her from duplicating the received particle into several in the same state. Actually, as Bob, Eve has one chance out of two to choose the wrong basis. If for instance Alice used the first basis and Eve the second, Eve will receive a state $|H\rangle$ or $|V\rangle$ but resend a particle in one of the states $|A\rangle$ or $|B\rangle$, which changes the flux of particles received by Bob and creates anomalies. Since there is a chance in two that Alice will choose the wrong basis, and also a chance in two that this will change the result observed by Bob, the perturbations created by Alice introduce an anomaly in 25% of the cases.

Alice's polarizations	↕ V	↗ A	↕ V	↔ H	↖ B	↖ B	↔ H	↕ V
Bits sent	1	0	1	0	1	1	0	1
Bob's measurement polarisation basis	↕↔	✕	✕	↕↔	↕↔	✕	✕	↕↔
Bob's results	1	0	0	0	1	1	0	1
Bits of common key	1	0		0		1		1

Figure 8.2 A typical run for a secret key exchange experiment. Alice sends her photons one by one by choosing randomly among four polarizations H, V, A, and B; then Bob measures the received polarization, either in the (H,V) basis or in the (A,B) basis. After the measurement is performed, they publicly communicate the choice of basis, either (H,V) or (A,B), for each pair, but not the results. The common key is then obtained from the results for which the two bases turned out to be the same; all results where the basis are different are ignored. In the case shown, only five bits are retained, but by repeating the experiment many times one can obtain a key of arbitrary length.

The strategy used by Alice and Bob is therefore to sacrifice some of the "good" bits that they have obtained when their bases coincide, to exchange them publicly, and to check if they are always identical as expected (Figure 8.2). It does not matter if Eve learns the values of these bits, which will not be used anymore. Alice and Bob compare their results and, if they differ by roughly 25%, know that someone has been spying on their communication. Whatever Eve can do inevitably modifies the correlation properties of the photons measured at the two ends of the line, so that Alice and Bob can become aware of the problem – provided they take the trouble to carefully compare their data and correlation rates.

Of course, our discussion is simplified, and only gives the general principles. In practice, when Alice and Bob choose the same basis, the coincidence between their results is not perfect, even in the absence of Eve. The reason is that various perturbations may occur during the propagation of the particles and during the measurements. If the corresponding error rate does not exceed a given value, one can use the techniques associated with classical error correction codes (as those used when reading CD or computer hard disks, for instance parity controls); the size of the key is then reduced even more, but random errors are very efficiently reduced. For a discussion of various methods of "privacy amplification" and "quantum secret growing", see [393].

The general scheme works with any sort of quantum particles but, for practical reasons, they are often photons propagating inside optical fibers. A difficulty is that the fibers may change the polarizations of the photons and destroy the correlations used to build the key. To overcome this problem, Bennett [394] has proposed another technique where time resolution is used, and where both Alice and Bob use a Mach–Zehnder interferometer with adjustable phase shift in one arm. Each photon emitted in Alice's laboratory can reach Bob's detectors through several different paths, two of which interfere in the time pulse he measures. The quantum interference effect that then takes place is very similar to the interference between different polarizations in the previous scheme, and a similar protocol of quantum measurements (involving noncommuting operators) can be used to transmit secure keys. This scheme was realized in an experiment in 1993 [395] and is still in use today.

8.2.2.b EPR protocol

In the preceding protocol, the quantum particles carrying the information are sent by one of the two partners and received by the other. It is also possible to design protocols that are more symmetric, where both receive particles belonging to pairs emitted by a common source. A particularly interesting case occurs when these pairs are in an entangled state such as that used in the Bell theorem; this makes a connection between the completeness of quantum mechanics (or its nonlocality), as discussed by EPR, and the quantum distribution of keys [390]. To emphasize the similarity with the discussion of the Bell inequalities in §4.1, we now come back to a situation with two spin $1/2$ particles, initially in the singlet state written in (4.1); but the transposition in terms of vertical or tilted polarizations of photons is straightforward.

The protocol to be used in this case remains rather similar to the BB84 protocol; it also gives a central role to the random choice of basis, but this time both Alice and Bob make a completely random choice of the components of the spins they measure. One can assume that they limit themselves to two directions of measurement, either along axis Oz, or along a perpendicular axis Ox. As above, two situations may occur: either they make different choices and their results have no correlation; or they make the same choice and, for an ideal experiment, the correlations are perfect. In this case also, Alice and Bob communicate only after the reception and measurement of all particles, and only about the choice of directions they have made – never about the results obtained. Each partner only retains the cases where it happens that the two choices have coincided. With this protocol, often called the "EPR protocol", the cryptographic key that Alice and Bob extract from their measurements is completely random: none of them can choose the series of bits of the key (while, in the BB84 protocol, Alice may do it), but this difference has no effect on the subsequent transmission of messages.

The effects of Eve's intrusions are similar to those in the BB84 protocol: since she does not know which base is used by Alice and Bob, any action from her implies a significant probability that the correlations observed by the two partners will be changed (in the cases when they have chosen the same basis). More precisely, when Eve performs a measurement on a spin and finds a result, the other spin is projected onto an eigenstate of the spin component that she has chosen, with the opposite eigenvalue. If Eve sends on the line another spin having the polarization that she has measured, and if by chance Alice and Bob choose a direction of measurement that is parallel to Eve's, they will observe a perfect anticorrelation of their results, exactly as if Eve had done nothing; her action is not visible. But if she chooses a perpendicular direction to measure the spin, there is one chance in two that they will observe the same result, which would be impossible in the absence of Eve's perturbation; an error rate of 25% then appears. As in the BB84 protocol, the result is that, if Alice and Bob decide to sacrifice a fraction of their results to check that no perturbation has occurred, they can easily verify that their communication has remained confidential.

Moreover, they can use a broader range for the directions of measurement, in particular those leading to a quantum violation of the BCHSH inequalities [139]; this provides another test to check that no action has been performed on the spins between the source and the measurements. In fact, the direction chosen by Eve and the result she observes play a role that is very similar to an additional variable λ, which determines the polarization properties of the particles received by Alice and Bob. The average over λ then leads very naturally to the BCHSH inequalities (in fact, one is precisely in the case studied in Appendix C). Checking that significant violations of the inequalities are still obtained is therefore a good criterion for the absence of Eve's intermediate measurements; this illustrates an interesting relation between the Bell theorem and quantum cryptography.

Conceptually, the EPR protocol nevertheless remains rather different from BB84. In the EPR protocol, when Eve can attempt to intrude into the system, the information that Alice and Bob will use to build the key still does not exist: all possible results of all spin components are still potentially present in the singlet entangled state, since no measurement has been performed – in other words, the useful information does not exist yet. If Eve attempts to acquire some knowledge, it is (so to speak) she who takes the responsibility of projecting the state vector and attributing definite polarizations to the spins; it is rather intuitive that this operation should be detectable by Alice and Bob.

One can go even further in this direction and conceive of schemes where, for the events that are useful for creating the key, no particle at all propagates along the line between Alice and Bob where Eve could intervene. Such a scheme has already been discussed at the end of §2.4 – see also [83]. This particular protocol involves one single particle only, and constructs the key from events where the particle has remained in Alice's laboratory; it never propagated between Alice's and Bob's locations, because of a destructive quantum interference effect. This is an extreme case where, for all events that are retained for constructing the key, no particle has propagated between the two partners, which prevents Eve from obtaining any useful information.

Other protocols have been proposed. It is possible for instance to use protocols making use of six quantum states, or individual quantum systems having a space of states with dimension larger than 2; again, for more details, we refer the reader to the review article [393].

8.3 Teleporting a quantum state

The notion of quantum teleportation [396] is also related to quantum nonlocality; the idea is to take advantage of the correlations between two entangled particles, which initially are for instance in an "all-or-nothing" state (6.16) (for $N = 2$), in order to reproduce at a distance the arbitrary spin state of a third particle. The scenario is the following (Figure 8.3): initially, two entangled particles propagate towards two remote regions of space; one of them reaches the laboratory of the first actor, Alice, while the second reaches that of the second actor, Bob; a third particle in an arbitrary quantum state $|\varphi\rangle$ is then provided to Alice in her laboratory; the final purpose of all this scenario is to put Bob's particle into exactly the same state $|\varphi\rangle$, whatever it is (without, of course, transporting the particle itself). One then says that state $|\varphi\rangle$ has been teleported.

What procedure can be followed to realize teleportation? One could naively think that the best strategy for Alice is to start by performing a measurement on the particle in state $|\varphi\rangle$ to be teleported. But this is not true: one can show that it is more efficient to perform a "combined measurement" that involves at the same time this particle as well as her particle from the entangled pair. In fact, for the teleportation process to work, an essential feature of this measurement is that no distinction between the two particles used by Alice should be made. With photons one may for instance, as in [397], direct the particles onto opposite sides of the same optical beam splitter, and measure on each side how many photons are either reflected or transmitted; this device does not allow one to decide from which initial direction the detected photons came, so that the condition is fulfilled. Then, Alice communicates the result of the measurement to Bob; this is done by some classical method such as telephone, e-mail, etc., in other words by a method that is not instantaneous but submitted to the limitations due to the finite velocity of light. Finally, Bob modifies the state of his particle by applying a local unitary transformation that depends on the classical information he has received. This operation puts his particle into exactly the same state $|\varphi\rangle$ as the initial state of the third particle, which realizes the "teleportation" of the state. The whole scenario is "hybrid" because it involves a combination of transmission of quantum information (through the entangled state) and classical information (the phone call from Alice to Bob). Here we discuss only the general ideas, without explicitly writing the calculations that show how the measurement process modifies the state vector, and which unitary transformation Bob should apply, even if these calculations are not very complicated. A more detailed description can be found in the original reference [396], or for instance in §9.8 of [398], or again in §6.5.3 of [399].

Teleportation may look either magical, or trivial, depending how one looks at it. The possibility of reproducing at a distance a quantum state from classical information is not in itself a big surprise. Suppose for instance that Alice decided to choose what the teleported state should be, and filtered the spin (she sends particles through a Stern–Gerlach system until she gets a $+1$ result[2]); she could then ask Bob by telephone to align his Stern–Gerlach filter in the same direction, and to repeat the experiment until he also observes a $+1$ result. This might be called a trivial scenario of teleportation, based only on the transmission of classical information. But real quantum teleportation does much more than this! First, the state that is transported is not necessarily chosen by Alice, but can be completely arbitrary. Second, a message with only binary classical information is used (the result of the combined experiment made by Alice[3]), which certainly does not provide sufficient information to reconstruct a quantum state. In fact, a quantum state depends on continuous parameters, while results of experiments correspond to discrete information only. Somehow, in the teleportation process, the finite binary information has turned into continuous information! The latter, in classical information theory, would correspond to an infinite number of bits.

From Bob's point of view, the amount of received information has two components: classical information sent by Alice, with a content that is completely "un-

---

[2] For filtering a spin state, one obviously needs to use a nondestructive method for detection after the Stern–Gerlach magnet. One could for instance imagine a laser detection scheme, designed in such a way that the atom goes through an excited state, and then emits a photon by returning to the same internal ground state (closed optical pumping cycle – this is possible for well-chosen atomic transition and laser polarization).

[3] Alice sends to Bob two classical bits corresponding to the result of a measurement performed on two particles.

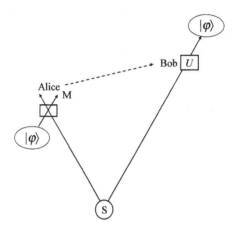

Figure 8.3 A source S emits two spin 1/2 entangled particles, which fly to Alice and Bob. In addition, Alice receives a particle in an arbitrary state $|\varphi\rangle$, which she does not know. She performs a measurement M involving both this particle and one of the particles of the pair emitted by S. She then sends a message to Bob by a classical communication channel (shown by a broken line in the figure) containing the result of her experiment (two bits of classical information since two particles are measured). Bob uses this information to perform a unitary transformation U on his particle, which, then, is transferred into exactly the same state $|\varphi\rangle$ as the initial remote particle. This process is often called "quantum teleportation".

controlled", since it is not decided by her, but just describes the random result of an experiment; quantum information is contained in the teleported state itself (what we will call a "qubit" in the next section). If the teleportation is repeated many times with the same state prepared by Alice, by successive measurements on the teleported particles, Bob will be able to determine its quantum state with arbitrary accuracy, including the direction that was chosen by Alice; he therefore receives a real message from her (for a discussion of the optimum strategy that Bob should apply, see [400]).

If one wishes to describe teleportation in a sensational way, one could explain that, even before Bob receives any classical information, he has already received "almost all the information" on the quantum state, in fact all the controllable in-formation (the content of the classical message is random); this "information" has come to him instantaneously, exactly at the time when Alice performed her combined experiment, without any delay proportional to the distance covered. The rest of information, which is the "difference" between continuous "information" and discrete information, comes only later and is, of course, subject to the minimum delay associated with relativity. But this is based on an intuitive notion of the "difference between quantum/controllable and classical/noncontrollable information" that we have not precisely defined; needless to say, this should not be taken as a real violation of the basic principles of relativity!

Finally, has really "something" been transported in the teleportation scheme, or just information? Clearly, what is transported is not a particle, but a quantum state. The essence of the teleportation process is completely different from any scenario of classical communication between human beings. The relation between quantum teleportation and Bell-type nonlocality experiments is discussed in [401]; see also [402] as well as [403] for a review of recent results, as well as [404] for a teleportation experiment of an N-photon GHZ state with the help of only one pair of entangled particles. Recently, teleportation of photon states over a distance of 16 kilometers in free space has been reported [405].

8.4 Quantum computation and simulation

What we said in this chapter of cryptography and teleportation is even more true concerning quantum computing. This field of research gives rise to a large activity, with new results constantly appearing in an abundant literature; giving an up-to-date report is out of the question here. We will thus limit ourselves to a brief introduction to some of the major ideas, and refer the interested reader who wishes to learn more to References [406–408] or to books [409–412].

8.4.1 General idea

The general idea of quantum computing [413] is to base numerical calculations, not on classical "bits", which can be only in two discrete states (corresponding to 0 and 1 in usual binary notation), but on quantum bits, or "qubits". These qubits are quantum systems, each having access to a two-dimensional space of states; this means that qubits, not only can be in one of the two states $|0\rangle$ and $|1\rangle$, but also in any linear superposition of them. For a single qubit, it is already clear that a continuum of states is a much "larger" ensemble than two discrete states only. For a large collection of bits, the difference is even more pronounced: for classical bits, the dimension of the state space increases linearly with the number of bits (for instance, the state of three classical bits defines a vector with three components, each equal to 0 or 1); for qubits, the dimension increases exponentially (this is a property of the tensor product of spaces; for instance, for three qubits the dimension of space is $2^3 = 8$). If one assumes that a significant number of qubits is available, one gets access to a space state with an enormous "size", where many kinds of interference effects can take place.

Now, if one could somehow make all branches of the state vector "work in parallel" to perform independent calculations, it is clear that one could perform much faster calculations, at least in theory. For instance, suppose that one wishes to calculate the solutions of a system of equations that depend on a parameter; one could imagine an algorithm where the system of qubits goes to a superposition of states associated with all possible solutions of the equations, simultaneously for all values of the parameter. But then the difficulty is to have access to these components: one cannot directly measure the state vector as a classical variable. Nevertheless, appropriate quantum measurement processes can be designed to make some use of this "quantum parallelism", which in fact opens up many possibilities. For instance, the notion of unique computational complexity of a given mathematical problem, which puts limits on the efficiency of classical computers, no longer applies in the same way.

A summary of the history and prehistory of quantum computing can be found for instance in [414]. Feynman, in a conference given in 1981 at MIT, remarked that it seems to be in general impossible to simulate the evolution of any quantum system on a classical computer with reasonable efficiency. This led him to propose a basic model for a quantum computer that would perform this task. In 1985, David Deutsch [413] described a "universal quantum computer", or "quantum Turing machine", that can simulate any other quantum computer, as a classical Turing machine can simulate any classical computer. The whole field, which is presently a domain of very active research, expanded rapidly in the decade 1990–2000 with the introduction of quantum gates and algorithms.

8.4.2 *Quantum gates and algorithms*

A "quantum computer" is generally considered as a combination of basic elements called "logical quantum gates", which are connected with each other in a way that is appropriate for the calculation to be performed. The simplest quantum gates act on a single qubit:

– the "X gate", which transforms $|0\rangle$ into $|1\rangle$, and $|1\rangle$ into $|0\rangle$; its action is represented by the σ_x Pauli matrix.

– the "Hadamard gate H", which acts according to:

$$H\,|0\rangle = \frac{1}{\sqrt{2}}\,[|0\rangle + |1\rangle]$$
$$H\,|1\rangle = \frac{1}{\sqrt{2}}\,[|0\rangle - |1\rangle] \tag{8.14}$$

(If the two basis states $|0\rangle$ and $|1\rangle$ correspond to photon states that are polarized horizontally and vertically, the effect of H is to turn these linear polarizations by 45°). Quantum gates acting on two qubit states also exist, such as the "cNOT gate" (for "controlled not"), which acts on both a control qubit and a target qubit. For an introduction to various gates and a rapid discussion of how they can be used to implement quantum algorithms, see for instance §6.5 of [399] or chapter 8 of [415].

Among quantum algorithms, one often cites first the Shor algorithm [416]. Shor noticed that the factorization of large integer numbers into prime factors may become enormously faster than with classical methods; an introduction to the relation between quantum mechanics and the factorization of integer numbers is given in [417], and with more details in chapter 3 of [410]. The Grover algorithm [418] is a quantum algorithm that allows searching an unsorted data base in a much more efficient way than classical computation (the gain is quadratic as a function of the number of objects in the base); the algorithm is probabilistic since it gives the correct answer with high probability, but the probability of failure can be decreased at will by repeating the algorithm. The Deutsch–Josza algorithm [419] provides another case where the gain offered by the quantum computation is exponential. Similar enhancements of the speed of computation are expected in the simulation of many-particle quantum systems [420]. Depending on the problem, the theoretical gain in speed provided by quantum algorithms is polynomial or exponential, but sometimes there is no gain at all. More recently, a new algorithm has been proposed to obtain useful information concerning the solutions of very large systems of linear equations, by providing approximate values of mathematical quantities depending on the solution [421].

Fundamentally, there are many differences between classical bits and quantum qubits. We have already mentioned that classical bits have two reference states that are fixed once and for all, while qubits can use any orthogonal basis in their space of states, but this is not the only difference by far. Classical bits can be copied at will and ad infinitum, while the no-cloning theorem mentioned in the preceding section (see also Appendix F) applies to qubits. On the other hand, classical bits can be transmitted only into the forward direction of light cones, while the use of entanglement and teleportation may remove this limitation for qubits. Another difference is that the information is less directly coded in quantum qubits than in classical bits: in order to transmit and receive usable information from qubits, one has to specify what kind of measurements should be made with them (this is related to the flexibility concerning the previously mentioned space state basis). Since the quantum measurement process involves randomness, often the algorithm will provide results with randomness also, so that repeating it may be necessary. In the end, as for all human beings, Alice and Bob can communicate only at a classical level, for instance by adjusting the macroscopic settings of their measurement apparatuses and observing the red and green light flashes associated with the results of measurements. Paraphrasing Bohr (see the end of §1.2.3), we could say that "there is no such concept as quantum information; information shared by humans is inherently classical, but may be transmitted through quantum qubits"; nevertheless, the whole field is now sometimes called "quantum information theory".

One of the first proposals of a scheme of a quantum computer involved cold trapped atomic ions, in particular the realization of a cNOT gate; see [332]. Since then, many laboratories have realized experiments to demonstrate the feasibility of quantum computations with ions [333]. In 2001, a group at IBM [422] used the techniques of nuclear magnetic resonance to implement the Shor algorithm, and to factorize the number $N = 15$ into 3×5. In 2011, another group at the Center for Quantum Photonics in Bristol used a quantum optical system to factor $N = 21$ [423]. More recently, a Chinese group in Hefei and Shanghai succeeded in factoring $N = 143$ [424]. Progress is constant, but we are still pretty far from practical applications!

8.4.3 Quantum error correction codes

Decoherence is the big enemy of quantum computation, since it constantly tends to destroy the useful coherent superpositions [408, 425]; this reduces the full quantum information to its classical, Boolean, component (made of diagonal matrix elements only). It is now perfectly clear that a "crude" quantum computer based on the naive use of nonredundant qubits will never work, at least with more than a very small number of them. It has been remarked that this kind of quantum computer would simply be a kind of resurgence of the old analog computers (errors in quantum information form a continuum), in an especially fragile version!

But it has also been pointed out that an appropriate use of quantum redundancy may allow one to design efficient error correcting schemes [426, 427]. Decoherence can be corrected by using a system containing more qubits, and by projecting its state into some subspaces in which the correct information about the significant qubit survives without error [428]; the theoretical schemes involve collective measurements of several qubits, which give access to some combined information on all of them, but none on a single qubit. In other words, the two states of a qubit are no longer merely the plus or minus state of a single spin, but now correspond to two entangled states of many spins; a coherent superposition is not stored in a single quantum system, but in a coherent superposition of these entangled states. By making use of the redundancy contained in the entangled states, it turns out that it is theoretically possible to "purify" these quantum states by applying to them local operations, in order to extract a smaller number of systems in nonperturbed states [429]. Chapter 5 of [410] discusses in more detail how this "miracle" can happen and how error correcting systems can be built from the basic quantum logic gates. For reviews of the subject, see [430, 431].

The "quantum threshold theorem", sometimes also called the "quantum fault-tolerance theorem", indicates the threshold under which the noise of a quantum computer should be in order to be able to simulate a perfect quantum computer [432, 433]. Quantum error correction codes apply to various situations, including quantum computation as well as communication or cryptography [434]. Similarly, the notion of "quantum repeaters" [435] has been introduced to correct for the effect of imperfections and noise in quantum communication. Another very different view of quantum computation has been proposed, based on a semiclassical concept where qubits are still used, but communicate only through classical macroscopic signals, which are used to determine the type of measurement performed on the next qubit [436]. This kind of computer should be much less sensitive to decoherence.

Another approach is to accurately control the coupling of a quantum system to its environment. While, for many forms of dissipation, decoherence tends to destroy the quantum effects that are used for quantum computation, dissipation may also sometimes have the opposite effect. With a good control of the coupling of the system to the outside reservoir, the environment may actually drive the system to a steady state in which the outcome of the quantum computation is encoded [437].

Generally speaking, whether or not it will be possible one day to beat or control decoherence in a sufficiently large system for practical quantum computing still remains to be seen. Moreover, although the factorization into prime numbers is an important question (in particular for cryptography), as well as the many-body quantum problem, it would be nice to apply the principles of quantum computation to a broader scope of problems! The question as to whether or not quantum computation will become a practical tool one day remains open to debate. In any case, this is an exciting new field of research.

8.4.4 Quantum simulation

The purpose of another related field of research is, not to use quantum systems to perform abstract calculations (such as factoring large numbers), but perform simulations. The idea, introduced by Feynman in famous lectures in 1959 and 1981 [438], is to simulate a complex quantum physical system of interest with another quantum system that is more easily accessible. Directly calculating the properties of large physical systems in quantum mechanics, for instance those frequently encountered in condensed matter physics, is known to raise very difficult computational problems. One can then replace the large system with another that behaves in the same way but is more conveniently accessible to experiments and controllable, in order to obtain useful information about the behavior of both systems. In general, two kinds of quantum simulations are possible: digital simulation, where the calculation is made with a quantum computer; analog simulation, where no calculation is made, but just the observation of the first quantum system that mimics the second. For reviews of this field, see [439], [440] and [441]; a detailed discussion of the impact of disorder and noise on the reliability of quantum simulation is given in Reference [442].

A brilliant success of the latter approach has been the direct observation of the phase transition of a quantum gas from a superfluid state to a Mott insulator [443]. In this case, the simulator was a gas submitted to the potential created by an optical lattice, but many other quantum systems have been tested for quantum simulation: trapped ions [333], nuclear spins [444] with very long coherence times, superconducting circuits [445], photons [446], etc. Interesting perspectives are opened by the possibility to simulate field theories, in particular by creating artificial gauge potentials for neutral atoms [447].

9

Quantum Measurement

The process of measurement plays an important role in quantum theory. Measurements can be direct, if the physical system S interacts directly with the measurement apparatus M (as we have assumed until now), or indirect. In the latter case, the physical system first interacts with an ancillary system B, which may have a space of states that is very different from that of S, for instance much larger; after this interaction has finished, M is used to perform a measurement on B, without any direct interaction with A. Because S is then "protected" from any direct interaction with the measurement apparatus, the state of S is not necessarily strongly modified, and may even be only weakly affected. In both cases, the process implies entanglement between several physical systems.

In this chapter, we study how this entanglement is created and used for measurements, as well as the notion of weak and continuous measurements. These questions play an important role in several of the interpretations of quantum mechanics that we discuss in Chapter 11. In §9.1, we study direct measurements and introduce the classical von Neumann model of ideal measurements, which provides a general frame and gives a way to take into account the effects of environment; in §9.1.4, we also discuss the Wigner–Araki–Yanase theorem. In §9.2, we study indirect measurements as well as the important notion of POVM observables. Finally, in §9.3, we study conditional (weak) measurements and continuous measurements, and the special dynamics (Wiener processes) they may introduce.

9.1 Direct measurements

The von Neumann model of quantum measurement [4] provides a general framework for describing the process in terms of correlations appearing (or disappearing) in the state vector associated with the whole system $S + M$. In this model, the two systems S and M are initially described by a product state $|\Psi_0\rangle$ and interact during the time of measurement, so that they become entangled; they then reach a final state $|\Psi'\rangle$ and do not interact anymore.

9.1.1 Ideal measurement, von Neumann model

We assume that the physical quantity measured on S is described by a Hermitian operator A acting in its space of states $\mathcal{E}_S$. This operator has normalized eigenvectors $|a_n\rangle$ associated with eigenvalues a_n (for the sake of simplicity, we assume that they are nondegenerate):

$$A\,|a_n\rangle = a_n\,|a_n\rangle \tag{9.1}$$

Initially, the state $|\varphi_0\rangle$ of S is any linear combination of the $|a_n\rangle$:

$$|\varphi_0\rangle = \sum_{n=1}^{N_S} c_n\,|a_n\rangle \tag{9.2}$$

with arbitrary complex coefficients c_n; the only condition is that the sum of their squared modulus is equal to 1 (normalization condition). As for the measurement apparatus M, we assume that before measurement it is always in the same quantum state $|\Phi_0\rangle$. The initial state of the whole system is then:

$$|\Psi_0\rangle = |\varphi_0\rangle \otimes |\Phi_0\rangle \tag{9.3}$$

9.1.1.a Basic measurement process

Let us first study the special case where, initially, S is in one of the eigenstates associated with the measurement: $|\varphi_0\rangle = |a_n\rangle$. The initial state of the whole system is then:

$$|\Psi_0\rangle = |a_n\rangle \otimes |\Phi_0\rangle \tag{9.4}$$

After the measurement, system S is left in the same state $|a_n\rangle$. Nevertheless, the measurement apparatus reaches a state $|\Phi'_n\rangle$ that is different from $|\Phi_0\rangle$ and depends on n; it has to be so since, to make the result accessible experimentally, the position of its "pointer" used for reading the result on the apparatus (macroscopic pointer moving in front of a scale, state of light emitting diodes, writing of the result into a computer memory, etc.) must depend on n. The various states $|\Phi_n\rangle$ are actually mutually orthogonal, since the pointer necessarily involves a very large number of atoms reaching a different state to make the reading by a macroscopic observer possible[1]. The effect of measurement is therefore summarized by:

$$|\Psi_0\rangle = |a_n\rangle \otimes |\Phi_0\rangle \implies |\Psi'_n\rangle = |a_n\rangle \otimes |\Phi'_n\rangle \tag{9.5}$$

where $|\Phi'_n\rangle$ is a normalized state. No correlation or entanglement has then appeared between the measurement apparatus and the measured system; this corresponds to the simple situation where the result of the measurement is certain.

In the general case, the initial state of system S is a superposition (9.2) of eigenstates associated with the measurement. The state replacing (9.4) is the linear combination with the same coefficients:

$$|\Psi_0\rangle = \sum_n c_n |a_n\rangle \otimes |\Phi_0\rangle \tag{9.6}$$

The linearity of the Schrödinger equation then implies that:

$$|\Psi_0\rangle \implies |\Psi'\rangle = \sum_n c_n |a_n\rangle \otimes |\Phi'_n\rangle \tag{9.7}$$

which is a state where the measurement apparatus is entangled with the measured system S. After the measurement, one can then no longer attribute a state vector to system S (pure state), but only a density operator obtained by partial trace. Since the $|\Phi'_n\rangle$ are mutually orthogonal and normalized, this density operator is given by:

$$\rho'_S = Tr_M \left\{ |\Psi'\rangle \langle \Psi'| \right\} = \sum_n |c_n|^2 |a_n\rangle \langle a_n| \tag{9.8}$$

This result looks very natural: it expresses the fact that the measured system has a probability $|c_n|^2$ to be in each of the states $|a_n\rangle$ associated with the results a_n, which corresponds exactly to Born's probability rule. It is indeed a very useful formula that summarizes some features of the measurement postulate of quantum mechanics. Nevertheless, as we have already emphasized (for instance in §§2.1 and 2.2), it does not contain one essential component of this postulate: the uniqueness of the result of measurement (emergence of macroscopic uniqueness). Indeed, all possible results are still contained in the partial trace, considered as equally possible even after the measurement. The system is described mathematically by a sum containing all possible results of the experiment simultaneously, which seems to be contradictory with the observation of a single result observed in a given realization. This was expected: equation (9.8) is only a consequence of the Schrödinger equation, which is unable to stop the endless progression of the von Neumann chain and to ensure macroscopic uniqueness.

As we have seen in §1.2.2.a, to solve this problem von Neumann introduced

---

[1] This is a necessary but not sufficient condition for M to constitute a good measurement apparatus. The states $|\Phi'_n\rangle$ should also provide a stable recording of the measurement result, which should not quickly be erased under the effect of the proper evolution of M or its coupling to the environment – we come back to this point in §9.1.2.

a specific postulate: the state reduction postulate, which forces the uniqueness of the result of the measurement. According to this postulate, a single result a_r is indeed observed after measurement and, in the summations of (9.7) and (9.8), all components except the component $n = r$ should be suppressed[2]. Under these conditions, the state vector after measurement becomes a product, entanglement has disappeared, and S is in a pure state again.

9.1.1.b Effects of the interaction, pointer observable

Relations (9.5) or (9.7) imply that some interaction has occurred between S and M. We can for instance assume that their interaction Hamiltonian H_{int} has the form:

$$H_{int} = g \, A \, P_M \qquad (9.9)$$

where A is the operator introduced in the preceding discussion (acting on S only), P_M an operator acting on M only, and g a coupling constant. We also assume that, in the space of states of M, operator P_M has a conjugate operator X_M:

$$[X_M, P_M] = i\hbar \qquad (9.10)$$

This means that P_M acts as the infinitesimal translation operator with respect to X_M. In other words, the action of its exponential on any eigenvector $|x_M\rangle$ of X_M:

$$X_M \, |x_M\rangle = x_M \, |x_M\rangle \qquad (9.11)$$

provides a translation of the eigenvalue x_M:

$$e^{-i\Delta x P_M /\hbar} \, |x_M\rangle = |x_M + \Delta x\rangle \qquad (9.12)$$

where Δx is any real number.

We now assume that $|\Phi_0\rangle$ (state of the apparatus before measurement) is an eigenstate of X_M with eigenvalue x_0, and ignore any other source of evolution[3] of the combined system other than the mutual interaction between S and M. The evolution operator between time $t = 0$ before measurement and time $t = \tau$ after the end of interaction is:

$$U(0, t) = e^{-ig\tau A P_M /\hbar} \qquad (9.13)$$

Its application to (9.4) then provides:

$$U(t) \, |a_n\rangle \otimes |\Phi_0(x_0)\rangle = |a_n\rangle \otimes |\Phi(x_0 + g\tau a_n)\rangle \qquad (9.14)$$

---

[2] Nonstandard interpretations of quantum mechanics (additional variables, modal interpretation, nonlinear dynamics, etc.) solve the problem in different ways (Chapter 11).

[3] This is a valid approximation if the interaction time τ is very short and g very large. One could include the independent evolutions of systems S and M by using the interaction representation associated with the sum of their Hamiltonians. Nevertheless, since here we are mostly interested in the effects of their mutual interaction, for the sake of simplicity we ignore the effects of the separate independent evolutions.

where the variables written between brackets in the state of the measurement appa-
ratus[4] refer to the eigenvalues of X_M. In terms of the kets $|\Phi'_n\rangle$ introduced in (9.5),
we therefore have:

$$|\Phi'_n\rangle = |\Phi(x_0 + g\tau a_n)\rangle \tag{9.15}$$

These relations show that, for the measurement apparatus M, the eigenvalue of X
has been translated by an amount $g\tau a_n$ that depends on the eigenvalue a_n of S.
For the measurement apparatus, the observable X_M therefore plays the role of the
position of a "pointer" which indicates the result of measurement after interaction
between both systems.

If, initially, the system is in an eigenstate of A, it will remain in the same eigenstate after measurement. One then says that the measurement is a "quantum nondemolition" measurement[5] (QND). Of course, in general, the system S is not initially
in an eigenstate of A, and the interaction with the measurement apparatus changes
its pure state into a complicated statistical mixture obtained by inserting (9.15) into
(9.7); the usual "von Neumann regress" (§2.1) takes place.

9.1.2 Effects of the environment, pointer states

The state (9.7) of the whole system after measurement is an entangled state, written
in the basis associated with the measurement. But we have noticed in §7.2.1 that,
for any entangled state, several such expansions with different basis are possible
(ambiguity of entanglement). Does this mean that the very nature of the measured
observable in a measurement process is ambiguous as well? This is not the case
because not any interaction process can lead to a measurement; several conditions
are required.

First, we have seen in §9.1.1.b that the measured observable depends on the form
of the interaction between S and the measurement apparatus; in this simple model,
the eigensates of measurement $|a_n\rangle$ are invariant under the effect of this interaction.
A second obvious condition is that the states $|\Phi'_n\rangle$ of the measurement apparatus
M should store the information related to the result of the measurement in a robust
way, and not destroy it immediately under the effect of the proper evolution of M.
This condition is fulfilled if X_M is a constant of motion of M, in other words if X_M
commutes with the proper Hamiltonian H_M of M.

Moreover, a measurement apparatus cannot remain completely isolated from
its environment, even at a microscopic level. This would require that none of its

---

[4] Needless to say, a measurement apparatus is macroscopic and has many more degrees of freedom than the
only position of its pointer. For the sake of simplicity, these degrees of freedom are not made explicit in our
notation.
[5] QND measurements can for instance be performed in quantum optics [448]. In Chapter 10, we discuss other
examples of QND measurements.

macroscopic number of atoms, electrons, etc., interacts with any particle of the environment to become correlated with it in some way (for instance, a photon that is scattered into a different state, depending on the position of the measurement pointer, as in the scheme discussed in §7.3.3.a). One can even remark that, because of its very purpose, the apparatus must be able to interact and correlate with the experimenter, when he acquires the knowledge of the result of measurement. This impossibility has stimulated Zeh [449] and Zurek [450, 451] to push further the analysis, and to include the environment of the measurement apparatus in the process of quantum measurement.

Models have been developed that, at the price of some simplification, lead to tractable calculations. This gives a good idea of the general nature of the physical phenomena occurring during the entanglement with the environment. In the Coleman-Hepp model [452, 453], the system under study is coupled to a semiinfinite linear array of N spin $1/2$ particles at fixed positions, representing the apparatus. The calculation then shows that, if "local" observables are built from a finite number of spin operators, their measurement cannot make the distinction between a coherent superposition of the various positions of the pointer and an incoherent statistical mixture. Following Leggett and Caldeira [358, 359], Zurek [451] considers an environment E made of a large number of harmonic oscillators [454]; as in [360], he also studies the case where the environment is made of an ensemble of spins $1/2$ ("central spin model") that are all coupled to the measurement apparatus M. The apparatus is simplified as well and treated as a two-level system (equivalent to a spin $1/2$). This coupling introduces a Hamiltonian $H_{ME}(i)$ between the ith spin of the environment and the spin of the measurement apparatus; $H_{ME}(i)$ is proportional to the product of the components of the two spins along Oz with a coupling constant g_i, which varies randomly as a function of i. One can then show that a preferred basis exists in the space of states of the measurement apparatus M (as simplified in this model), the "pointer states basis" in which the entanglement with E tends to destroy the coherences. What determines this basis, preferred to all the others, is the form of the coupling Hamiltonian between the measurement apparatus and its environment. For a more realistic measurement apparatus than a single spin, for instance an apparatus that includes a pointer moving in front of a scale, the preferred basis corresponds to states where the position of the pointer is well defined (as opposed, for instance, to the eigenbasis of its momentum, or to any other state that is significantly extended in space). Zurek calls this phenomenon "einselection" (for "environment induced selection").

This is an important idea: the very physical constitution of a measurement apparatus determines the way it is coupled to the environment. The Hamiltonian describing this coupling determines the basis of the pointer states [455]. Actually, if the proper dynamics of the apparatus is neglected, pointer states are merely the eigenvectors of the operator that commutes with the interaction Hamiltonian between the apparatus and the environment [450]; if this dynamics is included, the situation is more complicated. Several necessary conditions therefore determine if a device can be considered as satisfactory to give access to a physical quantity of S; first, the coupling between S and M must be appropriate to transfer the desired information from S to M; second, this information must be recorded in a way that makes it stable in time with respect to both the proper evolution of M and the coupling between M and E; finally, it should also be robust against small perturbations.

Needless to say, from a fundamental point of view, once more the same remark applies: uniqueness does not emerge from this theory. It just determines the basis of states of the apparatus that are sufficiently robust with respect to the environment to store the result of the measurement. An independent postulate is therefore still needed. Zurek [456] (see also §VI-D of [451] and §III-F of [346]) has nevertheless proposed to use the notion of "envariance" (environment-assisted invariance), a symmetry exhibited by correlated quantum systems and related to causality, to describe the nature of statistical ignorance and derive the Born rule and macroscopic classicality, without any notion of measurement or state vector collapse.

9.1.3 Hund paradox

A similar analysis applies to a problem in molecular physics, the so-called "Hund paradox" [457], namely the origin of the stability of chiral states of molecules. Chiral molecules in their ground state have access to two configurations, one left-handed configuration associated with quantum state $|L\rangle$, and one right-handed configuration associated with quantum state $|R\rangle$. A mirror image symmetry transforms each of these states into the other, but none of them is invariant: each has chirality, which means that a solution of molecules in one of these states exhibits "rotary power" (it rotates the plane of polarization of an optical beam that crosses the solution).

These two states are always coupled by the tunnel effect (as for the two symmetrical configurations of the Ammonia molecule), so that the real ground state of the molecule is the symmetrical combination:

$$|G\rangle = \frac{1}{\sqrt{2}} [|L\rangle + |R\rangle] \qquad (9.16)$$

This state is invariant under mirror symmetry, as expected: the nondegenerate ground state of any quantum system has the same symmetry as its Hamiltonian (in this case the Hamiltonian of the molecule, which is invariant under this symmetry). It has no chirality and therefore no rotary power.

Nevertheless, by measuring the rotation of the plane of polarization of light when it crosses the solution, one observes that molecules are "enantiomers": they are either in state $|L\rangle$ or $|R\rangle$, which indeed rotate polarizations in opposite directions, but they are not in the real ground state $|G\rangle$ with no rotary power. One can even sort out molecules of one configuration, $|L\rangle$ or $|R\rangle$ only, to enhance the rotation of the solution. The intriguing question is then: why aren't the molecules in a state with the same symmetry as the Hamiltonian of the molecule, such as state $|G\rangle$? More generally, why can't the molecules be found in any superposition:

$$|D_\pm\rangle = \frac{1}{\sqrt{2}}\left[|L\rangle \pm e^{i\xi}|R\rangle\right] \tag{9.17}$$

since this possibility is a direct consequence of the superposition principle?

The answer to the question involves the coupling to the environment, as before for pointer states. The molecule constantly collides with others or scatters photons so that it correlates with them in a way that is different for states $|L\rangle$ and $|R\rangle$ because interactions are local; those two states therefore provide the preferred basis with respect to the environment, while $|G\rangle$ is a combination of such states and therefore very fragile against decoherence [455, 458]. This explains why state $|G\rangle$ is not observed in the optical properties of the solution.

Actually, if the coupling with the other molecules is sufficiently fast and the tunnel effect relatively slow, the interaction with the environment completely blocks the tunnel effect. Assume a molecule is initially in state $|L\rangle$; if the molecule was completely isolated, it should oscillate between states $|L\rangle$ and $|R\rangle$ under the effect of its internal Hamiltonian, at a frequency fixed by the rate of the tunnel effect. Nevertheless, the effect of the environment blocks this oscillation, so that state $|R\rangle$ is never reached by the molecule (this is sometimes called the "quantum Zeno effect").

9.1.4 The WAY (Wigner–Araki–Yanase) theorem

Wigner showed that the measurement process has limitations in the presence of conservation laws [459] (this reference is translated into English in [460]). The question was later investigated in more detail by Araki and Yanase [461, 462], and for short the result is often called the WAY theorem. The idea is to examine the von Neumann model of measurement more precisely, and to show that it contains difficulties related to the conservation of additive constants of motion; more elaborate models are therefore needed. We begin with a simple case, a 1/2 spin measured in a Stern-Gerlach apparatus, and then generalize the discussion.

9.1.4.a Spin measurement in a Stern-Gerlach experiment

We assume that a spin $1/2$ particle is initially in an orbital state $|u_0\rangle$ describing its external variable and in a state $|+\rangle$ describing its spin the state (eigenstate of eigenvalue $\hbar/2$ of the S_z component of its spin along axis Oz). This particle moves towards a Stern-Gerlach magnet, which is part of a measurement apparatus of the Oz component of the spin. The initial state of the whole system is:

$$|\Psi_0\rangle = |+\rangle \otimes |u_0\rangle \otimes |M_0\rangle \tag{9.18}$$

where $|M_0\rangle$ is the initial state of the measurement apparatus, which in general contains a macroscopic number of variables (all the positions of the particles making the apparatus). It is convenient to introduce the ket $|\Phi\rangle$ describing both the orbital variables of the particle and the measurement apparatus; the initial value $|\Phi_0\rangle$ of this ket is given by:

$$|\Phi_0\rangle = |u_0\rangle \otimes |M_0\rangle \tag{9.19}$$

As in (9.3), the state describing the whole system is then:

$$|\Psi_0\rangle = |+\rangle \otimes |\Phi_0\rangle \tag{9.20}$$

Since the initial state is an eigenstate of the measured projection of the spin, after interaction with the measurement apparatus the state of the whole system is, according to the von Neumann model of §9.1.1:

$$|\Psi'_+\rangle = |+\rangle \otimes |\Phi'_+\rangle \tag{9.21}$$

In this ideal nonperturbing measurement, the spin has not changed its state, and both the orbital state of the particle and the measurement apparatus (including its "pointer" indicating the result of measurement) have reached state $|\Phi'_+\rangle$. Assume now that the initial state of the spin is $|-\rangle$ instead of $|+\rangle$ (a spin oriented in the opposite direction). The initial state is then:

$$|\Psi_0\rangle = |-\rangle \otimes |\Phi_0\rangle \tag{9.22}$$

and the final state:

$$|\Psi'_-\rangle = |-\rangle \otimes |\Phi'_-\rangle \tag{9.23}$$

where $|\Phi'_-\rangle$ is another final state for the orbital variables and the measurement apparatus. We have seen that, if the apparatus provides a readable result, $|\Phi'_-\rangle$ should be orthogonal to $|\Phi'_+\rangle$:

$$\langle \Phi'_+ | \Phi'_- \rangle = 0 \tag{9.24}$$

In general, the initial state of the spin is a linear combination of both spin states:

$$|\varphi_S\rangle = c_1 |+\rangle + c_2 |-\rangle \tag{9.25}$$

Then, the linearity of the Schrödinger equation provides the following value $|\Psi''\rangle$ of the state vector $|\Psi\rangle$ after measurement:

$$|\Psi''\rangle = c_1 |+\rangle \otimes |\Phi'_+\rangle + c_2 |-\rangle \otimes |\Phi'_-\rangle \tag{9.26}$$

The angular momentum $\mathbf{J}$ of the whole system is the sum of the spin angular momentum $\mathbf{S}$ of the particle, its orbital angular momentum $\mathbf{L}$, and the angular momentum $\mathbf{M}$ of the measurement apparatus:

$$\mathbf{J} = \mathbf{S} + \mathbf{L} + \mathbf{M} = \mathbf{S} + \mathbf{T} \tag{9.27}$$

where $\mathbf{T} = \mathbf{L} + \mathbf{M}$ is the angular momentum of the variables associated with the ket $|\Phi_0\rangle$. We assume that the interaction between the particle and the measurement apparatus is invariant under rotation, so that the total angular momentum $\mathbf{J}$ is conserved during evolution. When the state vector evolves from (9.20) to (9.21), because the state of the spin is unchanged, we must have:

$$\langle \Phi_0 | \mathbf{T} | \Phi_0 \rangle = \langle \Phi'_+ | \mathbf{T} | \Phi'_+ \rangle \tag{9.28}$$

Similarly, when the state vector evolves from (9.22) to (9.23), we must have:

$$\langle \Phi_0 | \mathbf{T} | \Phi_0 \rangle = \langle \Phi'_- | \mathbf{T} | \Phi'_- \rangle \tag{9.29}$$

The average values of the angular momentum in states $|\Phi'_+\rangle$ and $|\Phi'_-\rangle$ are therefore equal.

Assume now that the initial state of the spin is an eigenstate of the S_x component of the spin $\mathbf{S}$ along the transverse direction Ox:

$$|\varphi_S\rangle = \frac{1}{\sqrt{2}} [|+\rangle + |-\rangle] \tag{9.30}$$

with:

$$|\Psi_0\rangle = |\varphi_S\rangle \otimes |\Phi_0\rangle \tag{9.31}$$

The state of the whole system after measurement is then:

$$|\Psi''\rangle = \frac{1}{\sqrt{2}} \left[|+\rangle \otimes |\Phi'_+\rangle + |-\rangle \otimes |\Phi'_-\rangle \right] \tag{9.32}$$

In the initial state (9.31), the average value of the Ox component J_x of the total angular momentum is:

$$\langle \Psi_0 | J_x | \Psi_0 \rangle = \langle \Psi_0 | S_x | \Psi_0 \rangle + \langle \Psi_0 | T_x | \Psi_0 \rangle = \frac{\hbar}{2} + \langle \Phi_0 | T_x | \Phi_0 \rangle \tag{9.33}$$

where T_x is the Ox component of $\mathbf{T}$. In the final state $|\Psi''\rangle$, because of the orthogonality relation (9.24), we have:

$$\langle \Psi'' | S_x | \Psi'' \rangle = 0 \tag{9.34}$$

so that the transverse component of the spin has decreased by $\hbar/2$ during the interaction with the measurement apparatus. On the other hand, the average value of T_x is unchanged, since:

$$\langle\Psi''|T_x|\Psi''\rangle = \frac{1}{2}\left[\langle\Phi'_+|T_x|\Phi'_+\rangle + \langle\Phi'_-|T_x|\Phi'_-\rangle\right] = \langle\Phi_0|T_x|\Phi_0\rangle \quad (9.35)$$

where we have used (9.28) and (9.29) to obtain the second equality. The consequence is that:

$$\langle\Psi''|J_x|\Psi''\rangle = \langle\Phi_0|T_x|\Phi_0\rangle \quad (9.36)$$

This result which implies a contradiction with (9.33): the final transverse angular momentum is smaller (by $\hbar/2$) than the initial angular momentum. This is due to the disappearance of the coherence of the spin introduced by the orthogonality condition (9.24). In §9.1.4.c, we discuss how this contradiction can be solved.

9.1.4.b Generalization

We now generalize the preceding considerations to any measurement, using the method of Reference [463]. We call J an "additive conserved quantity" :

$$J = S + T \quad (9.37)$$

The word "additive" means that J is the sum of an operator S acting in $\mathcal{E}_S$ (the space of states of the measured system) and of an operator T acting in the space of states of the measurement apparatus. The word "conserved" means that, during the interaction between the system and the measurement apparatus, J is a constant of the motion: if $U(t)$ is the evolution operator associated with this interaction, the operator J obeys the relation:

$$J = U^\dagger(t)\,J\,U(t) \quad (9.38)$$

As previously, J may be for instance a component of the angular momentum or of the linear momentum, but also the conserved electric charge of the system, the baryon number (the third of the difference between the number of quarks and antiquarks), etc.

We call $\{|a_n\rangle\}$ (with $n = 1, 2, ..N_S$) an orthonormal basis of $\mathcal{E}_S$ formed with the eigenstates associated with the measurement. The state of the whole system before measurement is a linear combination of the states $|\Psi_n\rangle$ defined as:

$$|\Psi_n\rangle = |a_n\rangle \otimes |\Phi_0\rangle \quad (9.39)$$

where $|\Phi_0\rangle$ is normalized. Relation (9.5) expresses that this state evolves to a state $|\Psi'_n\rangle$ after measurement given by:

$$|\Psi'_n\rangle = U(t)|\Psi_n\rangle = |a_n\rangle \otimes |\Phi'_n\rangle \quad (9.40)$$

where, as we have seen in §9.1.1, the states of the measurement apparatus $|\Phi'_n\rangle$ are normalized (the evolution is unitary) and orthogonal for different results:

$$\langle \Phi'_n | \Phi'_p \rangle = \delta_{n,p} \tag{9.41}$$

Now, if we take the matrix element of the left-hand side of relation (9.38) between the bra $\langle \Psi_n |$ and the ket $| \Psi_p \rangle$, we obtain:

$$\langle \Psi_n | J | \Psi_p \rangle = \langle a_n | S | a_p \rangle + \delta_{n,p} \langle \Phi_0 | T | \Phi_0 \rangle \tag{9.42}$$

while the matrix element of the right hand side is:

$$\langle \Psi'_n | J | \Psi'_p \rangle = \delta_{n,p} \langle a_n | S | a_p \rangle + \delta_{n,p} \langle \Phi'_n | T | \Phi'_p \rangle \tag{9.43}$$

Equality (9.38) requires that these two matrix elements should be equal. It can therefore be obeyed only if:

$$\langle a_n | S | a_p \rangle \propto \delta_{n,p}$$
$$\langle \Phi'_n | T | \Phi'_n \rangle = \langle \Phi_0 | T | \Phi_0 \rangle \quad \text{for any } n \tag{9.44}$$

The second relation is the equivalent of equalities (9.28) and (9.29). The first relation indicates that the off-diagonal matrix elements of S must vanish in the basis $\{|a_n\rangle\}$ of the eigenstates of measurement. In other words, the two operators S and A (the measured observable) must commute. We therefore obtain the WAY theorem: in a perfect von Neumann measurement, the measured observable must commute with any additive conserved quantity.

9.1.4.c Discussion

We have now reached a situation that looks paradoxical: since the three components of the angular momentum are conserved, the WAY theorem seems to indicate that any measured quantity must commute with the three components of the spin. But no component of the spin is in this case, which would imply that the measurement of any component of a spin is impossible! Of course, this is not what the theorem means. It only means that the von Neumann scheme of a perfect measurement is an idealization. It assumes at the same time that the measurement is perfect (the measurement apparatus provides a perfectly accurate measurement, so that its states $|\Phi'_n\rangle$ after measurement must be perfectly orthogonal) and nonperturbing (the state of the measured system, if it is initially an eigenstate of the measurement, does not change at all during the process). Araki and Yanase [461, 462] as well as other authors [463, 464] have studied the conditions under which imperfect measurements (with apparatuses made of a very big number of particles) or disturbing measurements can remain compatible with conservation laws.

Assume for instance that (9.40) is replaced by:

$$|\Psi'_n\rangle = U(t)|\Psi_n\rangle = |a'_n\rangle \otimes |\Phi'_n\rangle \tag{9.45}$$

with:

$$|a'_n\rangle = |a_n\rangle + \varepsilon |u_n\rangle \tag{9.46}$$

where ε is a small number. We will limit the calculation to first order in ε; if the scalar product of $|u_n\rangle$ and $|a_n\rangle$ vanishes, $|a'_n\rangle$ is normalized to this order. Then:

$$
\langle \Psi'_n | J | \Psi'_p \rangle = \delta_{n,p} \langle a_n | S | a_p \rangle + \delta_{n,p} \langle \Phi'_n | T | \Phi'_p \rangle
$$
$$
+ \varepsilon \langle a_n | u_p \rangle \langle \Phi'_n | T | \Phi'_p \rangle + \varepsilon \langle u_p | a_n \rangle \langle \Phi'_p | T | \Phi'_n \rangle + \varepsilon^2 .. \tag{9.47}
$$

Because the matrix elements $\langle \Phi'_n | T | \Phi'_p \rangle$ and $\langle \Phi'_p | T | \Phi'_n \rangle$ relate to a macroscopic system, they can be much larger than those of S, so that the correction added by the terms in ε may be important. A very small nonorthogonality of the final states of S may bring a large correction to the right-hand side of (9.43), and allow a conservation of the mean value of J. General discussions of the consequences of the WAY theorem may be found in [465, 466].

9.2 Indirect measurements

To introduce the notions of indirect measurements and of POVM, we begin with a simple case: a two-level system coupled to another quantum system B with an arbitrarily large space of states.

9.2.1 A simple model: two-level system

We assume that the space of states $\mathcal{E}_S$ of S is two dimensional and call $|+\rangle$ and $|-\rangle$ two states forming an orthonormal basis in $\mathcal{E}_S$. Between times $t = 0$ and $t = \tau$, system S is coupled to another system B called "ancillary system", with space of states $\mathcal{E}_B$ spanned by an orthonormal basis $|\Phi_m\rangle$ (with $q = 0, 1, ..., N_B$; the dimension N_B may be much larger than 2). At time τ, a measurement is performed on the ancillary system, without any direct interaction with S; we ignore the proper evolution of S and B during this time interval, and take into account only their interaction (otherwise we should use the interaction representation with respect to the sum of the proper Hamiltonians of both systems, which would complicate the calculations).

9.2.1.a Interaction and entanglement

The interaction Hamiltonian is:

$$H_{int.} = g\,\sigma_z(S)\,\Xi(B) \tag{9.48}$$

where g is a coupling constant, $\sigma_z(S)$ the operator acting in the space of states of S (defined by the third Pauli matrix in the $\{|\pm\rangle\}$ basis), and $\Xi(B)$ an arbitrary operator acting on B only. Initially, the ensemble of $S + B$ is in the state:

$$|\Psi_0\rangle = |\varphi_0\rangle \otimes |\Phi_0\rangle \tag{9.49}$$

where $|\varphi_0\rangle$ is any state in $\mathcal{E}_S$:

$$|\varphi_0\rangle = \alpha\,|+\rangle + \beta\,|-\rangle \tag{9.50}$$

Both $|\varphi_0\rangle$ and $|\Phi_0\rangle$ are normalized. Since the proper evolutions of S and B are ignored, the evolution operator between $t = 0$ and $t = \tau$ is:

$$U(0, \tau) = e^{-ig\tau\,\sigma_z(S)\,\Xi(B)/\hbar} \tag{9.51}$$

If S is initially in one of the states $|+\rangle$ or $|-\rangle$, relation (9.5) becomes:

$$|\Psi_0\rangle = |\pm\rangle \otimes |\Phi_0\rangle \implies |\Psi'\rangle = |\pm\rangle \otimes \left|\Phi_B^\pm\right\rangle \tag{9.52}$$

with:

$$\left|\Phi_B^\pm\right\rangle = e^{\mp ig\tau\Xi(B)/\hbar}\,|\Phi_0\rangle \tag{9.53}$$

The state of S is then unchanged, and will not be affected by whatever measurement is performed on B, since the two systems are not entangled. Note that $\left|\Phi_B^\pm\right\rangle$ are normalized vectors (they are obtained by the action of unitary operators onto $|\Phi_0\rangle$) but that they are in general not mutually orthogonal: for instance, if the coupling constant g tends to zero, they both tend to the same vector $|\Phi_0\rangle$ in the space of states of B.

In the more general case where S is initially in state (9.50), the ensemble $S + B$ after interaction is described by the entangled state:

$$|\Psi'\rangle = U(0, \tau)\,|\varphi_0\rangle\,|\Phi_0\rangle = \alpha\,|+\rangle\left|\Phi_B^+\right\rangle + \beta\,|-\rangle\left|\Phi_B^-\right\rangle \tag{9.54}$$

so that the effect of the measurement performed on B may change the state of S, as we now discuss.

9.2.1.b Measurement on the ancillary system

At some time $t \geq \tau$, a measurement is performed on system B; the possible results b_m, corresponding to the orthonormal kets $|\Theta_m\rangle$, are assumed to be nondegenerate for simplicity. Expanding over the corresponding basis allows us to write $|\Psi'\rangle$ as:

$$|\Psi'\rangle = \sum_{m=1}^{N_B} \left[\alpha \langle \Theta_m | \Phi_B^+ \rangle |+\rangle + \beta \langle \Theta_m | \Phi_B^- \rangle |-\rangle \right] \otimes |\Theta_m\rangle \qquad (9.55)$$

where N_B is the dimension of the space of states of system B. If the result of the measurement is b_q, the projection postulate reduces this ket to one of its components in the summation over m, so that it becomes the product:

$$|\Psi'_q\rangle = \left[\alpha \langle \Theta_q | \Phi_B^+ \rangle |+\rangle + \beta \langle \Theta_q | \Phi_B^- \rangle |-\rangle \right] \otimes |\Theta_q\rangle \qquad (9.56)$$

System S is then in a pure state again, state $|\varphi'_q\rangle$ given by:

$$|\varphi'_q\rangle = \alpha \langle \Theta_q | \Phi_B^+ \rangle |+\rangle + \beta \langle \Theta_q | \Phi_B^- \rangle |-\rangle \qquad (9.57)$$

or:

$$|\varphi'_q\rangle = M_q |\varphi_0\rangle \qquad (9.58)$$

where M_q is the operator acting in $\mathcal{E}_S$ defined by:

$$M_q = \langle \Theta_q | \Phi_B^+ \rangle |+\rangle\langle+| \; + \; \langle \Theta_q | \Phi_B^- \rangle |-\rangle\langle-| \qquad (9.59)$$

Here, q is the index labeling the result of the indirect measurement. Every M_q is diagonal in the basis $\{|+\rangle, |-\rangle\}$, but with different (complex) diagonal elements. In the general case, the state of S is indeed changed by the action of the indirect measurement.

In general, M_q is not a projection operator, since its square is not the same operator:

$$\left[M_q \right]^2 = \langle \Theta_q | \Phi_B^+ \rangle^2 |+\rangle\langle+| \; + \; \langle \Theta_q | \Phi_B^- \rangle^2 |-\rangle\langle-| \qquad (9.60)$$

(except in the special case where $|\Theta_q\rangle$ is equal to one of the $|\Phi_B^\pm\rangle$ and orthogonal to the other). For this reason, the operation is often called a "nonprojective measurement", as opposed to projective measurements, such as those discussed in §9.1.

A similar calculation gives:

$$M_q^\dagger M_q = M_q M_q^\dagger = \langle \Phi_B^+ | \Theta_q \rangle\langle \Theta_q | \Phi_B^+ \rangle |+\rangle\langle+| \; + \; \langle \Phi_B^- | \Theta_q \rangle\langle \Theta_q | \Phi_B^- \rangle |-\rangle\langle-| \qquad (9.61)$$

9.2.1.c Probabilities and sum rule

The probability $\mathcal{P}_q$ of obtaining result b_q is nothing but the square of the norm of ket (9.56), which by (9.58) and (9.59) can be simply expressed as an average value over the initial state $|\varphi_0\rangle$ of S:

$$\mathcal{P}_q = \left| \alpha \langle \Theta_q | \Phi_B^+ \rangle \right|^2 + \left| \beta \langle \Theta_q | \Phi_B^- \rangle \right|^2 = \langle \varphi_0 | M_q^\dagger M_q | \varphi_0 \rangle \qquad (9.62)$$

A summation over index q in (9.61) then introduces a closure relation over basis $\{|\Phi_q\rangle\}$, and one obtains:

$$\sum_{q=1}^{N_B} M_q^\dagger M_q = \left\langle \Phi_B^+ \left| \Phi_B^+ \right\rangle |+\rangle\langle+| + \left\langle \Phi_B^- \left| \Phi_B^- \right\rangle |-\rangle\langle-| \right. \right. = |+\rangle\langle+| + |-\rangle\langle-| \quad (9.63)$$

or:

$$\sum_{q=1}^{N_B} M_q^\dagger M_q = 1 \tag{9.64}$$

which means that the sum of probabilities gives 1, as expected for a series of Kraus operators introduced[6] in §7.4.1.a).

As a consequence, the series of operators:

$$\Pi_q = M_q^\dagger M_q \tag{9.65}$$

provides a decomposition of the identity operator in the space of states of S. In general, the Π_q are not mutually orthogonal (the traces of their binary products are not zero), as one can easily see from (9.61). With mutually orthogonal projectors in a two-dimensional space associated with a spin $1/2$, the identity could only be obtained by summing two projectors over two opposite spin directions; it is impossible to build more than two orthogonal projectors in $\mathcal{E}_S$. With the nonorthogonal Π_q, a decomposition of the identity is obtained with N_B operators, the dimension of the space of state of B, which can be much larger than 2. For instance, one can obtain the identity in the space of states of a spin $1/2$ by summing projectors over any series of K spin directions defined by vectors in space having a zero sum, and dividing them by $K/2$. Since, moreover, the Π_q are not necessarily projectors, there is a large flexibility for forming series of operators summing to 1 as in (9.64).

9.2.2 Generalization: POVM

We now assume that the dimension of the space of states $\mathcal{E}_S$ has any value N_S, instead of necessarily 2. We also assume again that the two systems evolve under the effect of their interaction only, with a unitary evolution given for instance by (9.13). The kets $|u_n\rangle$ are eigenstates of this evolution, which changes the state of the whole system according to:

$$|\Psi_0\rangle_n = |u_n\rangle \otimes |\Phi_0\rangle \implies |\Psi'\rangle_n = |u_n\rangle \otimes |\Phi_B^n\rangle \tag{9.66}$$

---

[6] Here, the kets $|\Theta_q\rangle$ are determined by the type of measurement performed on system B; their number, and therefore the number of Kraus operators, is the number of possible results of measurement. It is therefore not possible to reduce this number by the change of basis used in § 7.4.1.b.

If the initial state of S is not $|u_n\rangle$ but a general superposition:

$$|\varphi_0\rangle = \sum_{n=1}^{N_S} c_n |u_n\rangle \tag{9.67}$$

linearity requires that the state $|\Psi'\rangle$ after interaction is:

$$|\Psi'\rangle = \sum_{n=1}^{N_S} c_n |u_n\rangle \otimes |\Phi_B^n\rangle \tag{9.68}$$

9.2.2.a Measurement on the ancillary system

We now assume that the same measurement as before is performed on system B: the possible (nondegenerate) results b_m are associated with the orthonormal eigenvectors $|\Theta_m\rangle$. The ket $|\Psi'\rangle$ can be expanded as:

$$|\Psi'\rangle = \sum_{n=1}^{N_S} \sum_{m=1}^{N_B} c_n \langle\Theta_m|\Phi_B^n\rangle |u_n\rangle \otimes |\Theta_m\rangle \tag{9.69}$$

After a measurement providing result b_q, the projection postulates transforms this ket into the ket $|\Psi_q'\rangle$ given by a single term $m = q$ of the sum over m:

$$|\Psi_q'\rangle = \sum_{n=1}^{N_S} c_n \langle\Theta_q|\Phi_B^n\rangle |u_n\rangle \otimes |\Theta_q\rangle \tag{9.70}$$

Now, if we define the operators M_q acting on $\mathcal{E}_S$ by:

$$M_q |u_n\rangle = \langle\Theta_q|\Phi_B^n\rangle |u_n\rangle \tag{9.71}$$

(they are diagonal but not necessarily real in the basis of the $|u_n\rangle$) and insert (9.67), we obtain:

$$|\Psi_q'\rangle = M_q |\varphi_0\rangle \otimes |\Theta_q\rangle \tag{9.72}$$

9.2.2.b Sum rule and POVM operators

The probability $\mathcal{P}_q$ to obtain result b_q is given by the square of the norm of $|\Psi_q'\rangle$ in the expansion (9.69) of the state vector before measurement. According to (9.72), this probability is:

$$\begin{aligned} \mathcal{P}_q &= \langle\varphi_0| M_q^\dagger M_q |\varphi_0\rangle \langle\Theta_q|\Theta_q\rangle \\ &= \langle\varphi_0| M_q^\dagger M_q |\varphi_0\rangle \end{aligned} \tag{9.73}$$

Since the sum of probabilities is equal to unity, and since $|\varphi_0\rangle$ is any ket in $\mathcal{E}_S$, the sum rule (9.64) is obtained again from this relation. The M_q are Kraus operators (§7.4.1.a).

We define the operators Π_q as:

$$\Pi_q = M_q^\dagger M_q \tag{9.74}$$

The sum rule indicates that their sum is equal to the identity operator:

$$\sum_{q=1}^{N_B} \Pi_q = 1 \tag{9.75}$$

Moreover, the average of each Π_q in any state of S is always a positive number (a probability). The measurements associated with such an ensemble of operators are called a POVM (an acronym for "positive operator valued measure"); by extension, the operators themselves are often called POVM. As in the case where S is a spin $1/2$, the number of operators in a POVM is not equal to the dimension of space $\mathcal{E}_S$, but to the dimension of space for system B, which may be much larger. It may also be much smaller if, for instance, the ancillary system B is a spin $1/2$: in this case, the number of POVM operators acting in the space of states of S cannot exceed 2, whatever the dimension of this space is.

A well-known example of a two-operator POVM acting in a two-dimensional space of states $\mathcal{E}_S$ spanned by the two vectors $|\pm\rangle$ is given by the two Kraus operators:

$$M_1 = |-\rangle\langle-| \qquad M_2 = |-\rangle\langle+| \tag{9.76}$$

leading to the POVM:

$$\Pi_1 = |-\rangle\langle-| \qquad \Pi_2 = |+\rangle\langle+| \tag{9.77}$$

with:

$$\Pi_1 + \Pi_2 = 1 \tag{9.78}$$

If M_0 is a projector, M_1 is not such an operator (its square vanishes); the cross products of these operators are $M_0 M_1 = M_1$ and $M_1 M_0 = 0$.

9.3 Conditional and continuous measurements

Situations where the entanglement produced during the interaction stage between S and M remains weak are appropriate for performing conditional (weak) and continuous measurements.

9.3.1 *Weak values, conditional measurements*

The notion of "weak values" was introduced by Aharonov, Albert, and Vaidman [382] (see also chapter 16 of [381], [383] and [384]). One often speaks of "weak measurements" in this context, but the name "conditional measurements" seems more appropriate. As in §9.1.1.b, let us consider the evolution of the coupled system $S + M$, but in the opposite case where the indication readable on the pointer is inaccurate and provides little information on the state of S. This case is obtained if, in the expression (9.9) of the coupling Hamiltonian, the coupling constant g is small and can be treated to first order only. The initial state of the system is:

$$|\Psi_0\rangle = |\varphi_0\rangle\,|\Phi_0(x_0)\rangle \tag{9.79}$$

where $|\varphi_0\rangle$ has the general form (7.48), expressed in terms of the eigenvectors $|a_n\rangle$ of A. Instead of assuming that $|\Phi_0\rangle$ is an eigenstate of the pointer observable X_M, we now assume that it is a state $|\Phi_0(x_0)\rangle$ with Gaussian components on the eigenvectors $|x_M\rangle$ of this operator; we choose a broad Gaussian function centered around $x_M = x_0$:

$$\langle x_M\,|\Phi_0(x_0)\rangle = \left(2\pi\sigma^2\right)^{-1/4}\,e^{-(x_M-x_0)^2/4\sigma^2} \tag{9.80}$$

where σ is the mean-square fluctuation of the position of the pointer in this initial state. After interaction between time $t = 0$ and time $t = \tau$, the state vector of the whole system is obtained by action on $|\Psi_0\rangle$ of the evolution operator (9.13), and given by[7]:

$$
\begin{aligned}
|\Psi'\rangle &= \left(2\pi\sigma^2\right)^{-1/4} \sum_n c_n \int dx_M\, e^{-(x_M-x_0)^2/4\sigma^2}\, e^{-ig\tau a_n P_M/\hbar}\, |a_n\rangle \otimes |x_M\rangle \\
&= \left(2\pi\sigma^2\right)^{-1/4} \sum_n c_n \int dx_M\, e^{-(x_M-x_0)^2/4\sigma^2}\, |a_n\rangle \otimes |x_M + g\tau a_n\rangle
\end{aligned} \tag{9.81}
$$

With a change of the integration variable x_M into $x = x_M + g\tau a_n$, we obtain:

$$|\Psi'\rangle = \left(2\pi\sigma^2\right)^{-1/4} \sum_n c_n \int dx\, e^{-(x-x_0-g\tau a_n)^2/4\sigma^2}\, |a_n\rangle \otimes |x\rangle \tag{9.82}$$

or, to first order in $g\tau$ (we assume $g\tau a_n \ll \sigma$ for any a_n):

$$|\Psi'\rangle = \left(2\pi\sigma^2\right)^{-1/4} \sum_n c_n \int dx \left[1 + g\tau a_n \frac{d}{dx_0} + ...\right] e^{-(x-x_0)^2/4\sigma^2}\, |a_n\rangle \otimes |x\rangle \tag{9.83}$$

The summation over n of the term in 1 in the bracket reconstructs the ket $|\varphi_0\rangle$ written in (9.2), while the second term in the bracket introduces $A\,|\varphi_0\rangle$. We then obtain:

$$|\Psi'\rangle = \left(2\pi\sigma^2\right)^{-1/4} \left[1 + g\tau A \frac{d}{dx_0} + ...\right] |\varphi_0\rangle \otimes |\Phi_0(x_0)\rangle \tag{9.84}$$

Now, assume that a measurement is performed on S, projecting this system into

---

[7] As in §9.2.1, we consider that the only evolution of the whole system is due to the interaction Hamiltonian between S and M, ignoring the proper evolution of the two isolated systems.

state $\left|\varphi_f\right\rangle$ corresponding to a nondegenerate eigenvalue of some operator acting in $\mathcal{E}_S$. After this measurement, the state becomes:

$$
\begin{aligned}
\left|\Psi''\right\rangle &\sim \left[\left\langle\varphi_f\left|\varphi_0\right\rangle + g\tau\left\langle\varphi_f\left|A\right|\varphi_0\right\rangle\frac{d}{dx_0}\right]\right|\varphi_f\rangle\left|\Phi_0(x_0)\right\rangle \\
&= \left\langle\varphi_f\left|\varphi_0\right\rangle\left[1 + \frac{g\tau\left\langle\varphi_f\left|A\right|\varphi_0\right\rangle}{\left\langle\varphi_f\left|\varphi_0\right\rangle\right.}\frac{d}{dx_0}\right]\right|\varphi_f\rangle\left|\Phi_0(x_0)\right\rangle
\end{aligned}
\tag{9.85}
$$

If the ratio between the matrix elements $\left\langle\varphi_f\left|A\right|\varphi_0\right\rangle$ and $\left\langle\varphi_f\left|\varphi_0\right\rangle$ is real, we finally obtain the simple relation:

$$
\left|\Psi''\right\rangle \sim \left|\varphi_f\right\rangle\left|\Phi_0\left(x_0 + \Delta x_0\right)\right\rangle
\tag{9.86}
$$

with:

$$
\Delta x_0 = g\tau\frac{\left\langle\varphi_f\left|A\right|\varphi_0\right\rangle}{\left\langle\varphi_f\left|\varphi_0\right\rangle\right.}
\tag{9.87}
$$

Δx_0 is the average shift of the position of the pointer for this particular series of events; it can be large if $\left\langle\varphi_f\left|\varphi_0\right\rangle$ has a very small modulus. If the ratio between the matrix elements is complex:

$$
\frac{\left\langle\varphi_f\left|A\right|\varphi_0\right\rangle}{\left\langle\varphi_f\left|\varphi_0\right\rangle\right.} = R + iJ
\tag{9.88}
$$

we set $\Delta x_0 = g\tau R$, and one can show that the additional term in J changes only the phase of the state of M after the measurement. To first order, this corresponds to the substitution:

$$
e^{-(x-x_0-\Delta x_0)^2/4\sigma^2} \implies e^{i\xi(x)}e^{-(x-x_0-\Delta x_0)^2/4\sigma^2}
\tag{9.89}
$$

where $\xi(x)$ is a phase variable[8]. Measurements of the position of the pointer variable are not affected by this change (but a measurement of the conjugate variable P_M would give access to it). In what follows, we therefore limit ourselves to the case where the ratio between the matrix elements is real.

The experimental procedure is therefore the following: first, one prepares S in a state $\left|\varphi_0\right\rangle$ and M in a state $\left|\Phi_0(x_0)\right\rangle$ associated with a wide distribution of the pointer variable x_M; the two systems are left to interact for a short time τ; one then measures S with an observable that has $\left|\varphi_f\right\rangle$ as a nondegenerate eigenstate,

---

[8] To first order, the phase shift $\xi(x)$ is given by:

$$
\xi(x) = g\tau J\, e^{(x-x_0)^2/\sigma^2}\frac{d}{dx_0}\left[e^{-(x-x_0)^2/\sigma^2}\right] = g\tau J\frac{x-x_0}{2\sigma^2}
$$

and one selects only the events where S is found in this state (this is a postselection process); finally, one observes the average shift Δx_0 of the pointer variable of M. The interesting result is that, with this combined scheme of pre- and postselection performed on S, the observed shift can be much larger than any of the shifts $g\tau a_n$ associated with the eigenvalues of A; the amplification factor is the real part of the ratio written in (9.88). Δx_0 is called the "weak value" associated with this measurement [381, 382] – even if its value may be surprisingly large. Note that our first-order calculation implies that the weak value should always remain much smaller than the width σ associated with the initial state of the measurement apparatus[9]. To detect this amplified shift with a reasonable accuracy, a large number of successive measurements is therefore necessary in order to reduce the uncertainty by averaging the individual fluctuations of the observations.

A good amplification factor is obtained if $\langle \varphi_f | \varphi_0 \rangle$ is small but $\langle \varphi_f | A | \varphi_0 \rangle$ large. Situations where $|\varphi_0\rangle$ and $|\varphi_f\rangle$ are almost orthogonal are therefore favorable, but this implies that the postselection process will reject most events, which means that the experiment has to be repeated a very large number of times to provide a significant result. For a spin $1/2$ for instance, one can assume that:

$$|\varphi_0\rangle = |+\rangle \quad \text{and} \quad |\varphi_f\rangle = \varepsilon|+\rangle + \alpha|-\rangle \tag{9.90}$$

with ε real and very small, and α almost equal to 1. If the measurement apparatus M measures the spin Pauli operator $A = \sigma_x$, then:

$$A|\varphi_0\rangle = A|+\rangle = \frac{1}{\sqrt{2}}|-\rangle \tag{9.91}$$

The scalar product of this ket with $|\varphi_f\rangle$ is not infinitesimal when ϵ tends to zero, but almost equal to $1/\sqrt{2}$. As a consequence, the amplification factor (9.88) is $\sim 1/(\varepsilon\sqrt{2})$, which can be arbitrarily large if ε is very small.

Another interesting situation occurs when:

$$|\varphi_0\rangle = |+\rangle \quad \text{and} \quad |\varphi_f\rangle = \frac{1}{\sqrt{2}}[|+\rangle + |-\rangle] \tag{9.92}$$

The spins then start in an eigenstate of σ_z and reach an eigenstate of σ_x, both with eigenvalues 1. We assume that:

$$A = \frac{1}{\sqrt{2}}[\sigma_x + \sigma_z] \tag{9.93}$$

---

[9] In a standard von Neumann measurement (§ 9.1.1), the opposite is true, since one assumes that a single measurement can provide a nonambiguous result; as a consequence, the entanglement between the system S and the measurement apparatus M introduces a complete decoherence between the various eigenstates of S associated with the measured observable A. Here, by contrast, the system S may evolve during interaction as a superposition of these eigenstates that remains (almost) perfectly coherent. This coherence produces a constructive interference effect, which explains why the weak value may be much larger than any value obtained in a standard von Neumann measurement.

which corresponds to a spin measurement along a direction in the xOz plane at 45°
of both axes Ox and Oz. An easy calculation then shows that the amplification factor is $\sqrt{2}$. In other words, the spins interact with M as if their component at 45° was
not ± 1, as for any component of $\vec{\sigma}$, but the geometrical sum $1/\sqrt{2} + 1/\sqrt{2} = \sqrt{2}$
obtained by adding the initial value of the spin with the final value (as if they were
classical perpendicular vectors). A curious property is that this value exceeds the
largest eigenvalue $+1$ of any component of $\vec{\sigma}$. The authors of [384] propose to
interpret this situation in a nonstandard way, by considering that the preselection
and postselection processes, when combined, select spins having two orthogonal
components perfectly well defined (which is of course totally impossible in the
usual formulation of quantum mechanics, where two perpendicular spin components cannot have well-defined values at the same time).

Similar ideas apply to the time evolution of a quantum system S coupled to an
external system B through an Hamiltonian $gQ_S Q_B$, where Q_S acts in the space
of states of S and Q_B in the space of states of B. Preselecting the state of B in a
quantum superposition of eigenstates of Q_B and postselecting its state in another
superposition may lead to a quantum evolution of S that is equivalent to the superposition of different Hamiltonians, and of evolutions at different periods of time
[467].

The general idea of measuring weak values is not restricted to quantum physics,
but applies to any wave theory. In classical optics, for instance, when detecting
the weak field scattered by an object, it is well known that one can enhance the
contrast by inserting almost orthogonal polarizers on the input and output beams.
Reference [468] actually considers that weak values are not inherently quantum,
but rather a purely statistical feature of pre- and post-selection with disturbance.
By contrast, Reference [469] considers that the concept of weak values arises due
to wave interference, so that it has no analog whatsoever in classical statistics;
see also [470]. Another question is whether the amplification associated with conditional measurements can really be used as a practical method to improve the
sensitivity of experiments. Various points of view on this question have been expressed in the literature [471–475]. For instance, Reference [476] concludes that
the frequency dependence of the technical noise in experiments determines the
answer to the question; in experimental settings with time-correlated noise, weak-
value-amplification may outperform conventional measurements, and is actually
near optimal.

A variety of experiments have been made to measure weak values, either in a
regime of classical optics, or in a purely quantum regime involving entanglement
between particles [477–487]. The measurement of weak values has also been used
[488] to test the Leggett–Garg inequalities (§4.2.2) or to obtain a "direct measurement" of a quantum wave function [385].

9.3.2 **Continuous measurements**

We now consider a situation similar to that of §9.3.1 but, instead of assuming that S is postselected in a given state, we calculate the evolution of its state when the position of the pointer of M is measured, still ignoring the proper evolution of the isolated systems S and M. We start from the expression (9.82) of the quantum state of the combined system $S + M$, which yields:

$$|\Psi'\rangle = \left(2\pi\sigma^2\right)^{-1/4} \int dx \, e^{-(x-x_0-g\tau A)^2/4\sigma^2} \, |\varphi_0\rangle \otimes |x\rangle \qquad (9.94)$$

and we assume that a measurement of the position of the pointer of M has given a result in the interval:

$$\left[x_r - \frac{dx_r}{2} \, , \, x_r + \frac{dx_r}{2}\right] \qquad (9.95)$$

where $dx_r \ll \sigma$. After the measurement, the state $|\Psi''\rangle$ is obtained by projecting $|\Psi'\rangle$ over the eigenstates of X_M corresponding to this interval:

$$|\Psi''\rangle = \left(2\pi\sigma^2\right)^{-1/4} e^{-(x_r-x_0-g\tau A)^2/4\sigma^2} \, |\varphi_0\rangle \otimes \int_{x_r-dx_r/2}^{x_r+dx_r/2} dx \, |x\rangle \qquad (9.96)$$

Since this ket is a tensor product, we can attribute a ket $|\varphi''\rangle$ to system S:

$$|\varphi''\rangle \sim e^{-(x_r-x_0-g\tau A)^2/4\sigma^2} \, |\varphi_0\rangle \qquad (9.97)$$

In what follows, we assume that the dimensionless parameter:

$$\varepsilon = \frac{g\tau a}{\sigma} \qquad (9.98)$$

(where a is the largest of the modulus of all eigenvalues of A) is small, and perform a second-order calculation with respect to this parameter; the reason why a first-order calculation is not sufficient will become apparent at the end of §9.3.2.c.

9.3.2.a *Probability of the result*

The probability $\mathcal{P}(x_r)dx_r$ of obtaining a result in the interval (9.95) is given by the square of the norm of $|\Psi''\rangle$, which gives:

$$\mathcal{P}(x_r)dx_r = \left(2\pi\sigma^2\right)^{-1/2} \langle\varphi_0| e^{-(x_r-x_0-g\tau A)^2/2\sigma^2} \, |\varphi_0\rangle \, dx_r \qquad (9.99)$$

By expanding the exponential to second order in ε, we obtain:

$$\begin{aligned}
\mathcal{P}(x_r) &= \left(2\pi\sigma^2\right)^{-1/2} e^{-(x_r-x_0)^2/2\sigma^2} \times \\
&\times \left\{1 + \frac{g\tau}{\sigma}\frac{x_r-x_0}{\sigma}\langle A\rangle_0 + \left(\frac{g\tau}{\sigma}\right)^2 \langle A^2\rangle_0 \left[-\frac{1}{2} + \frac{(x_r-x_0)^2}{2\sigma^2}\right] + \ldots\right\}
\end{aligned} \qquad (9.100)$$

where the averages of A and its square in state $|\varphi_0\rangle$ are defined by:

$$\langle A \rangle_0 = \langle \varphi_0| A |\varphi_0\rangle \quad \text{and} \quad \langle A^2 \rangle_0 = \langle \varphi_0| A^2 |\varphi_0\rangle \tag{9.101}$$

To second order in ε, we then obtain:

$$\mathcal{P}(x_r) = \left(2\pi\sigma^2\right)^{-1/2} \times$$
$$\times \left\{ e^{-(x_r-x_0-\Delta x_0)^2/2\sigma^2} + \left(\frac{g\tau}{\sigma}\right)^2 e^{-(x_r-x_0)^2/2\sigma^2} F(x_r - x_0) + \dots \right\} \tag{9.102}$$

where:

$$\Delta x_0 = g\tau \langle A \rangle_0 \tag{9.103}$$

and where F is an even function[10] of $(x_r - x_0)$. To zero and first order in $\varepsilon = g\tau a/\sigma$, the distribution is given by the first term in the bracket of (9.102), that is a shifted Gaussian distribution, centered around:

$$\bar{x}_r = x_0 + \Delta x_0 \tag{9.104}$$

The second term in the bracket introduces no second-order correction to the average of $(x_r - x_0 - \Delta x_0)$: the term in $(x_r - x_0)$ vanishes because of the parity of F, and the term in Δx_0 gives a third-order correction (at least).

We conclude that the average value of the result of the measurement is not exactly the most probable initial position of the pointer x_0, but a value shifted by an amount Δx_0 given by (9.103); the shift depends on the initial state $|\varphi_0\rangle$ of S. We then introduce the dimensionless random variable ξ_r defined by:

$$\xi_r = \frac{x_r - x_0 - \Delta x_0}{\sigma} \tag{9.105}$$

which characterizes the random result of the measurement, has zero average value, and a unit mean-square deviation.

9.3.2.b Evolution of the state

With this notation, the state (9.97) of S after measurement becomes:

$$|\varphi''\rangle \sim e^{-[\xi_r - \frac{g\tau}{\sigma}(A-\langle A\rangle_0)]^2/4} |\varphi_0\rangle \tag{9.106}$$

[10] The expression of F is:

$$F = -\frac{1}{2}\langle A^2 \rangle_0 + \frac{(x_r - x_0)^2}{2\sigma^2}\left[\langle A^2 \rangle_0 - (\langle A \rangle_0)^2\right]$$

or, by expanding the exponential to second order:

$$|\varphi''\rangle \sim e^{-(\xi_r)^2/4} \times$$
$$\times \left\{ 1 + \frac{g\tau}{2\sigma}\xi_r [A - \langle A \rangle_0] + \left(\frac{g\tau}{2\sigma}\right)^2 \left[\frac{(\xi_r)^2}{2} - 1\right][A - \langle A \rangle_0]^2 + ... \right\} |\varphi_0\rangle$$

(9.107)

We can ignore the prefactor $e^{-(\xi_r)^2/4}$, and normalize this ket by calculating the square of the norm of the ket written in the second line of this expression, that is the product of this ket by the associated bra. This product introduces average values in the initial state $|\varphi_0\rangle$, which we note $\langle \ \rangle_0$ as previously. The first-order terms both contain $(\langle A \rangle_0 - \langle A \rangle_0)$ and vanish; the zero and second-order terms then give:

$$1 + \left(\frac{g\tau}{2\sigma}\right)^2 \langle [A - \langle A \rangle_0]^2 \rangle_0 \left[(\xi_r)^2 + 2\left[\frac{(\xi_r)^2}{2} - 1\right]\right] + ...$$

(9.108)

To obtain a normalized state, we must therefore multiply the ket by:

$$\left\{ 1 + \left(\frac{g\tau}{2\sigma}\right)^2 \langle [A - \langle A \rangle_0]^2 \rangle_0 \left[2(\xi_r)^2 - 2\right] + ... \right\}^{-1/2}$$

(9.109)

or:

$$\left\{ 1 - \left(\frac{g\tau}{2\sigma}\right)^2 \langle [A - \langle A \rangle_0]^2 \rangle_0 \left[(\xi_r)^2 - 1\right] + ... \right\}$$

(9.110)

If we multiply the curly bracket in the right-hand side (9.107) by this bracket, we obtain the normalized ket $|\widehat{\varphi}''\rangle$ after measurement:

$$|\widehat{\varphi}''\rangle = \left\{ 1 + \frac{g\tau}{2\sigma}\xi_r [A - \langle A \rangle_0] - \frac{1}{2}\left(\frac{g\tau}{2\sigma}\right)^2 (\xi_r)^2 [A - \langle A \rangle_0]^2 + \right.$$
$$\left. + \left(\frac{g\tau}{2\sigma}\right)^2 \left[(\xi_r)^2 - 1\right]\left[[A - \langle A \rangle_0]^2 - \langle [A - \langle A \rangle_0]^2 \rangle_0\right] + ... \right\} |\varphi_0\rangle$$

(9.111)

9.3.2.c Wiener process; stochastic differential equation

We now assume that the system undergoes a continuous series of measurements separated by a time interval δt, with measurement apparatuses that are all identical and all in the same initial state $|\Phi_0(x_0)\rangle$. Since we wish to obtain a continuous process, we assume that two parameters go simultaneously to zero: the time interval δt and the parameter $\varepsilon = g\tau a/\sigma$ characterizing the perturbation introduced by each measurement. At this stage, in order to introduce the appropriate continuous limit, we recall a few properties of Brownian motion.

Brownian motion Consider a particle moving on an axis Ox by random jumps occurring constantly with a time interval δt between them. Each jump changes the position by an amount $\pm\delta l$, the two opposite values having the same probability $1/2$. We are interested in the continuous limit where both δl and δt tend to zero. We consider a fixed time interval dt, which we divide into N smaller intervals $\delta t = dt/N$, corresponding to times $t_0, t_1, ..., t_r, ..., t_N$ at which jumps $\delta x_r = \pm\delta l$ occur. Each jump is characterized by the dimensionless variable $\xi_r = \delta x_r/\delta l = \pm 1$.

- We first assume that the ratio $\delta l/\delta t$ keeps a constant value:

$$\frac{\delta l}{\delta t} = c \tag{9.112}$$

Since, during the time interval dt, the particle makes jumps in both directions with equal probabilities, the average $\overline{dx}$ of the variation of its position x vanishes:

$$\overline{dx} = 0 \tag{9.113}$$

Moreover, the average of the square of this variation is given by:

$$\overline{dx^2} = \overline{[\xi_0 + \xi_1 + ... + \xi_N]^2} (\delta l)^2 = N (\delta l)^2 \tag{9.114}$$

This is because, in the square of the sum, all cross terms containing the product of two different ξ_r have zero average value, so that only N square terms remain. The mean-square deviation of the distance covered by the particle during dt is therefore:

$$\sqrt{\overline{dx^2}} = \sqrt{N}\delta l = \sqrt{\frac{dt}{\delta t}}\delta l = c\sqrt{dt\,\delta t} \tag{9.115}$$

which tends to zero in the limit $\delta t \to 0$. We then see that, in the continuous limit where $\delta l/\delta t$ is constant, the particle does not move anymore; the infinitesimal distance covered by each jump is too small.

- We now assume that the ratio $(\delta l)^2/\delta t$ keeps a constant value D in the continuous limit:

$$\frac{(\delta l)^2}{\delta t} = D \tag{9.116}$$

The average value of $\overline{dx}$ is still equal to zero, as in (9.113), but now we have:

$$\overline{dx^2} = N (\delta l)^2 = \frac{dt}{\delta t} (\delta l)^2 = D\,dt \tag{9.117}$$

During a time interval dt, the particle now covers an average distance $\sqrt{\overline{dx^2}}$ proportional to $\sqrt{dt}$. This regime is called a Brownian motion regime.

Note that, for the sake of simplicity, we have assumed that the jumps take only two opposite values $\pm\delta l$ so that $\xi_r = \pm 1$. However, the same results hold if the jumps take a continuous range of values, in other words if ξ_r is any stochastic variable with zero average value and unit mean-square deviation. For a more detailed study of this random motion and its various applications in physics, see for instance [489].

The square of the displacement dx^2 is a random variable with significant fluctuations. But if, instead of considering the square of the sum of the ξ_r as in (9.114), we introduce the sum of the squares of the ξ_r:

$$ds = \left[(\xi_0)^2 + (\xi_1)^2 + \ldots + (\xi_N)^2\right](\delta l)^2 \tag{9.118}$$

we obtain another variable with an average value:

$$\overline{ds} = N\,(\delta l)^2 = \frac{dt}{\delta t}(\delta l)^2 = D\,dt \tag{9.119}$$

If $\xi_r = \pm 1$, this variable does not fluctuate, and has no random character at all. If ξ_r takes a continuum of values, the fluctuations of ds are given by $\sqrt{N}\,(\delta l)^2 = D\sqrt{\delta t\,dt}$, which tends to zero in the limit $\delta t \to 0$. Although initially defined as a random function, ds actually becomes a deterministic function varying linearly in time in the continuous limit. Relation (9.119) will be useful in what follows.

If one considers an ensemble of particles defining a statistical distribution $\rho(x, t)$ at time t and each undergoing a Brownian motion, one can show that the evolution of this distribution obeys the "diffusion equation":

$$\frac{d}{dt}\rho(x, t) = D\frac{d^2}{dx^2}\rho(x, t) \tag{9.120}$$

which has applications in many domains of physics (heat conduction, transport theory, etc.).

Stochastic evolution of the state vector We now apply the previous considerations to the evolution of the state vector. For the sake of simplicity, we assume that system S evolves only under the effect of its coupling with the series of measurement apparatuses. We also assume that the change of state written in (9.111) occurs at small time intervals δt, and that the constant $g\tau/2\sigma$ characterizing the size of the jump of the state vector is related to δt by a relation:

$$\left(\frac{g\tau}{2\sigma}\right)^2 = D\,\delta t \tag{9.121}$$

This relation is similar to (9.116), with $g\tau/2\sigma$ playing the role of δl in Brownian motion. At the end of the calculation, we will take the limit $\delta t \to 0$.

By analogy with the position undergoing steps $\xi_r\delta l$, we define the variations of a random function W by:

$$\delta W_r = \xi_r\frac{g\tau}{2\sigma} = \xi_r\sqrt{D}\,\delta t \tag{9.122}$$

If a time interval dt is split into N smaller intervals δt, the variation dW of W during dt is given by:

$$dW = W(t + dt) - W(t) = \sum_{r=1}^{N} \delta W_r \qquad (9.123)$$

In the limit where δt and the coupling constant g tend to zero, at constant D, W is not a regular function. It is actually a highly singular function, with a derivative that is always infinite. W is an example of what is called a "Wiener process". We briefly come back on the properties of Wiener processes in §9.3.2.c.

The sum of the squares of the variations δW_r has the same properties as the function ds introduced in (9.118), which is not stochastic but merely equal to $D\,dt$:

$$\sum_{r=1}^{N} (\delta W_r)^2 = D\,dt \qquad (9.124)$$

We can now use (9.111) to express the variation $|\delta\varphi\rangle_r = |\widetilde{\varphi}''\rangle - |\varphi\rangle_0$ of the state vector $|\varphi\rangle$ of S during one single infinitesimal measurement process, in terms of the Wiener process W:

$$|\delta\varphi\rangle_r = \left\{ \delta W_r \left[A - \langle A\rangle_0 \right] - \frac{1}{2} (\delta W_r)^2 \left[A - \langle A\rangle_0 \right]^2 + \right.$$
$$\left. + \left[(\delta W_r)^2 - \left(\frac{g\tau}{2\sigma} \right)^2 \right] \left[\left[A - \langle A\rangle_0 \right]^2 - \left\langle \left[A - \langle A\rangle_0 \right]^2 \right\rangle_0 \right] + \dots \right\} |\varphi\rangle \qquad (9.125)$$

Consider a sufficiently short time interval dt during which the evolution of $|\varphi\rangle$ is very small. We sum the variations $|\delta\varphi\rangle_r$ during the N smaller intervals δt to obtain $|\delta\varphi\rangle$. The sum of the terms linear in δW_r introduces a term in dW, as in (9.123). The quadratic term in δW_r of the first line of (9.125) gives a term in Ddt, according to (9.124). As for the quadratic term in the second line, it gives zero when relations (9.124) and (9.121) are inserted.

Finally, we take the continuous limit where both δt and $g\tau/2\sigma$ tend to zero at constant D in (9.121) and obtain:

$$|d\varphi\rangle = \left\{ dW \left[A - \langle A\rangle \right] - \frac{1}{2} D\,dt \left[A - \langle A\rangle \right]^2 \right\} |\varphi\rangle \qquad (9.126)$$

with[11]:

$$\langle A\rangle = \langle \varphi | A | \varphi\rangle \qquad (9.127)$$

Note that this stochastic differential equation is significantly different from any normal differential equation, the Schrödinger equation for instance: the term in

---

[11] Using (9.124), we can check that $(\langle\varphi| + \langle d\varphi|)(|\varphi\rangle + |d\varphi\rangle) = \langle\varphi|\varphi\rangle + O(dt^2)$, in other words that (9.126) conserves the norm of the state vector.

dW is stochastic and singular, with an amplitude proportional to $\sqrt{dt}$ instead of dt as usual. We now understand why a second-order calculation with respect to ε was necessary: the term in $D\,dt$ arises from the square of the variations $(\delta W_r)^2$, which is proportional to ε^2, and would have been missed by a first-order calculation only[12]. This term is called an "Ito term", and the corresponding sum an "Ito integral" [490]. Although the stochastic term in dW is much larger than the Ito term at a given time, it is also stochastic and may have opposite effects on the evolution of the state vector at successive times. By contrast, the Ito term is nonstochastic and has cumulative effects, so that they are not negligible in the long run when compared to the much larger stochastic terms. Note also that equation (9.126) is nonlinear, since $\langle A \rangle$ depends on the state $|\varphi\rangle$, which may lead to evolutions that are very different from a usual Schrödinger evolution. As expected, we check that $|d\varphi\rangle$ vanishes if $|\varphi\rangle$ is any eigenstate of A with eigenvalue a_n: in this case all successive measurements provide the same result $x_r = x_0 + g\tau a_n$, and the process is no longer stochastic (actually no evolution takes place). This type of nonlinear stochastic differential equation has been studied by Gisin [491, 492] and discussed in the context of quantum measurements [493]. For introductory texts on stochastic evolutions of the state vector and continuous measurements, see for instance [494, 495].

Properties of the Wiener process Since ξ_r is a stochastic variable with zero average, the definition (9.122) implies that the statistical average of the variations of W vanishes:

$$\overline{\delta W_r} = 0 \quad \text{or} \quad \overline{dW} = 0 \tag{9.128}$$

(the upper bar denotes an ensemble average over many realizations of the process). Since the mean-square deviation of ξ is 1, we also have:

$$\overline{[\delta W_r]^2} = D\delta t \tag{9.129}$$

Since the successive results of measurements result from successive quantum processes, which are all independent and fundamentally random, the stochastic variables $x_r - x_0$ are independent for different values of r. The same is therefore true of the ξ_r and of the δW_r. Combining this property with (9.129), we obtain:

$$\overline{\delta W_r\, \delta W_{r'}} = \delta_{r,r'} D\delta t \tag{9.130}$$

where $\delta_{r,r'}$ is a Kronecker delta symbol.

We can also introduce the time derivative $W'(t)$ of the Wiener process. This derivative may be defined as a random process obtained by dividing the variation

---

[12] A second-order calculation is indeed sufficient since third-order terms give a contribution in $dt^{3/2}$, playing no role in the limit $dt \to 0$.

of W by the time interval:

$$W'_r = \frac{\delta W_r}{\delta t} = \xi_r \sqrt{\frac{D}{\delta t}} \qquad (9.131)$$

In the continuous limit, we then have:

$$\overline{W'(t)W'(t')} = D\,\delta(t - t') \qquad (9.132)$$

where $\delta(t - t')$ denotes the Dirac delta function of $(t - t')$. This equality can be checked by remarking that, if $t' \neq t$, the average of the product of two W'_r vanishes for different values of the index r (they contain independent random variables ξ_r). Moreover, integrating (9.132) over t' introduces an integral that is the continuous limit of the discrete sum:

$$\sum_{r'} \delta t \left[\frac{1}{\delta t^2} \overline{\delta W_r \delta W_{r'}} \right] = \sum_{r'} \delta t \frac{1}{\delta t^2} \delta_{r,r'} D \delta t = D \qquad (9.133)$$

Therefore, $\overline{W'(t)W'(t')}$ is entirely concentrated in an infinitesimal interval around $t' = t$, and has an integral equal to D, which defines a delta function multiplied by D. Equation (9.132) says that the time variations of the Wiener process have no memory. They are what is often called a "white noise" in physics, with completely independent values at different times, even if the times are almost equal.

Now, instead of writing the evolution of the state vector in terms of an unknown Wiener process as in (9.126), it may seem more natural to express it as a function of the results of measurements, which are directly observable. In a given run of the experiment, what is actually obtained is a whole string of results $(x_r - x_0)$ – it is convenient here to take x_0 as an origin for expressing the results. By analogy with (9.105), we define the stochastic variable ζ_r by:

$$\zeta_r = \frac{x_r - x_0}{\sigma} = \xi_r + \frac{\Delta x_0}{\sigma} \qquad (9.134)$$

and, to obtain a proper continuous limit as in (9.122), we define the stochastic function R by:

$$\delta R_r = \zeta_r \frac{g\tau}{2\sigma} = \delta W_r + \frac{g\tau}{2\sigma} \frac{\Delta x_0}{\sigma} \qquad (9.135)$$

R is called the "measurement record" for each realization of the series of measurements. Using expression (9.103) as well as (9.121), we then obtain:

$$\delta W_r = \delta R_r - 2 \left(\frac{g\tau}{2\sigma} \right)^2 \langle A \rangle = \delta R_r - 2 \langle A \rangle D \delta t \qquad (9.136)$$

and, inserting this result into (9.126) and going to the continuous limit:

$$|d\varphi\rangle = \left\{ dR\,[A - \langle A \rangle] - \frac{1}{2} D\,dt\,\left[[A + \langle A \rangle]^2 - 4\langle A \rangle^2 \right] \right\} |\varphi\rangle \qquad (9.137)$$

This equation provides the evolution of the state vector as a function of the measurement record R.

Stochastic evolution of the density operator We finally study the evolution of the density operator $\rho = |\varphi\rangle\langle\varphi|$. Its infinitesimal variation is given by:

$$\delta\left(|\varphi\rangle\langle\varphi|\right)_r = \left[|\varphi\rangle + |\delta\varphi\rangle_r\right]\left[\langle\varphi| + \langle d\varphi|_r\right] - |\varphi\rangle\langle\varphi| \tag{9.138}$$

where $|d\varphi\rangle_r$ is given by (9.126) and $\langle d\varphi|_r$ is the corresponding bra. Usually, the term in $|d\varphi\rangle_r \langle d\varphi|_r$ can be ignored to first order, since it is in dt^2. Here the situation is different since, in terms of the Wiener process, $|d\varphi\rangle$ contains a term in dW_r, with a square proportional to δt as shown by (9.124). This term then introduces a contribution:

$$Ddt\ [A - \langle A\rangle]\rho\ [A - \langle A\rangle] \tag{9.139}$$

When this term is added to the linear terms contained in $|d\varphi\rangle_r$ and the associated bra, the evolution of ρ is obtained in the form:

$$d\rho = \left\{dW\left([A,\rho]_+ - 2\langle A\rangle\rho\right) - \frac{1}{2}Ddt\left[A,[A,\rho]\right]\right\} \tag{9.140}$$

containing an anticommutator $[A,\rho]_+$ and a double commutator $[A,[A,\rho]]$. This equation is sometimes called a Belakvin equation [496]. As is the case for the state vector, one can replace dW by $dR - 2\langle A\rangle Ddt$ to express the variation of the density operator as a function of the measurement record, instead of the Wiener process. We can easily check that if ρ is a statistical mixture of projectors over eigenstates of A, since it commutes with A, no evolution takes place under the effect of the successive measurements, for the same physical reasons previously.

The mathematical tools we have discussed in this chapter (Wiener processes, stochastic differential equations, etc.) turn out to be useful in some of the interpretations of quantum mechanics we discuss in Chapter 11, in particular those involving a stochastic Schrödinger dynamics.

10

Experiments: Quantum Reduction Seen in Real Time

From a fundamental point of view, we must acknowledge that, since about 1935, our conceptual understanding of quantum mechanics has not progressed so much. Really new ideas are few and far between – except of course the major line initiated by the contribution of Bell [6]. This is in big contrast with the rest of physics, where new theoretical and experimental discoveries in many fields have flourished, very often with the help of the tools of quantum mechanics. The fantastic evolution of the experimental techniques has completely changed the situation. At the beginning of quantum mechanics, the observation of the tracks of single particles in Wilson chambers [497] played the essential role in the introduction of the postulate of state vector reduction, but otherwise it was impossible to observe continuously a single electron, atom, or ion; the experiments that theorists were proposing in discussions on the principles of quantum measurement were therefore "thought experiments" ("Gedanken Experiment"), as for instance in the famous Solvay meetings [1, 21]. But nowadays, after almost of century of experimental progress, experiments that were then unthinkable have become a reality.

A huge number of contemporary experiments involves the laws of quantum mechanics in general; several books would not be sufficient to describe all of them. Nevertheless, in the majority of experiments, what is really observed is the sum over a very large number of particles of one individual microscopic observable (sum of atomic dipoles, for instance), which is accurately described by the average value of this observable. The Schrödinger equation can then be used to calculate this average, which is subsequently treated as if it was a classical variable. No particular use of the Born rule or of the projection postulate is then necessary – a typical illustration is given by many nuclear magnetic resonance experiments in chemistry and physics. Similarly, in many coincidence experiments, the relevant physical observable is the product of quantum operators associated to individual counting rates, and the Schrödinger equation can be used again to obtain the average of this product. Of course, this does not necessarily mean that the projection postulate is completely irrelevant for these experiments! In quantum optics and atomic physics, for instance, optical detection with photomultipliers and diodes is often used; the projection postulate then determines the size and nature of the random "shot noise" observed in the experiments. This noise limits the accuracy by adding a fluctuating component to the signal, while the latter itself has a smooth deterministic behavior as a function of the experimental parameters; since very often what is studied in detail is the signal rather than the noise, the projection postulate then plays only an auxiliary role.

Here we will focus our interest only on the small fraction of experiments where the effect of the projection postulate is particularly visible, for instance those where a single quantum particle is observed, and where "quantum jumps are visible in real time". The observations then exhibit more quantum features than what is predicted by the continuous Schrödinger evolution, and in this sense go beyond this equation. Our purpose is not to give a complete review, but just to provide a few particularly illustrative examples.

10.1 Single ion in a trap

It is possible to observe a single barium ion contained in a radio-frequency trap by monitoring its fluorescence under laser irradiation [498, 499]; the lifetime of the excited resonance level is about 10^{-9} second, meaning that even if only a thousandth of the fluorescence light is collected (as is typical in these experiments), one can still detect a flux of a million of photons per second, directly visible with the naked eye. Under these conditions, the authors of [498] write that it is possible to directly "watch the reduction of the wave function by the measurement process on the oscilloscope screen".

The relevant energy levels of a Barium ion used in the experiment performed by Dehmelt and coworkers are shown in Figure 10.1: the ground state is g, the two excited states are e_1 and e_2, and m_1 and m_2 are two metastable states. The transition $g - e_1$ from the ground state g to a first excited state e_1 is strongly illuminated with a laser, while the fluorescence at the corresponding wavelength is constantly monitored. Level e_1 spontaneously decays, not only to the ground state g, but also to a metastable level m_1, which decays into the ground state much more slowly than the excited state e_1. The ion could therefore remain trapped in m_1, without interacting with the laser, and the fluorescence would immediately stop. To prevent this trapping, a second laser is used to constantly excite the transition from m_1 to e_1. The result is to create a $g - e_1 - m_1$ closed circuit from which the ion cannot escape; this is used for laser cooling of the ion. If no other excitation of the ion was used, it would constantly fluoresce. But another, much weaker, light source excites the $g - e_2$ transition to a second excited level e_2. Sometimes, after reaching level

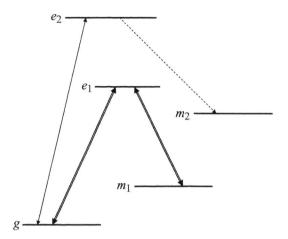

Figure 10.1 Energy levels of the barium ion involved in the experiment. The ground state of the ion is g, e_1 and e_2 are two excited states, and m_1 and m_2 two metastable states. Two lasers of high intensity (double arrows) excite, one the transition g-e_1, the other the transition m_1-e_1; they are used for laser cooling of the ion motion in the trap. The fluorescence emitted by spontaneous emission from level e_1 to g is constantly monitored. The transition g-e_2 is excited only weakly with a lamp; when the ion reaches e_2, it sometimes falls (dotted line) into the second metastable level m_2. It is then "shelved" and no longer fluoresces, until it falls spontaneously into the ground state g and starts fluorescing again.

e_2, the ion does not fall back into the ground state g, but into the metastable level m_2, where it can no longer be excited optically; it then stops fluorescing, and one then says that it has been "shelved" in the metastable state m_2. Nevertheless, since this level has a finite lifetime, the ion eventually falls back spontaneously into the ground state g, and starts fluorescing again.

If one applies the Schrödinger equation to this situation, one has to include in the quantum system the ion itself as well as the electromagnetic field, with several modes populated, and treat the effect of interactions (absorption, stimulated emission, spontaneous emission). The solution of the equation that is obtained in this way is, after some time, a superposition of two components: one where the ion is not shelved and strong spontaneous radiation is emitted, and another component where the ion is shelved and emits no radiation at the frequency of the $e_1 - g$ transi-

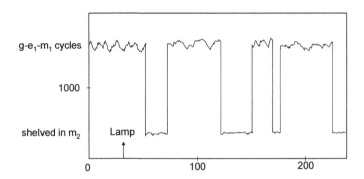

Figure 10.2 Fluorescence signal of a single ion at the wavelength of the transition $g - e_1$ as a function of time t in the experiment of [498]; the horizontal scale is seconds, and the vertical scale gives the number of counts per second delivered by a photomultiplier. Until $t = 30$s., the weak lamp remains off, so that the ion cycles between levels g, e_1, and m_1; it then constantly fluoresces (upper noisy line). The weak lamp is then switched on and may transfer the ion to metastable level m_2, where it stops emitting light. When this happens, the ion remains shelved for some time in the metastable level, and a dark period occurs (lower line) until the ion spontaneously returns to the ground state. The whole cycle then starts again from the beginning.

tion. The ion fluoresces and does not fluoresce at the same time, as the Schrödinger cat is alive and dead; the state vector provides an average fluorescence intensity that is intermediate between the two situations. In other words, within the Schrödinger equation, everything remains continuous, and "jumps" never occur.

But, since in the experiment the fluorescence intensity is continuously measured, the two components of the state vector also contain macroscopically different states of the measurement apparatus, and therefore the projection postulate applies: a fundamentally random process takes place, and spontaneously the system chooses one of the components only. Consequently, the fluorescence has either its maximal value (that corresponding to the ion circulating between the three levels g, e_1, and m_1), or zero; it does not vary continuously between these values.

The experimental results fully confirm this prediction: as shown in Figure 10.2, the observed fluorescence undergoes "jumps" between a fixed value and zero. The times at which these jumps occur are not controlled experimentally, since they are a consequence of the fundamentally random character of quantum measurement. Consequently, if many ions were fluorescing at the same time, each of them would switch between the fluorescing and the dark regime at different times, and on average one would observe only the average value of the fluorescence – the value predicted by the continuous Schrödinger equation. But, with a single ion the quantization of the signal is directly observed, and the system can undergo a very large number of jumps between the two values.

Experiments performed at the same time in Toschek's group [499] on the same ion have provided similar results in a case where only three levels are involved, also exhibiting very clear "quantum jumps". Similar experiments with other ions were performed in other laboratories, in particular with the mercury ion [500, 501]. In this case, three atomic levels are relevant: the ground state, a resonance excited level $^2P_{1/2}$ with short lifetime (2 ns.), and a metastable level $^2D_{5/2}$ with lifetime 0.1 s. When the transition between the ground state and resonance level is constantly excited, the ion fluoresces with no interruption. But when the transition to the metastable states is also excited, sometimes the ion is "shelved" in this state and stops fluorescing until it falls back to the ground state by spontaneous emission. Results are shown in Figure 10.3. As in Figure 10.2, the fluorescence of the ion exhibits discontinuities between two regimes, often called "quantum jumps" as a reminiscence of the historical Bohr theory of atoms (§1.1.1).

Long before these results were obtained, Schrödinger had analyzed a thought experiment where the light emitted by an atom was used in an interference experiment [502]; he had pointed out that quantum jumps, if they occur when an atom emits photons, cannot be instantaneous, but necessarily have a finite duration related to the radiative width of the atomic levels. Here, the situation is similar, since the change from the dark regime (no fluorescence) to the bright regime (fluorescence) is triggered by the spontaneous emission from level e_2 to m_2, while the transition from bright to dark is triggered by the transition from m_2 to the ground state g – in other words, two "jumps" play a role, instead of one, but Schrödinger's remark applies as well.

Then, how is this compatible with the observations of Figure 10.2? The explanation is that the signal of Figure 10.2 is averaged over time, so that it does not show the discontinuities that would appear under a more precise observation. At a much smaller time scale, the signal provided by the photomultiplier monitoring the fluorescence is a series of sudden "clicks" occurring at random times, corresponding to the detection of individual photons; Figure 10.2 actually shows nothing but the averaged frequency of these clicks. What is really observed is transitions between periods where the clicks occur frequently and others where they are rare; the transition time between these periods cannot be measured more precisely than the time between two consecutive clicks. As a consequence, the time at which the "jumps" occur can only be measured within some uncertainty, related to the emission rate and therefore to the radiative width of the levels, as implied by Schrödinger's argu-

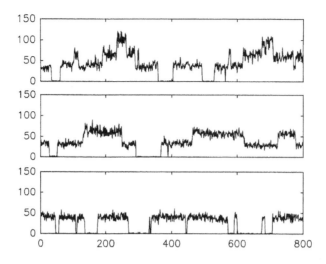

Figure 10.3 Lowest curve: fluorescence signal of a single Hg$^+$ ion at the resonance wavelength corresponding to the transition between the $^2S_{1/2}$ ground state of the ion and the $^2P_{1/2}$ excited state. The units of the horizontal axis are ms, and the vertical axis gives number of counts per ms. When the transition to a metastable $^2D_{5/2}$ level is simultaneously excited, quantum jumps are clearly visible, as in Figure 10.2. The middle curve is obtained when two ions sit in the trap; three cases may then occur, depending on the number (0, 1, or 2) of ions "shelved" in the metastable state, and resulting in three possible fluorescence intensities. The upper curve shows the fluorescence when three ions are trapped, and provide four possible levels of fluorescence at the resonance transition (figure kindly provided by D. Wineland and W. Itano).

ment – for a more precise discussion, see [503]. A more detailed theoretical study of the fluorescence intermittence phenomenon can be found in [504, 505].

10.2 Single electron in a trap

A single electron orbiting in a cyclotron trap can also give rise to observations reaching the quantum limit, as demonstrated by Gabrielse et al. [506]. In their experiment, a cold Penning trap was used to store one electron. Such a trap has a strong homogeneous axial magnetic field and a quadrupole electric field created by electrodes. In the magnetic field, the quantized levels of a charged particle are the

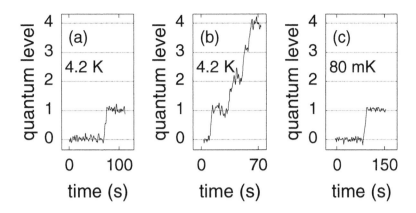

Figure 10.4 Real time detection of the quantum orbital number of a single electron in a Penning trap. In (a) and (b), the excitation between the Landau levels
is obtained under the effect of blackbody radiation in the experiment, at a few
degrees Kelvin. In (c), the temperature is much lower so that this process does
not occur anymore; the excitation of the electron towards excited Landau levels is then obtained by applying a microwave field (figure kindly provided by G.
Gabrielse).

equidistant orbital "Landau levels" with energy $nh\nu_c$, where ν_c is the cyclotron frequency and n an integer (the energy results from a combined effect of the coupling
of the magnetic field with the orbital and spin variables). What is measured is the
quantum number n characterizing this energy.

Two different motions of the electron in the trap are actually relevant in this
experiment. The main motion is the usual cyclotron motion, perpendicular to the
magnetic field, which is the central focus of the experiment. But there is also a longitudinal motion, parallel to the field, which has a frequency that depends weakly
on the energy of the cyclotron motion (experimentally, this coupling is introduced
by a small distortion of the homogeneous magnetic field). It is therefore possible to
have access to the cyclotron energy by a very precise measurement of the frequency
of longitudinal oscillations of the electron, which gives access to n. In the experiment, a single electron orbits in the magnetic field, while its longitudinal frequency
is constantly monitored with great accuracy, which amounts to a constant quantum
measurement of n. Figure 10.4 shows an example of the variation in time of this
frequency: it clearly exhibits steps occurring when the system is projected into an
energy eigenstate under the effect of quantum measurement. The general comments
of §10.1 apply to this experiment as well, which provides another illustration of a
"quantum nondemolition" measurement monitored in real time. More recently, a
new version of the cyclotron experiment has provided an extremely accurate mea-

surement of the magnetic moment of the electron and of the fine structure constant [507].

10.3 Measuring the number of photons in a cavity

In the experiments described in the two preceding sections, the observed quantum system was a material particle (particle with rest mass), either an ion or an electron. One could think that photons are less appropriate to be observed individually, since they are so easily absorbed in various processes, including their detection with photomultipliers or diodes. But recent experiments have obtained comparable results by observing photons in a cavity, illustrating the properties of quantum measurement in a way that is even more spectacular. Until relatively recently, the only way to detect a photon was to absorb it in a detector (a photomultiplier, for instance), so that further measurements on the same particle were impossible. But various "quantum nondemolition" methods now make it possible to measure the presence and the number of photons without destroying them [448, 508] – see also §6.2 of [374]. The combination of this possibility with the methods of cavity electrodynamics has resulted in experiments where the number of photons stored in a cavity is constantly monitored in real time [509].

Rydberg atoms are atoms in high energy levels, Rydberg levels, very close to the ionization threshold. Among them, the "circular Rydberg levels" (those where the rotation quantum number l takes its maximal value) interact with photons in a particularly simple and well-controlled way. These atoms have a very large dipole moment, which means that they are strongly coupled to the electromagnetic field; they can therefore be used as probes of its properties, even if it contains very few photons. Moreover, they can be ionized and detected with high efficiency and selective access to the various Rydberg levels. Current experimental techniques allow the production of beams of such atoms, which can be sent through a resonant electromagnetic cavity to test the number of photons it contains.

One could make use of the absorption of photons by the atoms, by sending atoms one by one through the cavity, controlling their energy level before the crossing of the field and measuring the energy level at the output – but obviously this process would absorb photons and not lead to a quantum nondemolition measurement. A much better technique is to choose a case where the frequency of the photons in the cavity differs significantly from the resonance frequency of the atoms (Bohr frequency associated with the transition between the two relevant Rydberg levels), so that the absorption probability of photons remains negligible. Due to nonresonant interactions between atoms and photons, the atoms create index (dispersive) effects for the photons and, conversely, the photons shift the atomic energy levels;

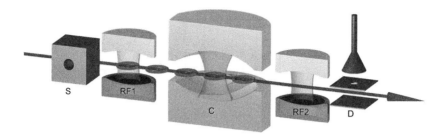

Figure 10.5 A beam of atoms in a given highly excited state (circular Rydberg
level) is created by a source S. Each atom crosses, first a region where a device
RF1 excites coherently a transition between the initial and another close Rydberg
level, then crosses a cavity C with resonant frequency shifted with respect to the
atomic transition, and a region where another device RF2 (locked in phase with
RF1) excites again the same transition. Devices RF1 and RF2 perform what is
called Ramsey spectroscopy on the atom: the first interaction creates a coherent
dipole associated with the two relevant Rydberg levels, and the second detects
the phase of this dipole, including its evolution between the two regions and the
effect of cavity C. Since the dipole accumulates a dephasing when it crosses C,
and since this dephasing depends on the number of photons contained in C, the
final population of the Rydberg level depends on this number of photons. One
then has direct access to it by measuring this population in detector D. Since the
cavity is not resonant at the atomic resonant frequency, the number of photons in
the cavity is not changed by the interaction; the method thefore provides quantum
nondestructive (QND) measurement (this figure has been kindly provided by J.M.
Raimond).

one can show that the frequency of the transition between the two atomic levels is
increased by an amount that is proportional to the number of photons in the cavity.

The idea is then to send into the cavity atoms that are in a coherent superposition
of two Rydberg states, let them interact with the photons inside the cavity, and at
the output measure the change of phase of the coherent superposition induced by
this interaction. Since the phase change is proportional to the number of photons
in the cavity, one obtains in this way a measurement of their number. This purely
dispersive technique is nondestructive: it changes neither the energy of the electromagnetic field (that is, the number of photons) nor the energy of the atom at the
output of the cavity. It provides quantum nondemolition measurements (§9.1.1.b).
The experiment is shown schematically in Figure 10.5.

One major experimental difficulty is that a sufficient number of atoms must cross
the cavity in order to give a reasonably accurate measurement, before the electromagnetic field has decayed by absorption in the walls of the cavity. As a consequence, one has to use a cavity with an extremely high quality factor Q. This has
been achieved with the help of a high quality superconducting cavity so that, in the

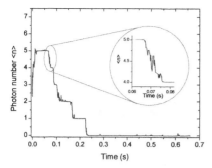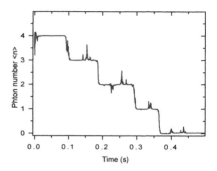

Figure 10.6 Continuous monitoring of the number of photons contained in an electromagnetic cavity, obtained by measuring the phase shift accumulated by Rydberg atoms crossing the cavity. Initially, the cavity is loaded with a coherent field of small intensity which, under the effect of quantum measurement, is projected to a state with a well-defined number of photons (five in the left part of the figure, four in the right part). Then the photons are progressively absorbed by the wall of the cavity and disappear, one by one. This process is constantly observed in real time (figure kindly provided by J.M. Raimond).

experiment, hundreds of atoms can cross the cavity before the number of photons changes. By constantly monitoring the state of the outgoing atoms, one has access to a really continuous and nondestructive measurement of the number of photons. Initially, the cavity is fed with a coherent state that does not have a well-defined number of photons; nevertheless, as soon as a few atoms have crossed the cavity and their phase has been measured at the output of the cavity, the system is projected into a state with fixed number of photons. Then, the photons disappear one by one as they are absorbed in the walls of the cavity, which occurs at a slow rate since the superconducting cavity has a very high quality factor. This produces a series of quantized changes in the observed signal until no photon remains in the cavity. Two examples of recordings made in this way are shown in Figure 10.6.

In this case, the discussion concerning the time it takes to obtain this projection (see end of §10.1) can be given with particular clarity, in terms of the number of atoms that are required before one can ascertain with good probability that a jump has occurred [509]. What one observes is clearly not the continuous evolution predicted by the Schrödinger equation, which has the same solution for all identical realizations of the experiment; in each experiment, one sees marked steps, which fall at times that are different for each realization. The Schrödinger equation provides only the average of the observations over many experiments; the steps seen in single experiment cannot be explained within this equation only, but require an additional ingredient. This is another particularly spectacular case where one can see "the effects of the reduction of the state vector occurring directly under your eyes".

10.4 Spontaneous phase of Bose–Einstein condensates

The notion of "identical particles" plays a much more important role in quantum mechanics than in classical physics. It leads to the introduction of a special "symmetrization" postulate, which determines the possible forms of the state vector of a system of identical particles, with a distinction between two classes of particles: bosons and fermions. The latter obey the Pauli exclusion principle: two fermions can never occupy the same individual quantum state. By contrast, an arbitrary number of bosons can occupy the same quantum state, as for instance photons do in a monomode laser beam. Another example is given by Bose–Einstein condensates; in a very dilute gas at very low temperatures, such condensates can be represented by states where all N bosons are in the same individual state $|\varphi\rangle$:

$$|\Psi\rangle = |1 : \varphi\rangle |2 : \varphi\rangle ... |N : \varphi\rangle \tag{10.1}$$

In this case, as opposed to the situation where one single particle is in state $|\varphi\rangle$ (§8.1.2), it becomes possible to determine the state, or equivalently its wave function $\varphi(\mathbf{r})$; one sometimes says that this wave function becomes analogous to a classical field. But, in classical physics, two wave packets of the same field may give rise to interference effects. By analogy, one can ask the question: if two Bose–Einstein condensates made of identical atoms are prepared independently, is it possible to observe interference effects between them? The answer to the question involves interesting quantum measurement effects, and has been clarified by Javanainen and Yoo [510] as well as by other authors (see for instance references given in [132]).

10.4.1 Interferences between independent condensates

Assume that N identical bosons occupy state $|\varphi\rangle$ and P other bosons (identical to the former) state $|\chi\rangle$. When wave functions $\varphi(\mathbf{r})$ and $\chi(\mathbf{r})$ overlap, the first idea that comes to mind is to assume that the wave functions interfere exactly as a classical field, so that the probability to find a particle at point $\mathbf{r}$ is proportional to:

$$|\varphi(\mathbf{r}) + \chi(\mathbf{r})|^2 \tag{10.2}$$

In classical optics for instance, the intensity at point $\mathbf{r}$ contains an interference term between the electric fields of two overlapping beams; here we have crossed interference terms in $\varphi^*(\mathbf{r})\chi(\mathbf{r})$ and $\varphi(\mathbf{r})\chi^*(\mathbf{r})$. But, in quantum mechanics, the phase of each wave function is arbitrary: no physical property associated with the wave functions is changed if they are replaced by $e^{i\alpha}\varphi(\mathbf{r})$ and $e^{i\beta}\chi(\mathbf{r})$, while this obviously changes the crossed terms in (10.2). This argument shows that the quantum probability cannot be given by this classical formula. An explicit quantum calculation confirms this point; when one detects the first particle, its probability of presence is given by:

$$|\varphi(\mathbf{r})|^2 + |\chi(\mathbf{r})|^2 \tag{10.3}$$

The probability is therefore merely the sum of the probabilities corresponding to a detected particle being initially either in state $|\varphi\rangle$ or in state $|\chi\rangle$, without any interference term. As far as the detection of the first particle is concerned, the relative phase of the two condensates is therefore nonobservable, because actually it does not exist: nothing in the definition of the initial state can be used to determine this phase. The same result can also be interpreted in terms of an uncertainty relation between the phase and the number of particles (similar to the time-energy uncertainty relation): if the number of particles is perfectly determined, as is the case here, the phase is completely undetermined.

   But, as the authors of [510] have shown, the situation becomes more interesting when the positions of several particles are measured. For a given realization of the experiment, one can study the correlations between these positions. The quantum prediction is then that, when more and more particle positions are measured, the relative phase becomes better and better defined. Already the first position measurement creates some information on this phase[1], which plays a role for the second measurement; the phase distribution, instead of being completely independent of the phase, is given by a sinusoid having maxima and minima at positions depending on the measured position. Then, while measurements accumulate, the phase distribution becomes a product of more and more sinusoids and exhibits a sharper and sharper maximum, making this phase better and better determined; one finally practically reaches a classical situation where the phase is perfectly determined. This provides an interesting process where, initially, the phase did not exist, but where the successive projections due to quantum measurement make it emerge, and give it a better and better defined value. Interestingly, the observations would be exactly the same if the phase had existed from the beginning for each realization, but was totally unknown. Nevertheless, from one realization of the experiment to the next, the new phase value that emerges is totally different, with no correlation with the preceding value.

   Experiments performed in Ketterle's group at MIT with condensates made of sodium atoms have confirmed these predictions [511]. Two condensates were independently prepared, and then released from their separate traps to allow them

---

[1] For instance, the measurement shows that the phase cannot take a value that puts a destructive interference antinode at the measured position (we assume that the two intensities are the same).

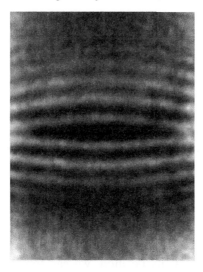

Figure 10.7 Observation of interference fringes when two Sodium condensates are prepared separately, and then released from their traps to spatially overlap. For each run of the experiment, the fringes have a well-defined phase, but the phase is completely uncorrelated from one run to the next (figure kindly provided by W. Ketterle).

to overlap spatially. Optical absorption was then used to measure the positions of atoms in the overlap region. Figure 10.7 shows the result obtained in a given run of the experiment: indeed, one observes fringes with a well-defined phase, even if quantum mechanics predicts that it did not exist before measurements; this provides a case where the state vector reduction is very directly visible. From one experiment to the next, the phase is completely uncorrelated; in other words, if one sums the results obtained in several runs, the fringes disappear, as predicted by theory. A specific property of the experiment is that the physical quantity that emerges from the measurement (the phase) directly fixes the value of a macroscopic quantity (the density of the gas at each point of the interference region), which provides a direct observation of the effects of state vector projection.

10.4.2 An additional variable?

The question that then arises very naturally is the same as for any quantum measurement: should we really think, as standard quantum mechanics invites us to do, that it is the act of measurement that creates the value of the phase? Or should we consider, on the contrary, that it only reveals a value that existed before the measurement? When a variable is macroscopic, as can be the case for the phase of an interference pattern, it does not seem very natural to consider that it can be created by the measurement, due to some uncontrolled perturbation of the measurement apparatus. This raises once more the question of the existence of additional variables, this time in a macroscopic context. Moreover, this experiment has some analogy with the discussion of the Schrödinger cat, where quantum uncertainty is transferred to the macroscopic world.

Leggett and Sols [512] have discussed a similar situation, where a Josephson current appears between two superconductors; the value of this current is fixed by the difference between the two superconducting phases. These authors ask whether such a phase, with all its consequences on a macroscopic current, can really appear under the influence of a measurement apparatus, which can be very small: "Does the act of 'looking to see' whether a Josephson current flows itself force the system into an eigenstate of current and hence of relative phase? ... Can it really be that by placing, let us say, a minuscule compass needle[2] next to the system, with a weak light beam to read off its position, we can force the system to 'realize' a definite macroscopic value of the current? Common sense certainly rebels against this conclusion, and we believe that in this case common sense is right".

10.4.3 Nonlocality of phase

One can add another component to the argument by assuming that the interference takes place, not in a single region of space, but in two remote regions where Alice and Bob operate. The situation is then analogous to that discussed with measurements performed on two remote spins (§4.1.1), but the correlations now occur between the phases of the interference patterns instead of the directions of spins. To push the analogy further, it is more convenient to assume that the condensates are spin condensates[3], with a relative phase controlling the average value of a transverse spin component in all overlap regions. The corresponding situation is schematized in Figure 10.8: two condensates, one containing spin + particles, the other spin − particles, extend in space and overlap in two remote regions, where Alice and Bob perform measurements of the transverse components of the spins. The predictions of quantum mechanics in such a case are a direct generalization of what was described in §10.4.1: when Alice performs a first measurement, the result is completely random; but, while measurements are accumulated, the successive results make a relative phase emerge with a better and better definition − in other

---

[2] The purpose of this needle is to measure, through its rotation, the magnetic field created by the macroscopic Josephson current.

[3] The "spins" we discuss here are not necessarily real spins. In quantum mechanics, any two-level system is equivalent to a spin 1/2; this provides a convenient way to describe the states and observables associated with the two states. Here, the two states in question may be for instance any pair of sublevels of the ground state of atoms. Longitudinal spin measurements then refer to measurements associated with observables represented by 2x2 diagonal matrices in the basis of the single-particle-states populated in the condensates; transverse spin measurements correspond to matrices that are off-diagonal

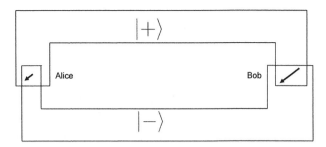

Figure 10.8 Two condensates, one corresponding to particles with spin state $|+\rangle$, the other to particles with spin state $|-\rangle$, have wave functions (represented in the figure) that overlap in two regions of space. In these two regions, two operators Alice and Bob measure a transverse component (in the plane that is perpendicular to the quantization axis) of the spins of the particles. Initially, the relative phase of the two condensates is completely undetermined, so that Alice's first measurement gives a completely random result. But, when she accumulates measurements, their effect on the quantum system is to determine the relative phase better and better; after some time, the phase is practically fixed, and one can say that the system has given transverse spin polarization in the overlap regions.

What is surprising is that, under the effect of Alice's measurements, the same direction of spin polarization appears in Bob's region, whatever the distance between the two regions is and without any mutual interaction. This is very similar to the usual situation of the EPR argument, but now with macroscopic EPR elements of reality if the number of particles in each region is macroscopic.

words a transverse spin polarization emerges in a well-defined direction. After a sufficient number of results, the results of spin measurements along this direction are practically certain. If the overlap region in which Alice operates contains a macroscopic number of particles, the emerging spin polarization is macroscopic.

Still, according to standard quantum mechanics, another effect of Alice's measurements is the appearance of a parallel spin orientation in Bob's region, without any time of propagation of the influence between the two regions. But is it really possible that a measurement performed by Alice on a small number of spins, a hundred for instance, will create an arbitrarily large spin orientation in Bob's region, at an arbitrary distance, especially if the distance between the two regions is extremely large? The question is strongly reminiscent of the EPR argument, but now for a case where the "elements of reality" are macroscopic quantities [132], which makes the argument even stronger. Can we really imagine that a macroscopic spin orientation is created in Bob's laboratory without any local interaction, and appears (so to say) from nothing? Or should we follow EPR and say that this orientation existed from the beginning of the experiment, so that the measurements do nothing but reveal the existence of a phase that was initially fixed at a random value? If so, the phase becomes an additional variable that should be inserted into quantum mechanics to make it complete. Since the discussion deals with macroscopic quantities, which presumably are directly accessible to human experience, it is difficult to guess precisely what Bohr would have replied to this form of the argument.

Moreover, the conservation of angular momentum raises delicate questions in such a case. How can the angular momentum contained in Bob's region be changed instantaneously as a result of measurements performed in an arbitrarily remote region[4]? Of course, if one considers only the region where Alice performs her experiment, no particular difficulty occurs. The measurement apparatus she uses, in order to be able to measure the angular momentum of the particles, must interact with them with a coupling Hamiltonian that contains their angular momentum, as well as its own angular momentum; this makes a transfer between them possible. One can then assume that the transfer depends on the result of measurement in a way that ensures perfect angular momentum conservation. Any paradox is avoided with the help of, so to say, a recoil effect of the measurement apparatus. But the coupling Hamiltonian in question certainly commutes with all operators associated with physical quantities localized in Bob's region; its action cannot change angular momentum in that region of space. It is therefore difficult to understand how Alice's measurement apparatus can create such an angular momentum at a distance. Moreover, if Alice performs her measurement on a small number of spins it seems paradoxical to consider, in order to conserve total angular momentum, that her apparatus gets a recoil effect that corresponds to the very large number of spins in Bob's region. But if we decided, rather, that her measurement apparatus cannot get a larger recoil than the maximal angular momentum allowed by quantum for a small number of spins, then we would have to give up angular momentum conservation. Here also, it is tempting to follow EPR and to avoid any difficulty by considering that the angular momentum in Bob's region already existed before Alice's measurement, which amounts to complete quantum mechanics with the addition of a phase.

Even if one modifies the theory to include this phase, this is not sufficient to obtain a completely classical behavior and to restore locality in the quantum predictions. Actually, the situation is relatively similar to the usual case with two spins, where the Bell theorem forces the EPR elements of reality to have a nonlocal evolution if one wishes to reproduce the predictions of quantum mechanics in all cases. Here also, when the relative phase is measured in two different regions, a model

---

[4] Here we discuss a single realization of the experiment. With a large number of measurements, the angular momentum may take all transverse directions, so that its average vanishes; no paraxox then occurs.

where this phase existed from the beginning (while being unknown) cannot reproduce all predictions of quantum mechanics [134]. This impossibility appears when
the spin components along several different directions are measured, and takes the
form of BCHSH inequalities that are violated by quantum mechanics. In addition,
and in opposition to what one could have expected, these violations do not tend to
zero for systems with a very large number of particles[5], but remain constant.

Indeed, Fock states (or Bose–Einstein condensates) provide interesting transpo-
sitions of the EPR argument and the Bell theorem to macroscopic systems; this
sheds a new light onto the arguments.

---

[5] The angles between the directions of the spin components to be measured become smaller and smaller when
the number of particles increases.

11

Various Interpretations and Reconstructions of Quantum Mechanics

Long ago, and almost in parallel with the "orthodox" Copenhagen/standard interpretation, other interpretations of quantum mechanics were proposed. Giving an exhaustive discussion of all points of view that have been put forward since then would probably be an impossible task. Moreover, while one can distinguish big families among the interpretations, it is also possible to combine them in many ways, with an almost infinite number of nuances. The Copenhagen/standard interpretation itself, as we have seen, is certainly not a monolithic construction, but can be declined in various forms. In this chapter, we will therefore limit ourselves to a general description of the major families of interpretations.

We will begin with a brief description of some frequent attitudes observed among scientists in laboratories, who do not necessarily pay much attention to the foundations of quantum mechanics, even when they do experiments in quantum physics. In practice, they often use pragmatic rules, which are sufficient to interpret all their experiments, and prefer to avoid difficult questions about the very nature of the measurement process. For instance, one popular point of view is the "correlation interpretation", which can be considered as a "minimal interpretation" – minimal but sufficient for all practical purposes; it is accepted as a valid rule by a large majority of physicists, even those who prefer to supplement it with other elements in order to reach a more precise interpretation for the whole theory. We will then proceed to discuss various families of interpretations that are less common, including additional variables, modified Schrödinger dynamics, consistent histories, Everett interpretation, etc. All of them tend to change the status of the postulate of state vector reduction; some interpretations incorporate it into the normal Schrödinger evolution, others make it a consequence of another physical process that is considered as more fundamental, still others use a formalism where the reduction is hidden or even never takes place. But the general purpose remains always the same: to solve the problems and questions that are associated with quantum measurement, the coexistence of two postulates for the evolution of the state vector, and the emergence of macroscopic uniqueness.

11.1 Pragmatism in laboratories

Until now, the fundamental difficulties of quantum mechanics have had little impact in the practical application of quantum mechanics in laboratories. Fortunately, in a sense, physicists know how to avoid these difficulties by just applying common sense (or physical intuition), even if the logic behind the method may sometimes remain rather vague at a fundamental level. This attitude has proved fruitful, opening the way to many discoveries, which would not necessarily have been possible if the physicists had remained blocked at the level of the foundations; applying the theory as it is to concrete examples can be more productive. Moreover, for reasons that we have given in the introduction of Chapter 10, in practice there is often no need to apply the quantum postulates of measurement (Born rule and state vector reduction), since the calculation of quantum averages with the Schrödinger equation is sufficient. It is then rather natural that, in laboratories, the problems associated with the conceptual difficulties of quantum mechanics should not be given first priority.

Conceptually, of course, even very pragmatic physicists would not mind having some explanation for the uniqueness of macroscopic observations; logical consistency requires us to be able to somehow incorporate it into quantum mechanics. In a first step, we will limit ourselves to the description of two popular strategies. The first (§11.1.1) is to break the von Neumann chain by hand, when it "obviously" goes too far; the appearance of uniqueness is applied, so to say, above the formalism of the theory. Another popular strategy (§11.1.2) is to use the "correlation interpretation", a point of view where macroscopic events are indeed considered as unique by definition (or by common sense), and where the role of the theory is just to give probabilities relating preparation events to measurement and observation events. Other scientists prefer to emphasize the central role of information in a theory such as quantum mechanics (§11.1.3).

11.1.1 Common sense: breaking the von Neumann chain by hand

When physicists put an end to the infinite von Neumann chain, they often do it in an implicit way, based on physical intuition. Here we attempt to make the process explicit by discussing two examples: modified macroscopic decoherence, and the effect of consciousness.

11.1.1.a Modified decoherence

One possible empirical rule to apply is to consider that, as soon as "significant" de-coherence takes place, the von Neumann chain automatically stops: all its branches spontaneously disappear, except one. Nature chooses this branch by some un-known physical process. In other words, one systematically associates emergence of uniqueness with macroscopic decoherence. For instance, as soon as a measuring apparatus is inserted in the experiment, if it is able to register results, one consid-ers that it actually registers a single result, whether or not a human being observes it. As we have already seen (Chapter 7), there is no hope of ever observing the physical effects of coherent superposition when they have propagated too far in the environment; assuming that they have merely vanished therefore creates no risk of contradiction with experimental observations. The difficulty is, of course, to define the precise meaning of the word "significant" in this context.

Breaking the von Neumann chain in this way "by hand" is, after all, not very different from applying the state vector reduction postulate in a slightly different way: instead of the conscious act of measurement, one considers that decoherence at some scale triggers the reduction. In other words, one believes in the Schrödinger equation, but not too far: only until it contains correlations with the environment that become too macroscopic. Somewhere, decoherence sets the border between Schrödinger's world and Born's (introduction of Chapter 2). The "postulate of macroscopic decoherence" is not very different from Bohr's point of view, since it also involves a macroscopic world that is accessible to our human experience and is unique. It contains inherent nonlocality to explain Bell-type experiments: when two spins in a single state are measured at very distant places, the correlation with the macroscopic world and the modified decoherence will occur simultaneously at these places in a way that is nonlocal.

One could nevertheless see this view more as phenomenology than as a full in-terpretation of the theory; questions such as "Precisely at what degree of entangle-ment is the von Neumann chain resolved into a single branch?" are left to common sense and personal judgment; in this sense, the theory is not complete. Theories with modified Schrödinger dynamics (§11.10) are precisely built to rationalize this approach by introducing a precise physical mechanism to stop the von Neumann chain. They provide precise answers to such questions, but the mechanism allow-ing the emergence of a single branch is more intrinsic to the system (it may involve the masses of the particles) than necessarily related to the environment.

11.1.1.b Effect of consciousness

We have already mentioned in §2.3 what is sometimes called the "Wigner inter-pretation" of quantum mechanics [59], where the origin of state vector reduction is related to human consciousness. The idea is similar to breaking the von Neumann chain when it reaches a measurement apparatus, but here the chain stops as soon as the linear superposition involves different states of consciousness. Similar views have been discussed by London and Bauer in 1939 [57], who emphasize that state vector reduction restores a pure state from a statistical mixture of the measured subsystem (see §7.3.3), as well as "the essential role played by the consciousness of the observer in this transition from a mixture to a pure state"; they then explain this special role by the faculty of introspection of conscious observers. Others prefer to invoke "special properties" of the electrical currents that correspond to perception in a human brain.

In any case, common sense suggests that consciousness provides at least an upper limit, a boundary beyond which the von Neumann chain certainly does not propagate; whether or not this propagation stops exactly at this limit, or before, is another question. Indeed, Wigner may have seen the introduction of an influence of consciousness only as an extreme case (exactly as the Schrödinger cat was introduced by Schrödinger as a "ridiculous case"), just for illustrating the necessity of a nonlinear ingredient in order to predict definite results of experiments (we come back to modified Schrödinger dynamics in §11.10). In any event, the merit of the idea is also to show how the logic of the notion of measurement in the standard interpretation can be pushed to its limits: indeed, how is it possible to ascribe such special properties to the operation of measurement without considering that the human mind also has very special properties?

11.1.2 Correlation interpretation

As already noted, the correlation interpretation is, more or less, a component that is common to all other interpretations. Those who would consider that it is not really sufficient to be a full interpretation may even call it "minimalistic". But, precisely, the fact that it is a sort of common core makes it especially interesting. After all, anyone who finds it insufficient may add more elements to it, for instance more realism. Here we will use the words "correlation interpretation" to describe this point of view, since it puts the emphasis on the correlations between successive results of experiments.

In this interpretation, the uniqueness of macroscopic events (including acts of measurements) is not questioned, but postulated; one also postulates that they occur in a stochastic way. The purpose of the theory is just to provide probabilities associated with all possible situations of preparation, evolution, and measurements performed on the quantum system.

11.1.2.a Calculating the probability for a sequence of results in successive measurements

We now evaluate the probability associated with any sequence of measurements, performed at different times. Let us assume that a measurement[1] of a physical quantity associated with operator M is performed at time t_1, with possible results m_i, m_j, ...; this is followed by another measurement of observable N at time t_2, with possible results n_k, n_l, ..., of observable O at time t_3, etc. The system is initially described by a normalized pure state $|\Psi(t_0)\rangle$. Between time t_0 and time t_1, this state evolves from $|\Psi(t_0)\rangle$ to $|\Psi(t_1)\rangle$ according to the Schrödinger equation:

$$|\Psi(t_1)\rangle = U(t_1, t_0)|\Psi(t_0)\rangle \tag{11.1}$$

where $U(t', t)$ is the evolution operator between times t and t'; we can expand this new state into its components corresponding to the various results that can be obtained at time t_1:

$$|\Psi(t_1)\rangle = \sum_m |\Psi_m(t_1)\rangle \tag{11.2}$$

where $|\Psi_m(t_1)\rangle$ is obtained by applying to $|\Psi(t_1)\rangle$ the projector $P_M(m)$ on the subspace corresponding to result m obtained by measuring M (§1.2.2.a):

$$|\Psi_m(t_1)\rangle = P_M(m)|\Psi(t_1)\rangle \tag{11.3}$$

The terms inside the sum over m of (11.2) are all mutually orthogonal; they will never give rise to interference effects in the future, since they correspond to different results of measurement. Actually, each term becomes correlated to an orthogonal state of the environment (the pointer of the measurement apparatus for instance) and decoherence will cancel any interference effect (cf. §7.3.3). Instead of making a complete calculation including the state of the environment, here we take the simpler point of view where we ignore the environment and consider the various orthogonal components $|\Psi_m(t_1)\rangle$ of $|\Psi(t_1)\rangle$ as independent from each other.

From time t_1 to time t_2, state $|\Psi_m(t_1)\rangle$ evolves under the effect of the Schrödinger equation and becomes a state $|\Psi_m(t_2)\rangle$ given by:

$$|\Psi_m(t_2)\rangle = U(t_2, t_1)|\Psi_m(t_1)\rangle \tag{11.4}$$

For the second measurement, the procedure repeats itself; we expand this new state according to:

$$|\Psi_m(t_2)\rangle = \sum_n |\Psi_{m,n}(t_2)\rangle \tag{11.5}$$

---

[1] We assume that all measurements are ideal; if nonideal measurements are considered, a more elaborate treatment is needed.

where $|\Psi_{m,n}(t_2)\rangle$ is obtained by the action of the projector $P_N(n)$ on the subspace corresponding to result n obtained by measuring N:

$$|\Psi_{m,n}(t_2)\rangle = P_N(n)\,|\Psi_m(t_2)\rangle \qquad (11.6)$$

Again, we consider that the evolution of each $|\Psi_{m,n}(t_2)\rangle$ is independent and, if a third measurement is performed at a later time t_3, it will generate one more decomposition, and so on. One can then check (Appendix G) that the probability of any given sequence of measurements m, n, p, etc., is nothing but the square of the norm of the final state vector:

$$\mathcal{P}(m, t_1; n, t_2; p, t_3; \ldots; q, t_r) = \left\langle \Psi_{m,n,p,\ldots,q}(t_r) \,\middle|\, \Psi_{m,n,p,\ldots,q}(t_r) \right\rangle \qquad (11.7)$$

The probability is therefore merely the square of the norm of the "branch" of the state vector, after the action of all projections at various times corresponding to the whole series of results.

This formula can also be written in terms of the initial density operator $\rho(t_0)$:

$$\rho(t_0) = |\Psi(t_0)\rangle\,\langle\Psi(t_0)| \qquad (11.8)$$

In the Heisenberg picture (for more details, see Appendix G), the projectors $P_M(m)$ and $P_N(n)$ become time-dependent operators $\widehat{P}_M(m, t)$ and $\widehat{P}_N(n, t)$. For two measurements, the probability of obtaining result m followed by result n can then be written as[2]:

$$\mathcal{P}(m, t_1; n, t_2) = Tr\left\{\widehat{P}_N(n, t_2)\widehat{P}_M(m, t_1)\rho(t_0)\widehat{P}_M(m, t_1)\widehat{P}_N(n, t_2)\right\} \qquad (11.9)$$

Relation (11.9) is sometimes called the Wigner formula[3]. It can easily be generalized to more than two measurements by inserting additional projectors on both sides in the reverse time order, and to situations where $\rho(t_0)$ describes a statistical mixture instead of a pure state.

11.1.2.b Getting rid of state vector reduction

Equations (11.7) and (11.9) can be seen as a consequence of the postulate of state vector reduction. Conversely, it is also possible to take either of these equations as a starting point, which then becomes a postulate in itself giving the probability of any sequence of measurements in a perfectly unambiguous way. The postulate of state vector reduction then becomes superfluous, since the generalized (multitime) Born rule given by (11.7) or (11.9) is sufficient to predict the probabilities of any sequence of measurements – one could still argue that the state vector reduction is

---

[2] Using circular permutation under the trace, one can in fact suppress one of the extreme projectors $\widehat{P}_N(n_2; t_2)$ in formula (11.9), but not the others.

[3] See equation (12) of [65, 513].

also contained in some way in the trace operation of (11.9), but no explicit reference to it is indispensable. From this perspective, the projection of the state vector resulting from measurement is no longer a postulate, but just seen as a convenient rule to make calculations, which can be derived from another postulate. As for the Schrödinger evolution, it is contained in the Heisenberg evolution of projection operators, so that the evolution of $| \Psi >$ is no longer directly apparent.

It is of course still necessary to postulate that the results of any measurements can give only one of the eigenvalues of the corresponding operator, and that the result is fundamentally random, which is essentially the content of the Born rule. The advantage is that, if one just uses formula (11.9) and ignores state vector reduction, the problems associated with a difficult coexistence of two different evolution postulates disappear; no discontinuous jump of any mathematical quantity ever occurs in the formalism. Why not then give up entirely the other postulates and just use this single formula for all predictions of results?

This is indeed the best solution for some physicists: if one accepts the idea that the purpose of physics is only to correlate the preparation of a physical system, contained mathematically in $\rho(t_0)$, with all possible sequence of results of measurements (by providing their probabilities), it is true that nothing more than (11.9) is needed. Why then worry about which series of results is realized in a particular experiment? There is no need in physics to do more than just give rules for the calculation of probabilities associated with the various preparation and measurement procedures (see for instance Dirac's quotation in §2.5). No attempt is made to describe the physical system itself during each realization of an experiment.

The "correlation interpretation" is perfectly consistent; it goes well with the view where a state vector expresses a preparation procedure (§1.2.3) rather than a physical property of the measured system. On the other hand, it is the complete opposite of the EPR reasoning, since it shows no interest whatsoever in questions related to physical reality as something "in itself". Questions such as "How should the physical system be described when one first measurement has already been performed, but before the second measurement is decided" are dismissed as superfluous or meaningless[4]. Needless to say, the notion of the EPR elements of reality itself becomes completely irrelevant to physics, which automatically solves all potential problems related to Bell-, GHZ-, and Hardy-type considerations. The same is true of the emergence of a single result in a single experiment; in a sense, the Schrödinger cat paradox is eliminated by putting it outside of the scope of physics,

---

[4] Suppose for instance that the polarization of a photon is measured nondestructively somewhere in Europe, that it propagates along a polarization preserving optical fiber, and reaches America where another polarization measurement is performed. In practice, most physicists consider intuitively that "something having physical properties" has propagated from one site of measurement to the other, and that physics has something to say about these properties. In the pure correlation interpretation, this intuition should be rejected: nothing real that physics could describe propagates along the fiber.

because the paradox is not expressed in terms of correlations. An interesting feature of this point of view is that the boundary between the measured system and the environment of the measuring devices is flexible; an advantage of this flexibility is that the method is well suited for successive approximations in the treatment of a measurement process, for instance the tracks left by a particle in a bubble chamber as discussed by Bell [55].

11.1.2.c Discussion

In practice, most physicists who favor the correlation interpretation do not feel the need for making it very explicit. Some nevertheless do; see for instance the article by Mermin [514], which starts from the statement: "Throughout this essay, I shall treat correlations and probabilities as primitive concepts". In a similar context, see also a recent "opinion" in *Physics Today* by Fuchs and Peres [110], who emphasize "the internal consistency of the theory without interpretation". On the other hand, the correlation interpretation is seen by some physicists as minimalistic because it leaves aside, as irrelevant, questions they find important; the best example is probably the notion of physical reality seen as an entity that should be independent of measurements performed by human beings. As we have already mentioned, this interpretation can easily be supplemented by other elements to make it more specific. Experience shows that defenders of the correlation point of view, when pressed in a discussion to describe their point of view with more accuracy, often express themselves in terms that come very close to the Everett interpretation (§11.12); in fact, they may sometimes be proponents of this interpretation without realizing it!

Let us finally mention in passing that formula (11.9) may be the starting point for many interesting discussions, taking it as a convenient formula, not necessarily basic in the interpretation. Suppose for instance that the first measurement is associated with a degenerate eigenvalue of an operator, in other words that $\widehat{P}_M(m; t_1)$ is a projector over a subspace of more than one dimension:

$$\widehat{P}_M(m; t_1) = \sum_{i=1}^{n} |\varphi_i\rangle \langle\varphi_i| \qquad (11.10)$$

(For the sake of simplicity we assume that $t_1 = t_0$, so that no time dependence appears in this expression). Inserting this expression into (11.9) immediately shows the appearance of interference terms (or crossed terms) $i \neq j$ between the contribution of the various $|\varphi_i\rangle$. Assume, on the other hand, that more information was actually obtained in the first measurement, so that the value of i was also determined, but that this information was lost, or forgotten; the experimenter ignores which of two or more i results was obtained. Then, what should be calculated is the sum of the probabilities associated with each possible result, which is a single sum over i from which all crossed terms $i \neq j$ have disappeared. In the first case, interference terms arise because one has to add probability amplitudes; in the second, they do not because one has to add the probabilities themselves (exclusive events). The contrast between these two situations may be understood in terms of decoherence: in the first case, several states of the system correlate to the same state of the measuring apparatus, which plays here the role of the environment; they do not in the second case, so that by partial trace all interference effects vanish. This remark is useful in the discussion of the close relation between the so-called "Zeno paradox in quantum mechanics" [515] and decoherence; it is also basic in the definition of consistency conditions for decoherent histories, to which we will come back later (§11.7).

11.1.3 Emphasizing information

Another popular point of view is to emphasize the role of information [398, 516–521, 523]; with the rise of the field of quantum information, the informational point of view is natural. Information can be about the whole experimental setup, including preparation and measurement devices; in this case, the idea becomes very similar to Bohr's emphasis on the relevance of the whole experimental setup (§3.3.2). Information can also include components that are acquired when experimental results become known, and undergo sudden changes when the results are observed; this property can be invoked to explain the von Neumann state vector reduction, seen as a "purely mental process" (Appendix A).

An interesting point is that any measurement process gives rise to a flow of information from the location where the measurement takes place to the environment. This is because the interaction between the measured system and the measurement apparatus, and then the environment, initiates a von Neumann chain, during which entanglement progresses further and further in the environment [521]. Consider a fixed volume containing all the measurement apparatus; as long as the entangled chain does not leave this volume, since any Hamiltonian evolution preserves pure states as well as entropy, the amount of entropy contained in the volume remains constant; but, as soon as entanglement proceeds beyond the fixed volume, the properties of the physical system contained inside the volume must be obtained by a partial trace (over the system outside of the volume), and the entropy contained in the volume increases (§7.2.3). To the local observer, the effect of this leak of entanglement towards the outside world appears as an entropy production – as an illustration, see for instance Peres's quotation in §11.3.2 and his discussion of the flux of information in an EPR experiment, and [524].

This informational point of view can be applied in a more or less strict way. One may focus the interest on the information content of the state vector[5], or consider

---

[5] Here is for instance how Fuchs [519] describes the program of informational quantum theory: "The quantum

that the nature of the state vector itself is mostly informational, or even take the more extreme view where it is purely informational (§11.3.2). The PBR theorem (§6.5.2) provides constraints about the relations between an informational content of the state vector and the possible existence of an underlying physical reality.

This point of view has obvious relations with the correlation interpretation, and can actually be used to complement it. It seems to explain the discontinuities arising from state vector reduction rather naturally, but of course the difficulties already discussed in §§1.2.3.a and 11.1.2 remain (for instance, questions about the division of the world into systems providing information and those on which information is acquired, or the description of independent reality during the experiment). Wigner's friend paradox is not problematic since, as long as the friend outside the closed laboratory has less information than the other, he continues to use an unreduced state vector, while the friend inside has already reduced his.

11.2 Ensemble interpretations

The terms "ensemble interpretation" of quantum mechanics and "statistical interpretation" are synonymous; both can be found in the literature. This class of interpretations does not discard the idea of describing physical systems, but specifies that the description given by the state vector applies only to ensemble systems prepared in identical conditions, not single systems (or single experiments). Einstein supported this point of view; for instance, in a letter to Schrödinger [62] written in 1932, he writes: "The Ψ function does not describe the state of a single system but (statistically) an ensemble of systems". The function Ψ would then contain an information that is similar to the information contained in the phase space distribution of classical statistical physics – which is not the most precise possible description within such a theory. As we have seen in Chapter 3, the EPR argument leads to the ensemble interpretation. A review of these interpretations has been given in 1970 by Ballentine [525], who writes: "Several arguments are advanced in favour of considering the quantum state description to apply only to an ensemble of similarly prepared systems, rather than supposing, as is often done, that it exhaustively represents an individual physical system. Most of the problems associated with the quantum theory of measurements are artifacts of the attempt to maintain the latter interpretation". This author distinguishes two classes of interpretations:

(i) "The statistical interpretation ... according to which a pure state ... need not provide a complete description of an individual system.

system represents something real and independent of us; the quantum state represents a collection of degrees of belief about *something* to do with that system ... The structure called quantum mechanics is about the interplay between these two things – the subjective and the objective. The task before us is to separate the wheat from the chaff. If the true quantum state represents subjective information, then how much of its mathematical support structure might be of the same character? Some of it, maybe most of it, but surely not all of it."

(ii) Interpretations which assert that a pure state provides a complete and exhaustive description of an individual system. This class contains ... several versions of the Copenhagen interpretation".

He then adds that "hypothesis (ii) is unnecessary for quantum theory, and moreover leads to serious difficulties". Other authors have expressed similar points of view; see for instance [122], where an alternative to the orthodox interpretation of quantum theory is proposed in the same line. For an example of explicit disagreement with the statistical interpretation, see for instance [45].

Within the ensemble interpretation, two different attitudes become logically possible:

(i) either one considers that the fact that a pure state description is valid for ensembles only, but not for a single system, signals that this description cannot be complete; more variables are needed to specify which particular system is considered inside the ensemble. This leads to introducing new elements of description that come in addition to the state vector[6], which leads to theories with additional (hidden) variables.

(ii) or one considers a theory that describes only ensembles of systems as perfectly satisfactory. If a single experiment is performed, one decides that a fundamentally random process takes place and makes a single result appear; no attempt is made to describe this process. This is, in a sense, a modern version of the historical "quantum jumps". Pushing this attitude to the extreme would lead to what Leggett [526] calls the "extreme statistical interpretation, according to which no physical meaning should be attached to the QM formalism either at the microscopic or at the macroscopic level".

Not all physicists favoring this interpretation make an explicit choice among these different possibilities. Nevertheless, with or without explicit reference to additional variables, one can find a number of authors who support the idea that the quantum state vector should be used only for the description of statistical ensembles. This general discussion is of course related to the status of the wave function, which, as we have discussed in §1.2.3 of Chapter 1, can be seen as describing either the quantum system itself, or only our knowledge about this quantum system. An interesting discussion of the subject is given by Aharonov et al. in [306], under the title "Meaning of the wave function". After remarking that "Since the discovery of quantum theory, a very fundamental question has haunted physicists: what is the physical meaning of the wave function?" the authors explain how it is possible to determine the time evolution of a state vector by actual measurements. This is done by considering a series of measurement that last a long time, named "protective measurements", during which the wave function is prevented from changing

---

[6] In his article, Ballentine remarks that "the introduction of hidden variables is fully compatible with the statistical predictions of quantum theory", and discusses the properties of these variables at the end of his article.

appreciably by means of another interaction it undergoes at the same time. To obtain a complete determination of the state vector, the method requires to perform measurements under various experimental conditions; it is therefore necessary to have access to a large sample of identically prepared quantum systems.

The authors of Reference [527] have developed a theory of ideal quantum measurements relying on subensembles: inside the whole ensemble of all realizations of a given measurement experiment, as soon as the measured system S and the pointer of the measurement apparatus M have become strongly correlated according to the von Neumann model, they distinguish subensembles of realizations. The smallest subensembles are of course individual realizations. It is assumed that these subensembles can be described by the same formalism and dynamic equations than the whole ensemble. From a detailed study of the relaxation created by the interaction between S and M, and from series of adequate interpretative principles, the authors of Reference [527] propose a progressive introduction of the von Neumann postulate of projection; it is then expressed in terms of subensembles of realizations of the experiment, and connected to precise properties of the dynamics of the interaction between S and M. No additional variables are introduced, but an ensemble of additional density operators obeying the standard equation of evolution; they actually have the same purpose: providing a more precise description of a single experiment than the single standard density operator.

11.3 Relational interpretation, relative state vector

Other possible interpretations of quantum mechanics emphasize the relative character of the state vector: several observers may use different state vectors to describe the same sequence of events, even if they use the same Galilean frame.

11.3.1 Relational interpretation

The relational point of view on quantum mechanics, introduced by Rovelli in 1996 [528, 529], is inspired by relativity, where different observers may use different times if they use different Galilean frames. In the relational interpretation, the notion of an absolute, observer-independent, state of a physical system is rejected: different observers may give different accounts of the same series of events by using different state vectors. But the difference does not arise from the use of different Galilean frames; it arises from different informations available to the observers, or on a more concrete basis from the use of different measurement apparatuses to make observations. Physical properties of systems are then not seen as absolute, but as depending on the apparatus used to measure the property. Quantum mechanics is considered as "a theory about information". In the words of Rovelli (§2.3

of [528]): "A quantum description of the state of a physical system S exists only if some system O, considered as an observer[7], is actually 'describing' S or, more precisely, has interacted with S. The quantum state of a system is always a state of that system with respect to other systems".

The emphasis is clearly put on the relations between physical objects, not the physical objects themselves. In an EPR experiment for instance, if the spin of one particle only is measured, this has no effect whatsoever on the state of the other spin, since it is not in relation with any measurement apparatus; nothing like a nonlocal effect of state vector reduction happens. In a second experiment, if the spins of the two particles are now measured along the same direction, one can then observe a property of the pair, namely that the results are always opposite; but this cannot be related to any property of the system in the first experiment, since the physical properties of the system with respect to a different apparatus may have completely changed. If finally the two spins are measured along different directions, one observes still other physical properties of the pair, with no special relation to those observed in the two first experiments.

In this interpretation, quantum state vector reduction (or collapse) becomes very different from what it is in the standard view. For instance, for a given quantum system, some observers may consider that state collapse has already occurred while, for others, collapse has not yet taken place – for them the system remains in a coherent superposition. Consider again an EPR experiment [530]: when Alice performs her experiment, a reduction occurs for her state vector describing her spin in relation with her measurement apparatus, and of course herself having observed a single result. Nevertheless, Bob, who is far away, knowing that she has performed an experiment but not knowing the result, describes the same ensemble as a coherent superposition involving all possible results. In other words, for Bob, Alice has become a Schrödinger cat. Similarly, if Bob also performs a measurement along a different spin direction, this reduces his own state vector, but Alice does not know the result and considers that he has put himself into a coherent superposition with respect to her and her apparatus. Each experimenter considers that the other is a Schrödinger cat, and this remains true until they communicate to exchange their results; the cats then disappear and both operators use the same state vector again. This example illustrates that, in this interpretation, the state vector does not directly describe reality, but rather the information that is available on this reality (not necessarily available to someone, but possibly relative to a measurement apparatus for instance).

In Bohr's point of view, the role of the experimental apparatus is also empha-

---

[7] System O can be understood as a measurement apparatus. This sentence then implies that (in a given Galilean frame) a particle can have a definite position, or velocity, or spin direction, etc. only with respect to the specific apparatus used to measure this property, never in an absolute way.

sized, but it is essential to consider it as a whole. Here, various parts of the apparatus may be distinguished, and may actually introduce different state vectors. Moreover, in the Copenhagen interpretation, measurement apparatuses and observers play a central role, as well as differences between the microscopic world and the world that is directly accessible to our human experience. In the relational interpretation, this becomes unnecessary. All physical systems are considered equal, with no mention of the necessarily macroscopic character of measurement apparatuses: the measurement process is just an ordinary interaction process, with no special role in the theory. The process can even take place in the absence of any observer, since the state vector and its reduction can be defined in relation with the measurement apparatus only.

As Rovelli summarizes in [528], "Quantum mechanics is a theory about the physical description of physical systems relative to other systems, and this is a complete description of the world". Since the relational interpretation considers that state vectors provide a complete description of physical reality, it is of course in complete opposition with the EPR argument, additional variables, etc.

11.3.2 *Pure informational point of view*

As we saw in §11.1.3, some physicists emphasize the informational content of the state vector. One can go further in this direction and take an extreme informational point of view, where $|\Psi\rangle$ is completely disconnected from physical reality; what is then considered is only the information content of the state vector $|\Psi\rangle$, which describes the information that one observer has concerning a given physical system (§11.1.3), but says nothing about the system itself. The state vector $|\Psi\rangle$ then relates to mental information only (Appendix A) and may take almost arbitrary values for the same system, depending on the amount of knowledge one ascribes to an arbitrary observer. This is analogous to a classical probability distribution, which also expresses a relation between some amount of knowledge and some independent reality; for instance, such a distribution may sometimes describe the system with perfect accuracy, but sometimes also contains no information at all (if the observer knows nothing about the system).

11.3.2.a *Discussion*

Consider an experiment that, in ordinary language, one would describe as: "A device at point A produces a particle, which propagates in space to another remote point B, where its interactions with a target are observed and recorded" – it can be a proton that is injected into the big CERN accelerator and collides with a target kilometers away. During the experiment, a wave packet associated with the state vector propagates from A to B. Nevertheless, from a strict informational point of view, one should consider that "nothing real" travels along within this wave packet: it is just a transfer of abstract information. The real world includes only the experimental devices and the observations; what propagates in between is just information about what the future experimental observations with the measurement apparatuses can be.

In [118], the words of Peres nicely illustrate this point of view, applied to the discussion of EPR experiment: "When Alice measures her spin, the information she gets is localized at her position, and will remain so until she decides to broadcast it. Absolutely nothing happens at Bob's location ... It is only if and when Alice informs Bob of the result she got ... that Bob realizes that his particle has a definite pure state. Until then, the two observers can legitimately ascribe different quantum states to the same system ... Quantum states are not physical objects: they exist only in our imagination ... The question raised by EPR 'can quantum mechanical description of physical reality be considered complete' has a positive answer. However, reality may be different for different observers".

Most physicists are prepared to accept the idea of observer-dependent density operators: if different observers have different amounts of information about the same physical system, they may describe it with different statistical mixtures, and therefore by different density operators ρ. The analogy with a classical distribution function is then obvious, since ρ does not correspond to the most precise possible description of a system within quantum mechanics; it actually attributes probabilities to several such descriptions (with state vectors, or pure states). But an observer dependence of state vectors $|\Psi\rangle$, or equivalently with density operators ρ associated with pure states, seems much more delicate to accept: pure states do not leave any room for a more precise quantum description. Moreover, most physicists would probably agree that, at least sometimes, the wave function does contain elements of reality; for instance, a system described by a ground state wave function is really in the ground state and has the properties of this state, not only in the mind of humans. The electrons of a superconductor at very low temperature are really in a BCS (Bardeen–Cooper–Schrieffer) ground state; the BCS wave function is not a mental process, but a description of the system itself. Similarly, chemists think that molecular orbitals describe real properties of molecules such as shape, size, etc., and not just knowledge of these properties. The standard view is rather that the Schrödinger evolution contains at the same time a physical evolution for those properties that quantum mechanics does attribute to the system (observables that have $|\Psi\rangle$ as an eigenvector), but also an evolution of probabilities (for all other observables) that represent only our knowledge of the system, and therefore can indeed be seen as mental.

We have already discussed the difficulties of the pure informational point of view in §§1.2.3.a and 11.1.3. If $|\Psi\rangle$ is only information, what is this information about?

How can we distinguish between systems that produce information and the systems on which information is acquired? If we accept the idea of an independent physical reality, how should we then describe the system itself? Does this mean that this very idea should be abandoned, and that physics completely gives up any idea of saying something about independent reality? In some early writings, Heisenberg seemed not to be very far from the pure information point of view, but later he took a more moderate point of view. As for Bohr, he was not a positivist but a realist in his own way (see §2.5, for example Von Weizsäcker's quotation at the end); the pure information point of view is usually considered as beyond the standard interpretation (§1.1.3).

11.3.2.b QBism

A recent version of this class of interpretations is QBism, an abbreviation for Quantum Bayesianism [531–533], which is based on the use of Bayesian probabilities. The most frequent definition of probabilities is the frequentist, or ensemble, definition: one assumes the existence of an ensemble of identically prepared physical systems, which provides many realizations of the same initial situation. The probability of a given event is then defined as the proportion of realizations, among all possible in the whole ensemble, where this particular event occurs; in other words, it is the relative frequency of occurrence of this event. By contrast, in the subjective Bayesian definition, an "agent" assigns a probability according to his/her reasonable expectation that the event will occur; this definition is more natural for events that cannot occur more than once (for instance, the probability that it will rain tomorrow). In QBism, the quantum state associated with a physical system characterizes the probabilities that this agent assigns to the results of possible future measurements. The quantum state is therefore not associated with the physical system itself only, or its preparation, but with a couple agent + system; two different agents may then assign two distinct quantum states to the same physical system. The quantum state is subjective and summarizes the knowledge of agents and their beliefs about the content of their future experience. In the words of Fuchs [532]: "For the Quantum Bayesian, quantum theory is not something outside probability theory ... but rather it is an addition to probability theory itself".

Under these conditions, discontinuous jumps of the state vector are of course perfectly natural, as natural as sudden jumps of classical probabilities when some new information is taken into account. The observers (called agent in this context) play a completely central role, since the whole quantum formalism deals only with their knowledge and assignments, having very little to do with the real world. There is still a frontier (a split) between different kinds of situations : ordinary Schrödinger evolution and situations where information is gathered by observers. Fuchs [532] notes that "In contemplating a quantum measurement, one makes a conceptual split in the world: one part is treated as an agent, and the other as a kind of reagent or catalyst". But this frontier is no longer objective; as Mermin writes [54] "The splits reside, not in the objective world, but at the boundaries between that world and the experiences of the various agents who use quantum mechanics". State vector reduction can only occur when an agent has the "mental facility to use quantum mechanics to update its state assignments on the basis of its subsequent experience". We then recover some analogy with Wigner's suggestion to take into account the effects of the consciousness of the observer (§11.1.1.b). Similar to Bohr, QBism considers that "The primitive concept of *experience* is fundamental to an understanding of science"; nevertheless, "Unlike Copenhagen, QBism takes the 'subjective' or 'judgmental' or 'personalist' view of probabilities" [534].

11.4 Logical, algebraic, and deductive approaches

In this section, we discuss a family of quantum theories where the state vector $|\Psi\rangle$ is not considered as the starting point for describing a quantum physical system. The state vector is rather seen as a mathematical tool that can be derived from more basic considerations concerning the possible statements that can be made about the properties of a physical system. We will briefly review two classes of theories: quantum logic, which is related to logic as a branch of philosophy applied to quantum phenomena, but also involves mathematics; and algebraic, C* algebra, and axiomatic theories, which are more purely mathematical, and also borrow from both logic and mathematical formalism. In the last part of this section, we discuss a general theorem, the Gleason theorem, which has interesting applications in these theories and, more generally, in all quantum mechanics.

The purpose of these approaches is to construct quantum mechanics ab initio from basic and very general principles, without for instance starting from classical mechanics and applying quantization procedures to it. The strong interest of such a logical construction is that it clarifies the deep structure of quantum mechanics. It remains nevertheless somewhat abstract and axiomatic; in fact, what it provides is more another method to introduce quantum mechanics than a better understanding of its physical content. We will therefore limit ourselves to a brief description, which remains superficial, but includes many references for the reader who wishes to learn more on the subject.

11.4.1 Quantum logic

A general historical presentation of quantum logic can be found in Chapter 8 of Jammer's book *The Philosophy of Quantum Mechanics* [58], with many references. Von Neumann, in his famous 1932 treatise [4], remarked that "the relations between a physical system on one hand, and the projections on the other, make possible a sort of logical calculus with these. However, in contrast to the concepts of ordinary logic, this system is extended by the concept of 'simultaneous decidability', which is characteristic of quantum mechanics". In 1933, an astrophysicist, Fritz Zwicky, suggested that nonclassical logic should be used in microphysics [535]; he proposed a "principle of flexibility of scientific truth", which "must be many-valued". In 1936, Birkhoff and von Neumann attempted to reconcile the apparent inconsistency of classical Boolean logic with the rules applying to the measurement of incompatible observables in quantum mechanics [536]; they pointed out that the characteristic difference between the logical structures of classical and quantum mechanics is that distributive identities hold in the former, but not in the latter[8]. Jordan also published a few articles on the subject around 1950 [539].

To understand in simple terms why, we have to study the role of projectors in the quantum formalism. Consider for instance an observable A (a Hermitian operator acting in the space of states) and one of its eigenvalues a_i. We call P_i the projector over the eigensubspace $\mathcal{E}_i$ associated with this eigenvalue (if the eigenvalue is denegerate, this subspace has a dimension larger than 1). This projector is Hermitian, and obeys the relation $(P_i)^2 = P_i$; its eigenvalues are 1 and 0. If the quantum state of the system is an eigenstate of P_i with eigenvalue 1, a measurement of A provides result a_i with certainty; one can then consider that the proposition "the value of observable A is a_i" is a true proposition. But, if the quantum system is an eigenstate of P_i with eigenvalue 0, a measurement of A will never provide result a_i; one can consider the proposition "the value of observable A is a_i" a false proposition[9].

We can therefore associate with every projector P_i a test made on the physical system in order to determine whether a proposition concerning the system is true or false. A correspondence is established in this way between projectors P_i and logical propositions π_i. It is easy to see that the supplementary projector $1 - P_i$ over the supplementary subspace of $\mathcal{E}_i$ is associated with the opposite proposition "not π_i"; in this way, we introduce the logical negation of a proposition.

Now, if we consider two commuting projectors P_i and P_j, one can also see that the projector over the intersection of the corresponding subspaces corresponds to the proposition "π_i and π_j"; the projector over the direct sum of the subspaces corresponds to the proposition "π_i or π_j". With three commuting projectors, one obtains the standard distributivity of logical propositions:

$$\pi_i \text{ and } (\pi_j \text{ or } \pi_k) = \left(\pi_i \text{ and } \pi_j\right) \text{ or } (\pi_i \text{ and } \pi_k) \tag{11.11}$$

---

[8] Soon after the publication of the Birkhoff and von Neumann article, Strauss criticized it in his Ph.D. dissertation (1938); he then introduced another version of logic, "complementarity logic" [537]. Later, in 1968, Popper also criticized the Birkhoff and von Neumann article, but this time by considering that it is logically inconsistent [538]; a debate ensued, in particular with Ramsay and Pool, which will not be reported here.

[9] We will come back to the relation between measurements and projectors in the discussion of the Gleason theorem (§11.4.3).

By contrast, if the projectors do not commute, this distributivity law is not obeyed in general, which completely changes the way propositions can be used. Quantum logic therefore studies the formal structure of deductive reasoning and of propositional statements obeying the same rules as projectors, in order to provide a set of rules that are compatible with the principles of quantum mechanics. This leads to the study of the "lattice of propositions" and "orthomodularity". For more details, we refer to a clear and concise introduction to the subject given in chapter 4 of the recent book *The Formalisms of Quantum Mechanics: An Introduction* by David [540]. For a simple introduction of quantum logic, see for instance [541]. More specialized books are for instance *Quantum Logic* by Mittelstaedt [542] or *The Logic of Quantum Mechanics* by Beltrametti and Casinelli [543]; many useful references can also be found in [544]. For a general and recent review on quantum logic and its relations to the modal interpretation of quantum mechanics (§11.9), see [545]. In [546], Bell briefly reviews the origins of quantum logic, and discusses the fundamental problem of its meaning; he proposes the introduction of "proximity spaces", which leads to the introduction of "attributes "defined over these spaces.

Other versions of quantum logic have been proposed. Many-valued logic corresponds to a version of quantum logic that has been proposed by Reichenbach in the early 1940s [547]. In his "probability theory of meaning", he considers that "a proposition has a meaning if it is possible to determine its weight, i.e. a degree of probability, for the proposition", as well as similar rules for groups of propositions. Later, Von Weizsäcker [548] proposed a different version of complementarity logic, where "every elementary proposition can have, apart from 1 and 0, a complex number as its truth value". The square modulus of this value gives, as in standard quantum mechanics, the probability that an experimental test will verify the logical proposition; the phase of the complex number is related to the complementary logical alternative. This logical construction is analogous to an infinite valued logic.

11.4.2 Algebraic theory, formal axiomatic theory

In quantum logic, we have seen that the purpose is to show how the structure of observables (Hermitian operators acting in the space of states) can emerge from basic assumptions concerning the operations of measurement performed on quantum physical systems. Once this structure of observables is obtained, one can go one step further and build the space of states itself, introduce the notion of physical state, etc. By contrast, in the algebraic formulation of quantum mechanics, the starting point is directly the algebra structure of observables, which is postulated, and one studies the properties of the space in which they act and of the physical states inside this space. A standard reference book on the algebraic approach to quantum theory is *Local Quantum Physics: Fields, Particles, Algebras* by R. Haag [549]; chapter 3 of the book by David [540] provides a heuristic, but also more physical, introduction to the subject for the interested reader. As before we only give a very brief, mostly historical, summary here.

The axiomatic approach and the "C* algebra theory" is a mathematical view of quantum mechanics involving algebras of operators and their representations. Initially, it arose from the need to develop a quantum theory treating rigorously systems with an infinite number of degrees of freedom, especially in quantum field theory but also in statistical mechanics. Von Neumann also initiated the field by considering algebras of operators [4], and continued his work with Jordan and Wigner [550]. In addition to the "von Neumann algebras", Gelfand and Naimark [551] introduced another class of algebras, now named "C* algebras", which were also used in the contributions of Segal [552] and later of Haag and Kastler [553]. Murray and von Neumann had given a classification of the various types of algebras [554], with type I, type II, etc. Type I algebras are appropriate to quantum mechanics with a space of states of finite dimension only. Von Neumann used type II algebras, but it turns out that the properties of III are actually necessary for quantum field theory (for a brief review, see for instance §§9.2 and 8.3 of [544] and §3.7.2 of [540]). A important clarification in this field was introduced by Connes, who gave a classification of algebras of type III [555] and further introduced the notion of noncommutative geometry [556].

In the line initiated by von Neumann, Hilbert, and Jordan, Mackey [557] provided a formalism for relating observables, system preparation filters, and states, and introduced an axiomatization of quantum mechanics based on two fundamental notions, observables and states. This was the first complete system of axioms for a formal theory. The work was then expanded by Piron [558] who, with Jauch [559], showed that one can reconstruct a Hilbert lattice (and not only an orthomodular lattice) by introducing an appropriate "covering property". For a more detailed review of the role of the Geneva school in this field, see §8.5 of [58]. A more recent and important contribution is the Solèr theorem [560], which, on the basis of algebraic assumptions, proves that the Hilbert space must be built either from real numbers or complex numbers (as in standard theory), or quaternions. An introduction to axiomatic quantum field theory is given in §3.8 of [540].

The precedings theories are not free of difficulties, sometimes illustrated by counterexamples. One of them is that, while the structure of a Hilbert space provides a natural way to treat the composition of several physical systems, with the notion of the tensor product (§12.2), one can not merely multiply Hilbert lattices. In fact, combining two or more physical systems may introduce new mathematical structures. To treat this problem, it has been proposed to introduce convex sets and the language of categories [561–563]; for a review, see [564].

11.4.3 Gleason theorem

A theorem due to Gleason [565] is useful in the context of this family of presentations of quantum mechanics. Consider a space of states $\mathcal{E}$ and the ensemble of the orthogonal projectors P_j onto all its subspaces[10]. Any orthonormal basis $\{|u_i\rangle\}$ of $\mathcal{E}$ immediately provides examples of such projectors, for instance projectors onto single states:

$$P_j = |u_j\rangle\langle u_j| \tag{11.12}$$

A sum of an arbitrary number of these projectors (without including the same operator twice or more) also gives a possible P_j:

$$P_j = |u_i\rangle\langle u_i| + |u_k\rangle\langle u_k| + .. |u_k\rangle\langle u_k| \tag{11.13}$$

since the $|u_i\rangle$ are mutually orthogonal. Moreover, any basis $\{|u_i\rangle\}$ can be chosen, so that it is clear that the ensemble of the P_i is large; it is neither limited to projectors onto single states, nor to projectors that are mutually orthogonal.

As we recalled in §11.4.1, each P_j is an Hermitian operator with two eigenvalues, 0 and 1. Physically, one may associate a measurement to P_j, with results 1 or 0. If the state $|\Psi\rangle$ of the system is invariant under the action of P_j, the result 1 is obtained with certainty; if $|\Psi\rangle$ is cancelled under the action of P_j, the result 0 is obtained with certainty. In all other situations, namely when the state is changed but not cancelled under the action of P_j, the result is randomly either 1 or 0; the probability $\mathcal{P}_j(1)$ of obtaining 1 is given by the average of P_j in $|\Psi\rangle$:

$$\mathcal{P}_j(1) = \langle\Psi| P_j |\Psi\rangle \tag{11.14}$$

A particular operator P_j is the identity operator $I(\mathcal{E})$ in $\mathcal{E}$, corresponding to a measurement having a result 1 with certainty, whatever the state of the system is.

The sum of operators P_j is not necessarily another operator P_n. It is a projector if all the P_j project over subspaces that are orthogonal, or equivalently if all the binary products P_iP_j vanish[11]. We then call them "mutually exclusive projectors", or just "exclusive projectors" for simplicity. Mathematically, they are exclusive because it

---

[10] To define P_j, one chooses a subspace $\mathcal{E}_S^j$. Any ket $|\Psi\rangle$ of $\mathcal{E}$ can be written as the sum of a vector $|\Psi\rangle_j$ belonging to $\mathcal{E}_S^j$ and a vector $|\Psi\rangle_j^\perp$ orthogonal to $\mathcal{E}_S^j$. The definition of P_j is then $P_j|\Psi\rangle = |\Psi\rangle_j$.
Orthogonal projectors are Hermitian. If the product P_jP_k of two operators vanishes, since the Hermitian conjugate P_kP_j also vanishes, P_j and P_k are commuting operators. Orthogonal projectors are a special case of the ensemble of projectors acting inside a space (oblique projectors), satisfying the relation $P^2 = P$.

[11] If two operators P_i and P_j project over orthogonal subspaces, one can choose orthonormal bases in each of these subspaces, and complete an orthonormal basis in $\mathcal{E}$ by adding a sufficient number of normalized vectors that are mutually orthogonal and orthogonal to the two subspaces. All these vectors are common eigenvectors

is impossible to find any ket $|\Psi\rangle$ that does not vanish under the successive action of the two orthogonal projectors. Physically, the projectors correspond to mutually compatible observables that can both be measured at the same time; when they are exclusive, no state exists for which two results 1 are certain.

Adding a sufficient number of orthogonal P can then lead to a point where their sum projects over the whole space $\mathcal{E}$ and gives the identity operator:

$$P_{j_1} + P_{j_2} + \ldots + P_{j_p} = I(\mathcal{E}) \tag{11.15}$$

One then says that this series of projectors provides a "resolution of unity". An obvious example is provided by the series of projectors over all kets $|u_i\rangle$ of any orthonormal basis of $\mathcal{E}$, but one can also group these projectors in any way to obtain other decompositions of identity.

Suppose now that a real function f is defined over all possible P_j: a real number is assigned to each of these projectors. Moreover, we assume that f has the following properties:

$$
\begin{aligned}
&0 \leq f(P_j) \leq 1 &&\text{for any } P_j \\
&f[I(\mathcal{E})] = 1 \\
&f(P_{j_1} + P_{j_2} + \ldots + P_{j_m}) \\
&\quad = f(P_{j_1}) + f(P_{j_2}) + \ldots + f(P_{j_m}) &&\text{for exclusive } P
\end{aligned}
\tag{11.16}
$$

(The second and third conditions imply immediately that the probability associated to any decomposition of identity is 1). These mathematical conditions have a simple physical interpretation: they mean that $f(P_j)$ can be considered the probability of obtaining result 1 in the quantum measurement associated with any P_j. The third relation expresses that the probability of the union of exclusive events is the sum of their probabilities.

The Gleason theorem then states that, if the dimension of $\mathcal{E}$ is more than 2, there exists a nonnegative Hermitian operator with unit trace ρ acting in $\mathcal{E}$ such that:

$$f(P_j) = Tr\{\rho P_j\} \qquad \text{for any } P_j \tag{11.17}$$

In other words, all the values of f can be derived from a single operator ρ by a simple product and trace operation[12]. The mathematical proof of this theorem is

---

of P_i and P_j with eigenvalues 1 or 0. Since by construction none of them has eigenvalue 1 for both P_i and P_j, the product $P_i P_j$ vanishes.

Conversely, if this product vanishes, the projectors necessarily commute (footnote [10]), and again one can build a common basis of eigenvectors in $\mathcal{E}$. The product $P_i P_j$ vanishes only if none of these eigenvectors has a double eigenvalue 1, which means that P_i and P_j project onto orthogonal subspaces.

In both cases, it is then easy to see that the sum $P_i + P_j$ is the orthogonal projector over the subspace given by the direct sum of the two initial projection subspaces. By recurrence, the generalization to more than two orthogonal projectors is straightforward.

[12] There is a one-to-one correspondence between the projectors and the subspaces over which they project, so that it is equivalent to consider that the function f is defined over the projectors or over the subspaces of $\mathcal{E}$. In mathematics, one often defines f as a "measure over all possible subspaces of $\mathcal{E}$".

not trivial, and we refer the reader to the original publication [565]. A detailed summary of the proof can be found in §4.4.3 of Reference [540].

If f is interpreted as giving the probabilities associated with the measurements defined by the projectors, this theorem shows that these probabilities can all be obtained from one single operator ρ, which can then be considered as the density operator of the system. This is an interesting result, since it derives mathematically the quantum formalism of the density operator and of the trace from general considerations concerning the necessary conditions satisfied by probabilities.

The Gleason theorem requires a lower limit of 3 for the dimension of the space of states; it does not apply to a spin 1/2. Nevertheless, Bush [566] has shown that, if one extends conditions (11.16) to POVM (see §9.2.2), this limitation of the theorem can be removed. POVM define a broader class of operators, and lead to more possible decompositions of identity, than orthogonal projectors. As a consequence, the Bush form of the theorem requires more assumptions than the Gleason form. This is the price to pay to obtain a more general result (also valid in two dimensions), as well as a simpler mathematical proof.

11.5 Veiled reality

The "veiled-reality" interpretation was introduced by d'Espagnat [24, 25]. As far as the mathematical formalism is concerned, it does not differ from the standard interpretation, but it proposes a conceptual and philosophical framework that is different. This framework is realist, as for Bohr, but here the definition of reality does not involve human perception relayed by measurement apparatuses; the existence of reality is considered fundamental, with no need to refer to humans and their perceptive structure. Analyzing the theory of measurement, the EPR and Bell arguments, a study of the relation between counterfactuality and realism, and the consequences of the intersubjective agreement, d'Espagnat concludes that quantum mechanics cannot lead to descriptive interpretations of individual objects. He distinguishes between independent reality, of which at best the general structures are accessible, and empirical reality (perceived phenomena). This leads him to conclude that the ultimate reality is a "veiled reality", only marginally accessible to discursive knowledge.

Other interpretations of quantum mechanics also distinguish between two levels of reality. This is for example the case of the de Broglie–Bohm theory (§11.8.1), with a physical field that can be manipulated (the wave function) and the particle positions (which are observable, but not manipulable). But it remains a very different approach, in particular because the two levels of reality appear in the mathematical formalism of the de Broglie–Bohm theory itself.

11.6 Contextual quantum reality

The authors of Reference [567] propose a completely different definition of reality, which is expressed in terms of an association between a quantum object and the environment provided by the ensemble of its measurement apparatuses. They start from three postulates:

(i) For a physical system, they define a "modality" as the values of a complete set of physical quantities that can be predicted with certainty and measured repeatedly on this system[13]. This complete set of physical quantities is called a context; it corresponds to a given experimental setting. The modality is attributed jointly to the system and the context.

(ii) For any given context, N different modalities exist, but they are mutually exclusive : if one set of predictions is true, the others are wrong. The value of N is the same in all relevant contexts; it is a characteristic property of a given quantum system, called the dimension.

(iii) For a given quantum system, the various contexts are related by transformations g that have the structure of a continuous group G.

Postulate (i) puts the emphasis on contexts, systems, and modalities (hence the acronym CSM given to this approach), postulate (ii) on quantization, and postulate (iii) on the relations between different contexts. The authors of Reference [567] then show that a theory that is compatible with these postulates is necessarily probabilistic; the introduction of probabilities is then no longer a postulate per se, but a consequence of other postulates. Moreover, by various reasonings, they are able to derive the whole quantum formalism. The structure of quantum mechanics then appears as a consequence of an interplay between the quantized number of "modalities" accessible to a quantum system and the continuum of "contexts" that define these modalities.

This approach is reminiscent of the Gleason theorem (§11.4.3), but independent of it. It is also very similar to Bohr's point of view, where physical reality can only be defined in terms of the whole experimental setup (*cf.* §2.5, part (i), and Jammer's quotation at the end of that section). A important difference, however, is the central role of quantization, as expressed by postulate (ii).

11.7 History interpretation

The interpretation of "consistent histories"[14] is also sometimes called "decoherent history interpretation", or just "history interpretation" – here we will use the latter. As we will see, it proposes a logical framework to discuss the evolution of a closed

---

[13] The word "complete" refers to the largest possible set compatible with certainty and repeatability.

[14] The notion of consistency is essential at the level of families of histories, rather than at the level of individual histories.

quantum system, without any reference to measurements. The general idea was introduced and developed by Griffiths [17]; Omnès, Gell-Mann , and Hartle also contributed to it and sometimes adapted it [568–570]. The history interpretation is probably the most recent among those that are discussed in this book. We will remain within the limits of a non-specialized introduction; the reader interested in more precise information on the subject should go to the provided references – for a general presentation, see also an article in *Physics Today* [571], as well as the references contained therein, or the introductory review article by Hohenberg [572].

11.7.1 Histories, families of histories

Consider any orthogonal projector P on a subspace $\mathcal{F}$ of the space of states of a system; it has two eigenvalues, $+1$ corresponding to all the states belonging to $\mathcal{F}$, and 0 corresponding to all states that are orthogonal to $\mathcal{F}$ (they belong to the supplementary subspace, which is associated with the projector $Q = 1 - P$). One can associate a measurement process with P: if the state of the system belongs to $\mathcal{F}$, the result of the measurement[15] is $+1$; if the state is orthogonal to $\mathcal{F}$, the result is 0. Assume now that this measurement is made at time t_1 on a system that is initially (at time t_0) described by a density operator $\rho(t_0)$; the probability for finding the state of the system in $\mathcal{F}$ (result $+1$) at time t_1 is then given by formula (11.9), which in this case simplifies into:

$$\mathcal{P}(\mathcal{F}, t_1) = \mathrm{Tr}\left\{\widehat{P}(t_1)\rho(t_0)\widehat{P}(t_1)\right\} \tag{11.18}$$

where $\widehat{P}(t)$ is the projector P in the Heisenberg picture at time t. This result can obviously be generalized to several subspaces $\mathcal{F}_1$, $\mathcal{F}_2$, $\mathcal{F}_3$, etc., and several measurement times t_1, t_2, t_3, etc. (we assume $t_1 < t_2 < t_3 < ...$). The probability for finding the state of the system at time t_1 in $\mathcal{F}_1$, then at time t_2 in $\mathcal{F}_2$, then at time t_3 in $\mathcal{F}_3$, etc. is, according to the Wigner formula (§11.1.2.a):

$$\mathcal{P}(\mathcal{F}_1, t_1; \mathcal{F}_2, t_2; \mathcal{F}_3, t_3...) = \mathrm{Tr}\left\{...\widehat{P}_3(t_3)\widehat{P}_2(t_2)\widehat{P}_1(t_1)\rho(t_0)\widehat{P}_1(t_1)\widehat{P}_2(t_2)\widehat{P}_3(t_3)...\right\} \tag{11.19}$$

where, as before, the $\widehat{P}_i(t_i)$ are the projectors over subspaces $\mathcal{F}_1$, $\mathcal{F}_2$, $\mathcal{F}_3$ in the Heisenberg picture.

We can now associate a "history" of the system with this equation: a history $\mathcal{H}$ is defined by a series of arbitrary times t_i, each of them associated with an orthogonal projector P_i over some subspace; its probability is given by (11.19), which,

---

[15] Alternatively, we can consider an observable M having an eigenvalue m associated with eigenspace $\mathcal{F}$. We will then concisely call "the probability of finding the state of the system in $\mathcal{F}$" the probability of obtaining result m in the measurement, so that the state of the system is projected onto $\mathcal{F}$.

for simplicity, we will write as $P(\mathcal{H})$. In other words, a history is the selection of a particular path, or branch, for the state vector in a von Neumann chain, defined mathematically by a series of times and projectors. In this point of view, no reference is made to measurements: a history describes inherent physical properties of the system itself, without having to invoke interactions with external systems. Needless to say, there are an enormous number of different histories, which can have all sorts of properties; some of them are accurate because they contain many times associated with projectors over small subspaces $\mathcal{F}$; others remain very vague because they contain only a few times with projectors over large subspaces $\mathcal{F}$ (one can even decide that $\mathcal{F}$ is the entire space of states, so that no information at all is contained in the history at the corresponding time).

There are in fact so many histories that it is useful to group them into families, or sets, of histories. A family is defined again by an arbitrary series of times t_1, t_2, t_3, ..., but now we associate to each of these times t_i an ensemble of orthogonal projectors $P_{i,j}$ that, when summed over j, restore the whole initial space of states. For each time, we then have a series of mutually orthogonal projectors that provide a decomposition of the unity operator:

$$\sum_j P_{i,j} = 1 \tag{11.20}$$

For each time t_i this gives the system a choice, so to say, among many projectors, and therefore a choice among many histories of the same family. It is actually easy to see from (11.20) and (11.19) that the sum of probabilities of all histories of a given family is equal to 1:

$$\sum_{\text{histories of a family}} P(\mathcal{H}) = 1 \tag{11.21}$$

which we interpret as the fact that the system will always follow one, and only one, of them.

The simplest case occurs when a family is built from a single history: a trivial way to incorporate one history into a family is to associate, at each time t_i ($i = 1, 2, ..., N$), in addition to the projector P_i, the supplementary projector $Q_i = 1 - P_i$; the family then contains 2^N individual histories. Needless to say, there are many other ways to complement a single history with "more accurate" histories than those containing the Q; this can be done by decomposing each Q into many individual projectors, the only limit being the dimension of the total space of states.

11.7.2 Consistent families

The preceding definitions are in general not sufficient to ensure a satisfactory logical consistency in the reasonings. Having chosen a given family, it is very natural to also enclose in the family all those histories that can be built by replacing any pair or projectors, or actually any group of projectors, by their sum; this is because the sum of two orthogonal projectors is again a projector (onto a subspace that is the direct sum of the initial subspaces). The difference introduced by this operation is that, now, at each time, the events are no longer necessarily exclusive[16]; the histories incorporate a hierarchy in their descriptive accuracy, including even cases where the projector at a given time is just the projector over the whole space of states (no information at all on the system at this time).

Consider the simplest case where two projectors only, occurring at time t_i, have been grouped into one single projector to build a new history. The two "parent" histories then correspond to two exclusive possibilities (they contain orthogonal projectors), so that their probabilities add independently in the sum (11.21). What about the daughter history? It is exclusive of neither of its parents and, in terms of the physical properties of the system, it contains less information at time t_i: the system may have either of the properties associated with the parents. But a general theorem in probability theory states that the probability associated with an event that can be realized by either of two exclusive events is the sum of the individual probabilities. One then expects that the probability of the daughter history should be the sum of the parent probabilities. But, in quantum mechanics, relation (11.19) shows that this is not necessarily the case; since any projector, $\widehat{P}_2(t_2)$ for instance, appears twice in the formula, replacing it by a sum of projectors introduces four terms: two terms that give the sum of probabilities, as expected, but also two crossed terms[17] (or "interference terms") between the parent histories, so that the probability of the daughter history is in general different from the sums of the parent probabilities. This difficulty was to be expected: we know that quantum mechanics is linear at the level of probability amplitudes, not probabilities themselves; interferences occur because the state vector at time t_i, in the daughter history, may belong to one of the subspaces associated with the parents, but may also be any linear combination of such states. As a consequence, a sum rule for probabilities is not trivial.

One way to restore the additivity of probabilities is to consider only families

---

[16] For these nonexclusive families, relation (11.21) no longer holds since it would involve double counting of possibilities.

[17] These crossed terms look very similar to the right-hand side of (11.19), but their trace always contains at some time t_i one projector $\widehat{P}_{i,j}(t_i)$ on the left of $\rho(t_0)$ and one orthogonal projector $\widehat{P}_{i,k}(t_i)$ on the right.

where the crossed terms vanish, which amounts to the condition:

$$\mathrm{Tr}\left\{...\widehat{P}_{3,j_3}(t_3)\widehat{P}_{2,j_2}(t_2)\widehat{P}_{1,j_1}(t_1)\rho(t_0)\widehat{P}_{1,j_1'}(t_1)\widehat{P}_{2,j_2'}(t_2)\widehat{P}_{3,j_3'}(t_3)...\right\}$$
$$\propto \quad \delta_{j_1,j_1'} \times \delta_{j_2,j_2'} \times \delta_{j_3,j_3'} \times ... \tag{11.22}$$

Because of the presence of the product of δ in the right-hand side, the left-hand side of (11.22) vanishes as soon as at least one pair of the indices (j_1, j_1'), (j_2, j_2'), (j_3, j_3'), etc., contains different values; if they are all equal, the trace merely gives the probability $\mathcal{P}(\mathcal{H})$ associated with the particular history of the family. In this way, we introduce the notion of a "consistent family": if condition (11.22) is fulfilled for all projectors of a given family of histories, we will say that this family is logically consistent, or consistent for short. Condition (11.22) is basic in the history interpretation of quantum mechanics; it is sometimes expressed in a weaker form, as the cancellation of the real part only; this, as well as other points related to this condition, is briefly discussed in Appendix K. We now discuss how consistent families can be used as an interpretation of quantum mechanics.

11.7.3 Quantum evolution of an isolated system

Let us consider an isolated system and suppose that a consistent family of histories has been chosen to describe it; any consistent family may be selected but, as soon as the choice is made, it cannot be changed (we discuss later what happens if one attempts to describe the same system with more than one family). This unique choice provides us with a well-defined logical frame, and with a series of possible histories that are accessible to the system and give information at all intermediate times t_1, t_2, ... Which history will actually occur in a given realization of the physical system is not known in advance: we postulate the existence of some fundamentally random process of Nature that selects one single history among all those of the family. The corresponding probability $\mathcal{P}(\mathcal{H})$ is given by the right-hand side of (11.19); since this formula belongs to standard quantum mechanics, this postulate ensures that the standard predictions of the theory are automatically recovered. For each realization, the system will then possess at each time t_i all physical properties associated to the particular projectors $P_{i,j}$ that occur in the selected history.

This provides a description of the evolution of its physical properties that can be significantly more accurate than that given by its state vector; in fact, the smaller the subspaces associated with the projectors $P_{i,j}$, the more accuracy is gained[18]. For instance, if the system is a particle and if the projector is a projector over some region of space, we will say that for a particular history the particle is in this region at the corresponding time, even if the whole Schrödinger wave function extends

---

[18] Obviously, no information is gained if all $P_{i,j}$ are projectors over the whole space of states, but this corresponds to a trivial case of little interest.

over a much larger region. Or, if a photon strikes a beam splitter, or enters a Mach–Zehnder interferometer, some histories of the system may include information on which trajectory is chosen by the photon[19], while standard quantum mechanics considers that the particle takes all of them at the same time. Since histories contain several different times, one may even attempt to reconstruct an approximate trajectory for the particle, while this is completely excluded in standard quantum mechanics (for instance, for a wave function that is a spherical wave); but of course one must always check that the projectors that are introduced for this purpose remain compatible with the consistency of a single family.

The physical information contained in the histories is not necessarily about position only: a projector can also project over a range of eigenstates of the momentum operator, or include mixed information on position and momentum (subject, of course, to Heisenberg relations, as always in quantum mechanics), information on spin, etc.. There is actually a huge flexibility in the choice of projectors; for each choice, the physical properties that may be ascribed to the system are all those that are shared by all states contained in the projection subspace, but not by any orthogonal state. A frequent choice is to assume that, at a particular time t_i, all $P_{i,j}$ are the projectors over the eigenstates of some Hermitian operator H: the first operator $P_{i,j=1}$ is the projector over all the eigenstates of H corresponding to the eigenvalue h_1, the second $P_{i,j=2}$ the corresponding projector for the eigenvalue h_2, etc. In this case, all histories of the family include exact information about the value of the physical quantity associated at time t_i to H (for instance, the energy if H is the Hamiltonian). But, as already mentioned, we are not free to choose any operator H_i at any time t_i: in general, there is no reason why the consistency conditions should be satisfied by a family built just by choosing physical quantities in an arbitrary way.

Using histories, we obtain a description of the properties of the system in itself, without any reference to measurements, conscious observers, etc. This does not mean that measurements are excluded; but they can be treated merely as particular cases, by incorporating the corresponding physical devices in the system under study. Moreover, one attributes properties to the system at different times; this contrasts with the orthodox interpretation, where a measurement does not reveal any pre-existing property of the physical system, and moreover projects it into a new state that may be totally independent of the initial state. It is easy to show that the

---

[19] Assume that, with a Mach–Zehnder interferometer, the family provides information about the path of the photon inside the interferometer. Then consistency requires that it gives no information on its output path (after the last beam splitter), and on which output detector it triggers at the end. This is because the probability of having a particle in each output path contains interference terms between the paths inside the interferometer, but these interference terms vanish by sum over the two outputs.

One can also build other consistent families where the output channel is specified, as well as the excited output detector, but then consistency requires that no information about the path inside the interferometer is available. This is an illustration of how quantum complementarity applies within the history interpretation.

whole formalism of consistent families is invariant under time reversal, in other words that it contains no difference between the past and the future (instead of the initial density operator $\rho(t_0)$, one may use the final density operator $\rho(t_N)$ and still use the same quantum formalism [573]) – for more details, and even an intrinsic definition of consistency that involves no density operator at all, see §III of [574]. In addition, one can develop a relation between consistent families and semiclassical descriptions of a physical system; see [569] for a discussion of how classical equations can be recovered for a quantum system provided sufficient coarse graining is included (in order to ensure, not only decoherence between the various histories of the family, but also what the authors of this reference call "inertia" to recover classical predictability). See also chapter 16 of [570] for a discussion of how classical determinism is restored, in a weak version that ensures perfect correlations between the values of quasi-classical observables at different times (or course, there is no question of fundamental determinism in this context). The history point of view undoubtedly has many attractive features, and seems to be particularly clear and easy to use, at least as long as one limits oneself to one single consistent family of histories.

11.7.4 Incompatibility of different consistent families

How does the history interpretation deal with the existence of several consistent families? They are a priori all equally valid, but they will obviously lead to totally different descriptions of the evolution of the same physical system; this is probably the most delicate aspect of the interpretation (we will come back to this point in the next subsection). The history interpretation considers that different consistent families should be regarded as mutually exclusive (except, of course, in very particular cases where the two families can be embedded into a single large consistent family). Any family may be used in a logical reasoning, but not combined together with others in general. The physicist is then free to choose any point of view in order to describe the evolution of the system and to ascribe properties to the system. In a second independent step, another consistent family may also be chosen in order to develop other logical considerations within this different frame; but it would be totally meaningless (logically inconsistent) to combine considerations arising from the two frames. This very important fundamental rule, somewhat reminiscent of Bohr's complementarity, must be constantly kept in mind when one uses this interpretation. We refer the reader to [574] for a detailed and systematic discussion of how to reason consistently in the presence of disparate families, and to [575] for simple examples of incompatible families of histories (photon hitting a beam splitter, §II) and the discussion of quantum incompatibility (§V); various classical analogies are offered for this incompatibility, including a two-dimensional representation of a three-dimensional object by a draftsman, who can choose many points of view to make a drawing, but can certainly not take several at the same time – otherwise the projection would become inconsistent.

11.7.5 Comparison with other interpretations

In the history interpretation, there is no need to invoke conscious observers, measurement apparatuses, etc. The system has properties in itself, as in the Bohmian or modified Schrödinger dynamics interpretations. When compared to the others, a striking feature of the history interpretation is the enormous flexibility that exists for the selection of the points of view that can be chosen for describing the system; we have seen that all the times t_1, t_2, ... are arbitrary (actually their number is also arbitrary) and, for each of them, many different projectors P may be introduced. One may even wonder if the interpretation is sufficiently specific, and if this very large number of families of histories is not a problem.

What is the exact relation between the history interpretation and the orthodox theory? There is certainly a close relation, but several concepts are expressed in a more precise way with histories. For instance, complementarity stands in the Copenhagen interpretation as a general, almost philosophical, principle. In the history interpretation, it is related to mathematical conditions, consistency conditions, and orthogonality of projectors. It is impossible to come back to classical physics and to a simultaneous definition of all its observables: no projector can be more precise than the projector over a single quantum state $|\varphi\rangle$, which is itself obviously subject to the uncertainty relations because of the very structure of the space of states. Of course, it is still possible to make Bohrian considerations on incompatible measurement devices, or distinctions between the macroscopic and microscopic worlds, but with histories they lose part of their fundamental character. The history interpretation allows a quantum theory of the universe (compare for instance with quotation (x) of §2.5); we do not have to worry about dividing the universe into observed systems and observers.

The more important difference between the orthodox and the history interpretations is probably the way they describe the time evolution of a physical system. In the usual interpretation, we have two different postulates for the evolution of a single entity, the state vector, which may sometimes create difficulties; in the history interpretation, the continuous Schrödinger evolution and the random evolution of the system among histories are put at very different levels, so that the conflict is much less violent. Actually, in the history interpretation, the Schrödinger evolution plays a role only at the level of the initial definition of consistent families (through the evolution operators contained in projectors in the Heisenberg picture) and in the calculation of the probability $\mathcal{P}(\mathcal{H})$; the real time evolution takes place between the times t_i and t_{i+1} and is purely stochastic. A completely nondeterministic evolution has now become the major source of evolution! There is a kind of inversion of priorities with respect to the orthodox point of view, where the major source of evolution is deterministic under the effect of the continuous Schrödinger equation. Nevertheless, and despite these differences, the decoherent history interpretation remains very much in the spirit of the orthodox interpretation; indeed, it has been described as an "extension of the Copenhagen interpretation", or as "a way to emphasize the internal logical consistency of the notion of complementarity". Gell-Mann and Hartle [576] prefer a more general point of view on the history interpretation, which makes the Copenhagen interpretation just "a special case of a more general interpretation in terms of the decoherent histories of the universe. The Copenhagen interpretation is too special to be fundamental ...".

What about the relation with the "correlation interpretation"? It also seems to be very close, since both points of view give a central role to the Wigner formula. In a sense, this minimal interpretation is contained in both the Copenhagen interpretation and in the history interpretation. Some physicists favoring the correlation interpretation would probably argue that adding a physical discussion in terms of histories to their mathematical calculation of probabilities does not add much to their point of view: they are happy with the calculation of correlations and do not feel the need for making statements on the evolution of the properties of the system itself. Moreover, they might add that they wish to insert whatever projectors correspond to a series of measurements in (11.9), and not worry about consistency conditions: in the history interpretation, for arbitrary sequences of measurements, one would get inconsistent families for the isolated physical system, and one has to include the measurement apparatuses to restore consistency. We have already noticed in §11.1.2 that the correlation interpretation allows large flexibility concerning the boundary between the measured system and the environment. For these physicists, the history description appears probably more as an interesting possibility than as a necessity, but without introducing any contradiction either.

Are there also similarities with theories with additional variables? To some extent, yes. Within a given family, there are many histories corresponding to the same Schrödinger evolution and, for each history, we have seen that more information on the evolution of physical reality is available than from the state vector (or wave function) only. Under these conditions, the state vector appears as an incomplete description of reality, and one may even argue that the histories themselves are indeed additional variables[20]. In a sense, histories provide a kind of intermediate view between an infinitely precise Bohmian trajectory for a position and a very delocalized wave function. In the Bohm theory, the wave function pilots the po-

---

[20] They would then be family dependent, and therefore not EPR elements of reality, as we discuss later.

sition of the particles; in the decoherent history interpretation, the propagation of the wave function pilots rather the definition of histories (through a consistency condition) as well as a calculation of probabilities, but not the evolution between times t_i and t_{i+1}, which is supposed to be fundamentally random.

Two theories, one deterministic, the other completely indeterministic, are necessarily different conceptually. Nevertheless, it is not impossible to transpose some of the Bohmian ideas to the history interpretation, and make it deterministic. Consider a given consistent family for which, at time t_1, the number of different projectors P_{j1} is Q_1; at time t_2 the number of different projectors is Q_2; etc. One could for instance introduce an additional variable $x(t_i)$ that, by definition, always belongs to the $[0, 1]$ interval, and postulate that this variable provides a criterion to decide, among all projectors P_{ji} at each time t_i, which is "realized" (which projector defines physical properties that are associated with this history at this time). Mathematically, this can be done by dividing the $[0, 1]$ interval into Q_i smaller intervals I_{ij} (with $j = 1, 2, ..., Q_i$) and associating the values of $x(t_i)$ within the interval I_{ij} to the realization of projector P_{ji}. In this way, a "trajectory", defined by the values of $x(t)$ at all discrete times $t_1, t_2, ..., t_i,...$ defines a single history of the family. At the initial time t_1, as in Bohmian theory one then could assume a probability distribution that reproduces the quantum predictions, and finally define a law of motion of the additional variable that ensures compatibility with the predictions of standard quantum mechanics[21] – in this case relation (11.19). By adding this variable $x(t)$ to the history interpretation, the latter could be made deterministic and, to some extent, compatible with the Bohmian ideas. Still, with the present states of the two theories, we can probably safely conclude this comparison between the Bohmian and history interpretations by saying that they give very different points of view on quantum mechanics.

Finally, what is the comparison with theories using a modified Schrödinger evolution? In a sense, they correspond to completely opposite strategies, since they introduce into one single equation the continuous evolution of the state vector as well as a mechanism simulating the state vector reduction (when needed); by contrast, the history interpretation puts on different levels the continuous Schrödinger evolution and a fundamentally random selection of history selection by the system. One might venture to say that the modified nonlinear dynamics approach is an extension of the purely wave program of Schrödinger, while the history interpretation is a modern version of the ideas put forward by Bohr. Another important difference is, of course, that a theory with modified dynamics is not strictly equivalent to standard quantum mechanics, and could lead to experimental tests, while

---

[21] This could be done either continuously by postulating some equation of evolution of $x(t)$, or by discrete steps by postulating a mapping of the interval $[0, 1]$ over itself corresponding to the translation in time from t_i to t_{i+1} (we note in passing a similarity with the notion of Poincaré maps in classical mechanics).

the history interpretation is built to reproduce exactly the same predictions in all cases – even if it can sometimes provide a useful point of view that allows one to grasp its content more conveniently [436].

11.7.6 A profusion of histories: discussion

We finally come back to a discussion of the impact of the profusion of possible points of view, which are provided by all the families that satisfy the consistency condition. We have already remarked that there is, by far, no single way in this interpretation to describe the evolution of properties of a physical system – for instance, all the complementary descriptions of the Copenhagen interpretation appear at the same level. This is indeed a large flexibility, much larger than in classical physics, and much larger than in the Bohmian theory, for instance. Are the rules that we have defined ("no combination of points of view") really sufficient to ensure that the theory is completely satisfactory? The answer to this question is not so clear, for several reasons. First, for macroscopic systems, one would like an ideal theory to naturally introduce a restriction to sets corresponding to quasi-classical histories; unfortunately, the number of consistent sets is in fact much too large to have this property [454]. This is the reason why more restrictive criteria for mathematically identifying the relevant sets are (or have been) proposed, but no complete solution or consensus has yet been found; the detailed physical consequences of consistency conditions are still being explored, and actually provide an interesting subject of research. Moreover, the historical paradoxes are not all solved by the history interpretation. Some of them are, for instance the Wigner friend paradox, to the extent where no reference to observers is made in this interpretation. But some others remain unsolved, just with a reformulation in a different formalism and vocabulary.

Let us for instance take the Schrödinger cat paradox, which initially arose from the absence of any ingredient in the Schrödinger equation able to create the emergence of a single macroscopic result – in other words, to exclude impossible macroscopic superpositions of an isolated, nonobserved, system. In the history interpretation, the paradox translates into a choice of families of histories: the problem is that there is no way to eliminate the families of histories where the cat is at the same time dead and alive; actually, most families that are mathematically acceptable through the consistency condition contain projectors on macroscopic superpositions, and nevertheless have exactly the same status as the families that do not. One would much prefer to have a "super-consistency" rule that would eliminate these superpositions; this would really solve the problem, but such a rule does not exist for the moment. At this stage, one can then do two things: either consider that the choice of sensible histories and reasonable points of view is a matter of common sense – a case in which one returns to the usual situation in the traditional interpretation, where the application of the postulate of wave packet is also left to the good taste of the physicist; or invoke decoherence and coupling to the external world in order to eliminate all these unwanted families – a case in which one returns to the usual situation where, conceptually, it is impossible to ascribe reasonable physical properties to a closed system without referring to the external world and interactions with it[22].

Finally, one may notice that, in the history interpretation, there is no attempt to follow "in real time" the evolution of the physical system; one speaks only of histories that are seen as complete, "closed in time", almost as histories of the past in a sense. Basic questions that were initially at the origin of the introduction of the state reduction postulate, such as "how to describe the physical reality of a spin that has already undergone a first measurement but not yet a second", are not easily answered. In fact, the consistency condition of the whole history depends on the future choice of the observable that will be measured, which does not make the discussion simpler than in the traditional interpretation, maybe even more complicated since its very logical frame is now under discussion. What about a series of measurements that may be, or may not be, continued in the future, depending on a decision that has not yet been made? As for the EPR correlation experiments, they can be reanalyzed within the history interpretation formalism [577] (see also [241] for a discussion of the Hardy impossibilities and the notion of "consistent counterfactuality"); nevertheless, at a fundamental level, the EPR reasoning still has to be dismissed for exactly the same reason already invoked by Bohr long ago: it introduces the EPR notion of "elements of reality", or counterfactual arguments, that are no more valid within the history interpretation than in the Copenhagen interpretation (see for instance §V of [577] or the first letter in [576]). We are then brought back to almost the same old debate, with no fundamentally new element. We have nevertheless already remarked that, as with the correlation interpretation, the history interpretation may be supplemented by other ingredients, such as the Everett interpretation[23] or, at the other extreme, EPR or deterministic ingredients, a case in which the discussion would of course become different.

For a more detailed discussion of this interpretation, see the references given at the beginning of this section. For a discussion of the relationship with decoherence,

---

[22] For instance, in the context of histories, one sometimes invokes the practical impossibility to build an apparatus that would distinguish between a macroscopic superposition and the orthogonal superposition; this would justify the elimination of the corresponding histories from those that should be used in the description of reality. Such an argument reintroduces the notion of measurement apparatus and observers in order to select histories, in contradiction with the initial motivations of this point of view – see Rosenfeld's quotation in §2.5. Moreover, this immediately opens again the door to Wigner friend-type paradoxes, etc.

[23] Nevertheless, since the Everett interpretation completely suppresses from the beginning any specific notion of measurement, measuring apparatus, etc., the usefulness of completing it with the history interpretation is not obvious.

the notion of "preferred (pointer) bases", and classical predictability, see [454]. For a critique of the decoherent history interpretation, see for instance [578], where it is argued that consistency conditions are not sufficient to predict the persistence of quasi-classicality, even at large scales in the universe. See also [579], which claims that these conditions are not sufficient either for a derivation of the validity of the Copenhagen interpretation in the future; but see also the reply to this critique by Griffiths in [575]. The history interpretation is related to the theory of continuous measurements in quantum mechanics [580] (§9.3.2) as well as to the technique of stochastic quantum theories (§11.10.3). Finally, another interesting reference is an article published in 1998 in *Physics Today* [18], which contains a discussion of the history interpretation in terms that stimulated interesting reactions from the proponents of the interpretation [576].

11.8 Additional ("hidden") variables

With additional/hidden variables, elements that clearly do not belong to the orthodox interpretation are explicitly introduced. Additional variables are added to the quantum state vector in order to obtain a more precise description of a single system. We have already mentioned that the EPR theorem itself can be seen as a strong argument in favor of the existence of additional variables. These variables are sometimes still called "hidden", even if this word is somewhat paradoxical, since they are often more visible than the complex state vector[24]; Bell proposed to use the word "beable" instead [582], which tends to be more and more used in recent articles. Here we will use the word "additional", because of its generality.

Theories with additional variables are often built mathematically to reproduce the predictions of orthodox quantum mechanics exactly. If they give the same probabilities for all possible measurements, it is clear that there is no hope to disprove experimentally orthodox quantum mechanics in favor of these theories, or the opposite. In this sense, they are not completely new theories, but rather variations on a known theory (an exception is mentioned in §11.8.1.g). They nevertheless have a real conceptual interest since they propose a description and an explanation of quantum phenomena that differ from those of standard theory. They can restore realism and solve the difficulties related to the presence of two kinds of evolution of the state vector (Schrödinger cat paradox). They can also restore determinism, although not necessarily: one can also build theories with additional variables that remain fundamentally nondeterministic.

---

[24] They appear directly in the results of measurements so that, instead of being hidden, the additional variables are actually visible. In [581], Bell writes "Absurdly, such theories are known as 'hidden variables' theories. Absurdly, for there it is not in the wave function that one finds an image of the visible world, and the results of the experiments, but in the complementary 'hidden' (!) variables".

11.8.1 De Broglie-Bohm (dBB) theory

In standard quantum mechanics, a single quantum object may sometimes behave as a particle, sometimes as a wave, sometimes combine both, depending on the experiment considered. In the de Broglie–Bohm theory (dBB theory for short), the particle and the wave always coexist; they actually go together, and the particle is constantly guided by the wave. The elementary system called "quantum particle" in standard theory is replaced by an inseparable couple made of a particle and a field. After a historical introduction, we begin with a brief discussion of the general framework of this theory, and then discuss trajectories for one or two particles as well as quantum measurements. We will then outline a few objections that have been made to the theory, and conclude with a discussion putting the achievements of the theory into perspective.

11.8.1.a Historical introduction

Theories with additional variables started in 1926–1927 with the early work of de Broglie [583, 584], along the lines of his thesis [30]. L. de Broglie first elaborated his "theory of the double solution", which he named so since the same wave equation has two solutions: the usual continuous wave function $\Psi(\mathbf{r})$, and a solution with mobile singularities $u(\mathbf{r})$ representing the physical particle itself. The particle is then considered as an energy concentration in the singularity region of the new field $u(\mathbf{r})$; it therefore remains perfectly localized as in classical physics. It is also considered as a small clock having internal vibrations, and a motion determined by a synchronization condition between these vibrations and the external wave: "the particle glides on its wave in such a way that its internal vibrations always remain in phase with the vibration of the wave at the point where it sits"[25]. Mathematically, this leads to a "guidance equation" for $u(\mathbf{r})$ in terms of the wave $\Psi(\mathbf{r})$. Since the latter is subject to diffraction effects on external obstacles, one can then recover the predictions of standard quantum mechanics, provided one assumes that the statistical distribution of singularities is given by $|\Psi(\mathbf{r})|^2$.

Nevertheless, when invited by Lorentz to report on his work at the 1927 Solvay meeting, it seems that de Broglie was too worried about the mathematical difficulties of this theory to report on it in front of this audience [101, 585]. He preferred to discuss a simpler version[26], the "pilot wave theory", where the localized singu-

---

[25] In de Broglie's view, a single particle is really represented by a wave $u(\mathbf{r})$, while the wave function $\Psi(\mathbf{r})$ gives only statistical information on an ensemble of particles, so that it does not have the same status of reality. He describes this as a "curious mixture of single events and statistics " [101].

[26] At the Solvay meeting, Pauli made objections to the pilot wave theory, in particular that it could not reproduce the results of standard theory for inelastic collisions. We now know that it is perfectly possible to build a pilot wave theory that provides exactly the same predictions as standard quantum mechanics, so that this objection is not valid.
  Later, de Broglie regretted [101] having "weakened his point of view" by presenting a "truncated version" of his double solution theory, from which the singularity of the wave had been suppressed and useful features

larity of $u(\mathbf{r})$ is directly replaced by a position of the particle, as in classical theory, while its motion remains totally different from classical physics since it is determined by the pilot wave. The general idea is then that the wave function does not directly represent a quantum particle but, rather, is a wave that guides the motion of the particle.

Actually, also in 1926, Born had envisaged the possibility of introducing "additional parameters" to the theory, in the course of the famous article where he introduced the probabilistic interpretation of the wave function [34]; he even worked on this subject in more detail with his assistant Frenkel [58] but, unfortunately, this work seems to have been lost. Chapter 8 of [586] discusses early attempts at causal theories in more detail.

In 1952, Bohm [11, 587] independently elaborated a more complete version of the pilot wave theory. As in the de Broglie pilot wave theory, he added positions to the wave function and assumed that the motion of the particle in space is guided by the gradients of the wave function, but also introduced new elements as we will see later. Both points of view thus share common basic concepts, and one often speaks of the "de Broglie–Bohm" theory. Bohmian theory is probably the best known among those that add variables to standard quantum mechanics. Another well-known example is nevertheless the work of Wiener and Siegel [12], who elaborated a mathematical formulation of quantum mechanics in terms of probabilities (or probability densities), instead of probability amplitudes, while the results remain exactly equivalent to those of the standard calculation of probabilities. Such a point of view completely eliminates the need for a special postulate concerning measurement processes.

11.8.1.b General framework

We first introduce the essential components of the de dBB theory, the guiding formula and the quantum equilibrium condition. References [588–592] provide general books introducing and discussing this theory.

Guiding formula None of the usual ingredients of standard quantum mechanics disappears in theories with additional variables such as the dBB theory. Instead of an object with intermediate status of reality (see §1.2.3), the wave function becomes a field that acts directly on the real position of the particle through the gradient of its phase. The Schrödinger equation itself remains strictly unchanged, but a completely new ingredient is added to it: in addition to its wave function/field, each particle now gets an additional variable λ, more precisely three additional

disappeared. He was particularly interested in the similarities with the theory of general relativity, where the motion of singularities of the gravitational field follows geodesics of space-time. He also considered the idea of introducing a nonlinear equation of motion for the wave $u(\mathbf{r})$, which spreads the singularity over a tiny mobile region of space.

variables that are the three components of a vector $\mathbf{Q}$. The evolution of $\mathbf{Q}$ is coupled to the wave function field through a "guiding formula" (or "quantum velocity term") which, for a single particle with wave function $\Psi(\mathbf{r}, t)$, is given by[27]:

$$\frac{d}{dt}\mathbf{Q} = \frac{1}{m\,|\Psi(\mathbf{Q}, t)|^2}\,\mathrm{Re}\left[\Psi^*(\mathbf{Q}, t)\frac{\hbar}{i}\boldsymbol{\nabla}_{\mathbf{r}}\Psi(\mathbf{Q}, t)\right] = \frac{\hbar}{m}\boldsymbol{\nabla}_{\mathbf{r}}\xi(\mathbf{Q}, t) \qquad (11.23)$$

where m is the mass of the particle and $\xi(\mathbf{r}, t)$ the phase of the complex number $\Psi(\mathbf{r}, t)$; in the right-hand side of this equation, the gradients of these functions are first taken with respect of the current variable $\mathbf{r}$, which is then replaced by the Bohmian position $\mathbf{Q}$. This expression is nothing but the ratio between standard probability current:

$$\mathbf{J}(\mathbf{r}, t) = \frac{1}{m}\,\mathrm{Re}\left[\Psi^*(\mathbf{r}, t)\frac{\hbar}{i}\boldsymbol{\nabla}\Psi(\mathbf{r}, t)\right] \qquad (11.24)$$

and the density of probability $|\Psi(\mathbf{r}, t)|^2$, both taken at point $\mathbf{r} = \mathbf{Q}$; in other words, the time derivative of $\mathbf{Q}$ is nothing but the local velocity of the fluid of probability, in the hydrodynamic version of the Schrödinger equation introduced by Madelung in 1927 [593]. The vector $\mathbf{Q}$ is called the Bohmian position of the particle and its time derivative (11.23) the Bohmian velocity, but for simplicity the word "Bohmian" will sometimes be omitted.

For a system made of N particles, the additional variables are the $3N$ components of $\mathbf{Q}_1$, $\mathbf{Q}_2$, ..., $\mathbf{Q}_N$, which together define a vector in configuration space evolving according to a direct generalization of (11.23): the time derivative of each $\mathbf{Q}_i$ is obtained by replacing the gradient $\boldsymbol{\nabla}$ by the gradient $\boldsymbol{\nabla}_{\mathbf{r}_i}$ of the partial derivative with respect to the corresponding variable in the wave function – see for instance (11.26). An important point is that there is no retroaction of the additional variables onto the wave function; the coupling goes one way only. From the beginning, the theory therefore introduces a marked asymmetry between the two mathematical objects that are used to describe the system; we will see later that they also have very different physical properties.

The quantum velocity term depends only on the gradient of the phase of the wave function, not on its modulus. Therefore, vanishingly small wave functions may have a finite influence on the position of the particles. With a Gaussian wave packet, for instance, the influence of the wave packet on the velocity of the particle is comparable near the center of the wave packet or at arbitrarily large distances, where the wave function is vanishingly small. Of course, situations where the position of the particle is extremely far from the center of the wave packet are very

---

[27] We assume that the potential vector $\mathbf{A}(\mathbf{r}, t)$ vanishes. In the presence of a potential vector, and for a particle of charge q, in the right-hand side of (11.23) the derivation $(\hbar/i)\boldsymbol{\nabla}$ should be replaced by $(\hbar/i)\boldsymbol{\nabla} - q\mathbf{A}(\mathbf{r}, t)$, so that $\hbar\boldsymbol{\nabla}\xi$ becomes $\hbar\boldsymbol{\nabla}\xi - q\mathbf{A}(\mathbf{r}, t)$. The time derivative of $\mathbf{Q}$ is then gauge invariant.

rare, but when they occur by chance, the position is guided exactly with the same efficiency in all space.

In order to make his theory more similar to classical mechanics, initially Bohm did not start directly from (11.23), but introduced the notion of a "quantum potential", defined as:

$$V_{\text{quantum}}(\mathbf{r}) = -\frac{\hbar^2}{2m} \frac{\Delta \, |\Psi(\mathbf{r}, t)|}{|\Psi(\mathbf{r}, t)|} \tag{11.25}$$

(where Δ is the Laplacian); the quantum potential depends on the wave function. To understand its role, one can take the time derivative of relation (11.23) to obtain the acceleration of the Bohmian position $\mathbf{Q}(t)$, and use the Schrödinger equation to calculate the time derivative of the right-hand side. One then sees that the product of this acceleration by the mass m of the particle is nothing but the gradient of the sum of two potentials: the usual external potential $V(\mathbf{r})$, and the quantum potential. The first term corresponds of course to the classical Newton law, the second introduces quantum effects. One then has the choice between two equivalent points of view: either directly postulate that (11.23) is always valid, or assume that it is valid at time $t = 0$, and use Newton's law to obtain the velocity at any time t after adding $V_{\text{quantum}}(\mathbf{r})$ to the usual potential. Subsequent versions of Bohmian mechanics have nevertheless often discarded the quantum potential in favor of a quantum velocity term only.

Quantum equilibrium condition The theory also postulates an initial random distribution of the position variables $\mathbf{Q}_1, \mathbf{Q}_2,..., \mathbf{Q}_N$ that reproduces exactly the initial quantum probability distribution $|\Psi(\mathbf{Q}_1, \mathbf{Q}_2, ...)|^2$ of standard theory for position measurements. This distribution is often called the "quantum equilibrium distribution". The distribution is not due to a preparation of the system that is insufficiently accurate, but postulated as fundamental. For a given realization of an experiment, there is no way to select which value of the position is actually realized inside the distribution; from one realization to the next, a new completely random choice of position is spontaneously made by Nature. This assumption conserves the fundamentally random character of the predictions of quantum mechanics [594].

The additional variables then depend on the wave function in two ways, through both their initial values and their evolution. Combining the Schrödinger equation with the form of the "quantum velocity term" (11.23), one can show that, if at time t the distribution of positions is equal to $|\Psi(\mathbf{Q}_1, \mathbf{Q}_2, ...)|^2$, the equality also holds at time $t + dt$. This ensures that the property continues to be true for any time, and automatically provides a close contact with all the predictions of quantum mechanics, since all predictions concerning the probabilities of position measurements are then identical [28]. In particular, under the effect of the quantum velocity term, the Bohmian positions are constantly dragged by the wave function and can never move away from it; they remain in regions of space where it does not vanish, which ensures that neither the guiding formula (11.23) nor the quantum potential (11.25) become indeterminate. Another important consequence of this postulate is to avoid a conflict with relativity [595], since arbitrary distributions would introduce the possibility of superluminal signaling (Appendix H). Since the Born rule is a consequence of the quantum equilibrium, one can then consider that this rule is not an independent postulate of quantum mechanics, but merely a consequence of the relativistic impossibility of instantaneous signal transmission.

One can then restore determinism[29], and assume that the results of measurements merely reveal the initial pre-existing value of the positions, chosen among all possible values in the initial probability distribution (in §11.8.1.d, we come back in more detail to the measurement process in dBB theory). This assumption solves many difficulties, for instance those related to understanding why quantum systems can manifest both wave and particle properties in interference experiments. The system always contains two indissociable objects, a wave and a particle; the wave may produce interference effects and guide the particle in a way that forces its trajectory to reproduce the interference pattern – nothing especially mysterious conceptually (later we study Bohmian trajectories in more detail). Similarly, in the negative experiment with a Mach–Zehnder interferometer discussed in §2.4, in all events a wave propagates in the two arms, uninfluenced by the position of the particle; the interference effect it produces at the output beam splitter is different, whether one of its components in one of the arms is absorbed by the object, or whether it is not, and the particle is guided accordingly. The final outcome of the experiment is just a result of an initial choice for the path of the particle and of its guiding by the wave at the output beam splitter; no paradox at all! The same is also true for the Schrödinger cat: depending on the exact initial position of a many-dimension variable λ, which belongs to an enormous configuration space (including the variables associated with the radioactive nucleus as well as all variables associated with the cat), the cat remains alive or dies. Nevertheless, decoherence will act in exactly the same way as in standard quantum mechanics, and make it impossible in practice to observe interferences with macroscopic objects in very different states (this is related to the notion of "empty waves", see §11.8.1.d); the reason is that the theory is built to be equivalent to standard quantum mechanics.

---

[28] Nevertheless, one can also build modified version of the dBB theory that are not equivalent to standard quantum mechanics; see §11.8.1.g

[29] At least to some extent, see the discussion in §11.8.1.i.

Description of physical reality For anyone who is very familiar with quantum mechanics, but not with the concept of additional variables, at first they may look somewhat mysterious, since they require a drastic change in the way we usually reason in quantum mechanics. This may explain why they are sometimes called "hidden", but this is only a consequence of our much better familiarity with standard quantum mechanics! In fact, these variables are less abstract than wave functions. The additional variables are those that are directly "seen" in a measurement, as opposed to the state vector, which remains invisible; it actually plays an indirect role, through its effect on the additional variables. In the example of a particle creating a track in a bubble chamber, what we see directly on the photograph is the successive values of an additional variable, which is actually nothing but the position of the particle. By contrast, who has ever obtained a single shot photograph of the wave function?

Two different attitudes are possible to describe physical reality within the dBB theory. In the first, the Bohmian positions and the guiding wave function are both considered as physically real, therefore with a similar status. In the second, the wave function is rather considered only as a mathematical tool allowing one to obtain the velocities of the real positions. We now discuss these two possibilities.

(i) Physically real wave function

Bell wrote [596]: "No one can understand this theory (the dBB theory) until he is willing to think of Ψ as a real objective field rather than just a probability amplitude". In this point of view, a "particle" always involves a combination of both a position and the associated field, which is physically real. Since this field can extend at arbitrarily large distance from the Bohmian position, it is then by no means surprising that two particles should influence each other even if their positions are very far away: the influence results from the interaction between two physical objects, a position and a field, as for instance in classical electromagnetism.

This point of view may look unified, but still implies a description of physical reality at two different levels:

– First, one corresponding to the elements associated with the wave function (or state vector), which are not directly visible but can be influenced in experiments by applying fields (or moving walls), since the evolution of the state vector depends on a Hamiltonian that can be controlled by applying fields (and on boundary conditions). This evolution takes place in a space with high dimensionality: for instance, for a system made of N spinless particles, the wave function evolves in a space with dimension $3N$. The state vector is not sufficient to give a complete description of a physical system.

– Second, another corresponding to the additional variables, the Bohmian positions, which are visible in experiments and move in ordinary three dimension space. It is nevertheless impossible to manipulate them directly, and for instance to change their distribution from quantum equilibrium. The reason for this impossibility is fundamental, otherwise it would become possible to send superluminal signals, in contradiction with relativity [595]. The Bohmian positions can only be manipulated indirectly, through actions on the state vector, which then guides their velocities.

(ii) Wave function as a mathematical tool

In another point of view, the wave function is not a real physical field: it is a mathematical function that has the role to provide, through its partial derivatives, the velocities of the Bohmian positions. Since the velocities directly provide the time evolution of the positions, they should also be considered as part of the description of physical reality, according to the principle (expressed in familiar terms) "anything that acts on something real has to be real too". In a experiment, the wave function characterizes the preparation procedure of the quantum physical system (§1.2.3.b). Its role is similar to that of an Hamiltonian or a Lagrangian in classical mechanics: these functions are mathematically defined in configuration space, but only their values at the positions of the particles (or in their infinitesimal vicinity), are physically relevant. An illustration of this analogy is the effect of a change of the electromagnetic gauge to describe an externally applied magnetic field: the classical Lagrangian changes, but not the forces applied to the particles; the wave function changes, but not the Bohmian velocities of the particle positions. One can then consider that physical reality exists and propagates in ordinary three-dimensional space, while it is influenced by a mathematical object propagating in multidimensional configuration space. Reference [592] discusses this second point of view in more detail; see in particular Remarks 5 and 6 of §5.1.1, where a comparison is made between the properties of the wave function and those of the electric and magnetic field in classical electromagnetism. For a more extended discussion on "Reality and the role of the wave function", in particular the role of the wave function of the universe, see for instance Reference [597].

In the two points of view, two sorts of variables are necessary (and sufficient) to obtain a complete description of reality of a physical system and of its time evolution. We have already mentioned that there is no retroaction of the additional variables onto the state vector, which creates an unusual situation in physics: usually, when two physical quantities are coupled, they mutually influence each other[30]. Another unusual feature is that the effect of the field on the particle position does not depend on the intensity of the field, but only on its relative variations on space. Interestingly, we are now facing another sort of duality, which distinguishes between direct action on physical systems (or preparation), directly related to the

---

[30] Reference [598] discusses the possible effects of a back-action term onto the wave function. A variant of the dBB theory where this back-action plays an important role will be discussed in §11.10.4.

state vector, and results of observations performed on them (results of measurements), determined by the additional variables.

11.8.1.c Bohmian trajectories

As soon as particles get a time-dependent position, they also get a velocity, an acceleration, etc., and a trajectory as in classical physics. By studying these trajectories, one obtains a variety of interesting, sometimes unexpected, results. For instance, even in the simplest case, a single particle in free space, the trajectories are not necessarily simple straight lines [588, 599]; they may be curved. We now study a few situations leading to characteristic Bohmian trajectories.

One particle A first remark is that (11.23) gives zero velocity for any wave function that is real; a nonzero velocity can occur only if the phase of the wave function varies in space. As a consequence, for instance in the ground state of the hydrogen atom, the Bohmian position of the electron does not revolve around the proton as one could expect, but remains static. The effect of the quantum potential exactly cancels the attraction of the proton when the electron is in its ground state, so that it experiences no force. Similarly, for an harmonic oscillator in the ground state (or in any stationary state), the position of the particle does not oscillate in the potential, but remains at the same place.

This property is general: each time the Hamiltonian is invariant under time reversal, one can choose a basis of stationary wave functions that are real, and for which the corresponding Bohmian velocity always vanishes[31]. Of course, with arbitrary wave functions that are coherent superpositions of stationary states, the situation is different: under the effect of the phase changes induced by the Schrödinger equation, the position and the associated velocities become functions of time. An example is an oscillator in a coherent quasi-classical state, where the time evolution of the position reproduces the classical oscillation in the potential well. Moreover, even in real stationary states, the correlation functions of positions are time dependent because, in dBB theory, the effect of the measurement on the wave function has to be included (we come back to correlations between measurements at different times in the discussion of §11.8.1.h).

We now consider a typical interference experiment, as shown schematically in Figure 11.1: a source S emits, one by one, a series of single particles that reach a screen D through which two apertures have been created, so that interferences

---

[31] The Hydrogen atom also has stationary wave functions that are not real, with a phase factor $e^{im_l\varphi}$, where φ is the azimuthal angle and m_l the quantum number associated with one component of the orbital momentum. For these states, the Bohmian trajectories do revolve around the proton, as in the classical picture. Nevertheless, in the absence of a magnetic field, the values $\pm m_l$ have the same energy and, by linear combination, one can build a basis of stationary wave functions that are real and correspond to zero Bohmian velocity. So, for a given degenerate energy level, depending on the wave function one chooses, this velocity vanishes or not.

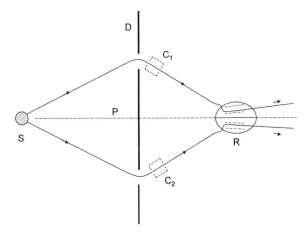

Figure 11.1 Bohmian trajectories associated with a particle emitted by a source S and crossing a screen D with two holes. In §11.8.1.h, two cavities C_1 and C_2 will be added into the scheme, but for the moment they can be ignored. Quantum mechanics predicts that the particle interferes with itself and produces an interference pattern in the observation region R; the dotted lines in this region symbolize the bright fringes. In Bohmian mechanics, the particle trajectories are not necessarily straight lines in free space; they bend inside the two holes under the effect of diffraction, and moreover oscillate in the interference region. The final result is that the predictions of quantum mechanics are exactly reproduced. One notices a "no-crossing rule": the trajectories never cross the symmetry plane P (central horizontal dashed line). More details about the shape of the trajectories in the interference region can be seen in the right part of Figure I.3 of Appendix I.

can be observed in region R on the other side of the screen. The experiment is symmetric with respect to an horizontal plane drawn as a broken line in the figure. We know that Bohmian mechanics is built to reproduce exactly the predictions of standard quantum mechanics concerning the measurement of the positions of particles; how will it manage to reproduce the interference pattern observed on the other side of the screen if one measures the position of the particle?

The dBB theory solves the problem because (11.23) predicts that free particles may have curved trajectories; this feature is actually indispensable for the statistics of the positions to reproduce the usual predictions of quantum mechanics [600]. Two typical trajectories are shown in the figure. A first interesting effect occurs when the particle crosses the screen through a hole that is sufficiently small to diffract the wave function (its size is smaller than, or comparable to, the de Broglie wavelength): without touching the walls, the particle is deflected because the wave function that pilots its velocity is subject to diffraction. Moreover, after the screen and in region R where the states coming from the two holes overlap, bending effects again occur, since the velocity of the particle is affected by interference effects. The trajectories are then modified in a way that ensures that, when the experiment is repeated many times, the series of measured positions reproduces the quantum interference pattern exactly.

One can easily see by symmetry that, at points in the plane of symmetry of the experiment (central horizontal dashed line of Figure 11.1), the Bohmian velocity always lies within this plane: no trajectory of the particle can cross the symmetry plane. This result is sometimes called the "no-crossing rule" of Bohmian trajectories. Now consider two wave packets, each coming from a different hole, and crossing each other in the region of the symmetry plane. The no-crossing rule makes the trajectories "bounce" on the symmetry plane so that, after the wave packets have crossed each other, the trajectories that followed one wave packet have "jumped" to the other. As a consequence, after the screen, any trajectory that is above the symmetry plane necessarily went through the upper hole, and conversely.

See [55] for a discussion of the trajectory of a particle that can be reconstructed from the observation of its successive positions in a cloud chamber, and Chapter 5 of [588] for a discussion of Bohmian trajectories through a potential barrier (tunnel effect). Similar effects occur when particles propagate inside a Mach–Zehnder interferometer; in the absence of a final beam splitter, their position jumps from one wave packet to the other at the ouptut, making their trajectory look like a zig-zag [601].

One may find these trajectories very strange, and be tempted to reject the dBB interpretation for this reason. It should nevertheless be clear that the trajectories are not introduced arbitrarily by this interpretation: they actually already exist in standard theory, where they are the trajectories of the probability current. The only addition of the dBB interpretation is to give them a stronger physical content, since the velocity of a probability fluid becomes the velocity of a point position.

Two particles; conditional wave function For systems of two particles or more, the situation becomes even more interesting[32]. The velocities of the two particles are

---

[32] A feature of the dBB theory is that, even for interacting particles, their two positions are not directly coupled. The interaction Hamiltonian affects the wave function, which in turn guides the positions, but the world of Bohmian positions remains completely free of direct mutual interactions.

given as a function of the two-particle wave function $\Psi(\mathbf{r}_1, \mathbf{r}_2; t)$ by:

$$\frac{d}{dt}\mathbf{Q}_1 = \frac{1}{m_1 \, |\Psi(\mathbf{Q}_1, \mathbf{Q}_2; t)|^2} \, \mathrm{Re}\left[\frac{\hbar}{i}\Psi^*(\mathbf{Q}_1, \mathbf{Q}_2; t) \, \nabla_{\mathbf{r}_1}\Psi(\mathbf{Q}_1, \mathbf{Q}_2; t)\right]$$

$$\frac{d}{dt}\mathbf{Q}_2 = \frac{1}{m_2 \, |\Psi(\mathbf{Q}_1, \mathbf{Q}_2; t)|^2} \, \mathrm{Re}\left[\frac{\hbar}{i}\Psi^*(\mathbf{Q}_1, \mathbf{Q}_2; t) \, \nabla_{\mathbf{r}_2}\Psi(\mathbf{Q}_1, \mathbf{Q}_2; t)\right]$$

(11.26)

where m_1 and m_2 are the masses. As in (11.23), the gradients in the right-hand side are first taken with respect of the current variables $\mathbf{r}_1$ and $\mathbf{r}_2$, and these variables are then replaced by the Bohmian positions $\mathbf{Q}_1$ and $\mathbf{Q}_2$.

Since the Schrödinger equation remains unchanged in the dBB theory, as in standard quantum mechanics the wave function still propagates in the six-dimensional configuration space; it pilots a six-dimensional vector obtained by grouping the components of $\mathbf{Q}_1$ and $\mathbf{Q}_2$. The fact that the velocities of the positions should be calculated in configuration space is a crucial element of the dBB theory, with many consequences, as we see later. On the other hand, each of the two Bohmian positions, when seen separately, propagates in ordinary three-dimensional space. As a result, nonlocal propagation effects may appear: the velocity of each particles depends, not only on its own position, but also on the position of the other particle, even if it is at a very large distance.

A convenient notion is that of "conditional wave function". If the motion $\mathbf{Q}_2(t)$ of particle 2 is known, one can define the conditional wave function $\psi_{\mathrm{cond}}(\mathbf{r}_1, t)$ of particle 1 as:

$$\psi_{\mathrm{cond}}(\mathbf{r}_1, t) = \Psi\left(\mathbf{r}_1, \mathbf{r}_2 = \mathbf{Q}_2(t); t\right)$$

(11.27)

If we replace Ψ by ψ_{cond} in the single-particle guiding equation (11.23), we obtain the equation of motion of $\mathbf{Q}_1(t)$ given by the first line of (11.26). The conditional wave function of particle 1 therefore provides the motion of this particle by exactly the same equation as for an isolated particle. Note that $\psi_{\mathrm{cond}}(\mathbf{r}_1, t)$ depends on the time for two reasons: because of the direct dependence of Ψ on time, and because of the time dependence of $\mathbf{Q}_2(t)$. Of course, if $\mathbf{Q}_1(t)$ is known, one can in the same way define a conditional wave function of particle 2. These conditional wave functions are a specific notion of dBB theory: in standard quantum mechanics, one can trace out the variables of one particle, but then the remaining particle is in general described by a density operator, not a pure state.

Now, if the wave function $\Psi(\mathbf{r}_1, \mathbf{r}_2; t)$ is a product:

$$\Psi(\mathbf{r}_1, \mathbf{r}_2; t) = \varphi(\mathbf{r}_1, t) \, \chi(\mathbf{r}_2, t)$$

(11.28)

it is easy to see that (11.26) simplifies into:

$$\frac{d}{dt}\mathbf{Q}_1 = \frac{1}{m_1\,|\varphi(\mathbf{Q}_1,t)|^2}\,\mathrm{Re}\left[\frac{\hbar}{i}\varphi^*(\mathbf{Q}_1,t)\,\mathbf{\nabla}_{\mathbf{r}_1}\varphi(\mathbf{Q}_1,t)\right]$$

$$\frac{d}{dt}\mathbf{Q}_2 = \frac{1}{m_2\,|\chi(\mathbf{Q}_2,t)|^2}\,\mathrm{Re}\left[\frac{\hbar}{i}\chi^*(\mathbf{Q}_2,t)\,\mathbf{\nabla}_{\mathbf{r}_2}\chi(\mathbf{Q}_2,t)\right]$$

(11.29)

Each particle then propagates independently, guided locally by its own wave function. The two conditional wave functions are merely the functions $\varphi(\mathbf{r}_1,t)$ and $\chi(\mathbf{r}_2,t)$ contained in the product (11.28).

If the wave function is not a product, the velocities have to be evaluated at a point of six-dimensional configuration space that depends on the positions of both particles; the result for the velocity of particle 1 may then depend explicitly on the position of particle 2, and conversely. Suppose for instance that the two-particle wave function can be written as a sum of two products:

$$\Psi(\mathbf{r}_1,\mathbf{r}_2;t) = \alpha\,\varphi(\mathbf{r}_1,t)\,\chi(\mathbf{r}_2,t) + \beta\,\varphi'(\mathbf{r}_1,t)\,\chi'(\mathbf{r}_2,t) \qquad (11.30)$$

and let us study the Bohmian velocity of particle 1; when do nonlocal effects occur? This depends on the overlap of the single-particle wave functions in the two terms of (11.30).

(i) Consider a particular point $(\mathbf{Q}_1,\mathbf{Q}_2)$ of a trajectory in configuration space. If, at time t, one of the two wave functions $\chi(\mathbf{Q}_2,t)$ or $\chi'(\mathbf{Q}_2,t)$ vanishes, only one term of (11.30) is relevant in the expression of the velocity of particle 1 at point $\mathbf{Q}_1$. The wave function of particle 2 disappears from this expression, as in (11.29); particle 1 then propagates locally.

(ii) If both wave functions $\chi(\mathbf{Q}_2,t)$ and $\chi'(\mathbf{Q}_2,t)$ are nonzero, then the same simplification does not take place. nonlocal effects take place in general, and the velocity of particle 1 is explicitly dependent on the position of the other particle. Of course, this is not always the case; if, for instance, it turns out that at point $\mathbf{Q}_1$ the two wave functions of the first particle have the same relative variation:

$$\frac{\mathbf{\nabla}_{\mathbf{r}_1}\varphi(\mathbf{Q}_1,t)}{\varphi(\mathbf{Q}_1,t)} = \frac{\mathbf{\nabla}_{\mathbf{r}_1}\varphi'(\mathbf{Q}_1,t)}{\varphi'(\mathbf{Q}_1,t)} = \mathbf{W}(\mathbf{Q}_1,t) \qquad (11.31)$$

again a simplification occurs[33] and the velocity at point $\mathbf{Q}_1$ is independent of position $\mathbf{Q}_2$. But, in general, relation 11.31 does not hold.

To summarize, nonlocal effects of particle 2 on particle 1 may occur when the wave functions $\chi(\mathbf{r}_2,t)$ and $\chi'(\mathbf{r}_2,t)$ overlap, and if the position $\mathbf{Q}_2$ falls in the overlap region. If the two wave functions do not overlap, whatever trajectory is

---

[33] If, in (11.26), we replace Ψ by its expression (11.30), $\mathbf{\nabla}_{\mathbf{r}_1}\varphi(\mathbf{Q}_1,t)$ by $\varphi(\mathbf{Q}_1,t) \times \mathbf{W}(\mathbf{Q}_1,t)$ and $\mathbf{\nabla}_{\mathbf{r}_1}\varphi'(\mathbf{Q}_1,t)$ by $\varphi'(\mathbf{Q}_1,t) \times \mathbf{W}(\mathbf{Q}_1,t)$, we obtain $d\mathbf{Q}_1/dt = \hbar\,\mathrm{Im}\,[\mathbf{W}(\mathbf{Q}_1,t)]\,/m_1$, which is independent of Q_2.

selected, one of the two components always vanishes at the Bohmian positions of the particles; this component then plays no role whatsoever in the evolution of the positions, and the particles move independently. The vanishing component[34] of the wave function is what Bohm calls an "empty wave" [11].

Empty waves and effective waves In the dBB theory, empty waves do not occur only when two particles are entangled as in the preceding example. In the single-particle experiment described in Figure 11.1, for instance, each time a particle is emitted by the source, the initial position of the Bohmian variable determines which path it will follow and which hole it will cross. One component of the wave function then guides the particle while the other becomes an empty wave, playing no role as long as the two components do not overlap. Reference [594] calls the active component of the wave function (the nonempty part) the "effective wave function". Nevertheless, when later both waves recombine in the interference region R, they both contribute to the guiding of the particle and create the observed fringe pattern. During the interference phenomenon, the empty wave has, so to say, recovered its particle, so that it becomes effective. Moreover, as we have seen in §11.8.1.c, the empty wave "catches" the particle while the wave packets cross; after the crossing, the wave packet that was empty becomes effective, and conversely.

Something similar may happen with wave function (11.30). When two particles are entangled but functions $\chi(r_2)$ and $\chi'(r_2)$ have no overlap, we have seen that one of the components is necessarily empty: particle 2 is guided by one wave only so that, for particle 1, no interference effect between $\varphi(r_1)$ and $\varphi'(r_1)$ can occur, even if these wave functions overlap[35]. But suppose that the future evolution of the system creates an overlap between $\chi(r_2, t)$ and $\chi'(r_2, t)$, while $\varphi(r_1)$ and $\varphi'(r_1)$ continue to overlap. If Q_1 and Q_2 fall inside their respective overlap region, the empty component of the two-particle wave function becomes effective, so that interference effects can occur again: in general, the velocity of each particle depends (nonlocally) on the position of both particles. A complicated motion of two coupled positions can then occur. This situation happens in two-particle-interference experiments [602, 603], where the Bohmian trajectories reproduce the results of standard quantum mechanics [604].

If particle 2 now becomes entangled with a large number of other particles, the von Neumann chain propagates too far and it becomes impossible in practice to restore interference effects. When decoherence becomes irreversible, empty waves remain empty forever. This is the case when the other particles are those contained in a measurement apparatus (§11.8.1.d). For more discussion of the role played

---

[34] Of course, depending on the trajectory considered, it may be one or the other component that is empty.

[35] In standard quantum mechanics, this absence of interference also occurs: by partial trace over particle 2, the coherences of particle 1 disappear.

by empty waves in the context of Bohmian theory or other interpretations, see for instance [71] and [605].

Other possible definitions of Bohmian velocities The agreement between the predictions of standard quantum mechanics and the statistics over Bohmian trajectories rests on a relation where only the divergence of the velocity defined in (11.23) plays a role. If one adds to the velocity any vector with no divergence, any curl, the agreement remains unaffected. This remark gives the impression of an almost too large flexibility for Bohmian theories, and even to some extent of an insufficiently defined framework. But in fact more constraints exist than this simple agreement to build reasonable trajectories. Reference [606] provides a discussion of the requirements introduced by Galilean invariance, which creates strong limitations, while still leaving several possibilities for defining the velocities.

It turns out that one can show that, in Einsteinian relativity, the conserved current is indeed unique [607, 608], so that the Bohmian velocity can be defined without any ambiguity. This can be shown by using the Dirac equation, which means that one has to consider particles with spin, as we will do in §11.8.1.e. It is then sufficient to take a nonrelativistic limit to obtain a unique Galilean value for the velocity in the guiding equation.

Moreover, another argument in favor of the usual definition of the Bohmian velocity has been proposed in Reference [609]. This article shows that the usual Bohmian velocity can be measured by an appropriate combination of a strong (projective) measurement and a weak measurement (§9.3) of the position at different times, when their difference is divided by the time difference. Proposing a technique of measurement that provides the value of a variable experimentally is, of course, a good way to attribute more physical value to it.

11.8.1.d Quantum measurement, projection of the conditional wave function, nonlocality

We now study in more detail how the mechanism of empty waves ensures that the uniqueness of results of quantum measurements appears naturally in dBB theory. This success is obtained at the price of an explicit nonlocality, which we will also discuss in this section.

Measurement in the dBB theory In the dBB theory, there is no need to introduce any special postulate for state vector reduction; it is already contained in the theory as a consequence of the mechanism of "empty waves", which in turn is a consequence of the equations of motion and of the notion of trajectory. This solves the difficulties related to the definition of the border between the Schrödinger equation and state reduction, which disappears. In other words, the Schrödinger equation applies equally well during all stages of a measurement process and at any other time.

To understand in more detail why, we consider a physical system after it has interacted with a measurement apparatus. It is then necessarily entangled with the apparatus (§9.1), so that the wave function of both systems takes the form:

$$\Psi(\mathbf{r}, \mathbf{r}_1, \mathbf{r}_2, ..., \mathbf{r}_N; t) = \sum_j \varphi_j(\mathbf{r}, t) \chi_j(\mathbf{r}_1, \mathbf{r}_2, ..., \mathbf{r}_N; t) \qquad (11.32)$$

where the $\varphi_j(\mathbf{r}, t)$ are mutually orthogonal states of the measured system associated with the various possible results. The $\chi_j(\mathbf{r}_1, \mathbf{r}_2, ..., \mathbf{r}_N; t)$ are the corresponding states of the measurement apparatus (including the "pointer"), which are also orthogonal. Actually, they are not only orthogonal but also without any spatial overlap – otherwise the observation of the position of the pointer would not provide a result of measurement. This form of the state vector is similar to (11.30), in the case where there is no overlap, so that the same discussion applies. We have seen (§11.8.1.b) that the Bohmian variables cannot "leave" the wave function (they cannot reach points of configuration space where it vanishes). After measurement, the variables $\mathbf{Q}_1$, $\mathbf{Q}_2$, ..., $\mathbf{Q}_N$ associated with the measurement apparatus necessarily belong to one of the domains of configuration space where one of the wave functions χ_j does not vanish; since these functions have no overlap, they cannot belong to more than one of these domains.

Therefore, for any possible trajectory, only one term of the sum (11.32) can play a role, all the others being "empty waves". As for the measured system itself, its conditional wave function (§11.8.1.c) is only one of the $\varphi_j(\mathbf{r}, t)$: it is reduced to the component associated to the observed result, which is just equivalent to what is predicted by the projection postulate. The von Neumann prescription is now a consequence of the dBB dynamics, no longer a postulate!

As mentioned at the end of §11.8.1.c, because of the very large number of variables associated with a measurement apparatus and its environment, it is impossible in practice to restore an overlap of all the components and to drive each of the many Bohmian positions to its specific region of overlap of the wave function. These empty waves can therefore no longer influence the guiding of the position of the particles, and have become completely irrelevant to determine the results of future position measurements; they remain empty forever. It is then possible to retain only this nonempty term, without affecting in any way the future dynamics of the combined system. Moreover, one can normalize this term since, in Bohmian theory, the velocities are unchanged if the wave function is multiplied by a constant.

The net result is that, for any realization of the experiment, corresponding to a given trajectory:

– a single result is obtained

– only one component of (11.32) plays a role in the calculation of the future motion of the Bohmian positions in configuration space (as well as in the calculation of probability of any possible future measurement performed on the system). State vector reduction is then reconstructed and gets a dynamics.

Within dBB theory, the result of any experiment is just a consequence of the random initial position of the system in configuration space. The uniqueness of this result follows from the impossibility of a single point in configuration space to belong at the same time to more than one of the domains associated with the components of (11.32). This automatically makes all components of the wave function vanish, except the one corresponding to the result of measurement. Determinism is restored in principle but, since this additional variable cannot be controlled in the preparation stage of the experiment – see the preceding section and point (ii) in §11.8.1.b – in practice quantum experiments give random results. While in standard quantum mechanics, the mechanism of decoherence (correlation with the environment) is not sufficient to explain the emergence of a single result in a single experiment, in the dBB theory it is, thanks to the introduction of the Bohmian variables of the measurement apparatus and the mechanism of empty waves. This is a great achievement!

Position measurements and Bohmian trajectories are compatible If a series of position measurements are performed at various times on a particle, one may wonder if the results of all these measurements reconstruct the Bohmian trajectory – otherwise, of course, the physical interpretation of Bohmian trajectories would become delicate. Why this is indeed the case is discussed in §2 of Appendix I. Actually, the proof is simple only in the case where the successive measurement apparatuses are sufficiently fast to signal the passing of the particle rapidly. When the apparatuses are slow and provide only delayed information, the problem is more delicate, as we will see in §2 of Appendix I.

Nonlocality In § 11.8.1.c, we have seen that Bohmian positions evolve according to equations (11.26), which are explicitly nonlocal. Nonlocality is of course a surprising feature for any physical theory, but we should keep in mind that this feature is not artificially added into a theory that otherwise would be completely local: standard quantum theory is not perfectly local either. Actually, in configuration space, the equations of both theories are local, but nonlocalities may appear when one comes back to ordinary three-dimensional space.

Consider for instance two spinless particles and their six-dimensional configuration space. The formalism of standard quantum mechanics offers no local description of the physics taking place at the position of one single particle, even if the particles are contained in remote regions of space. The standard description is actually contained in the wave function[36], which has values that depend on the positions $\mathbf{r}_1$ and $\mathbf{r}_2$ of the two particles. If the wave function is not a product, the only proper characterization of the system therefore occurs in configuration space. For instance, the time evolution of the phase of this wave function depends on the values of the single particle potentials $V(\mathbf{r}_1)$ and $V(\mathbf{r}_2)$. The same is true of the probability current $\mathbf{J}$, which is defined in the same configuration space, and has a time derivative containing both potentials in general (nevertheless, for both theories, nonlocality disappears from the equations as soon as a partial trace over the other particle is taken; in Bohmian theory, this involves an integration over all possible positions of the second particle).

The difference between the two theories is therefore not so much in the mathematical equations, which in neither case are really local. It is more that, in Bohmian theory, one ascribes physical reality to variables that evolve nonlocally (positions), while in standard quantum mechanics nonlocal evolutions occur only for physical quantities that are more indirectly related to physical reality.

Persistent empty waves Consider a series of experiments providing a chain of successive results for arbitrary observables (for instance, spins, as discussed in §11.8.1.e). For each realization of the series, the chain of results that is actually observed is tagged by the Bohmian positions of the pointers; it corresponds to a nonempty component of the state vector, the "effective wave function" (§11.8.1.c). All other chains of results (those that have not been obtained) are associated with empty waves, which still exist, but can have no effect whatsoever on the future evolution of any Bohmian position. These empty waves never disappear from the solution of the Schrödinger equation (as in the Everett interpretation, §11.12), but their fate is to remain forever inactive, without any observable effect. If the number of successive experiments is increased, the number of empty waves becomes very large, while only one wave remains nonempty (the effective wave function) and may have future physical effects. If for instance we consider the state vector of the universe, we clearly have to face an absolutely fantastic number of empty waves. Of course, they can be discarded without affecting the physical predictions in any way, but their status of reality is not obvious (we come back to this point in §11.8.1.i).

---

[36] This is true in the Schrödinger picture. In the Heisenberg picture, the dynamics is contained in the evolution of operators; their matrix elements belong to a space with dimension equal to the square of that of the configuration space, which is even larger. If one uses second quantization or field theory, the dimensions become infinite (in field theory, coordinates in ordinary space time appear in the formalism as parameters defining the operators, but the space in which these operators act has infinite dimension).

11.8.1.e Spin and field theory

Bohmian theory can include treatments of spin and of quantum fields. Here we give only a brief summary and limit ourselves to a few simple examples.

Pauli spinors Spins can easily be included in the dBB theory within a nonrelativistic treatment based on Pauli spinors. A simple method does not add any specific Bohmian variable associated with the spin, but just keeps the usual Bohmian position of the particle. Even within this simple context, interesting effects occur, but still in a way that avoids any contradiction with the predictions of standard quantum mechanics.

For a single particle with spin S, the quantum state can be defined in terms of a spinor having $2S + 1$ components $\Psi_\mu(\mathbf{r}, t)$. If $S = 0$ (spinless particle), the single component of the spinor is the wave function $\Psi(\mathbf{r}, t)$ already appearing in (11.23); if the particle has a spin $1/2$, it is described by two components $\Psi_\pm(\mathbf{r}, t)$ obeying the Pauli equation, etc. At each point of space, the Bohmian velocity of the particle is defined as the ratio of the local current of probability by the local density of probability, which are both obtained by a trace over the spin variable (sum over index μ). Equation (11.23) is then replaced by:

$$\frac{d}{dt}\mathbf{Q} = \frac{1}{m\sum_\mu |\Psi_\mu(\mathbf{Q}, t)|^2} \, \mathrm{Re} \sum_\mu \left[\frac{\hbar}{i} \Psi_\mu^*(\mathbf{Q}, t) \nabla \Psi_\mu(\mathbf{Q}, t) \right] \tag{11.33}$$

Instead of using the Pauli nonrelativistic spinor, one can start from the relativistic Dirac equation and use its probability density current. This turns out to define the conserved current and therefore the local Bohmian velocity in an unique way [607, 608]. One can then take the nonrelativistic limit. The interesting point is that the Bohmian velocity is the sum of the right hand-side of (11.33) and of an additional contribution containing a curl:

$$\frac{1}{m\sum_\mu |\Psi_\mu(\mathbf{Q}, t)|^2} \nabla \times \sum_\mu \Psi_\mu^*(\mathbf{Q}, t) [\sigma \, \Psi(\mathbf{Q}, t)]_\mu \tag{11.34}$$

By calculating the local average of the spin orientation from the spinor in standard quantum mechanics, one can also define a local direction of the spin, which leads to an even more visual representation of the propagation of the particle. The evolution of the spin contains a term called the "quantum torque" term (§9.3.2 of [588]); its effect is visible in the spin trajectories shown in Figure 11.2. Interestingly, this term also appears in the kinetic theory of a Bose–Einstein condensed gas at very low temperatures [610], where it introduces a rotation of the spins that is independent of the interactions.

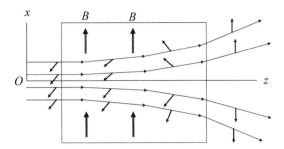

Figure 11.2 In a Stern–Gerlach experiment, atoms propagating along direction Oz enter a region (rectangle) where a high magnetic field B along direction Ox is applied. Its strong gradient acts on the magnetic moments of the atoms and creates a spin-dependent deviation. Initially, the atoms have a transverse spin direction along Oy. While they progress in the gradient, the direction of their spin changes; when they exit the apparatus, two separate beams are formed with spins either parallel or antiparallel to Ox. Four examples of Bohmian trajectories are shown (the directions of the arrows show the changes of the polar angle between the spin and Ox; for clarity, the rapid precession of the azimuthal angle around B is not included).

A few examples The effects taking place for a spinless particle have their counterpart for particles with spin. We now discuss a few examples.

(i) The direction of the spin does not necessarily remain constant along a trajectory, even in free space – for more details, see for instance chapters 9 and 11 of [588], and [611]. As in [612], consider an atom entering a Stern–Gerlach apparatus that measures the component of the spins along Ox. If, initially, the spins of the particle point in a perpendicular direction Oy, their directions change progressively when the trajectory propagates in the magnetic field gradient, as shown schematically in Figure 11.2 (for clarity, the uniform precession of the spins around the homogeneous part of the magnetic field has been removed). At the output of the device, the spins are now either parallel or antiparallel to Ox. The former case corresponds to the trajectories leading to spin result $+1$, the latter to those giving result -1.

Spin rotations along trajectories occur even for free particles; there is an equivalent of the "no-crossing rule": if two wave packets associated with opposite spin directions cross in free space, along any Bohmian trajectory the direction of spin turns during the crossing and, after the position has jumped from one wave packet to the other, ends up in the opposite direction – an analogous effect to the exchange of wave packets by trajectories that we discussed previoulsy.

(ii) An interesting thought experiment considers one spin particle propagating in direction Oz that is sent through a series of spin analyzers (Stern–Gerlach systems for instance) with various orientations; for instance, the first analyzer separates the two values of the spin component along Ox, the second along a perpendicular direction Oy, the third along Ox again, etc. (at each step, the direction of measurement changes by 90°), as shown in Figure 11.3. The first analyzer divides the trajectories into two groups, those going to positive directions along Ox, and those going to negative directions, depending on the initial position of the particle. The second analyzer divides again each of these two groups of trajectories into two subgroups, depending again on the initial position of the trajectory. The same phenomenon repeats when more and more measurements are added. One could then hope to reach a situation where the determination of the initial position becomes sufficiently accurate to provide a determination of the trajectory of the particle in the next analyzer. Actually, this never happens: whatever number of analyzers is used, one never reaches a situation where the initial position becomes sufficiently well known to be able to predict the deviation observed in the next measurement with certainty; the result of the spin measurement remains completely random. The reason why this is the case is that repeated measurements introduce an exponentially increasing sensitivity to the initial position of the particle, in analogy with chaotic situations in classical physics. At each step, the distribution of the Bohmian variables remains exactly the quantum equilibrium distribution. There is no way, even with many measurements, to determine the position sufficiently well to eliminate the fundamentally random character of quantum mechanics.

Wigner has pointed out the time irreversibility of the evolution of the additional variables in such situations [613]; Clauser has remarked that reversibility can be restored if one takes into account the spin polarization variables [614]. In addition, it is also possible to assume the existence of additional variables associated with the successive measurement apparatuses, as in "von Neumann's informal hidden variable argument"; then, more randomization takes place at each measurement, so that it is no longer necessary to assume this extreme sensitivity to the initial conditions.

(iii) For two particles, consider an EPRB experiment of the type described in §3.3.1 and the evolution of the spins and positions of the two particles when they are far apart. If particle 1 is sent through a Stern–Gerlach analyzer oriented along direction a, the evolution of its Bohmian position is obviously affected in a way that depends on a: since the position has to follow the quantum wave functions, it has the choice between two wave packets that are separated along axis a. This situation is that illustrated in the right part of Figure 11.2, were the interaction between the particle and the magnet has oriented the spin in a direction that is parallel or antiparallel to a, depending on the wave packet. Assume now, as in

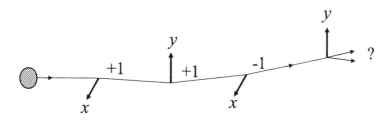

Figure 11.3 A particle with spin 1/2 propagating in direction Oz crosses several Stern–Gerlach analyzers, which measure its spin component in succession along directions Ox, Oy, Ox, etc. Result +1 for the spin component is obtained if the deviation of the trajectory of the particle occurs in one direction, result −1 if it occurs in the opposite direction. These results are completely random, and provide better and better information about the initial Bohmian position of the particle inside the wave packet associated with the wave function at the input of the system. Nevertheless, whatever the number of measurements is, this information is never sufficient to give any information on the result of the next measurement.

References [615, 616], that during this first measurement particle 2 is still flying in free space towards the second Stern-Gerlach magnet. Then, according to dBB theory (see also Appendix I, §I.1.2), this particle acquires a spin orientation that is antiparallel to that taken by particle 1. Interestingly, before the two wave packets of particle 1 separated in the magnet, particle 2 had no Bohmian spin orientation (in agreement with the zero quantum value of the average of each spin vector in a singlet state), but it acquires one under the effect of this separation. The conclusion is that, with a singlet spin state, the selection of an orbital wave packet for the first particle produces a projection of the spin state of the second particle onto a state that depends on a. This nicely illustrates the fact that the result of a measurement does not reveal the value of some pre-existing quantity attached to the particle, but is created during the interaction of the whole measurement apparatus with the whole quantum system.

Now, if the two particles penetrate the two Stern-Gerlach magnets simultaneously, the orbital wave function that determines the motion of the point in the configuration space is not a product. The position of each particle 1 has the choice between two wave packers, and this choice changes the position $(\mathbf{Q}_1, \mathbf{Q}_2)$ of the point representing the system in the six-dimensional configuration space. As we have seen previously, this immediately changes the quantum velocity term of the other particle, in a way that then depends explicitly on direction of measurement. No wonder if such a theory has no difficulty in reproducing the nonlocal effects associated with violations of the Bell inequalities!

In this case, the advantage of introducing additional variables is to emphasize the effects of nonlocality, while these effects often remain relatively hidden in the orthodox formalism (one more reason not to call these variables "hidden"!). Bell for instance wrote (last sentence of [617]): "it is a merit of the de Broglie–Bohm interpretation to bring this (nonlocality) out so explicitly that it cannot be ignored" – in fact, historically, he came to his famous inequalities precisely through this channel. As mentioned in §4.3.2, it is a matter of debate whether the standard interpretation of quantum mechanics just hides the nonlocality it actually contains, or can be seen as a local theory despite the violations of the inequalities.

Relativistic treatment of spin We have already mentioned that the dBB theory can be extended to the relativistic Dirac equation, which leads for instance to the additional term (11.34) in Bohmian velocities. This is possible since this equation, as the Schrödinger equation, can be reformulated in terms of local densities and currents, as remarked by des Cloizeaux in 1983 [618]. Chapter 12 of the book by Holland [588]) discusses various aspects of the dBB continuation of the Dirac equation. This equation raises specific questions because it predicts positive and negative energies; one generally considers that the levels of negative energies are filled by fermions, which form what is usually called the "Dirac sea". It turns out that it is actually possible to ascribe particle trajectories to all fermions, including those in the Dirac sea, and to develop a theory that is deterministic [619].

dBB field theory In the second of his initial articles (appendix A of [120]), Bohm included a brief discussion of field quantization, in the context of a study of the Compton effect. The close analogy between an harmonic oscillator and the dynamics of a free mode of the electromagnetic field in a cavity is a useful guide for introducing Bohmian trajectories within quantum electrodynamics – see also §10.6 and chapter 12 of [588]. Bohm introduces an additional variable $\mathcal{A}$ associated with the quantum operator A describing the vector potential, which plays the role of the "position" of the harmonic oscillator; the "wave function" in the representation where A is diagonal then guides $\mathcal{A}$ according to the usual pilot equation. As often in quantum electromagnetism, this is done in the Coulomb gauge, where the vector potential is transverse and in direct relation with the magnetic field; the electric field is proportional to its time derivative, that is to the momentum associated with the potential vector (it is possible to show that the theory remains gauge invariant).

Instead of introducing a Bohmian variable for the potential vector, one can also directly associate a Bohmian variable $\mathcal{E}$ to the electric field. This variable is guided by the "wave function" in the representation where the electric field is diagonal. Then $\mathcal{E}$ and its time derivative $d\mathcal{E}/dt$ directly describe the electric and magnetic (within a constant factor) fields associated with the mode considered. An inverse Fourier transform provides the values of the electromagnetic fields at every point inside the cavity. The same is of course true for the magnetic field $\mathcal{B}$.

Similar methods may be applied to other bosonic fields. Massive bosons have position operators (as opposed to photons, which have no rest mass). For states containing a fixed number of nonrelativistic bosons, a wave function can be defined in the configuration space of all positions. One can then directly apply the same method as in §11.8.1.b to introduce Bohmian positions: the pilot equation remains exactly the same as for distinguishable particles, all Bohmian positions being guided by a fully symmetric wave function. This method is nevertheless not applicable when the state vector is a superposition of states containing different total numbers of bosons. One can then proceed as before for the electromagnetic field, and introduce Bohmian variables for the fields[37]. It is even possible to introduce creation and annihilation operators [620, 621], which create or annihilate Bohmian positions, so that trajectories can begin or end at points. Altogether, the situation for bosons is that there seems to be a choice between field and position Bohmian variables.

For fermionic fields, if the state contains a fixed number of particles, the situation is the same as before; the Bohmian positions are guided by a fully antisymmetric wave function. Introducing Bohmian variables for fields associated with anticommuting operators is more complicated than for bosons; a possibility is to decide that the Bohmian variables of fermionic systems are only positions. Combining this approach with the previous description of bosonic fields, one then obtains a description of reality where bosons and fermions are treated in a different way, the former having Bohmian field variables and the latter only position variables; the usual symmetry of second quantization is lost. But, after all, boson and fermion systems are indeed very different physically; there is no special reason why similar Bohmian variables should be used for both. Moreover, composite bosons (atoms, for instance) made of an even number of fermions have a Bohmian position variable in this theory, the center of mass of the positions of the fermionic constituents.

Bell discussed a method to attribute Bohmian variables (called "beables") to quantized fermionic fields on a lattice. In this model, discussed in §11.8.2, the de-

---

[37] The difficulties associated with divergences in standard quantum field theory, and the necessity of introducing renormalization procedures, still occur in Bohmian theory, as one could expect.

terministic guiding equation of dBB theory is replaced by a stochastic evolution process, but a conjecture is that this stochastic feature should go away in the continuum limit. For reviews on dBB field theories, see for instance [620] and [622], or [623–625] for relativistic theory.

11.8.1.f Applied dBB mechanics, quantum cosmology

Since Bohmian mechanics is strictly equivalent to nonrelativistic standard quantum mechanics, it can also be used as a convenient tool to address practical problems in this domain, independently of any interpretation. For instance, computing trajectories can be a convenient way to study problems in molecular dynamics [626]; quantum Monte Carlo methods and conditional wave functions are useful to obtain simulations of atoms in ultra-strong laser fields [627]; the behavior of electrons in nano-electronics circuits can also be studied [628, 629]. A general review of many possible applications is given in [630].

In physical cosmology, one problem is to study the evolution of the structure of the early universe during the initial "big bang" period. Quantum mechanics plays an important role in this problem, but also gravity. General relativity introduces an equation giving the metric, where the general relativistic scale factor appears as a quantum operator. As for the quantum particles, when the temperature is sufficiently high, they are ultrarelativistic and they can be considered as forming a relativistic fluid (similar to a radiation field); this leads to a generic form of the action describing all these microscopic degrees of freedom. Both systems have quantum fluctuations, which can create small initial perturbations growing in time and eventually leading to the formation of stars and galaxies; but this phenomenon obeys complicated coupled quantum equations. A perturbation expansion can be introduced with respect to small inhomogeneous and anisotropic fluctuations of the metric, but even then the solution still remains very difficult, because of the presence of the quantum operator associated with the relativistic scale factor.

It is then usual to treat the scale factor approximately as a classical number; of course, conceptually, this is not very satisfactory. But one can also develop a Bohmian theory for quantum gravity just along the same lines as for the electromagnetic field. There is then just a Bohmian variable for the metric. It has been shown in References. [631] and [632] that, after an unitary transformation has been performed to transform the Hamiltonian into a sum of two terms, it is useful to introduce the Bohmian trajectory of the scale factor. The trajectory can be calculated to successive orders, which makes it possible to inject the corresponding values into the successive perturbation equations of the other degrees of freedom. This by no means implies that quantum effects are ignored: as we have seen, Bohmian trajectories are sensitive to interference or tunneling, and their use does not suppress any quantum result. For instance, the Bohmian trajectory can be used in theories where the universe collapses and expands, but where the collapse takes place only until a small (but nonzero) value of the scale factor is reached, where the quantum effects of gravity become so strongly repulsive that the universe rebounds back out ("Big Bounce") [633]. The method is particularly well suited to define a global time in cosmology; various interesting results in quantum cosmogenesis have been obtained in this way. Of course, one may, at will, interpret the Bohmian trajectory as defining a real physical phenomenon, or just as a convenient intermediate mathematical tool that facilitates a perturbation calculation.

11.8.1.g Modified dBB theory

Already in his initial articles published in 1952 (§9 of the first part of Reference [11]), Bohm emphasized that the dBB theory could be a good starting point to introduce modifications of the standard theory, for instance to solve difficulties appearing at very short distances. These modified dBB theories are no longer completely equivalent to standard theory, which means that they become experimentally testable against it; they therefore introduce new physics. For instance, Bohm discussed the introduction of a relaxation term into the expression of the acceleration of the particle (this changes the guiding formula and the quantum equilibrium condition), or into the equation of evolution of the state vector (this changes the quantum dynamics). He came back to the problem in 1953 to show in a simple model that, under the effect of random collisions, an arbitrary probability density will ultimately tend to a distribution given by $|\Psi(\mathbf{r}_1, \mathbf{r}_2, ...)|^2$, so that its difference with quantum equilibrium is likely to remain very small at all times. In 1954, he and Vigier treated a more general case, by assuming irregular fluctuations of the fluid that guides the positions of the particles [634].

In 1991, Valentini took a different approach [635] inspired from the H-theorem in statistical mechanics, and introduced a "subquantum entropy" H to characterize any distribution differing from the quantum equilibrium condition; he then showed that a course-grained value of H can only increase in time, and that the maximal value is obtained for the quantum equilibrium. The latter then appears as an attractor, and the reasoning can be seen as a derivation of the Born rule, which no longer appears as an independent postulate, but as a consequence of the dynamics. See [636, 637] for more recent numerical simulations along the same line.

Another type of modified dBB theory is discussed in §11.10.4 (page 388). See also Reference [638], which introduces a variant of the dBB theory where additional variables are introduced in phase space, instead of the configuration space; if a particle is in a stationary bound state, it can then move, as opposed to what happens with the dBB theory.

11.8.1.h Objections and solutions

It is not possible here to discuss all aspects of the dBB theory. It is not a "main-stream" theory, and has certainly not been applied to all problems that have been treated successfully within standard quantum mechanics. These problems range from usual applications in condensed matter physics, quantum optics, etc., to quantum chromodynamics (the theory of the strong interaction involving quarks), an essential component of our present physical description of the world. Since our purpose here is rather to focus on the interpretation, we will just discuss a few examples of applications of the dBB theory. They have been selected because they involve situations that help clarify the physical content of the Bohmian point of view, as well as some frequent misunderstandings concerning this theory. A review of various objections, some common misunderstandings, and their answers, can be found in §15 of [589].

Are Bohmian trajectories real? We have seen that a past Bohmian trajectory can be reconstructed from a single observation of the position of the particle, which allows us to calculate preceding positions and therefore the trajectory. For the future, it is nevertheless impossible to make more accurate predictions than those of standard quantum mechanics; the same fundamental indeterminism applies. In this sense, the explanations of quantum phenomena given by the dBB theory are more retrodictive than predictive.

The authors of References [611] propose to study an interference experiment in an interesting case (see also the discussion of Reference [639]). The usual interference device is supplemented by electromagnetic cavities, which can store the energy of photons and be used as a "Welcher Weg" device (a "which way" device that tells the experimenter which hole the particle went through in an interference experiment). The first particle (test particle) goes through a screen having two holes, with two cavities sitting near the holes (C_1 and C_2 in Figure 11.1). If the particle goes through the upper hole, it leaves a photon in the first cavity[38]; if it goes through the lower hole, it leaves it in the second cavity. Probing the number of photons in one of the cavities then allows one to know which hole the particle went through. A way to obtain the information is to observe the trajectory of a second particle, which is sent afterwards through this cavity, experiences the field it contains, and takes a trajectory that depends on the number of photons[39]. At the end, one can compare the trajectories of the two particles, which can interact indirectly through a photon in a cavity, and investigate the conditions under which they can influence each other.

---

[38] We may assume that the particles are atoms in high Rydberg states, which can have very large dipole moments and can easily emit or absorb photons in superconducting cavities – see §10.3.

[39] A similar scheme is used in the experiment described in §10.3.

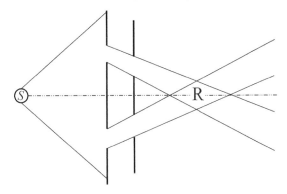

Figure 11.4 An interference experiment with a source S and screens designed in order to create wave packets that cross in the interference region R.

For the discussion, it is more convenient to assume that the wave packets of the test particle originating from the two holes cross each other, defining a finite interference region, instead of overlapping until an infinite distance from the screen under the effect of diffraction. This simpler situation is schematized in Figure 11.4. The experiment is symmetric with respect to the horizontal plane P shown in the figure. In standard quantum mechanics, the two wave packets cross each other without any mutual effect (the Schrödinger equation is linear). Therefore, the wave packet that crosses the upper slit, and may leave in passing a photon in C_1, then propagates downwards in the figure; after the interference region, a particle in this wave packet can only be found below the symmetry plane P. In the same way, the wave packet that may leave a photon in C_2, propagates upwards and corresponds to particles that can be detected above the symmetry plane after the interference region. But we have seen in §11.8.1.c that, in the absence of the photon cavities detecting the test particle, its dBB trajectories can never cross the symmetry plane ("no-crossing rule"): in the interference region they "jump" from one wave packet to the other and avoid crossing the plane. Let us assume for a moment that this is still true in the presence of the which way detectors of the photon (we discuss this point later). Then, finding the particle under the symmetry plane P would necessarily mean that the corresponding trajectory went through the lower hole, even if a photon is left in the upper cavity. This would then lead to a retrodictive contradiction between standard quantum mechanics and dBB theory: they consider that the particle went through different holes. The test particle has left a photon in the cavity that its Bohmian trajectory never crossed! Does this mean that the results of quantum position measurements are contradictory with Bohmian trajectories? If this is the case, do the trajectories have any physical meaning?

The contradiction looks even stronger if the photon left in the cavity is later detected by its influence on the trajectory of a second particle: the net result is that the two Bohmian trajectories influence each other without ever coming close to each other. From this strange conclusion, the authors conclude that the Bohmian retrodictive trajectories may be incompatible with the results of measurements, and are therefore "surrealistic"; we now investigate this subject.

Remarks:

(i) Before we enter a more detailed discussion of the paradox, we note that it relies on the idea that, in dBB theory, the particles should be identified with their positions and trajectories only. This would correspond to a truncated form of dBB theory in which only the positions of the particles have physical reality, with the exclusion of any element of the wave function. But in the dBB theory, positions and guiding waves are completely indissociable, as we have seen in §11.8.1.b, and there is no reason whatsoever why the wave could not also influence its surrounding. For instance, if one takes the point of view where the wave function becomes a real classical field, this field exists as objectively as an electromagnetic field for instance (see Bell's quotation in §11.8.1.b). The simplest way to solve the apparent paradox is therefore to consider that the real field associated with the first particle can interact locally with the electromagnetic field in the cavity, leaving a photon in it; later this photon acts on the trajectory of the second particle.

Equivalently, if one prefers to use the notion of quantum potential and nonlocality, one can express the same idea by saying, as the authors of [640]: "the energy has been transferred nonlocally to the cavity-field mode by the action of the quantum potential ... a phenomenon no less real than the nonlocal correlations observed in EPR effects ..." – see the figures of this reference for examples of plots of the trajectories.

(ii) Another preliminary remark is that, within the standard interpretation of quantum mechanics, "interaction-free measurements" are described in a way that is actually very similar to this Bohmian description. In §2.4, we have described an experiment where some events can occur only because the Schrödinger wave is absorbed in the arm of an interferometer, while the particle had only access to the other arm to propagate; the propagations of the wave and of the particle were also to some extent dissociated in these events. Standard interaction-free quantum measurement processes are not less "surreal" than Bohmian trajectories!

Effects of entanglement on the trajectories   In order to correctly interpret the indications provided by a measurement apparatus concerning the trajectories of a test particle, we have to study the dynamics of the interaction between the two physical systems. As we will see, the Bohmian positions associated with the particles contained in the measurement apparatus play an essential role to determine the trajectories of the test particle when it reaches the interference region. Moreover, in the limit where the apparatus contains a very large number of particles, surrealistic trajectories completely disappear.

We will first study the case where the measurement apparatus is made of massive particles, and then the situation where the Welcher Weg apparatus contains zero mass particles, as the photons of a micromaser in Reference [611].

(i) Entanglement with massive particles

We assume that the wave of the test particle crosses the two holes of a screen, and then reaches a region of space where quantum interference effects may take place, as illustrated in Figure 11.4. A second particle plays the role of the pointer of a measurement apparatus: if the test particle crosses the upper hole, the quantum state of the pointer particle is $|\chi_1\rangle$, and another orthogonal quantum state $|\chi_2\rangle$ if the test particle crosses the lower hole. The wave function of the whole physical system then has the form (11.30), where the functions $\varphi(\mathbf{r}_{1,2})$ and $\chi(\mathbf{r}_{1,2})$ are time-dependent since both particles propagate in space. Each particle has a Bohmian position, $\mathbf{Q}$ for the test particle, and $\mathbf{Q}_1$ for the pointer particle; they are coupled by the nonlocal guiding equation 11.26 (with the substitutions of $\mathbf{Q}$ instead of $\mathbf{Q}_1$ and of $\mathbf{Q}_1$ instead of $\mathbf{Q}_2$). One can also generalize this discussion by assuming that the pointer contains more than one particle, and that these particles with positions $\mathbf{Q}_1, \mathbf{Q}_2, .., \mathbf{Q}_N$ are entangled with the test particle; the quantum state of the whole system is then a simple generalization of (11.30).

An initial remark is that the "no-crossing rule" (trajectories can never cross the horizontal symmetry plane P of Figure 11.1) does not hold in this case. In Reference [611], the existence of the paradoxical trajectories was derived from this rule. It certainly applies for a single particle propagating in a symmetrical environment with a symmetrical wave function, but no longer in the presence of the pointer particle. This is because the symmetry is in general broken by the choice of the initial value of the pointer position $\mathbf{Q}_1$; thus, there is no special reason why trajectories should necessarily bounce on the symmetry plane.

Reference [294] gives a general discussion of the trajectories of the test and pointer particles, and discusses how the Bohmian positions of the pointer play an essential role. Here, we just summarize the results, but more details are given in §I.3 of Appendix I. If the wave packets of the pointer particles are already well separated when the test particle reaches the interference region ("fast pointer"), the situation is very simple: the trajectories of the test particle in the interference region are then practically straight lines. Therefore, the no crossing rule never applies and no surrealistic trajectory occurs. More interesting effects occur with a "slow pointers" (the wave packets of the pointer particles are not yet separated when the test particle reaches the interference region) [641]. If the number of pointer particles is limited to one or a few units, quantum nonlocal effects may take place, originating from simultaneous "jumps" in configuration space of all particles from one wave packet to the other. This introduces the surrealistic trajectories of Reference [611], which can be seen, either as a nonlocal influence between remote particle positions, or as a local influence of the field of one particle on the position of the others. Nevertheless, if the number of pointer particles becomes large, as is the case with a macroscopic pointer, these nonlocal effects completely disappear and no surrealistic trajectory occurs.

(ii) Entanglement with photon modes in a micromaser

We now assume that, after the particle has crossed the cavities, the test particle is entangled with the two fields in the cavities; again it is not possible to treat it as an isolated quantum system. To study the micromaser within dBB theory, we must introduce position variables associated with the modes of the field in each cavity. We treat the modes of the field with a simple model: harmonic oscillators associated with massive particles in a potential. The position variable of the wave function associated with cavity C_1 is $\mathbf{r}_1$, that associated with cavity C_2 is $\mathbf{r}_2$, corresponding respectively to position variables $\mathbf{Q}_1$ and $\mathbf{Q}_2$ in Bohmian mechanics. We call $\mathbf{r}$ the variable of the wave function of the test particle, with Bohmian position $\mathbf{Q}$. The entangled situation is then described by a wave function:

$$\Psi(\mathbf{r}, \mathbf{r}_1, \mathbf{r}_2) = \frac{1}{\sqrt{2}} \left[\varphi_{\text{upper}}(\mathbf{r}) \chi_e(\mathbf{r}_1) \theta_g(\mathbf{r}_2) + \varphi_{\text{lower}}(\mathbf{r}) \chi_g(\mathbf{r}_1) \theta_e(\mathbf{r}_2) \right] \qquad (11.35)$$

where $\varphi_{\text{upper}}(\mathbf{r})$ and $\varphi_{\text{lower}}(\mathbf{r})$ are the wave functions of the particle that originate from the upper and lower hole respectively, $\chi_{g,e}(\mathbf{r}_1)$ the wave functions associated with the first harmonic oscillator in either its ground or first excited state, and $\theta_{g,e}(\mathbf{r}_2)$ the similar wave functions for the other oscillator.

The most probable locations of the positions $\mathbf{Q}$, $\mathbf{Q}_1$, and $\mathbf{Q}_2$ are those where $|\Psi(\mathbf{Q}, \mathbf{Q}_1, \mathbf{Q}_2)|^2$ takes relatively large values, which implies that at least one of the two components of (11.35) should have a significant modulus. For instance, a maximal value of the first component requires that $\mathbf{Q}_1$ should be chosen to give a maximal value to $|\chi_e(\mathbf{Q}_1)|^2$, while the second component requires that $|\chi_g(\mathbf{Q}_1)|^2$ should be maximal. But the functions $\chi_{e,g}(\mathbf{r})$ are different (they are actually orthogonal), and their maxima do not coincide; whatever the random value of $\mathbf{Q}_1$ is, at most one of these functions can be relatively large, while the other remains rather small[40]. The same is also true for variable $\mathbf{Q}_2$. Therefore, for a random choice of $\mathbf{Q}_1$ and $\mathbf{Q}_2$, one component of (11.35) dominates over the other. In other words, for a particular trajectory of the system (in nine-dimensional configuration space), there is a large probability that the two components should have different mod-

---

[40] For instance, if $\mathbf{Q}_1$ happens to be at the maximum of the probability distribution for the ground state, the wave function of the excited state vanishes exactly, so that one of the terms in (11.35) completely disappears.

ulus. This breaks the initial symmetry and changes the interference pattern: the no-crossing rule is no longer valid. Some paradoxical trajectories (those where the trajectory goes through one cavity and excites the field in the other) still exist, but they require particular configurations of Q_1 and Q_2, which significantly reduces their probability.

In addition, if we require some information on the state of the field in the cavities to be obtained, we need to couple it to other physical systems, a second particle for instance. In turn, the position of this particle will be scattered in directions that depend on the state of the field in the cavities. The superposition written in (11.35) will then contain four functions and four variables, and by the same effect the unbalance between the two terms will be even more reduced for most trajectories (in 12-dimensional space). As we discuss in Appendix I, when this second particle reaches a measurement apparatus, the number of entangled degrees of freedom becomes so large that one reaches a situation where one of the waves becomes empty forever. The paradoxical (surreal) trajectories then completely disappear (a similar situation occurs if the test particle leaves a trace in a bubble chamber).

(iii) Conclusion

Our conclusion will be that, if one single particle with one single Bohmian position plays the role of the measurement apparatus and its pointer, some trajectories of the test particle cross the interference region, some others bounce and therefore exhibit quantum nonlocality (surrealistic trajectories). Nevertheless, if the pointer is macroscopic, a very large number of Bohmian positions must be included in the analysis, so that all trajectories cross the interference region: surrealistic trajectories then disappear. One should remember that, when entanglement occurs, consistency in dBB theory requires that the position variables associated with all entangled subsystems should be included.

At a microscopic level and with slow pointers, the fact that trajectories of two particles can influence each other, even if they never come close to each other (in an EPR type experiment, for instance) should not be considered as absurd, but just an illustration of the explicit character of nonlocality in dBB theory – see the discussion of the same thought experiment by Griffiths [642]. There are many other examples illustrating that the quantum phenomena are indeed local in configuration space, but not necessarily in ordinary space. Several articles [643, 644], have studied similar problems in the context of weak (§9.3.1), continuous, or protective measurements.

Remark:

There is actually a connection between this analysis and one of the famous paradoxes discussed by Einstein and Bohr at the Solvay meetings. Einstein raised an objection against the uncertainty principle, considering an interference experiment where the momentum recoil of the screen with the two interference holes is measured, so that "Welcher Weg" (which way) information is obtained (see for instance complement D_I of [645]). Bohr [646] replied that one should take into account the quantum character of the screen, which is also subject to uncertainty relations; a proper calculation then shows that, if an interference can be observed, the state of the screen is necessarily such that the determination of the path of the particle is impossible. Einstein then agreed that the paradox was lifted. Here, we have a very similar situation: momentum is also transferred from the particle to the screen, in particular while the trajectories are bent inside the holes; this has consequences on the Bohmian positions associated with the screen. Treating the two quantum systems differently, one with Bohmian positions, the other without, would be inconsistent. The fact that quantum mechanics applies in the same way to all systems was exactly the conclusion of the Einstein–Bohr discussion concerning the combination of an interference experiment and of a which way detector. It is amusing to see that this historical argument should find another application in the context of dBB theory!

Correlation between measurements at different times Until now, we have only considered measurements made at one time, but quantum mechanics also gives predictions about measurements performed at different times; it is interesting to compare them to those of dBB theory [647]. In fact, at first sight, it is not obvious that the results will coincide; for instance, if one considers a system in a stationary state with a real wave function, the Bohmian velocity vanishes, so that the position $\mathbf{Q}$ remains constant. On the other hand, in quantum mechanics, the two-time correlation function in a stationary state has no reason to be time independent: for a harmonic oscillator in its ground state, for instance, it is well known that the two-time correlation function oscillates in time. One could then naively think that the dBB theory predicts time-independent correlation functions, in contradiction with the well-known time dependence obtained within standard quantum mechanics.

Actually, again a proper use of Bohmian mechanics shows that this contradiction does not exist. Within this theory, the effect of the first measurement must be taken into account – this is indispensable even if the two measured observables commute, and even if they correspond to independent systems. The first measurement correlates the position of the particle with that of a pointer; it becomes then necessary to take into account this second Bohmian position as well. Because the wave function of the whole system is no longer a product, situations similar to those discussed for two particles in §11.8.1.c then occur: the position of the measured system takes a velocity that depends on the position of the pointer. This completely changes its future evolution, so that the correlation function is also modified. At the end of a correct calculation within full dBB theory, one recovers exactly the same results as in standard quantum mechanics – for more details, see Appendix I, which also discusses other experiments in the context of Bohmian mechanics [648–651].

The rules of standard and dBB quantum theory are similar The conclusion of this brief review of various objections to Bohmian mechanics is that, after all, its structure is more similar to that of standard quantum mechanics than one might think; all standard rules that are applied to the state vector have their direct counterpart with the Bohmian velocities and positions. We should therefore not repeat mistakes that we have learned to avoid in standard quantum mechanics. In standard theory, we know that, when the system becomes quantum correlated with others, in particular measurement apparatuses, we must take this correlation into account by using an entangled state vector. Similarly, in the dBB theory, we must also take into account the correlations between the corresponding Bohmian variables – this introduces the phenomenon of "empty waves" (§11.8.1.d). We have learned from Bohr's historical arguments that the consistency of quantum mechanics requires that all parts of the experiments be treated as quantum systems, and for instance that one cannot ignore the recoil effect of parts of the apparatus (the screen). Similarly, the Bohmian description also requires that the effects of this recoil should be taken into account in terms of positions.

The same rule applies to correlation functions. In standard quantum mechanics, correlation functions are calculated by taking into account the perturbation of the first measurement on the evolution of the system until the second measurement; one can for instance use Wigner's formula (11.9) for this purpose. In dBB theory, the effect of this perturbation must also be included when calculating the evolution of the Bohmian positions between the two measurements.

Generally speaking, many objections to dBB theory can be raised if this theory is truncated to some of its Bohmian variables. A full dBB theory requires that all position variables of the full ensemble of entangled systems should be included. If this necessity is ignored, various inconsistencies appear; they can nevertheless be solved by the proper introduction of these variables.

11.8.1.i Summary and discussion

One could summarize the main features of the de Broglie–Bohm version of quantum theory as follows:

– It describes the same events in a richer way than standard theory, but without introducing any contradiction with its predictions. It often brings unexpected views on phenomena by introducing intricate arrays of trajectories, particles jumping from one wave packet to the other[41], etc., and leads to representations that may sometimes be found esthetically pleasing. It makes nonlocality completely explicit

---

[41] Of course, these pictures are not really specific to dBB theory. They can also be obtained within standard

but, as we have seen (§11.8.1.d), this nonlocality is not totally absent from standard quantum mechanics either. This explicit character of nonlocality may be the starting point of interesting new discoveries, as the Bell theorem has illustrated.

– It has often been rejected for bad reasons, starting from Pauli's objections at the 1927 Solvay conference (inability to treat nonelastic collisions), which does not hold. Its predictions are indeed equivalent to those of standard quantum theory, but when entanglement is involved (or partial trace operations), for a correct treatment of the problem one must include the associated Bohmian positions; omitting some of them leads to incorrect predictions.

– It brilliantly succeeds in eliminating the role of observers during measurements. With the mechanism of empty waves, measurements are considered as ordinary interaction processes; the uniqueness of the final result emerges without any external observation. State vector reduction is contained in the dynamics of conditional wave functions, and therefore no longer a postulate.

– Concerning determinism, the situation is less clear-cut: the theory provides a scheme where the measurement process appears as deterministic, but where an average over an initial and uncontrollable "quantum equilibrium" distribution is necessary. The measurement reveals some pre-existing property of the Bohmian positions, but only in a retrodictive way; from the result, one can obtain information about the past trajectory of a particle, but it is not possible to make more precise predictions than standard quantum mechanics for the future. This is because there is no possibility to control these positions in order to suppress the random character of the event, or even only partly reduce the indeterminism of standard theory. If one wishes, one can consider that the result of all experiments in physics is a consequence of the initial Bohmian positions of all particles of the universe, at the big bang, but one cannot say that indeterminism has really disappeared[42]; it just gets a more concrete basis.

– Concerning realism, we have seen in §11.8.1.b that there is a choice between two possibilities: either the wave function is considered as a real objective field, as proposed by Bell, or it is seen as a mathematical tool providing the evolution of the real positions. In both cases, positions alone are not sufficient to obtain a full description of the physical system and of its evolution; the wave function is needed to characterize the conditions under which the quantum system was prepared, which are real, and to determine the evolution of the positions, which are also real. The consequence is that the dBB theory does not succeed in re-introducing a simple, naive, realism: if one wishes to avoid incompatibility with relativity (avoid the pre-

theory by drawing figures of the probability current. The difference is that the dBB theory attributes more physical reality to these figures.

[42] In standard quantum mechanics, random events occur all the time – at least each time a measurement is performed somewhere in the universe. In Bohmian mechanics, a big single random event occurred at the creation of the universe; it is the source of all probabilities for future measurements.

diction of possible superluminal signaling), one is forced to accept that Bohmian positions cannot be manipulated. This implies the existence of two levels of description of reality, as noted at the end of §11.8.1.b (see also Appendix H):

• One level associated with the preparation of quantum systems; the fact that an atom or an ion is prepared in a given electronic configuration, a system of two spins in a singlet state, etc., is a part of objective reality that cannot be described in terms of Bohmian positions only. This part requires a description by a field (wave functions) or a state vector; it can be manipulated directly in experiments (by changing potentials, moving walls, etc.), but is not directly observable. It propagates in configuration space, which is much larger than the usual three-dimensional space (except for a single spinless particle). This description includes empty waves, with a reality status that is intermediate and somewhat delicate[43] [605].

• Another level of reality that is associated with observations and described by Bohmian positions, which are variables propagating in the usual three-dimensional space and directly observable. This level cannot be manipulated directly – it is actually completely interaction free (Bohmian positions do not interact with each other). The corresponding variables are only guided by the field, which allows manipulating them indirectly, but without ever changing their quantum equilibrium distribution.

A sort of dualism is then reintroduced in this way, but an asymmetrical dualism with two levels; one level of reality (preparation) influences the latter (positions), but not the reverse. Moreover, while positions propagate naturally in ordinary three-dimensional space, they are guided only by a field propagating in a completely different space, the abstract configuration space. The very fact that the two levels of reality are distinguished by the possible influence of experimental manipulations means that the observer, whose role was supposed to disappear, actually reappears in the physical description of phenomena. In this sense, one could consider that the interpretation does not succeed in bringing the complete simplification of the interpretation that was initially intended. Moreover, we observe that the tension (without contradiction) between relativity and the standard theory has its counterpart in the dBB interpretation: relativity is the source of the fundamental impossibility to manipulate Bohmian variables [595], which in turn ensures that the nonlocal evolution of these variables cannot be used for superluminal signaling.

Accepting or rejecting the dBB interpretation therefore remains a matter of per-

---

[43] As already discussed at the end of §11.8.1.c, one question is what level of reality should be attributed to empty waves associated with the results that have not been obtained. These waves tend to constantly proliferate; for instance, the wave function of the universe develops an enormous number of empty branches, with a single nonempty branch that is the only one having any influence on future positions (the effective wave function). It is then tempting to decide that only the nonempty component of the state vector is a real Bohmian field. But this would amount to reintroducing the postulate of state vector reduction, while its disappearance was one main motivation of the theory; moreover, it would raise delicate questions concerning the exact stage at which an empty wave becomes empty forever.

sonal taste. It is not so surprising that even its inventors, de Broglie and Bohm, have themselves sometimes changed their views on the subject during their lives. A point of comparison might be the theory of path integrals introduced by Feynman [51, 52], which is usually considered as a branch of standard quantum mechanics (§ 1.3). Most quantum physicists do not see any contradiction between this point of view and the use of the standard Hamiltonian quantification procedure and of the Schrödinger equation (or of the Heisenberg point of view); they are just complementary (actually, path integrals are more general) and any of them may be used according to what is more convenient in every context. In the same way, Bohmian theory may be seen as still another possible point of view, complementary but not contradictory.

11.8.2 BBB theory

Bell proposed to generalize the dBB theory to any quantum system, particles, or fields, and to introduce what he called "beables" [170] (see also §4.2.2.b) corresponding to real properties possessed by the system at any time; beables are similar to the Bohmian positions of material particles, which are assumed to be really at these positions. In his article [582], he takes the example of fermionic fields defined on a discrete space lattice, but his method can easily be applied to any physical system described by a state vector $|\Psi(t)\rangle$. This method can be called the "BBB theory", for de Broglie, Bohm, and Bell.

11.8.2.a Quantum calculation

Consider one "observable" A with discrete eigenvalues a_k and eigenstates:

$$|a_k, \alpha\rangle \tag{11.36}$$

When a_k is degenerate, index α distinguishes between the several orthonormal eigenstates associated with this eigenvalue. In fact, operator A is not necessarily a single operator, but may symbolize a series of commuting operators B, C, D, ... , while a_k symbolizes their eigenvalues b_{k_1}, c_{k_2}, d_{k_3}, ..; if this ensemble of operators is a "complete set of commuting operators ", this series of eigenvalues is sufficient to specify the eigenstate, and index α may be suppressed. We call $\langle n_k \rangle$ the population of all the eigenstates associated with eigenvalue a_k:

$$\langle n_k \rangle = \sum_\alpha \langle \Psi(t) | a_k, \alpha \rangle \langle a_k, \alpha | \Psi(t) \rangle \tag{11.37}$$

According to the Schrödinger equation, its time evolution is given by:

$$\frac{d}{dt}\langle n_k\rangle = 2\,\text{Re}\sum_\alpha \langle\Psi(t)\,|a_k,\alpha\rangle\langle a_k,\alpha|\frac{H}{i\hbar}|\Psi(t)\rangle$$

$$= 2\,\text{Re}\sum_{\alpha,\beta,m}\langle\Psi(t)\,|a_k,\alpha\rangle\langle a_k,\alpha|\frac{H}{i\hbar}|b_m,\beta\rangle\langle b_m,\beta\,|\Psi(t)\rangle \qquad (11.38)$$

where H is the Hamiltonian operator and Re the real part. We therefore have:

$$\frac{d}{dt}\langle n_k\rangle(t) = \sum_m Q_{km}(t) \qquad (11.39)$$

with:

$$Q_{km}(t) = 2\,\text{Re}\sum_{\alpha,\beta}\langle\Psi(t)\,|a_k,\alpha\rangle\langle a_k,\alpha|\frac{H}{i\hbar}|b_m,\beta\rangle\langle b_m,\beta\,|\Psi(t)\rangle \qquad (11.40)$$

If we interchange k and m, the expression under the summation becomes the opposite (because of factor i) of its complex conjugate. Therefore:

$$Q_{km}(t) = -Q_{mk}(t) \qquad (11.41)$$

It is also easy to check on (11.40) that $Q_{kk} = 0$. Therefore, the evolution of each population $\langle n_k\rangle$ is the sum of the evolutions associated with each pair of different eigenspaces k and m.

Now consider a given pair of such eigenspaces with indices i and j, where i is the index of the pair for which $Q_{ij} \geq 0$ (therefore Q_{ij} negative). If $\langle n_j\rangle(t) = 0$, relation (11.37) shows that all coefficients $< a_j,\alpha|\Psi(t) >$ vanish, so that $Q_{ij}(t)$ defined by (11.40) also vanish; we can then merely ignore the evolution introduced by this pair. If $\langle n_j\rangle(t) \neq 0$, we can then introduce the positive (or zero) coefficient T_{ij} by setting:

$$T_{ij}(t) = \frac{Q_{ij}(t)}{\langle n_j\rangle(t)} \qquad (11.42)$$

which allow us to write the evolution due to the pair i, j as:

$$\left.\frac{d}{dt}\right|_{ij}\langle n_i\rangle(t) = T_{ij}(t)\langle n_j\rangle(t)$$

$$\left.\frac{d}{dt}\right|_{ij}\langle n_j\rangle(t) = -T_{ij}(t)\langle n_j\rangle(t) \qquad (11.43)$$

We can interpret these relations as expressing a probability of transfer per unit time $T_{ij}(t)$ of any physical system belonging to eigenspace j into a state of eigenspace i. Note that there microreversibility is not obeyed: transitions occur between j and i, but not the reverse; but, of course, the total population is conserved.

We can now proceed in the same way for all pairs of eigenspaces i and j, and rewrite (11.39) as:

$$\frac{d}{dt}\langle n_k\rangle(t) = \sum_{m\subset M_+} T_{km}(t)\,\langle n_m\rangle(t) - \sum_{m\subset M_-} T_{mk}(t)\,\langle n_m\rangle(t) \tag{11.44}$$

where M_+ is the range of values of m for which $Q_{km}(t) \geq 0$, and M_- the range of values of m for which $Q_{km}(t) < 0$.

11.8.2.b Introducing beables

We now assume that, for each realization of an experiment where the system is described by the quantum state $|\Psi(t)\rangle$, a beable (a Bohmian variable) $\mathcal{A}$ can be associated with the system; this beable is equal to one of the possible eigenvalues a_k. At any time, we therefore have:

$$\mathcal{A} = a_{k=k_\mathcal{A}} \tag{11.45}$$

where the value $k_\mathcal{A}$ of the index k may change in time. If A symbolizes a large number (or an infinity) of observables B, C, D, ..., the beable $\mathcal{A}$ symbolizes a long series of eigenvalues $b_{k_\mathcal{A}}$, $c_{k_\mathcal{A}}$, d_B, ... In his example, Bell assumes that A is the list of all operators associated with the occupation number of a fermionic field at all nodes of a discrete lattice; at each time, and for each realization, all these occupation numbers heve well-defined values specified by the value of $\mathcal{A}$.

If the same experiment is repeated many times, we assume that at time $t = 0$, the proportion of realizations $\mathcal{P}_\mathcal{A}$ where a given value $\mathcal{A}$ is obtained is equal to the quantum population $\langle n_{k=k_\mathcal{A}}\rangle$:

$$\mathcal{P}_\mathcal{A} = \langle n_{k=k_\mathcal{A}}\rangle \tag{11.46}$$

This is nothing but the initial quantum equilibrium condition. Moreover, we assume that the value of $\mathcal{A}$ evolves stochastically: if it is given at time t by (11.45), during the infinitesimal time dt it has a probability $T_{ik_\mathcal{A}}(t)\,dt$ to change to any other value $\mathcal{A} = a_{k=k_\rangle}$ for which $T_{ik_\mathcal{A}}(t) > 0$. This ensures that the various probabilities $\mathcal{P}_\mathcal{A}(t)$ will follow the same equation of evolution as populations $\langle n_k\rangle(t)$, so that equality (11.46) remains valid at any time.

Since a given value of $\mathcal{A}$ can change to many different other values, the evolution of the Bohmian variables is necessarily stochastic, while it is deterministic in standard dBB theory. Nevertheless, one can assume that, for each value of $\mathcal{A}$, another finer Bohmian variable $\mathcal{B}$ exists, which leads to the next value of $\mathcal{A}$ in a deterministic way. The next time step will require that, in turn, $\mathcal{B}$ is completed by a still finer Bohmian variable C, etc. At the end, one reaches a continuous limit for the Bohmian variables, where one can hope to restore determinism. The continuum limit of general BBB-type quantum field theories is discussed in [652]; see also Refs.[625, 653, 654] for recent discussions of these theories.

The description by the beable $\mathcal{A}$ is maximally precise if the ensemble of operators symbolized by A is a complete set of commuting operators. It is then equivalent to specify the value of $\mathcal{A}$ or the single corresponding state vector $|\Psi_{\mathcal{A}}(t)\rangle$, which can be used instead of $\mathcal{A}$ to describe a single realization. One then reaches a point of view where the physical system is described by two state vectors: the usual state vector $|\Psi(t)\rangle$ associated with the ensemble of realizations, and the beable state vector $|\Psi_{\mathcal{A}}(t)\rangle$ associated with a single realization. At each time, the usual state vector can be expanded on the various possible beable state vectors $|\Psi_{\mathcal{A}}(t)\rangle$:

$$|\Psi(t)\rangle = \sum_i c_i(t) \, |\Psi_{\mathcal{A}}(t)\rangle \qquad (11.47)$$

This point of view reached in this way is closely related to the modal interpretation of quantum mechanics (§11.9).

11.8.3 Nelson mechanics

Another formulation of quantum mechanics including additional variables was introduced in 1952 by Fenyes [655] and in 1966 by Nelson [656], in a somewhat different form. Instead of taking the Schrödinger equation for granted and adding positions to it as dBB theory does, the idea is to derive the equation by introducing stochastic motions at a lower level. One considers point particles with mass m, each subject to a Brownian motion (Wiener process) with diffusion coefficient $\hbar/m$ and no friction; the effect of an external force deriving from a potential is included by simply using Newton's law. Nonlinear equations are then obtained for two classical real variables, which can be grouped into one single complex variable. An appropriate change of variables[44] then results in a sort of "miracle": a linear equation is obtained at the end. For appropriate values of the parameters, assuming a universal Brownian motion therefore leads to a natural derivation of the time-independent and time-dependent Schrödinger equations, including the possible role of an external electromagnetic potential.

One may consider this procedure as a purely mathematical method for quantization, without any particular physical consequence. But one may also attribute to it a more fundamental character, by considering the random positions and the Brownian motions of all particles as real. One then obtains another interpretation of

---

[44] The two classical real variables are the "current velocity" and the "osmotic velocity", defined as the half sum and half difference of two velocities: the forward and backward time derivatives of the (nondifferentiable) random position of the particle. The change of variables introduces Ψ, defined with a modulus equal to the square root of the probability density of the position, and with a phase gradient that is proportional to the current velocity.

quantum mechanics with additional variables, which is reminiscent of the Bohmian theory. One difference with this (and standard) theory is that the evolution of the wave function is no longer given by a postulate, but actually derived from other postulates that are considered more fundamental, including a "universal Brownian motion". Another difference is that the Nelson point of view makes direct use of Newton's law, so that it remains much more classical than a theory where this classical law is modified. But, as in Bohmian theory, one introduces continuous trajectories for the particles, even if they are also not directly accessible by experimentation. Again, the formalism is built to lead exactly to the same predictions as standard quantum mechanics, so that a large part of our discussion concerning Bohmian theory can be transposed to Nelson mechanics. For the discussion of statistical mixtures in this context, see [657].

As in the de Broglie–Bohm theory, a central role is played by the Madelung hydrodynamic equations [593]. It has been pointed out [658] that these equations are not sufficient to ensure a single valued wave function, if this wave function has nodes; to ensure a single value, one has to add circulation quantization conditions around the nodes, in addition to the local equations of motion of the fluid. In stochastic mechanics as well, such additional constraints are necessary to ensure total compatibility with standard quantum mechanics, which makes the physical interpretation more complicated.

Wiener processes have been used in many other domains of physics – we have already seen applications to quantum measurement (§9.3.2.c). They can actually provide a general quantization method. For instance, stochastic quantization has been applied in quantum field theory [659, 660], in particular by Parisi and Wu, who have proposed a quantization method with one additional time variable and a stochastic Langevin equation, and allowing gauge-independent perturbation calculations [661, 662].

11.9 Modal interpretation

The words "modal interpretation of quantum mechanics" are used for a class of interpretations [663, 664], introduced by van Fraassen [665], which also attribute more properties to physical systems than standard quantum mechanics does – but, in this case, the position of particles plays no particular role. In fact, the additional variable of modal theories takes the form of another state vector, which is perfectly similar to the standard state vector (it belongs to the same space of states). This additional variable nevertheless plays a different role, since it directly characterizes the real physical properties of the physical system. As a consequence, in these interpretations, a system may for instance have a perfectly well-defined value for a physical observable, even if its standard state vector is not an eigenstate of this observable. The very fact that the additional ingredient is a state vector ensures, automatically, that the additional properties are sufficiently restricted to avoid brutal contradictions or inconsistencies with the predictions of standard quantum mechanics. On the other hand, they are sufficiently rich to lead to definite macroscopic events and results of measurement; the state vector reduction postulate is then no longer necessary.

Van Fraassen therefore associates two distinct states (kets) with any physical system:

(i) The "dynamical state", which is the usual quantum state (or density operator) of standard quantum mechanics. This state describes the evolution of the system, which, for an isolated system, is given by the Schrödinger equation. In the modal interpretation, the dynamical state never undergoes state vector reduction.

(ii) A "value state", which represents the physical properties of the system at any time (it may be tempting to call it "physical state", but the tradition is to use the words "value state" in this context).

Consider a system S that is part of a larger system T, which, for the sake of simplicity, is assumed to be isolated. We now discuss how the dynamical and the value states describing S may differ if the dynamical state of S (density operator obtained by partial trace) is not a projector over a pure state (at least two of its eigenvalues do not vanish). Various possibilities have been suggested to define the value state. Van Fraassen [665] initially proposed relatively general rules, assuming that the state value could be any state appearing in the decomposition of the dynamical state (it can be any linear combination of the eigenstates with nonzero eigenvalues of the partial density operator ρ_S defining the dynamical state of S). The definition of the value state then remains rather loose. Other authors have proposed to be more specific and to use the bi-orthonormal decomposition (Schmidt decomposition, §7.2.2) to write the entangled dynamical state of the whole system T as:

$$|\Psi\rangle = \sum_n c_n |\varphi_n\rangle \otimes |\Phi'_n\rangle \qquad (11.48)$$

where the $|\varphi_n\rangle$ are normalized and mutually orthogonal, as well as the $|\Phi'_n\rangle$. The $|\varphi_n\rangle$ are then the possible value states describing system S. Of course, if all c_n but one are zero, within standard quantum mechanics system S is already in a pure state, and there is nothing new. But the modal interpretation postulates that, even when several c_n do not vanish, system S has all the properties associated with a single $|\varphi_n\rangle$. This postulate remains fundamentally nondeterministic: the only possible prediction is given by the probability $|c_n|^2$ for S to be in state $|\varphi_n\rangle$, which reconstructs the Born rule (at this stage, the use of the dynamical state cannot be avoided). But, even if it is impossible in advance to predict which of the accessible states will be reached, when it is reached, all sets of propositions about S that would be true if S was in state $|\varphi_n\rangle$ within standard quantum mechanics are indeed true. This point of view, where the dynamical state does not determine value states in an univocal way, but just provides possible values for them, is called "modal" because it leads to a modal logic of quantum propositions (§11.4.1).

The result is that, in general, the value state of S may contain more information about the physical properties of S than its dynamical state. There is nevertheless a limit on the amount of information that may be added in this way: the set of physical properties resulting from this more precise description should not exceed the maximum that is permitted by a standard description with any quantum state. In other words, one may well attribute to S a pure state even if this system is entangled with another quantum system (i.e., even if this system cannot be associated with a single pure state in standard quantum mechanics), but it is not possible to go beyond this and obtain an even more precise description. One can then never attribute simultaneous sharp definitions to noncommuting operators, position and momentum for instance (as opposed to Bohmian mechanics, where a particle has a perfectly well-defined position and velocity at every time). This provides a description with two state vectors that is somewhat reminiscent of the Bohm–Bub theory of measurement [7], to which we come back in the next section.

Consider a measurement process involving system S and a measurement apparatus M. During the initial stage of interaction between S and M, both systems develop entanglement. This is sufficient to introduce a value state for S that differs from its standard description (with a density operator). From this moment, the interpretation guarantees that both subsystems have all the properties associated with the emergence of a single result from the measurement (macroscopic uniqueness).

This program was later extended by Kochen [666] with more emphasis on the relational character of properties of physical systems: he considers that the value states of S do not describe properties of S alone, but properties in relation to all the other systems that are entangled with it (including possibly measurement apparatuses). Dieks [667] introduces another view where systems do have intrinsic properties, and discusses how measurement processes and the existence of macroscopic behavior can be understood within the modal interpretation. He points out that this interpretation is realist in the sense that it assigns definite values to a set of magnitudes pertaining to a physical system ("beables" or "existents"), as opposed to results of measurement only in standard theory.

Healey [668] has proposed an "interactive interpretation", which has similarities with a modal interpretation. He introduces a distinction between:

(i) the "quantum state" of the system, which is the usual state of standard quantum mechanics.

(ii) the "dynamical state" of the system, which contains information about all its current dynamical properties. "The dynamical state of a system at an instant may be identified with a truth value assignment to all sentences ascribing a quantum dynamical property to that system at that instant". Nevertheless, the dynamical state does not generally suffice to determine the future behavior of the system, nor even the probabilities of future behaviors.

The function of the quantum state is precisely to generate probabilities. It characterizes a system's current probabilistic disposition, some of which will be manifested in future interaction, but does not suffice to determine its future behavior. Every quantum system always has a dynamical state, but not necessarily a quantum state (if it is entangled with another system). Note the difference in vocabulary: the dynamical state of the interactive interpretation is analogous to the value state of the modal interpretation, while the dynamical state of the modal interpretation is analogous to the quantum state of the interactive interpretation.

The role of environment is clearly essential in all these interpretations[45]. For a measurement apparatus, the role of "pointer states" is important (§9.1.2). Bacciagaluppi [669], Dickson [670], Berkovitz and Hemmo [671], and others have also made contributions to this interpretation, discussing for instance the role of imperfect measurements, the relation with the BKS theorem (§6.4, etc.). Nevertheless, this interpretation has some difficulties, in particular with models of environmental decoherence assigning definite values to pointer positions [672] when they are continuous [669], with Lorentz invariance and contextuality [671, 673], and with applications in relativistic quantum field theory [674].

11.10 Modified Schrödinger dynamics

Another way to resolve the coexistence problem between the two evolution postulates of quantum mechanics is to fuse them into a single postulate, by changing the Schrödinger equation itself in order to incorporate all features that are necessary to describe a measurement. One assumes that the equation of evolution of the wave function contains, in addition to the usual Hamiltonian terms, new terms (possibly nonlinear and/or stochastic), which introduce the equivalent of state vector reduction when necessary [7, 13, 15, 493, 675]. The main objectives of this approach are:

(i) Uniqueness of macroscopic reality: the new dynamics should be built in such a way that superpositions of states that are macroscopically distinguishable are never reached, in other words that Schrödinger cats (as defined in the original arti-

---

[45] Of course, the idea to create a correspondence between the physical properties of a system and its environment is not specific to this interpretation; it exists also, for instance, in the pragmatic views discussed in §§11.1.1 and 11.1.2.

cle) never occur. In particular, measurements made with macroscopic instruments must have definite outcomes.

(ii) Compatibility with standard quantum mechanics at the macroscopic level: the probabilities of the different outcomes of measurements must reproduce the usual Born probability law.

(iii) Compatibility with standard quantum mechanics at the microscopic level: the theory should remain consistent with the enormous number of known experimental results where quantum mechanics has already been applied to various microscopic systems successfully.

Condition (i) implies that, for macroscopic superpositions (involving for instance pointers of measurement apparatuses), the new terms select one branch of the superposition, and cancel all the others. The postulate of state vector reduction of standard quantum mechanics then becomes useless; a single law of evolution is sufficient. In other words, macro- and microdynamics are unified. Condition (ii) adds that the selection of one branch must be (or seem to be) random in a way that exactly reproduces the probabilistic rules of standard mechanics and the effect of state vector reduction. Condition (iii) implies that the consequences of the change in the Schrödinger equation must remain extremely small in all situations involving microscopic objects only (atoms, molecules, etc.); this will immediately ensure that the enormous amount of successful predictions of quantum mechanics is preserved.

Two extremes must therefore be avoided: either to perturb the linear Schrödinger equation too much, and make interference effects disappear while they are still needed (for instance, recombination of the two beams at the exit of a Stern–Gerlach magnet if no decoherence has occurred); or perturb it too little to ensure the complete disappearance of macroscopic indeterminacy (Schrödinger cats). The perturbation term must be efficient when (but not before) any microscopic system becomes strongly correlated to a macroscopic environment, which ensures that significant decoherence has already taken place; the recovery of interference effects is then already impossible anyway within standard theory. When necessary, the process then reproduces the effect of the postulate of the wave function collapse, which no longer appears as an independent postulate, but as a consequence of the "normal" evolution of the wave function.

Including the emergence of macroscopic uniqueness in the fundamental dynamical equation of all physical systems solves the major conceptual difficulties of quantum mechanics. Obviously, this suppresses the problem of specifying the limits of application of two different postulates: measurement processes become ordinary interaction processes between two physical systems. Moreover, the state vector can then be said to directly describe physical reality, instead of being just a computational tool. The theory contains the elements that explain [14] "why events occur"! General discussions of the necessity of introducing the notion of events in quantum physics, and the "transmutation from possibilities to facts", can be found in [676, 677].

To really define the theory, these conditions are of course not sufficient; it is necessary to specify exactly what form the modified Schrödinger equation has, with no special guidance for doing so except the "border lines not to cross" provided by conditions (ii) and (iii). It is therefore not surprising that several versions of theories with modified Schrödinger dynamics should have been proposed.

11.10.1 Evolution of the ideas

It is possible to build versions of modified Schrödinger dynamics by changing the Schrödinger equation either without introducing additional variables, or by introducing them at the same time; both methods have been used.

11.10.1.a Early work

In 1966, Bohm and Bub [7] proposed a theory of measurement that contains a mechanism leading to state vector reduction. It belongs to the second category (including additional variables) since these authors incorporate in their theory the additional variables previously considered by Wiener and Siegel [12], which are contained in a "dual vector". The latter is a mathematical object that is actually very similar to the usual state vector $| \Psi >$, but obeys an entirely different equation of motion – in fact, both vectors evolve with coupled equations. The probabilities of the results of measurement are just consequences of the initial random distribution of the dual vector. For some "normal" distribution of the new variables, the predictions of usual quantum mechanics are recovered; but it is also possible to assume the existence of "dispersion free" distributions that lead to more precise, nonorthodox, predictions. As a consequence, this theory leads to a statistics that is richer than usual quantum mechanics. The authors discuss under which circumstances the effect of the new variables, which have not been observed yet, could be detected in the future.

In 1976, Pearle [13] introduced a version of the theory that belongs to the first category, where additional variables are not necessary. He considers the state vector of the combined system containing both the measured system and the apparatus, just after their interaction, when they are strongly entangled in a coherent macroscopic superposition (§9.1). The idea is to assume that the result of a measurement is determined by the relative phases of the components of this state vector; since these phases are not controlled experimentally with perfect accuracy, the final result of measurement then takes values that seem to be completely random. This behavior is obtained by adding a new nonlinear term into the equation of evolution of the state vector, with a time rate constant γ; this term drives the state vector to one of its macroscopic components only, in a way that depends on the initial relative phases. From this point of view, the randomness, postulated by standard quantum mechanics as fundamental, appears to be just apparent, being a mere consequence of experimental conditions that are never perfectly reproducible. Relatively little is added to the usual formalism of standard quantum mechanics, but the new term added in the equation of evolution should obey necessary conditions listed by the author:

– Property 1: a single result is predicted by the new equation of evolution, meaning that, among a series of probability amplitudes, all of them go to zero after the interaction, except one that gets modulus unity – the equivalent of property (i) above.

– Property 2: if the experiment is repeated many times, the selection of a single nonvanishing probability amplitude happens in a way that is consistent with the probabilistic predictions of quantum mechanics – the equivalent of property (ii) above.

In both the preceding theories, the reduction of the state vector is seen as a dynamical process that, like any dynamical process, has a finite time duration. We will see in this section examples of theories (GRW, for instance, discussed in the next section) where this is not the case.

In 1979, Pearle proposed a different approach [14], this time by introducing stochasticity into the Schrödinger equation; the equation then becomes a stochastic differential equation with a term containing a randomly fluctuating operator, and acting on the state vector in a nonlinear way. The presence of this new operator puts the theory into the category with additional variables. By contrast with the 1976 version, the selection of the outcome of the experiment is no longer due to uncontrolled phases, but to a randomly fluctuating operator. The mathematical process through which the new equation reduces the state vector to one component associated with a single result is described by a Fokker–Planck equation; it is analogous to a random walk between absorbing barriers – the corresponding statistical properties are sometimes described as those of the "gambler's ruin game". This version of Pearle's theory is more general than the previous one; it nevertheless still has some problems, which he listed in a review article [678], including "preferred basis problem", "interaction problem", "trigger problem", etc. In the Bohm–Bub theory, the complete reduction time was infinite, but here, because the statistical process associated with the "gambler's ruin game" has a finite mean duration, this time is finite [679].

11.10.1.b GRW spontaneous localization ("hits")

In 1986, Ghirardi, Rimini, and Weber (GRW) introduced a new version of the theory, which provides an "unified dynamics for microscopic and macroscopic systems" [15]. They obtain this result by adding to the usual Schrödinger equation a random process, called "spontaneous localization" (SL), which suddenly changes the state vector by localizing its wave function – as if the positions of the particles were measured at random times, but not with perfect accuracy (the localization remains approximate). Some features of this theory are actually reminiscent of the effect of continuous observations and measurements in standard quantum mechanics [680, 681] (§9.3.2), or that of approximate measurements. Spontaneous localization occurs at random times with a time constant that is adjusted so that, for macroscopic systems, the occurrence of superposition of far-away states is destroyed – condition (i); but this is true for macroscopic systems only, not microscopic systems – condition (ii). Earlier (1984), Gisin had also introduced a modified Schrödinger dynamics with similar equations of evolution implementing the projection postulate [493], but where the collapse time of the state vector was infinite.

A remarkable property of the GRW approach is that it completely solves the "preferred basis problem" as well as the "trigger problem"; the basis necessarily corresponds to position localized states. In this model, for a single realization of an individual quantum system (as opposed to an ensemble of many realizations[46] of the same system), the localization processes are sudden; they are sometimes called "hitting processes". This of course makes them completely different from what happens within usual Schrödinger dynamics. Consider for instance a single particle described by a state vector $|\Psi(t)\rangle$. The effect of one of these processes is to suddenly replace $|\Psi(t)\rangle$ by a ket $|\Psi'(t)\rangle$ given by:

$$|\Psi'(t)\rangle = \frac{F_j |\Psi(t)\rangle}{\langle \Psi(t)| \left(F_j\right)^2 |\Psi(t)\rangle} \qquad (11.49)$$

(the numerator of this expression ensures the conservation of norm of the state vector) where F_j is a Hermitian operator diagonal in the position representation. GRW assume that F_j localizes the particle around point of space $\mathbf{r}_j$ with an accuracy characterized by a free parameter α of the theory ($\alpha^{-1/2}$ is a length):

$$F_j = c\, e^{-\alpha(\mathbf{R}-\mathbf{r}_j)^2/2} \qquad (11.50)$$

---

[46] For an ensemble of systems, the discontinuities are averaged, and one recovers continuous equations of evolution for the density operators. Since most of the discussion of [15] is given in terms of density operators/matrices, and of the appearance of statistical mixtures (decoherence), one may get the (incorrect) impression that individual realizations are not considered in this work; but this is in fact not the case and "hitting processes" are indeed introduced at a fundamental level.

In this equation, $\mathbf{R}$ is the position operator of the particle and c a real normalization factor.

A series of operators F_j is then introduced, with the following condition:

$$\sum_j \left(F_j\right)^2 = 1 \tag{11.51}$$

With operators F_j defined in (11.50), it is natural to assume that index j defines a continuous position variable $\mathbf{r}_j$; the sum over j then becomes an integral over d^3r_j, and it is easy to check that:

$$c \int d^3r_j \, e^{-\alpha(\mathbf{R}-\mathbf{r}_j)^2} |\mathbf{r}\rangle = c \int d^3r_j \, e^{-\alpha(\mathbf{r}-\mathbf{r}_j)^2} |\mathbf{r}\rangle = c \left[\frac{\pi}{\alpha}\right]^{3/2} |\mathbf{r}\rangle \tag{11.52}$$

Condition (11.51) is therefore obeyed if:

$$c = \left[\frac{\alpha}{\pi}\right]^{3/2} \tag{11.53}$$

The $\mathbf{r}_j$ may also define discrete positions over the nodes of a lattice with a unit cell that has a size much smaller than $1/\sqrt{\alpha}$; the value of c then depends on this size. For convenience, in this section we continue to write discrete sums with index j, but the transposition to continuous indices and integrals is straightforward.

Now assume that all localization processes corresponding to the various $\mathbf{r}_j$ act in parallel, constantly and randomly localizing the particle at these points, each with a probability per unit time given by:

$$\gamma \, \langle \Psi(t)| \left(F_j\right)^2 |\Psi(t)\rangle \tag{11.54}$$

where γ is another free parameter of the theory (a rate, the inverse of a time). Condition (11.51) ensures that the total probability for any sort of hit is independent of the initial state $|\Psi(t)\rangle$. The random effect of these localization processes is then added to the usual Schrödinger evolution. One then obtains a theory where, for each possible realization of the random hits at all times, the state vector follows a trajectory that is well defined, but different for each realization.

Consider the first localization process. If $\Psi(\mathbf{r}, t)$ is the wave function associated with state $|\Psi(t)\rangle$, according to (11.50) the probability of occurrence of the process with index j is:

$$\mathcal{P}_j = \gamma c^2 \int d^3r \, e^{-\alpha(\mathbf{R}-\mathbf{r}_j)^2} \, |\Psi(\mathbf{r}, t)|^2 \tag{11.55}$$

The localization processes are therefore more likely to be centered at values of $\mathbf{r}_j$ where the density of probability $|\Psi(\mathbf{r}, t)|^2$ is maximal; conversely, spontaneous localizations are very unlikely to occur around positions where the modulus of the wave function is tiny. Whatever $\mathbf{r}_j$ is selected in the first localization process, the wave function after this process is multiplied by $e^{-\alpha(\mathbf{r}-\mathbf{r}_j)^2/2}$, which tends to restrict the wave function to a neighborhood of $\mathbf{r}_j$ with spatial extension $\alpha^{-1/2}$.

For the sake of simplicity, let us first assume that the wave function does not have the time to evolve before a second localization process takes place; because of the effect of the first localization, the second is then more likely to be centered at a point not too far from $\mathbf{r}_j$. Similarly, the third localization is likely to select a point that is in the neighborhood of the two preceding localization points, and so on: after a few localization processes, the wave function is well localized around a point $\mathbf{r}$ that is random, but well defined. This creates the spontaneous spatial reduction process of the wave function. Now, if the wave function evolves between the localization processes, the succession of the points of localization reconstructs a trajectory for the particle, as the track created in a Wilson or cloud chamber.

The scheme can immediately be generalized to a system made of N particles, by assuming that all the particles independently undergo localization processes. The operators F_j are then replaced by Hermitian operators F_j^i acting on the i-th particle, and their effect is summed over both indices i and j in the Schrödinger equation, but again for each realization one obtains a single trajectory for the state vector $|\Psi(t)\rangle$ describing the physical system. The effect of the random processes is to spatially localize the wave function around a single point of configuration space with $3N$ dimensions, with a spatial extension $\alpha^{-1/2}$ in every direction.

Consider now an ensemble of realizations of the same physical system, described by an ensemble density operator ρ. The average over all realizations of the hits provides the time evolution of ρ in the form:

$$\frac{d\rho(t)}{dt} = \frac{1}{i\hbar}\left[H(t),\rho(t)\right] + \gamma\left[\sum_{j,i} F_j^i\,\rho(t)\,F_j^i - \rho\right] \quad (11.56)$$

In the right-hand side of this equation, the first term containing the Hamiltonian $H(t)$ gives the usual Schrödinger–von Neumann evolution. The second term describes the average effect of stochastic processes that replace the density operator by a new value $F_j^i\rho(t)F_j^i$ with a probability per unit time γ.

In quantum mechanics, localizing a particle is automatically associated with a change of its kinetic energy; the hitting process therefore "heats" the particle, necessitating a study of the role of energy conservation. In practice, this means that the time constant γ and localization constant α have to be adjusted to avoid the prediction of spontaneous heating effects that have never been observed. GRW propose choices of the parameters ($\gamma = 10^{-16}$ s^{-1} and $\alpha^{-1/2} = 10^{-7}$ m.) such that the heating rate of a mole of ideal gas is approximately 10^{-15} K. per year, which is indeed compatible with known results.

Benatti *et al.* have checked that the main effect of the new term is to produce an actual reduction of the state vector [682]. A more general study of the consequences of introducing non-Hamiltonian evolutions of the state vector was given in [683], discussing in particular the possibility of introducing position and momentum localizations simultaneously; the conclusion of the study is that, in order to avoid unacceptable modification of the predictions concerning microscopic processes, one should limit oneself to spontaneous localization of positions only. A problem, nevertheless, of the GRW approach is that it assumes independent localizations of all the particles, so that it does not preserve the symmetry requirements of quantum mechanics for identical particles, bosons, or fermions.

A general approach to derive spontaneous localization models has been proposed [684]: Event-Enhanced Quantum Theory (EEQT), where the quantum system S under study is coupled to a classical system C having a state described by a classical parameter λ. The standard quantum formalism is extended by replacing the usual space of states $\mathcal{E}$ by a family of λ-dependent spaces $\mathcal{E}_\lambda$; similarly, all quantum operators get an extra index λ. In the simplest picture, λ can take two values only, corresponding for instance to the two positions of the pointer of a measurement apparatus; but α may also take any series of discrete values, or even be a vector with many components having such discrete values. The value of λ determines the evolution of the state vector $|\Psi(t)\rangle$ of S, which obeys an equation containing λ-dependent nonHermitian terms Λ_λ (in addition to the usual Hamiltonian). The value of α changes randomly at times determined by the value of the average of Λ_λ in state $|\Psi(t)\rangle$; as a consequence, the evolution of $|\Psi(t)\rangle$ depends on λ and conversely. What the authors call an "event" is a change of λ. This model is generic and, in particular, contains the GRW theory as a special case. Nevertheless, inasmuch as λ is considered as describing a classical physical system, the model reintroduces the Bohrian notion of a frontier between classical and quantum worlds, while a motivation of theories with modified Schrödinger dynamics is precisely to suppress this frontier.

11.10.1.c Continuous spontaneous localization

In 1989, Pearle showed [685] that it is possible to solve this problem and to get rid of the discontinuous character of the hitting processes, while retaining the attractive features of the GRW model. This is obtained by adding terms of "continuous spontaneous localization" (CSL) to the usual Hamiltonian in the Schrödinger evolution. The new terms introduce Markov processes that depend on a series of random functions of time $w_j(t)$ with a broad frequency spectrum (white noise), and contain a time rate γ as well as a set of mutually commuting Hermitian operators A_j. For an appropriate choice of these operators, a full compatibility with the standard notion of identical particles in quantum mechanics is realized. One advantage of this point of view is that it introduces less radical changes with respect to the

standard theory, since the state vector still evolves according to a differential equation. Nevertheless, this equation includes random functions of time (Itô stochastic
differential equation) as well as antiHermitian terms; the norm of the state vector
$|\Psi(t)\rangle$ is no longer conserved.

This nonconservation of the norm actually plays an important role in the theory,
since the statistical properties of the random functions are defined precisely from
this norm: one postulates (CSL probability rule) that the probability[47] for realiz-
ing any time dependence $w_j(t)$ that leads to a given value of $|\Psi(t)\rangle$ is nothing but
$\langle\Psi(t)|\Psi(t)\rangle^2$. At this point, the nonlinear character of the theory becomes obvious:
the evolution of $|\Psi(t)\rangle$ depends on functions $w_j(t)$, which, in turn, have statistical
properties constantly following the norm of $|\Psi(t)\rangle$ itself. Such a postulate strongly
favors the realizations of the random functions that give a large norm to the state
vector, while reducing the effect of all the others that give exponentially small values to the norm (even if mathematically they correspond to many more possibilities). One can then check that this choice of statistical properties remains consistent
with the independent Markovian evolution of each realization of the state vector. A
Fokker–Planck equation may be obtained for the time evolution of the probability
densities.

In order to understand the mechanism of state reduction in this theory (described
in more detail in Appendix J), let us first consider a simple case where one single
operator A (with eigenvalues a_n) is introduced; we assume that A corresponds to
the observable measured in a quantum measurement process such as that discussed
in §9.1. Just after the end of the interaction between the measured system and
the apparatus, both systems are entangled and described by the ket (9.7) of §9.1.
Now, under the effect of antiHermitian terms controlled by the random functions
$w_j(t)$, the modulus of each probability amplitude $c_n(t)$ fluctuates in time, instead of
remaining constant. Among the large number of mathematically possible $w_j(t)$, according to the CSL probability rule, only a very small proportion may occur with a
nonnegligible probability: the proportion leading to a large sum over n of all $|c_n(t)|^2$.
It turns out that, among these special functions, the most effective in providing a
large norm for the state vector are those that give a large value to one $|c_n(t)|^2$ only.
This is because, during its fluctuations, $w_j(t)$ can favor one value of n, but not several at the same time; situations where the fluctuations of $w_j(t)$ successively favor
two (or more) of these coefficients lead to a dilution of the norm preservation effect
and to an exponentially smaller value of the total norm at the end. We then obtain a
process that selects a single outcome in a way that is reminiscent of the "gambler's
ruin game" of [14]. The fluctuations of the random functions break the symmetry

---

[47] This is somewhat reminiscent of postulating (11.54). Several realizations of the random function $w_j(t)$ may
lead to the same value of $|\Psi(t)\rangle$ at time t; the square norm of $|\Psi(t)\rangle$ then gives the probability of each of these
realizations.

between all possible measurement outcomes, so that this theory reproduces state vector reduction.

In the CSL theory, the operator A is actually not associated with any particular measurement process – this would make it measurement dependent and immediately destroy one of the achievements of the theory (giving no special role to measurements). One rather assumes that A is replaced by a series of position localization operators A_j, acting on all particles of the system, and localizing them at all possible positions in space (j then becomes an index for spatial positions, and thus may be continuous). As in GRW theories, perfect localizations of the particles would transfer an infinite amount of energy to them, which is physically unacceptable. One then postulates that the localization provided by each A_j is imperfect, over a spatial range $\alpha^{-1/2}$; all A_j are mutually commuting operators. Despite these changes concerning the definition of the operators, the essence of the localization process remains similar to that previously discussed. It introduces a selection that eventually localizes each particle into a single random region of space – a spatial reduction process of the quantum state. For small quantum systems (single particle, atoms, molecules, etc.), the probability of occurrence of any collapse process remains extremely low for a very long time (γ is small). For macroscopic systems in quantum superpositions of two spatially distinct states, collapse is very likely to take place rapidly and to cancel one of the components; this is because all the particles involved are constantly subject to localization, while the localization of a single particle is sufficient to destroy one of the two components.

The functions $w_j(t)$ are considered as fundamentally random; as the Bohmian position variables, they cannot be manipulated directly, but they cannot be detected directly either (they could be called "hidden functions" instead of "hidden variables"). However, if one changes by hand the external parameters controlling the Hamiltonian (changing the magnetic field, for instance), the state vector is affected, which may in turn change the statistical properties of the $w_j(t)$ and therefore affect these functions indirectly.

In 1990, the study of Markov processes and continuous spontaneous localization for identical particles was expanded by Ghirardi, Pearle, and Rimini [686]. They showed that, for an ensemble of systems, equation (11.56) is then replaced by a "Lindblad form" (§7.4.3) giving the time evolution:

$$\frac{d\rho(t)}{dt} = \frac{1}{i\hbar}[H(t),\rho(t)] + \frac{\gamma}{2}\sum_{j=1}^{N}\left[2A_j\rho A_j - \left(A_j\right)^2\rho(t) - \rho(t)\left(A_j\right)^2\right] \qquad (11.57)$$

where the A_j are mutually commuting Hermitian operators (they may be arbitrary operators but, in this context, they are position localization operators). In general, discrete Markov processes in Hilbert space can be reduced, in the limit of small hits with infinite frequency, to a continuous spontaneous localization. For instance, if one chooses operators such that $A_j = F_j$ in (11.57), and if one uses the sum relation (11.51), one recovers (11.56); this introduces an interesting correspondence between discrete (GRW type) and CSL models [686] in terms of the average evolution of density operators. But, conversely, the CSL theory does not require that the sum of squares of all operators should necessarily be 1, so that the most general Lindblad equation cannot be obtained from a hitting process. The CSL theories therefore offer a broader range of possibilities, which turns out to be useful for elaborating relativistic versions of the theory (§11.10.1.e) or in cosmology [687].

A problem nevertheless is that in CSL, again, a complete collapse of the wave function is never obtained in any finite time. Even for a macroscopic system, when most of the wave function goes to the component corresponding to one single outcome of an experiment, there always remain a tiny component on the others (even if extremely small and continuously decreasing in time). The existence of this component is not considered as problematic by the proponents of the CSL theory [688]. Nevertheless, Shimony has argued [689] that it is philosophically objectionable to associate physical reality with one single state, as long as the system remains in a quantum superposition of this state and another; he considers that this is true even if the weight of the former is enormous in comparison with that of the latter. This is often called the "tail problem" of modified Schrödinger dynamics; see [690] and for instance [691] for a proposed solution to the tail problem involving a combination of modified Schrödinger dynamics and Bohmian velocities, both stochastic.

An important common feature to all these theories is that they include new universal constants. They appear in the modified Schrödinger dynamics, and are adjusted to satisfy conditions (i) to (iii) of §11.10. For instance, we have seen that the GRW theory (SL) introduces a time scale γ^{-1} for the rate at which spontaneous localization takes place, and a length scale $\alpha^{-1/2}$ to characterize how narrow this localization is. These constants may in a sense look like *ad hoc* physical quantities, introduced only for technical reasons. But actually they have an extremely important conceptual role: they define the border between the microscopic and macroscopic world. The corresponding border line, which was ill-defined in the Copenhagen interpretation, is now introduced in a perfectly precise way. Conceptually, their role is therefore somewhat reminiscent of that of the Planck constant.

11.10.1.d Relation with gravity

In 1989, starting also from a study of quantum stochastic processes ("quantum Wiener processes" [12]) as models for state vector reduction [675], as well as from the treatment of continuous measurements [692], Diosi proposed an interesting modification of the GRW theory [693]. The new theory still provides a unification of micro- and macrodynamics, but without requiring the introduction of any new

parameter. The general idea is to replace the previous spontaneous position localization processes by a stochastic process of mass localization, with a strength that
is proportional to the Newton universal gravitational constant G; this is the only
parameter appearing in the new terms that are added to the usual Schrödinger dynamics. Diosi then introduced a treatment of the collapse of the wave function from
a completely general law of density localization, which results in a parameter-free
unification. One can then really speak of a universal mechanism for reduction!

Nevertheless, Ghirardi *et al.* [694] soon showed that this appealing approach
also introduces serious problems at short distances, for which completely unrealistic predictions are obtained; for instance, atomic nuclei would receive so much
energy that they could not stay in their ground state, but would rapidly be excited
(or even dissociated). Ghirardi et al. propose a modification of Diosi's theory that
solves these problems while still retaining the idea of implying gravitation. One
of the two parameters of the GRW theory is eliminated with the help of a relation
implying Newton's constant, but the other remains necessary, a constant having the
dimension of a length. In this sense, the "universal" character of the mechanism is
lost – the authors take this as a general indication that a new parameter is really
necessary to solve the problem of the quantum theory of measurement.

Penrose has often invoked a relation between gravity and uniqueness of macroscopic reality in quantum mechanics (state vector reduction). In [695], within a
more general and philosophical context including the notion of consciousness, he
proposes that "the quantum gravity threshold for self-collapse is relevant to consciousness". In [696], he considers quantum superpositions of the same massive
object at two different locations in space; in the absence of gravity, this superposition can have a very well-defined energy because of space translation invariance.
He then uses general relativity to show that, because masses curve space-time, this
necessarily implies a superposition of two different space-times. He then uses the
"principle of general covariance" to study the properties of the time translation
operator, and shows that the considered situation necessarily has an energy uncertainty ΔE. Using the energy–time uncertainty relation, he conjectures that the
inverse of ΔE corresponds to a finite lifetime of the initial superposition: the superposition is unstable, and spontaneously decays into one of its components. This
gives rise to state vector reduction and ensures macroscopic uniqueness of the position of all massive bodies. He nevertheless warns the reader that "this proposal
does not provide a *theory* of quantum state reduction. It merely indicates the level
at which deviations from the standard linear Schrödinger (unitary) evolution are to
be expected owing to gravitational effects".

11.10.1.e Relation with relativity

The localization process introduced in the GRW theory is not easy to describe in a relativistic way, since it assigns a special role to positions and their localization into a finite volume at a given time; the concept belongs more to Galilean, than Einsteinian, relativity. Nevertheless, relativistic versions of stochastic quantum dynamics and of the CSL theory have been developed [697, 698]. Moreover, in 2005 Pearle introduced a quantization of the classical random field of the CSL theory [699], leading to a "completely quantized theory of state vector collapse"; this resolves a problem related to the conservation of energy under the effect of the collapse mechanism where particles gain energy, by assigning an energy to the random field, which loses energy so that the total energy remains constant. See also [700] and [701].

A general remark is that nonlinearity and stochasticity must go together if one wants to avoid contradictions with relativity. For instance, if one introduces nonlinear changes in the Schrödinger equation (at a microscopic level) while keeping the standard reduction postulate of quantum mechanics (normally this postulate is no longer useful in these theories), faster than light signaling becomes possible, as pointed out for instance by Gisin [702] and by Polchinski [703]. Tumulka has proposed a relativistic version of the GRW theory [704], which he calls "flash ontology", where the "local beables"(§11.8) are given by a discrete set of space-time points, at which the collapses are centered; these points have a random distribution that is determined by the initial wave function [48]. For general or historical reviews on collapse and dynamical reduction theories, see [705] and [678].

11.10.1.f Relation with experiments

As we have seen, the fundamental motivation of theories with modified Schrödinger dynamics is to provide a unification of all kinds of physical evolution, including the emergence of a single result in a single experiment. They obtain this result by the introduction of new physical constants characterizing new physical mechanisms, which means that they are necessarily more specific about the conditions under which state vector reduction takes place. In other words, they are more predictive than standard quantum mechanics. This provides at the same time a strong constraint, namely compatibility with all known experimental results, but also interesting opportunities for testing new theory experimentally. This domain is therefore not limited to purely theoretical considerations.

Concerning possible conflicts with existing experiments, we have already mentioned the discussion of [694], which points out incompatibilities of Diosi's universal form of the theory with known properties of microscopic objects. A similar

---

[48] We recover one ingredient of the de Broglie–Bohm theory, but here we have collapse points in space-time instead of purely spatial position. Moreover, in Bohmian theory collapse does not occur.

case is provided by the generalization of quantum mechanics proposed by Weinberg [706], which he introduced as an illustration of a nonlinearity that may be incompatible with available experimental data; see also [707] for an application of the same theory to quantum optics and [702] for a proof of the incompatibility of this theory with relativity, due to the prediction of superluminal communication (the proof is specific to the Weinberg form of the nonlinear theory and does not apply to the aforementioned other forms).

Another possibility would be to detect the very weak spontaneous heating effect predicted by these theories. Under the effect of the collapse terms in the evolution equation, atoms and nuclei should become partly excited, and emit some weak radiation, which one could try to detect [708]. It turns out that several experiments of this sort have been performed for different purposes, for instance detecting radiation in Germanium crystals due to possible collisions with "dark matter", or two neutrino double-β decay [709]. The authors of Reference [710] have proposed to measure the heating effect of the localization term in gaseous Bose–Einstein condensates. For the moment, no experiment has been able to confirm or to exclude theories with modified Schrödinger dynamics.

Significant tests of these new theories could be given in the future by quantum interference experiments performed with objects containing many particles and going through two spatially distinct paths at the same time. While, in the experiments performed until now with small objects, the localization terms have a completely negligible effect, with larger objects they should reduce or even cancel the contrast of the observed fringes. With the presently proposed values of parameters γ and α, significant tests would be obtained with objects having 10^8 nucleons. Nevertheless, observing interference patterns with objects having the corresponding masses remains a very difficult experimental challenge with present technology. For a review of recent results obtained with clusters and molecules, and a discussion of possible future experiments, see [711]; various aspects of the theory and experimental prospects are discussed in [712].

11.10.2 Physical description of reality within modified dynamics

In these interpretations of quantum mechanics, the state vector no longer keeps the subtle intermediate status it had in standard quantum mechanics (§1.2.3): it now directly represents physical reality "in itself". This reality evolves according to a unified dynamical theory, whether or not measurements, human observations, etc., are involved. Of course, this does not mean that the theory becomes more or less similar to classical mechanics. For instance, since the state vector evolves in a complex state space (Hilbert space), the description of reality should be made in this space, instead of the usual three-dimensional space; the two spaces are very different (not only because of their dimensions; even a classical configuration space with many dimensions remains very different from a Hilbert space). Nevertheless, the unification of the theory and of its dynamics is perfectly reached, which brings an enormous conceptual simplification. When compared to the de Broglie–Bohm theory, this description of reality appears simpler, since it involves only the wave function and no position of particles. The presence of a mechanism of reduction acting directly on the state vector also suppresses the existence of "empty waves".

Similar physical descriptions are obtained, whatever specific form of the non-linear dynamics theory is preferred. For instance, when a particle crosses a bubble chamber, the new terms create the appearance (at a macroscopic level) of a particle trajectory; they also select one of the wave packets at the measurement output of a Stern–Gerlach analyzer (and eliminate the other), but not before these packets become correlated with orthogonal states of the environment (e.g., detectors). Of course, a localization process of the wave function that operates in the space of positions, rather than in that of momenta, destroys to some extent the usual symmetry between positions and momenta in quantum mechanics. But this is actually not a real problem, since one can easily convince oneself that, in practice, what is measured in experiments is basically the positions of particles or objects (pointers, etc.); momenta are only indirectly accessible.

What about the Schrödinger cat and similar paradoxes? If the added nonlinear term has all the required properties to properly mimic the state vector reduction, they are easily solved. For instance, a broken poison bottle must have at least some parts that have a different spatial localization (in configuration space) than an unbroken bottle; otherwise it would have all the same physical properties. It is then clear that the modified dynamics will resolve the components long before it even reaches the cat, which ensures the emergence of a single possibility. For a recent discussion of the effects of the modified dynamics on "all or nothing coherent states" (§6.1.3) in the context of quantum optics, and of the effects on perception in terms of the "relative state of the brain" (§11.12), see [713].

How is an EPRB experiment described in this context? In the case of Bohmian trajectories, we emphasized the role of the "quantum velocity term", which has a value defined in configuration space, not in ordinary space. Here, what is essential is the role of the localization terms added into the Schrödinger equation and their action on the state vector. Intrinsically, these terms remain perfectly local; in the GRW theory, for instance, at any time each particle may undergo a spontaneous localization process to any point of space where its probability density does not vanish. Consider two particles with spin propagating towards Stern–Gerlach magnets along orientations a and b. Nothing special occurs as long as each of the particles propagates towards a Stern–Gerlach analyzer, or even within its magnet, since the particles are microscopic and have an extremely weak localization probability. As in standard theory, they can perfectly well be in coherent superpositions, even if they are far away from each other. But as soon as particle 1 (for instance) hits a detector at the output of the magnet, the system develops correlations with the particles contained in the detector, an electronic current, the amplifier, etc. A von Neumann chain then develops with two branches, each arising from one of the two wave packets split by the magnet. A macroscopic level is reached and the localization term becomes very effective, acting almost instantaneously. Assume for instance that the localization selects the branch where particle 1 has a + component along direction a. Mathematically, this localization amounts to a projection of the two-particle state vector onto a component where the first particle is in the + state along a; because of the mathematical structure of the entangled state, this also projects particle 2 onto a spin state with component − along the same direction[49]. As a consequence, the spin state of the second particle is projected exactly as if the state vector projection postulate had been applied. Nonlocality is not introduced by the localization process alone, but by its action onto a state that already contains nonlocal entanglement. Since this point of view emphasizes the role of the detectors and not of the analyzers, it is clearly closer to the standard interpretation than the Bohmian interpretation. State vector reduction takes place later than in that interpretation, since entanglement with the experimental environment is required, and not only a magnetic splitting of one of the spin wave packets. Both interpretations emphasize the role of nonlocality in a very explicit way, but with a different mechanism.

The program can be seen as a sort of revival of the initial hopes of Schrödinger, for whom all relevant physics should be contained in the wave function and its progressive evolution (see the end of §1.1.2); this is especially true, of course, of the versions of nonlinear dynamics that are continuous (even if fluctuating extra quantities may be introduced), and less true of versions including "hits" that are somewhat reminiscent of the state vector reduction. Here, the state vector directly describes the physical reality, in opposition with our discussion of §1.2; we have a new sort of wave mechanics, where the notion of point particles is completely abandoned in favor of wave packets. The theory is different from theories with Bohmian additional variables because the notion of infinitely precise position in configuration space never appears. It is free of the difficulties mentioned in §11.8.1.i, since there is no need to introduce several levels of reality or a distinction between what can be observed and what can be manipulated; it nevertheless introduces in the evolution of the state vector stochastic functions (or stochastic hits) that cannot be controlled by human action. As we have seen, another important difference is that these theories with modified dynamics are really new theories: in some cir-

---

[49] The equations are similar to those written in the second part of Appendix I in the context of Bohmian theory, except that they involve entanglement with more than the other spin only.

cumstances, they lead to predictions that differ from those of orthodox quantum mechanics, so that experimental tests might be possible. We should finally emphasize once more that, in these theories, the wave function still cannot be considered as an ordinary field: it continues to propagate in a high-dimension configuration space instead of the usual three-dimensional space.

A conclusion of our discussion of modified Schrödinger dynamics may be given by quoting Bell's essay "Speakable and unspeakable in quantum mechanics" (Chapter 18 of [6]). Speaking of standard quantum mechanics, he writes: "The 'problem' then is this: how exactly is the world to be divided into speakable apparatus ... that we can talk about ... and unspeakable quantum system that we cannot talk about? How many electrons, or atoms, or molecules, make an 'apparatus'? The mathematics of the ordinary theory requires such a division, but says nothing about how it is to be made[50] ... Now in my opinion the founding fathers (of quantum mechanics) were in fact wrong on this point. The quantum phenomena do *not* exclude a uniform description of micro and macro worlds ... system and apparatus. It is *not* essential to introduce a vague division of the world of this kind". Modified Schrödinger dynamics does provide an answer to these questions, and shows that a theory where the state vector directly describes the physical reality is perfectly possible: quantum mechanics can indeed be modified without immediately introducing blatant contradictions with presently known results.

After all, if we truly believe that quantum mechanics is fundamentally nondeterministic, why should we request that its fundamental equation of evolution should remain deterministic, as is the standard Schrödinger equation? Is it not more natural to accept the existence of a small stochastic term in the evolution equation itself to account for this random character and unify all dynamics? The border between the situations where macroscopic uniqueness emerges, or does not emerge, is then no longer vague as in the standard interpretation, but precisely contained in the equations. Moreover, the theory opens the possibility for new physical phenomena occurring at this border. It is true that the precise form of the additional terms in the dynamics of quantum systems is not known; a whole class of possibilities exists, so that this is still a subject for exploration. But simply the fact that one can build a theory without giving up realism is important, if only just as a proof of existence.

In particular, physics teachers do not necessarily have to introduce quantum mechanics to students with all the conceptual difficulties associated with the standard interpretation. In fact, if they wish they could take a simpler point of view, retaining the familiar idea that the properties of the physical world exist independently of observations, and considering that a physical system is really described by a wave function (or state vector) propagating in configuration space. They should admit

---

[50] Compare, for instance, with the quotation by Landau and Lifshitz in §2.5.

that the evolution of this wave function obeys an equation that, for the time being, is still unknown in the most general case. What is nevertheless perfectly clear is that, in the limit of microscopic systems, this equation reduces to the Schrödinger equation, which describes these systems with an extreme accuracy. For larger systems, the equation contains terms that lead to the spontaneous emergence of macroscopic uniqueness; unfortunately, for the moment the precise form of these mathematical terms is still unknown. Several possibilities for these terms have been proposed, but the final form of the theory is still a subject of research and debate. But, after all, this situation is not worse than in the standard interpretation, where almost nothing can be said about the nature of state vector reduction! Eventually, the price to pay for this simplification of the general conceptual frame of quantum mechanics would be the introduction of more complicated mathematical equations, containing the reduction process. We can hope that, one day, experiments will be able to tell us which point of view is best.

11.10.3 Open quantum systems in standard quantum mechanics

One can also introduce stochastic terms into the Schrödinger equation without any fundamental purpose, just to obtain a convenient method of calculation [714–716]. These terms are used for the calculation of the evolution of a partial trace density matrix describing a subsystem, within an unchanged linear Schrödinger equation for the whole system. The method replaces a master equation for the partial density operator by the calculation of the evolution of a series of state vectors; these vectors are submitted to random perturbations, which may introduce sudden changes that mimic quantum jumps. Each time evolution of a state vector provides an individual quantum trajectory.

In some circumstances, it turns out that this approach saves computing time very efficiently. The method has found many useful applications, in particular in quantum optics. It is sometimes called the "quantum trajectory method", "Monte Carlo wave function", or "quantum jumps simulation"; for a review, see [717]. In the limit of very small jumps, it becomes the method of "quantum state diffusion" [718, 719].

11.10.4 Attractive Schrödinger dynamics

It has also been proposed to combine the ideas of modified Schrödinger dynamics with those of the dBB theory, in order to obtain a version of quantum mechanics that attempts to combine the advantages of the two points of view [720]. The Bohmian positions are still part of the dynamical equations: they are driven by the wave function (as in the dBB theory), but they also react to it (as opposed to what happens in the dBB theory). Their effect is to create a distribution of density in space, which constantly attracts the wave function. Since this distribution of density follows the Bohmian positions, for each realization of the experiment it selects a single result, as these positions do in dBB theory, and therefore selects a single branch of the state vector.

Consider a system made of N identical particles. The proposed model introduces the following modified Schrödinger equation of the normalized state vector $|\overline{\Phi}(t)\rangle$:

$$i\hbar \frac{d}{dt} |\overline{\Phi}(t)\rangle = \left[H(t) + \overline{H}_L(t) \right] |\overline{\Phi}(t)\rangle \qquad (11.58)$$

where $H(t)$ is the usual Hamiltonian, and where $\overline{H}_L(t)$ is the (antiHermitian) operator given by:

$$\overline{H}_L(t) = i\hbar\gamma_L \int d^3r \left[\Psi^\dagger(\mathbf{r})\Psi(\mathbf{r}) - D_\Phi(\mathbf{r},t) \right] N_B(\mathbf{r},t) \qquad (11.59)$$

In this equation, γ_L is a free parameter of the model, $\Psi(\mathbf{r})$ is the field operator of the identical particles, and $D_\Phi(\mathbf{r})$ is the standard quantum local density:

$$D_\Phi(\mathbf{r},t) = \langle \overline{\Phi}(t) | \Psi^\dagger(\mathbf{r})\Psi(\mathbf{r}) | \overline{\Phi}(t) \rangle \qquad (11.60)$$

We also introduce a function $N_B(\mathbf{r},t)$ in order to characterize the number of Bohmian positions $\mathbf{q}_n$ that are at a distance from $\mathbf{r}$ comparable to a distance α_L:

$$N_B(\mathbf{r},t) = \sum_{n=1}^{N} e^{-(\mathbf{r}-\mathbf{q}_n)^2/(\alpha_L)^2} \qquad (11.61)$$

where α_L is the second free parameter of the model. The averaged Bohmian density in space is then defined as:

$$n_B(\mathbf{r},t) = \frac{N_B(\mathbf{r},t)}{(a_0)^3} \qquad (11.62)$$

Equation (11.58) is nonlinear since $D_\Phi(\mathbf{r}')$ depends on the state vector $|\overline{\Phi}\rangle$. More details concerning the properties of this model are discussed in Appendix L.

As in GRW and CSL theories, the two parameters γ_L and α_L are chosen in a way that introduces a completely negligible effect for microscopic systems, but rapidly destroys superpositions of states that correspond to macroscopically different densities. The dynamics of this process is nevertheless very different from that of GRW and CSL, since it is deterministic; the random character of the result of measurement is just a consequence of the random value of the initial positions. The form of the attractive term added to the Schrödinger dynamics automatically preserves the exchange symmetry of the state vector for identical particles.

The difficulties related to the existence of never-ending empty waves are then suppressed. In the dBB theory, the particle positions provide a direct description of reality; here, they play the role of mathematical random variables that are used to determine the stochastic time evolution of the quantum density distribution $D_\Phi(\mathbf{r}, t)$ in space. The notion of physical point particles disappears in favor of a spatial density distribution $n_B(\mathbf{r}, t)$, similar to that of a fluid, which can be seen as providing the closest description of physical reality. A specificity of this model is that each particle is not constantly subject to a very fast random perturbation that tends to localize it; a smoother time dependence and space averaged effect are assumed. In a solid, for instance, as we discuss in Appendix L, the attraction occurs within the whole volume of the solid, so that the localization effect occurs over a much larger distance; it is therefore considerably smoother in space than that of GRW/CSL theories, so that the heating effects taking place inside a solid are significantly attenuated (the localization process actually occurs mostly at the surface of the solid where density gradients occur, acting as a sort of additional cohesive force).

In Appendix L, we show that this model introduces no significant change to the usual dynamics, except in a special case: when the state vector undergoes a branching towards a quantum superposition of states where a macroscopic object is located in different regions of space. In each branch of the quantum superposition, if the object is held together by cohesive forces between its constituent particles, all positions of these particles have to occupy the same regions of space. But, since Bohmian positions cannot "leave the wave function"[51] this means that, in a single realization of an experiment, the Bohmian positions also necessarily remain clustered together in one region only. All these positions then create a strong attraction in the region of space associated with one component of the wave function only, so that the attractive term projects the state vector in this particular region; this is a very efficient process if the object contains a large number of particles. In a measurement, the pointer of the apparatus can play the role of the macroscopic object, and the localization process creates exactly what is needed for a dynamical introduction of state vector reduction that reproduces the von Neumann postulate. By the same token, one immediately obtains the uniqueness of the space localization of any macroscopic object maintained together by cohesive forces (absence of Schrödinger cats). But macroscopic objects that are not in this case (Bose–Einstein condensed gases, for instance) may remain in quantum superpositions of different locations. In a world made only of substances without internal cohesion, gases for instance, state vector reduction should not take place, or be very slow; indeed, gaseous Schrödinger cats would last for a long time before being reduced!

---

[51] More precisely, the ensemble of Bohmian positions defines a point in configuration space where the wave function does not vanish.

11.11 Transactional interpretation

The transactional interpretation of quantum mechanics also considers quantum states as real, rather than a mathematical representation of knowledge. It was proposed by Cramer in 1986 [721] as a generalization of previous work he had made in 1980 [722] on the possible role of advanced and retarded waves in EPR situations. It is known that, in classical electromagnetism, advanced waves (waves propagating from the future to the past) are solutions of the Maxwell equations, as well as the usual retarded waves. In nonrelativistic quantum mechanics, since the Schrödinger differential equation is first order in time, it does not have this double type of solutions for a given energy. Nevertheless, in relativistic quantum mechanics, second-order time equations replace the Schrödinger equation, so that in this case also advanced waves coexist with retarded waves. In transactional quantum mechanics, the microscopic exchange of a single quantum between a present emitter and a future absorber is described in terms of exchange of retarded and advanced waves between them.

This exchange is called a "transaction", also described by Cramer as a "handshake" between the two participants of a quantum process. The emitter produces a retarded wave (called "offer wave"), which travels to the absorber, making it produce an advanced wave ("confirmation wave"), which travels back in time to the emitter and reacts to it. The cycle repeats itself until a standing wave regime is reached, in which destructive interference cancels the wave outside the time interval of the transaction (destructive interference between either the two retarded, or the two advanced, waves). The whole process has a finite extent in space and time, which means that in relativity it cannot be seen as an event (a point in space-time). As for the collapse of the state vector, it does not happen at any precise time in this interpretation either; it takes place during the whole transaction, in a symmetric emission and absorption process. It is clear that the theory is explicitly nonlocal, so that it has no problem in explaining Bell-type correlations [722].

The mathematical elegance of this point of view makes it attractive. It nevertheless remains uncommon among physicists, probably because it requires giving up intuitive ideas about the past influencing the future and not the reverse, which may be considered as too high a price for a better understanding of quantum mechanics.

More generally, the retro-causal interpretations of (time-symmetric) quantum mechanics and the "loophole" it can offer to escape the Bell inequalities have been the object of recent interest; see for instance References [723–725].

11.12 Everett interpretation

A now famous point of view is that proposed in 1957 by Everett [726], who named it the "relative state interpretation". Other names also appear in the literature, such as the "many-worlds interpretation" (MWI) [727], the "many-minds interpretation", the "universal wave function interpretation", or the "splitting universe interpretation" (the word "splitting", or "branching", refers here to the ramifications of the state vector of the universe into various branches, which we discuss below). These names correspond to different members of the same family of interpretations, which may differ slightly from each other, or even sometimes be significantly different. A common feature of all the family is that any possible contradiction between the two evolution postulates is removed by a simple but efficient method: the second postulate is merely suppressed.

11.12.1 No limit for the Schrödinger equation

In the Everett interpretation, the Schrödinger equation is taken even more seriously than in any other interpretation. One does not attempt to explain how, in a sequence of measurements, a well-defined result is obtained for each measurement. Instead, one merely considers that a single result never emerges: for each measurement, all possible results are realized and observed at the same time! The von Neumann chain is then never broken; its tree of possibilities is left free to develop its branches ad infinitum.

Everett considers the state of the composite whole system made of the observed microscopic system, the measurement apparatus, and the observer(s) after a measurement. To describe this reunion of correlated subsystems, he writes [726]: "there does not exist anything like a single state for one subsystem. Systems do not possess states that are independent of the states of the remainder of the system ... One can arbitrarily choose a state for one subsystem and be led to the relative state for the remainder" – this seems to be just a description of quantum entanglement, a well-known concept. But, now, the novelty is that the observer is merely considered as a physical system, to be treated within the theory on the same footing as the rest of the environment, microscopic or macroscopic. "As models for observers we can, if we wish, consider automatically functioning machines, possessing sensory apparatus and coupled to recording devices registering past sensory data and machine configuration". Everett adds that "Current sensory data, as well as machine configuration, is immediately recorded in the memory, so that all the actions of the machine at a given instant can be considered as functions of the memory contents only"; all relevant experience that the observer keeps from the past is then contained in this memory (magnetic tape, counter, even configurations of brain cells). From this, Everett concludes that "There is no single state of the observer ... With each succeeding observation (or interaction), the observer state branches into a number of different states ... All branches exist simultaneously in the superposition after any sequence of observations". He then checks that "Experiences of the observer" (registered in any memory, counter system, etc.) "are in full accord with the predictions of the conventional 'external observer' formulation of quantum mechanics". Much later, in a letter to D. Raub [728], he writes that this interpretation is "the only completely coherent approach to explaining both the contents of quantum mechanics and the appearance of the world".

Consider for instance a physical system made of one microscopic system (or several) under study, a measurement apparatus and the observer, and then assume that this ensemble is isolated from the rest of the universe. Its state vector then represents reality itself, not our knowledge of this reality. While measurements are performed inside the whole system, its state vector ramifies into branches corresponding to all possible results of measurements, without ever selecting one of these branches; all remain real after measurement. The observer is part of this ramification process, and his/her mind becomes captured in an entangled state with the experimental apparatus and all the environment, reaching simultaneously different states corresponding to the registration of several different results (hence the words "many minds"). In other words, the observer plays the role of the Schrödinger cat in the historical paradox. Nevertheless, one assumes that the observer cannot perceive several different results at the same time. Indeed, each "component of the observer" has no relation whatsoever with all the others, as well as with the state vectors that are associated with them (hence the name "relative state interpretation").

The emergence of macroscopic uniqueness is then just a delusion, a consequence of the way the observer perceives his own memory, of his abilities of introspection. This delusion appears as a consequence of the limitations of the human mind; since the universe itself evolves in a perfectly regular and deterministic way, randomness occurs only in the mind of the observer as a consequence of the specific properties of his memory (storage into the memory and reading its content). Independently of the observers, nothing puts a limit on the predictive deterministic power of the Schrödinger equation. In reality, the random process we usually call "quantum measurement" never takes place!

The "universal wave function" is the wave function of the universe, which contains many branches, in particular all those created by quantum experiments leading to several possible results. Nevertheless, a split of one branch into several others may also occur in interaction processes that were not designed specifically as measurements[52]. Different branches corresponding to macroscopically different situa-

---

[52] When Wigner's friend (§2.3) communicates the result he has obtained to the physicist outside of the laboratory, for instance by giving him a telephone call, more branching takes place; the physicist outside becomes

tions, involving for instance the position of the pointers and their consequences, have independent evolutions. The reason is that no interaction Hamiltonian has matrix elements coupling states where a macroscopic number of elementary quantum systems are in orthogonal states; they cannot be made to interfere in practice, since this would require acting coherently on the individual microscopic states of too many particles. This is why one sometimes considers these branches as different "worlds" existing in parallel, which is the origin of the acronym MWI; others prefer to consider that one still has a unique world in a quantum superposition of very different states, sometimes called "multiverse" [729]. In any case, the universe has a single quantum state, but its subsystems, including the observers, are simultaneously in many states.

11.12.2 Consistency of the interpretation

In the Copenhagen interpretation, the consistency of quantum mechanics requires the existence of a classical world where external observers make and record observations. In the Everett interpretation, observers are neither external nor classical; the postulate of state vector reduction disappears as well. Moreover, the Born probability rule is then seen, not as an independent postulate, but as a rule that should be derived from the superposition principle. This means that one has to explain, within the formalism of the space of states and of the linear Schrödinger equation, why:

(i) as we have already remarked, every observer has the impression that a single result has been obtained when he uses his memory to recall the experiments that were made in the past. Moreover, different observers of the same experiment agree when they compare the results contained in their memories.

(ii) why, when repeating the same experiment, the observer also has the impression that the frequency of occurrence of each possible result corresponds to the usual Born rule.

(iii) it is also necessary to define precisely the appropriate basis of the space of states in which the preceding properties are true, and the conditions under which the branching (or splitting) of the observers take place.

We now examine these three points in succession.

11.12.2.a Perception of results

Point (i) is a postulate related to the faculties of the human mind of registering results and of introspection. The observers belong to a perfectly deterministic universe, but, after they have performed experiments, they perceive the results regis-

entangled with the branches resulting from the experiment, while he was not before he learnt the result. There is a relation between branching and the flow of information.

tered in their memories as random events. The memory of the observer is considered as a quantum system that is entangled with others, and quantum correlations between them are calculated by the standard method; an essential component of the interpretation is that the various states of the memory are considered as completely independent and without any possible communication between them. As a consequence, in each state of his mind, the observer can have access only to the content of his memory in the corresponding branch of the state vector; there is no influence whatsoever of all the other components. Similarly, when exchanging information on results of experiments with other observers, each memory state has access to only one branch associated with one single well-defined content of memories. In other words, any group of observers is split into many completely independent components, without any possible exchange of information between these components.

Of course, this point of view is not straightforward to accept; it is not surprising that it should have been criticized by several authors. For instance Peres writes (§12.1 of [398]), speaking of this family of interpretations: "None is satisfactory because they merely replace the arbitrariness of the collapse postulate by that of the no communication hypothesis" (no-communication between branches of the state vector); see also [730] and [731].

11.12.2.b Born rule

Point (ii) requires that, without invoking the notion of external observations, probabilities should emerge from considerations on the mathematical structure of the entangled state vector, and that their value should be predicted [732, 733]. The Schrödinger equation is linear, and of itself does not imply any particular meaning to the square of the norm of its various components, which must be related to the frequencies of occurrence of the various results. When the same experiment is repeated many times, many branches appear in the state vector, corresponding to all possible series of results; the objective is to show that, in most cases, the observer is part of a branch where the relative frequency of the results corresponds to the Born rule. The difficulty is, of course, to give a precise meaning to the words "in most cases"; for this, we must introduce the notion of probabilities, in other words attribute a measure to each of the various "trajectories" of the observers and their memories. A first remark is that a natural way is to choose the usual norm of the state vector for this measure, since it conserves the total probability. Everett then proposes a more precise reasoning. He assumes that the probability attached to each branch of the state vector is a function of its norm only[53], and that this probability has a property of additivity (if the measurement results are grouped in subsets, the

---

[53] This is not a trivial assumption; for instance, it excludes a theory in which some additional parameter would determine the probabilities.

total probability is still the sum of that of the subsets). He then shows that the usual norm is the only measure that has the required property. The measure associated with any particular sequence of results stored in a memory is therefore given by the squared norm[54], and the Born rule is recovered. See also the derivations developed in pp. 71-78 and 183-186 of Reference [727].

Here we give a simplified version of the argument, assuming that the identical microscopic systems to be measured $S_1, S_2, .., S_N$ are N spins $1/2$, all initially in the same superposition $|\varphi_0\rangle$ of two states $|+\rangle$ and $|-\rangle$:

$$|\varphi_0\rangle = \alpha |+\rangle + \beta |-\rangle \tag{11.63}$$

with the normalization condition:

$$|\alpha|^2 + |\beta|^2 = 1 \tag{11.64}$$

Moreover, before the experiments are performed, the whole systems also contains N memory registers $R_1, R_2, ..., R_N$ that are initially in the same state $|M_0\rangle$. The purpose of the memory register R_1 is to store the result of the measurement performed on S_1, that of R_2 to store the result corresponding to S_2, etc. The initial quantum state of the whole system is:

$$|\Psi_0\rangle = |S_1 : \varphi_0\rangle |S_2 : \varphi_0\rangle .. |N : \varphi_0\rangle \otimes |R_1 : M_0\rangle |R_2 : M_0\rangle .. |R_N : M_0\rangle \tag{11.65}$$

Now, after measurement, each memory register reaches state $|M_+\rangle$ if the result is $+$, state $|M_-\rangle$ if the result is $-$. The state of the whole system after all measurements have been performed is then:

$$\begin{aligned}
|\Psi\rangle = {}& [\alpha |S_1 : +\rangle |R_1 : M_+\rangle + \beta |S_1 : -\rangle |R_1 : M_-\rangle] \\
& \otimes [\alpha |S_2 : +\rangle |R_2 : M_+\rangle + \beta |S_2 : -\rangle |R_2 : M_-\rangle] \otimes .. \\
& .. \otimes [\alpha |S_N : +\rangle |R_N : M_+\rangle + \beta |S_N : -\rangle |R_N : M_-\rangle]
\end{aligned} \tag{11.66}$$

Now consider all sequences of measurement results where result $+$ is obtained N_+ times, result $-$ is obtained N_- times (with $N_+ + N_- = N$). In the entangled ket (11.66), the number $n(N_+, N_-)$ of components containing these results is:

$$n(N_+, N_-) = \frac{N!}{N_+! N_-!} \tag{11.67}$$

and each of these components has a square norm $|\alpha|^{2N_+} |\beta|^{2N_-}$. The total square norm $Q(N_+, N_-)$ associated to results N_+ and N_- is therefore:

$$Q(N_+, N_-) = \frac{N!}{N_+! N_-!} |\alpha|^{2N_+} |\beta|^{2N_-} \tag{11.68}$$

---

[54] There is an interesting relation between this work and Gleason's theorem (§11.4.3), which was discovered independently and published almost simultaneously.

which is nothing but the usual binomial distribution. Since:

$$\frac{Q(N_+ + 1, N_- - 1)}{Q(N_+, N_-)} = \frac{N_-}{N_+} \frac{|\alpha|^2}{|\beta|^2} \tag{11.69}$$

it is easy to see that $Q(N_+, N_-)$ is maximal when:

$$\frac{N_+}{N_-} \simeq \frac{|\alpha|^2}{|\beta|^2} \tag{11.70}$$

Moreover, when N is very large, it is known that this binomial distribution is narrowly peaked around this maximum (the relative width of the distribution is proportional to $1/\sqrt{N}$).

Of course, if we wish to derive the Born rule within the Everett interpretation, it would be circular to assume that the norms we have calculated are equal to probabilities. But we can make a weaker postulate: we can assume, for instance, that events never occur when they are associated with a norm that is less than a small number ε; this number can be chosen arbitrarily small, 10^{-20} for instance. Then, since the binomial distribution becomes narrower and narrower when the number of measurements N increases, all events that are not rejected by this postulate satisfy relation (11.70) to a better and better approximation (Figure 11.5); in the limit of an infinite number of measurements, they satisfy it exactly.

The final result is that the Born rule has been derived, at the price of a weak postulate in order to introduce the norm into the reasoning. Of course, we have discussed only a simplified version, with only two possible results for each measurement (it can nevertheless be generalized without any basic difficulty, just at the price of a slightly more complicated notation). Our proof is valid only if $N \to \infty$, while a more general reasoning should include the derivation of the Born rule for small N. One may also consider situations where the nature of the experiment changes during the sequence, or for instance where several observers repeat different experiments in parallel, and reach compatible conclusions.

More recent versions of the Everett interpretation are formulated in terms of information and flow of information (§11.1.3). For instance, Deutsch [732] has proposed an information-theoretic derivation of the Born rule by combining the Everett interpretation with game theory; see also the work of Wallace [734] and Saunders [735]. Zurek [456, 736] has discussed the relations between the Born probabilities and "environment-assisted invariance", or "envariance", which he compares to Laplace's standard definition of probabilities based on the "principle of indifference". Envariance can also be used to derive the dynamical independence of the branches of the state vector. Nevertheless, whether or not one can derive the Born rule from the Everett theory in a way that is more economical than the rule itself remains somewhat controversial.

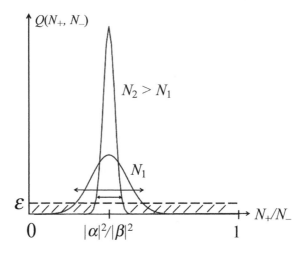

Figure 11.5 Variations of the total square norm $Q\,(N_+, N_-)$ associated with a sequence of N_+ results $+$ and N_- results $-$, as a function of the ratio N_+/N_- and at constant $N = N_+ + N_-$; the lower curve corresponds to $N = N_1$, and the upper and narrower curve to a larger value $N = N_2$. If one postulates that results corresponding to $Q\,(N_+, N_-) < \varepsilon$ are never observed (ε is an arbitrarily small number), then the possible observed values of the ratio N_+/N_- are restricted to a domain (double arrow in the figure) that decreases when N increases. In the limit $N \to \infty$, the ratio is equal to $|\alpha|^2 / |\beta|^2$, and the usual Born rule of quantum mechanics is recovered.

In any case, the Everett notion of probability is very different from any usual notion of probability, including that used within standard quantum mechanics. Normally, probability relates to a situation of uncertainty between a set of possible events, while only one becomes actual. It may happen that the event has already taken place in the past (the probability of an asteroid impacting the earth 65 millions years ago), but the important point is the exclusive character of these events (the collision took place, or it did not). This is not the situation in Everettian theory, where every outcome of a measurement actually occurs, so that there is no distinction between what actually happens and what does not. In reality, there is no uncertainty at all. Moreover, in the usual frequentist notion of probability, the observer samples an ensemble of possibilities among those of a statistical ensemble that is external to him/her; similarly, in a Bayesian view of probabilities, an agent estimates the probabilities associated to external events. By contrast, in the Everett interpretation, the observer contains all possibilities in the various branches and therefore plays himself/herself, in a sense, the role of the statistical ensemble. A new concept of probability is then introduced; see, for instance, the discussions of

References [737] and [738]. Curiously, instead of disappearing from the scene as one could have expected[55], the observer plays a central role in the theory again, by generating probabilities.

11.12.2.c Branching points

Point (iii) is also very important. In the Everett interpretation, macroscopic uniqueness is not a property of the outside world, but rather of the states of this world that become entangled with the relevant basis of states for the memory of the observers[56] – a basis in which they store permanent information about the past, and that they use to exchange consistent information on the results of the experiments with others. Each observer is always macroscopic, having no direct access to the microscopic world: an appropriate macroscopic device is necessary to transmit the microscopic information to his memory. Therefore, an observer can never become directly entangled with a microscopic system, for instance with the two trajectories that a single particle can take in an interferometer (which would destroy the interference effect). Such an entanglement can only occur indirectly, through an apparatus acting as an environment of the observed system. It then becomes natural to invoke macroscopic decoherence as the source of branching of the observer [739]; we have seen (§§7.3.3 and 11.1.1.a) that decoherence prevents the observation of any interference effect between the various components of a coherent superposition, whether the observer is involved or not in this superposition. Generally speaking, the "pointer state basis" (§9.1.2) created by the environment naturally appears as the best choice for expanding the Everett entangled state and applying properties (i) and (ii). This is a common point of view, even if it creates a somewhat indirect relation: both the memory of the observer and the external environment are coupled to the pointer, but the idea is to assume that the nature of the second coupling determines the relevant basis for the first interaction and entanglement. In other words, the relevant basis of the memory is not directly determined by its coupling with the observed object (pointer), but by its coupling with a third partner (considering that the memory of the observer is the only effective environment of the pointer seems rather unrealistic). One should then assume that both couplings share common properties concerning the preferred basis.

Another point of view is to consider (as is often the case in the variant called "many-world interpretation") that the splits of the state of the universe take place

---

[55] See for instance Rosenfeld's quotation in §2.5.

[56] Consider macroscopic subensembles of the universe that have interacted with observers. Macroscopic uniqueness is not an intrinsic property that could be attributed to them; rather, it corresponds to the perception of a property arising from the type of correlation occuring between them and the memory registers of the observers. In this interpretation, the observers therefore play a role that is at least as important as in the other interpretations of quantum mechanics. In the parts of the universe that have never interacted with observers, macroscopic uniqueness has no physical meaning at all.

only during the processes of measurement; the Copenhagian notion of measurement is then conserved. In the "many-mind interpretation", the split takes place at the level of the minds of the observers, a point of view that can be seen as an elaborate version of Wigner's interpretation involving human consciousness (§§11.1.1.b and 2.3). Since each branch of the state vector involves a different history of the world, it is also relatively natural to combine the Everett interpretation with the decoherent history interpretation (§11.7).

11.12.3 Discussion

We now discuss the application of the Everett theory in a few practical cases, which provide a better understanding of its content. We then summarize a few critical opinions that have been expressed on the subject.

11.12.3.a EPRB experiment, locality, external variables

How is an EPRB experiment seen in this theory? In the Bohmian interpretation, we emphasized the role of Stern–Gerlach analyzers; in the modified Schrödinger dynamics, the role of the detectors and decoherence; here we have to emphasize the correlations with the external world, which take place for the memories in the minds of the two human observers. The state vector actually develops its von Neumann chain through the analyzers and the detectors and, at some point, also includes the observers whose brains become part of the superposition. For each choice of the settings a and b, four branches of the state vector coexist; each branch depends explicitly on both parameters, and it is not possible to write mathematically a state vector describing locally the properties associated with a limited region of space and depending on a only, or on b only. Since in the Everett interpretation the state vector provides the most direct description of reality, the very expression of an entangled state vector automatically leads to a nonlocal description of reality. As for the observers, Alice and Bob, they are themselves embedded in this nonlocal von Neumann chain, with components that depend on both a and b; therefore, the choice of a has a distant influence on what Bob's registers, and conversely. Clearly, the Everett interpretation implies a strongly nonlocal description of reality[57].

A question that naturally then arises is whether or not the notion of "external variables" remains valid within the Everett interpretation (cf. discussion and Bell's quotation in §4.3.1.c). The memories of the observers are considered as registers functioning according to the deterministic Schrödinger equation. Should one then

---

[57] Nevertheless, if one considers the flux of information taking place during the Schrödinger (or Heisenberg) evolution of a system of qubits reproducing an EPR experiment, one is led to equations that remain perfectly local [740]; the authors of this reference conclude that a violation of the Bell inequalities means that one cannot attribute stochastic variables to experimental results, which is similar to rejecting counterfactuality (or realism).

consider that these observers retain free will, so that they can arbitrarily choose the settings *a* and *b*, or rather that these choices are causally determined, exactly as their memories? In the latter case, one arrives at superdeterminism, and the notion of free parameter disappears altogether; the proof of the Bell inequalities becomes impossible, since the settings *a* and *b* may be considered as consequences of common causes in the past; all the discussion of the experimental tests of locality we have given becomes meaningless.

11.12.3.b Cosmology

We have seen above that different branches of the state vector of the universe are independent. Then, by construction, the Everett interpretation leads to experimental predictions that are identical to those of the standard interpretation; in this sense, it is not falsifiable. DeWitt [16] even considers that this interpretation is a mere consequence of the formalism: after asking the question "Could the solution to the dilemma of indeterminism be a universe in which all possible outcomes of an experiment actually occur?" he states that "the mathematical formalism of the quantum theory is capable of yielding its own interpretation" – see also the interesting debate [19] that was stirred by the publication of this point of view. DeWitt also argues that the Everett interpretation is a necessity[58] in quantum cosmology: "Everett's ['many worlds'] interpretation has been adopted by the author (Bruce DeWitt) out of practical necessity: he knows of no other. At least he knows of no other that imposes no artificial limitations or fuzzy metaphysics while remaining able to serve the varied needs of quantum cosmology, mesoscopic quantum physics, and the looming discipline of quantum computation" (page 144 of [741]).

Considering the wave function of the universe is of course natural when studying it as a whole. Generally speaking, the idea of multiple universes is not uncommon in astrophysics and cosmology. It has sometimes been evoked to explain the existence of dark matter and dark energy; in the context of the Everett interpretation, one could propose an explanation involving the existence of some interaction between the branches of the state vector of the universe. Multiple universes have also been conjectured in the context of the "anthropic principle" (assuming that the state of the universe contains many branches where the conditions are such that the appearance of intelligent beings is not possible; nevertheless, mankind can only observe what is happening in the small proportion of branches where this appearance has been possible, which corresponds to very special universes). For a review of parallel universes and many worlds in the context of cosmology, see the articles by Tegmark [742], who emphasizes that "parallel universes are not a theory,

---

[58] Other interpretations, for instance those involving modified Schrödinger dynamics (§11.10), are nevertheless also perfectly suitable for introducing the state vector of the universe.

but a prediction of certain theories". A large proportion of physicists specialized in quantum cosmology seems to prefer the Everett interpretation [743].

11.12.3.c Critical opinions

At first sight, the Everett interpretation may look like a welcome unification in quantum physics, especially since it requires no change in the formalism. This makes it attractive to many physicists in a first contact; but on further study some find it difficult to really assimilate, and the interpretation has been criticized.

Consider for instance the "real-time photon counting experiment" described in §10.3, with the observations of horizontal steps and limited by sudden vertical changes of the measured number of photons. Of course, the solution of the Schrödinger equation does not exhibit any step or vertical discontinuity, but just a perfectly continuous decreasing exponential. Then, how does the Everett theory explain the presence of horizontal steps separated by vertical parts, as shown in Figure 10.5? In this theory, these observed steps do not really occur, since they do not occur in the whole state vector; they are only a property of the perception that the observers have of their own memory registers. At each time t, each observer actually divides into different components of the state vector. This happens as many times as an observation is performed, for instance as often as a Rydberg atom crosses the cavity. But the atoms are also quantum objects delocalized in space that arrive continuously, so that the time at which they cross the cavity and are detected is also a random result of quantum measurement. So, since time is continuous, this branching effect occurs constantly and an infinite number of times; in each branch of the state vector, the memory registers store the steps at different times. The curve of Figure 10.5 does not show what happened during the experiment, but just a particular perception of the content of the memory registers of the observers, among many other possible perceptions. Can this scheme be considered as a satisfactory description of the experiment? Should we consider that the observed quantum jumps do not really occur, but that they are only a perceptive effect? In any case, it is probably not in this way that most physicists usually describe the automatic acquisition of results of their experiments. The Everett interpretation seems to increase the distance between theory and experiments in physics.

As for Bell, in [744] he writes: "the elimination of arbitrary and inessential elements from Everett's theory leads back to, and throws new light on, the concepts of de Broglie". Indeed, while the Everett interpretation is often considered as the exact opposite of dBB interpretation, often seen as more naive, both are actually rather similar. The ensemble of the Bohmian empty waves, together with the nonempty wave, reconstructs the same state vector as in the Everett interpretation. Moreover, when an Everettian observer examines the content of his memory after performing a series of experiments, he has access to a single branch of the state vector selected along a sequence of ramifications. But this branch can also be defined by a Bohmian trajectory in configuration space; during each measurement, the trajectory directly indicates which branch has been selected by the Bohmian position to "surf" on it. The dBB theory is therefore nothing but the Everett theory supplemented by an indicator (the position) of the branch of the state vector that should be taken seriously. The only difference is that the sequence of Everett state vectors gives only a blurred view of the trajectory, while it is defined with infinite accuracy in Bohmian theory. Nevertheless, this difference has no consequence in practice since, in both cases, the observer remembers the same sequence of events, and nothing more. In [55], Bell adds that: "this multiplication of universes ... serves no real purpose in the theory, and can simply be dropped without repercussions[59]".

Peres calls it a "bizarre theory" [398] and considers the interpretation as noneconomical (quotation in §11.12.2). Leggett discusses this interpretation in the following terms [105]: "The branches of the superposition which we are not conscious of are said to be 'equally really', though it is not clear ... what these words, ostensibly English, are supposed to mean". One question is what we should expect from a physical theory; does it have to explain how we perceive results of experiments, and if so of what nature should such an explanation be? Since the emphasis is put, not on the physical properties of the systems themselves, but on the effects that they produce on our minds and memories, notions such as perception ([726] speaks of "trajectory of the memory configuration") and their description in neurosciences (the properties of introspection) become part of the theory. What is clear, anyway, is that the Everett interpretation is attractive aesthetically but remains somewhat mind-boggling. Since the human population of the earth is made of billions of individuals, and presumably since each of them is busy making quantum measurements all the time without even knowing it, should we see the state vector of the universe as constantly branching at an exponentially fantastic rate?

---

[59] If one cuts by hand all branches of the various universes but one, one reaches an interpretation that is very similar to the "pragmatic" interpretation discussed in §11.1.1.a.

Conclusion

Quantum mechanics is, with relativity, the essence of the big conceptual revolution of the physics of the twentieth century. The progress, in theoretical and experimental physics, has been extraordinary. New breakthroughs have constantly occurred, ranging from pure theory to practical applications; they have given access to situations and devices that would not even have been conceivable before the advent of quantum mechanics. The words "Second quantum revolution" have sometimes been used to characterize the incredible amount of ideas and results that have been obtained in the last few decades within the frame of quantum mechanics. Our present understanding of the universe is only possible with the use of quantum mechanics at almost every step.

Now, to what extent do we really understand this marvelous theory? We understand its machinery pretty well, and we know how to use its formalism to make predictions in an extremely large number of situations, even in cases that may be very intricate. Quantum mechanics has provided many counterintuitive predictions that, initially, seemed hard to believe, but turned out to be verified by experiments – even if sometimes the experimental verification came long after. One striking illustration is the observation of Bose–Einstein condensation in dilute gases, which was predicted by Einstein in 1925 [745], but observed only in 1995 [746, 747]. The phenomenon takes place at extremely low temperatures (much lower than that of the cosmic radiation background), so that, presumably, it had never occurred anywhere in the universe before 1995! It was just in limbo, waiting hidden in the equations of quantum mechanics, until it was realized one day. This illustrates the extraordinary predictive power of the theory. Heinrich Hertz, who played such a crucial role in the understanding of electromagnetic waves in the nineteenth century (Hertzian waves), remarked that, sometimes, the equations in physics are more intelligent than the person who invented them: "It is impossible to study this amazing theory without experiencing at some times the strange feeling that the mathematical formulas somehow have a proper life, that they are smarter than we, smarter than their author himself, so that we obtain from them more than what was originally put into them" [748]. The remark certainly applies to the equations of quantum mechanics and to the superposition principle: they contain probably much more substance that any of their inventors thought, for instance in terms of unexpected types of correlations, entanglement, etc. It is really astonishing to observe than, in all known cases, the equations have always predicted the correct results so exactly.

But, conceptually, whether or not we really understand the theory is less clear, as illustrated by the variety of interpretations used by physicists. True, the Schrödinger equation reaches the initial aim of quantum mechanics perfectly well: it explains why atoms, molecules, and matter in general are stable, which is impossible within classical physics. But the same equation, with its enormous number or successful predictions, seems to go too far when it extends its linear superpositions into the macroscopic world. It then raises difficult questions about the uniqueness of our observations and the very existence of any classical reality on a human scale. Some interpretations (Everett, for instance) consider that this uniqueness is not part of reality, but just a delusion arising from the way our memories register events. Others propose to change the Schrödinger equation to remove these problematic macroscopic superpositions and turn the wave function into a real wave, propagating in configuration space. Still other interpretations (dBB, for instance) propose to introduce additional variables to solve the problem of macroscopic uniqueness. In this context, a major issue is whether or not the present form of the theory of quantum mechanics is complete. If it is, it will never be possible in the future to give a more precise description of the physical properties of a single particle than its wave function; this is the position of the proponents of the orthodox interpretation. If it is not, future generations may be able to do better and introduce some kind of description that is more accurate.

We have seen why the EPR argument is similar to Gregor Mendel's reasoning, which led him from observations performed between 1854 and 1863 to the discovery of specific factors, the genes (the word appeared only later, in 1909). These genes turned out to be associated with microscopic objects hidden inside the plants that he studied. In such cases, one infers the existence of microscopic "elements of reality" from the results of macroscopic observations. Mendel could derive rules obeyed by the genes, when they combine in a new generation of plants, but at his time it was totally impossible to have any precise idea of what they really could be at a microscopic level (or actually even if they were microscopic objects, or macroscopic but too small to be seen with the techniques available at that time). It took almost a century before Avery *et al.* (1944) showed that the objects in question were contained in DNA molecules; later (1953), Franklin, Crick, and Watson discovered how subtle the microscopic structure of the object actually was, since genes corresponded to subtle arrangements of nucleic bases hidden inside the double helix of DNA molecules. We now know that, in a sense, rather than simple microscopic objects, the genes are arrangements of objects, and that all the biological machinery that reads them is certainly far beyond anything that could be conceived at Mendel's time. Similarly, if quantum mechanics is one day supplemented with additional variables, these variables will probably not be some trivial extension of the other variables that we already have in physics, but variables of a very different nature. But, of course, this is only a possibility, since the histories of biology and physics are not necessarily parallel!

A natural comparison is with special relativity, since neither quantum mechanics nor relativity is intuitive; indeed, experience shows that both, initially, require a lot of thought from each of us before they become intellectually acceptable. But the similarity stops here: while it is true that the more one thinks about relativity, the more understandable it becomes (at some point, one may even get the feeling that relativity is actually a logical necessity!), one can hardly make the same statement about quantum mechanics. Nevertheless, among all intellectual constructions of the human mind, quantum mechanics may be the most successful of all theories, since, despite all efforts of physicists to find its limits of validity (as they do for all physical theories), and many sorts of speculations, no one for now has been able to obtain clear evidence that these limits even exist. The future will tell us if this is the case; surprises are always possible!

12

Annex: Basic Mathematical Tools of Quantum Mechanics

This chapter gives a summary of the mathematical formalism used in quantum mechanics, with a short bibliography put directly at its end. It should be seen as a complement, to be used by readers who wish to know more than what has been recalled about the mathematical tools in the other chapters. Some results are given without explicit proofs; they can be found for instance in Chapters II and IV of [12 − 1] (this chapter has its own list of references on page 432, distinct from the main list, and designated as [12 − n]). Many quantum mechanics textbooks introduce its general formalism in a more complete way, for instance chapter VII of [12 − 2], chapter 3 of [12 − 3], or chapter 2 of [12 − 4].

We first summarize the general formalism of quantum mechanics for any physical system (§12.1), with the Dirac notation; we then study how this formalism treats the grouping of several physical systems into one single quantum system (§12.2); finally, we study a few simple special cases (§12.3), for instance a single particle in an external potential, with or without spin; the reader who prefers wave functions to more abstract reasonings in spaces of states may begin with this section.

12.1 General physical system

The general formalism of quantum mechanics applies to all physical systems, whether they contain a single particle, many particles of various sorts, one or several fields, etc.

12.1.1 Quantum space of states

The physical state of a system at each time is defined in quantum mechanics by a state vector which, in Dirac notation, is written $|\Psi\rangle$ – or $|\Psi(t)\rangle$ if one wishes to make the time dependence explicit. This vector $|\Psi\rangle$ belongs to a complex state vector space $\mathcal{E}$, which may have a more or less complicated structure depending on the system considered. In Dirac notation, the vectors of $\mathcal{E}$ are often called "kets". We explain below (§12.1.3) how the physical properties of the system can be calculated from $|\Psi\rangle$. The space $\mathcal{E}$ is called either the "space of states", or "Hilbert space" for historical reasons.

Consider any pair of vectors $|\Psi_1\rangle$, $|\Psi_2\rangle$, ... belonging to $\mathcal{E}$; by definition of a complex vector space, any linear combination with complex coefficients α, β:

$$\alpha\,|\Psi_1\rangle + \beta\,|\Psi_2\rangle + ... \tag{12.1}$$

is another vector $|\Psi\rangle$ belonging to $\mathcal{E}$. The "superposition principle" states that any linear combination of state vectors provides another possible state vector for the considered physical system; this principle has many physical consequences, several of which are discussed in this book.

In this space, one defines the scalar product of state vectors; the scalar product of vector $|\Psi\rangle$ by vector $|\Phi\rangle$ is a complex number that is usually written:

$$\langle\Phi\,|\Psi\rangle \tag{12.2}$$

This number depends linearly on $|\Psi\rangle$ (it is multiplied by λ if $|\Psi\rangle$ is multiplied by λ) and antilinearly on $|\Phi\rangle$ (it is multiplied by the complex conjugate μ^* of μ if $|\Phi\rangle$ is multiplied by μ). In Dirac notation, the mathematical object $\langle\Phi|$ is often called "bra".

A basis $\{|u_i\rangle\}$ of $\mathcal{E}$ is an ensemble of vectors $|u_i\rangle$ such that any vector of $\mathcal{E}$ can be written as a linear combination:

$$|\Psi\rangle = \sum_i x_i\,|u_i\rangle \tag{12.3}$$

in a unique way; the x_i are called the "components" of $|\Psi\rangle$ in this basis.

If the dimension of state space is finite and equal to P, the number of vectors $|u_i\rangle$ (and therefore the number of terms in this sum) is equal to P. Basis $\{|u_i\rangle\}$ is orthonormal if the scalar products between its vectors obey the relations:

$$\langle u_i\,|u_j\rangle = \delta_{i,j} \tag{12.4}$$

where $\delta_{i,j}$ is the Kronecker delta. With an orthonormal basis, one can show that the components x_i appearing in (12.3) are given by:

$$x_i = \langle u_i\,|\Psi\rangle \tag{12.5}$$

so that:

$$|\Psi\rangle = \sum_{i=1}^{P} \langle u_i\,|\Psi\rangle\,|u_i\rangle \tag{12.6}$$

We also have:

$$\langle \Psi | \Psi \rangle = \sum_i^P |x_i|^2 \qquad \langle \Phi | \Psi \rangle = \sum_i^P y_i^* x_i \qquad (12.7)$$

where the y_i are the components of ket $|\Phi\rangle$.

If the dimension of the state space is infinite, only normalizable kets correspond to a physical state, namely kets for which:

$$\sum_i |x_i|^2 = \text{finite number} \qquad (12.8)$$

Some "bases"are not only infinite, but also continuous[1], which means that the sums of (12.3), (12.6), and (12.7) are all replaced by integrals; it is also necessary to replace the Kronecker delta in (12.4) by a Dirac delta distribution.

12.1.2 Operators

One also defines the action of linear operators in the state space. An operator A is linear when the action of A on vector $\alpha |\Psi\rangle + \beta |\Phi\rangle$, where α and β are complex constants, is given by:

$$A\left[\alpha |\Psi\rangle + \beta |\Phi\rangle\right] = \alpha A |\Psi\rangle + \beta A |\Phi\rangle \qquad (12.9)$$

In any finite basis, operator A may be written as a $P \times P$ matrix with matrix elements $A_{i,j}$; the columns of the matrix contain the components of the transforms of the basis vectors under the action of A. In an orthonormal basis $\{|u_i\rangle\}$, the matrix elements are given by the scalar product of $A |u_j\rangle$ by $|u_i\rangle$:

$$A_{i,j} = \langle u_i | A | u_j \rangle \qquad (12.10)$$

One can easily define the product of an operator by any complex constant α, the sum of operators, or more generally the linear combination $\alpha A + \beta B$ of two operators A and B by:

$$\left[\alpha A + \beta B\right] |\Psi\rangle = \alpha \left(A |\Psi\rangle\right) + \beta \left(B |\Psi\rangle\right) \qquad (12.11)$$

12.1.2.a Product, commutator, eigenvectors

The product of two operators A and B is defined by its action on any ket $|\Psi\rangle$:

$$AB |\Psi\rangle = A |\Psi'\rangle \quad \text{where:} \quad |\Psi'\rangle = B |\Psi\rangle \qquad (12.12)$$

This product AB corresponds to the action of the B operator, followed by that of A. The matrix corresponding to the product operator $C = AB$ is merely given

---

[1] The vectors of a continuous basis are not normalizable, and therefore do not really belong to $\mathcal{E}$, which explains our quotation marks; see for instance the discussion in §A-3 of [12 − 1] .

by the product of the matrices associated with A and B, with the usual rule for multiplying matrices (multiplying the lines of the left matrix by the columns of the right matrix). In general, the product operator AB is different from the product operator BA in the reverse order; the operator $[A, B]$ defined by:

$$[A, B] = AB - BA \qquad (12.13)$$

is called the "commutator" between operators A and B. If this commutator vanishes, one says that operators A and B commute.

The eigenvectors $|a_k\rangle$ of A associated to the eigenvalues a_k are defined as vectors for which the action of A is simply a multiplication by a_k:

$$A|a_k\rangle = a_k|a_k\rangle \qquad (12.14)$$

A method to obtain the eigenvalues and the eigenvectors of A is to proceed to a diagonalization of the matrix of coefficients $A_{i,j}$; nevertheless, for some matrices, the complete diagonalization is not possible. If an eigenvalue corresponds to several eigenvectors that are not proportional, one says that the eigenvalue is "degenerate".

12.1.2.b Hermitian and unitary operators

(i) To each operator A one associates an Hermitian conjugate operator $A^\dagger$ with matrix elements obtained by complex conjugation of those of A, followed by a transposition of lines and columns:

$$\langle u_i | A^\dagger | u_j \rangle = \langle u_j | A | u_i \rangle^* \qquad (12.15)$$

An operator A is Hermitian if it is equal to its own Hermitian conjugate: $A = A^\dagger$; this means that all its matrix elements obey the relations[2]:

$$A_{i,j} = \left[A_{j,i}\right]^* \qquad (12.16)$$

In other words, matrix elements in symmetrical positions with respect to the first diagonal of the matrix are complex conjugate. One can show that any Hermitian operator can always be fully diagonalized, and that all its eigenvalues are real (which does not mean that they are all distinct: some may be degenerate).

(ii) A linear operator U is unitary if the product of its Hermitian conjugate by U is the unity operator:

$$U^\dagger U = 1 \qquad (12.17)$$

---

[2] One can show that, if this relation is satisfied in one basis $\{|u_i\rangle\}$, it is also satisfied in any other orthonormal basis.

Consider two kets $|\Phi\rangle$ and $|\Psi\rangle$ in the space of states and the action of U on these kets:

$$|\Phi'\rangle = U\,|\Phi\rangle \quad \text{and} \quad |\Psi'\rangle = U\,|\Psi\rangle \tag{12.18}$$

Relation (12.17) shows that:

$$\langle\Phi'\,|\Psi'\rangle = \langle\Phi|\,U^\dagger U\,|\Psi\rangle = \langle\Phi\,|\Psi\rangle \tag{12.19}$$

This result expresses that the action of a unitary operator does not change the scalar product (as a consequence, it does not change the norm of a ket either). A unitary operator transforms any orthonormal basis into another orthonormal basis of the space of states. Conversely, any orthonormal change of basis in the space of states defines a unitary operator. This property can be used as an alternative definition of unitary operators.

The matrix associated to a unitary operator is called a unitary matrix. Its columns contain the components of the kets $U\,|u_i\rangle$ in the initial basis $\{|u_i\rangle\}$. Unitarity can easily be checked by selecting any pair of columns i and j, taking the complex conjugate of the first column, and calculating the line-by-line product; the result should be the Kronecker delta $\delta_{i,j}$ for any choice of i and j.

A unitary operator can always be diagonalized in an orthogonal basis, as a Hermitian operator. The eigenvalues are all complex numbers of the form e^{ia}, where a is a real number. A consequence is that U can always be written in the form $U = e^{iA}$, where A is a Hermitian operator; unitary operators are, so to say, complex exponentials of Hermitian operators.

It is easy to show from (12.17) that the product of any number of unitary operators is another unitary operator; unitary operators can be chained without losing unitarity. Nevertheless, two unitary operators U_1 and U_2 do not commute in general.

12.1.2.c Trace

The trace of an operator is a number, the sum of all its diagonal matrix elements:

$$Tr\{A\} = \sum_i A_{ii} \tag{12.20}$$

One can show that the value of the trace is independent of the basis used to calculate it. Moreover, circular permutations of the operators inside a trace do not change the value of the trace; for any three operators A, B, and C:

$$Tr\{ABC\} = Tr\{BCA\} = Tr\{CAB\} \tag{12.21}$$

12.1.3 Probabilities

Hermitian operators play an important role in quantum mechanics: any physical quantity that can be measured on the system (such as its energy, momentum, angular momentum, etc.) corresponds to a Hermitian operator acting in its state space $\mathcal{E}$.

We have seen that, when A is Hermitian, it can be "diagonalized". One can then find a basis made of vectors of $\mathcal{E}$ that are all eigenvectors $|a_k\rangle$ of A, associated with real eigenvalues a_k; in this basis, the operator is represented by a diagonal matrix (having all zero matrix elements except the diagonal elements, which are the a_k).

A fundamental postulate of quantum mechanics is the Born probability rule. It is also discussed in the main text (§1.2.1.c) but, for completeness, we briefly recall it here. The rule says that, when the physical quantity associated to operator A is measured on a system in a normalized state $|\Psi\rangle$:

(i) the only possible results are the eigenvalues a_k,

(ii) the probability $\mathcal{P}(a_k)$ to find a particular result a_k is given by the squared modulus of the scalar product of $|\Psi\rangle$ by the corresponding normalized eigenvector:

$$\mathcal{P}(a_k) = |\langle a_k | \Psi \rangle|^2 \tag{12.22}$$

We have assumed that a single eigenvector corresponds to eigenvalue a_k (nondegenerate eigenvalue); if several distinct eigenvectors correspond to this eigenvalue, one must add the probabilities (12.22) corresponding to all orthonormalized eigenvectors associated with a_k, as in equation (1.4).

Assume we have an ensemble of systems that have been all prepared in the same state $|\Psi\rangle$. If one performs a large number of measurements of the physical quantity associated with operator A, since the Born rule is probabilistic, one generally obtains different values at each measurement. Nevertheless, one can calculate an average value $\langle A \rangle$ and obtain from this rule:

$$\langle A \rangle = \sum_k a_k \, \mathcal{P}(a_k) = \langle \Psi | A | \Psi \rangle \tag{12.23}$$

Consider the operator:

$$P_\Phi = |\Phi\rangle \langle \Phi| \tag{12.24}$$

where $|\Phi\rangle$ is normalized ($\langle \Phi | \Phi \rangle = 1$). Its action on an arbitrary ket $|\Psi\rangle$ is:

$$P_\Phi |\Psi\rangle = |\Phi\rangle \langle \Phi | \Psi \rangle = c |\Phi\rangle \tag{12.25}$$

where $c = \langle \Phi | \Psi \rangle$ is a number. Any ket is transformed under the action of P_Φ into a ket that is proportional to $|\Phi\rangle$; this indicates that P_Φ is the projector onto state $|\Phi\rangle$, obeying the characteristic projector relation $P^2 = P$. We note $P(a_k)$ the projector onto eigenvector $|a_k\rangle$:

$$P(a_k) = |a_k\rangle \langle a_k| \tag{12.26}$$

if a_k is nondegenerate; if it is degenerate, $P(a_k)$ is defined as the sum of projectors over an orthonormalized ensemble of eigenvectors associated with this same eigenvalue. Applying the Born probability rule shows that the probability of obtaining result a_k is then given by:

$$\mathcal{P}(a_k) = \langle \Psi | P(a_k) | \Psi \rangle = \langle P(a_k) \rangle \tag{12.27}$$

12.1.4 Time evolution

Until now we have considered the properties of the physical system at one time. When time evolution is taken into account, the state vector becomes a time-dependent ket $|\Psi(t)\rangle$, with an evolution given by the Schrödinger equation:

$$i\hbar \frac{d}{dt} |\Psi(t)\rangle = H(t) |\Psi(t)\rangle \tag{12.28}$$

where $H(t)$ is the Hamiltonian operator at time t (operator associated with the energy of the system) and $\hbar$ the Planck constant h divided by 2π. Since (12.28) is a first-order time differential equation, it provides a continuous time evolution of the state vector from any initial value. The Hamiltonian operator $H(t)$ is Hermitian but may take various forms, sometimes very complicated, so that it is not necessarily possible to solve equation (12.28) exactly; nevertheless, the equation is always valid.

The Schrödinger evolution does not change the norm of the state vector:

$$i\hbar \frac{d}{dt} \langle \Psi(t) | \Psi(t) \rangle = \langle \Psi(t) | \left[H(t) - H^\dagger(t) \right] | \Psi(t) \rangle = 0 \tag{12.29}$$

which means that the solution $|\Psi(t)\rangle$ of the equation at time t has the same norm as the initial state $|\Psi(t_0)\rangle$. The same reasoning immediately shows that the scalar product $\langle \Psi_1(t) | \Psi_2(t) \rangle$ of two different solutions of the Schrödinger equation does not change in time, whatever the initial values of these solutions were at time t_0. In other words, time propagation from t_0 to t corresponds to a unitary evolution (§12.1.2.b), which defines a unitary operator $U(t, t_0)$:

$$|\Psi(t)\rangle = U(t, t_0) |\Psi(t_0)\rangle \tag{12.30}$$

The Schrödinger equation (12.28) then becomes:

$$i\hbar \frac{d}{dt} U(t, t_0) = H(t) \, U(t, t_0) \tag{12.31}$$

with:

$$U(t_0, t_0) = 1 \tag{12.32}$$

Moreover, the evolution operator $U(t, t_0)$ between time t_0 and t obeys the chain relation corresponding to an evolution between times t and t'', followed by evolution between times t'' and t':

$$U(t', t) = U(t', t'') \, U(t'', t) \tag{12.33}$$

If the Hamiltonian H is independent of time, the evolution operator is given by a simple exponential:

$$U(t, t_0) = e^{-iH(t-t_0)/\hbar} \tag{12.34}$$

It is then easy to check that relations (12.31) and (12.32) are obeyed.

12.1.5 Schrödinger and Heisenberg pictures

The time evolution equations we have written above belong to what is often called the Schrödinger picture, where the state vector $|\Psi(t)\rangle$ is time-dependent. Nevertheless, we can cancel this time dependence by applying to $|\Psi(t)\rangle$ the inverse of the unitary transformation $U(t, t_0)$. This amounts to defining the state vector in the Heisenberg picture $|\Psi_H(t)\rangle$:

$$|\Psi_H(t)\rangle = U^{-1}(t, t_0) \, |\Psi(t)\rangle \tag{12.35}$$

which is indeed independent of time, since:

$$|\Psi_H(t)\rangle = U^{-1}(t, t_0)U(t, t_0) \, |\Psi(t_0)\rangle = |\Psi(t_0)\rangle \tag{12.36}$$

We can then simplify the notation $|\Psi_H(t)\rangle$ into $|\Psi_H\rangle$. Similarly, any operator A in the Schrödinger picture becomes in the Heisenberg picture a time-dependent operator $A_H(t, t_0)$ defined by:

$$A_H(t, t_0) = U^{-1}(t, t_0) \, A \, U(t, t_0) \tag{12.37}$$

Performing a unitary transformation on both the state vector $|\Psi(t)\rangle$ and the operators does not change the averages of these operators in $|\Psi(t)\rangle$, since:

$$\langle\Psi_H|A_H(t, t_0)|\Psi_H\rangle = \langle\Psi(t)| \, U(t, t_0) \, U^{-1}(t, t_0) \, A \, U(t, t_0) \, U^{-1}(t, t_0) \, |\Psi(t)\rangle$$
$$= \langle\Psi(t)| \, A \, |\Psi(t)\rangle \tag{12.38}$$

In particular, if we choose $A = P(a_k)$ as defined in (12.26), we see that the probability of observing any result a_k in a measurement of A can be obtained in the same way in the Heisenberg and in the Schrödinger pictures. In other words, we have a choice between a picture where the time dependence is contained in the state vector, and another picture where it remains perfectly constant but where the time dependence is transported to the observables A.

12.1.6 Density operator

Expressions such as (12.22), (12.23), and (12.27) are not linear with respect to the state vector $|\Psi\rangle$. One can nevertheless obtain linear expressions as functions of an operator ρ called "density operator", which replaces $|\Psi\rangle$.

12.1.6.a Definition

The density operator associated with a normalized state vector $|\Psi\rangle$ is the projector over this state, defined by:

$$\rho = |\Psi\rangle\langle\Psi| \tag{12.39}$$

The trace of this operator is one[3]:

$$Tr\{\rho\} = 1 \tag{12.40}$$

It is a Hermitian operator:

$$\rho = \rho^\dagger \tag{12.41}$$

One can then replace expressions (12.22) and (12.23) by:

$$\mathcal{P}(a_k) = Tr\{P(a_k)\rho\} \tag{12.42}$$

and:

$$\langle A\rangle = Tr\{A\rho\} \tag{12.43}$$

A useful property of these expressions is their linearity, which is convenient to combine the notions of classical and quantum probabilities. Assume that the state $|\Psi\rangle$ of a physical system is not known, but that the system is in normalized state $|\Psi_1\rangle$ with probability p_1, in normalized state $|\Psi_2\rangle$ with probability p_2, ... in normalized state $|\Psi_n\rangle$ with probability p_n. We define the density operator ρ as the sum of the density operators of these states with a weight equal to their probabilities:

$$\rho = \sum_n p_n |\Psi_n\rangle\langle\Psi_n| \tag{12.44}$$

where:

$$0 \le p_n \le 1 \quad \text{and} \quad \sum_n p_n = 1 \tag{12.45}$$

---

[3] This is a consequence of the normalization of $|\Psi\rangle$ since the calculation of the trace in any orthonormal basis $\{|u_i\rangle\}$ gives $Tr\{|\Psi\rangle\langle\Psi|\} = \sum_i \langle u_i |\Psi\rangle\langle\Psi |u_i\rangle$, which is equal to $\langle\Psi |\Psi\rangle = 1$.

Equations (12.40) to (12.43) then remain valid. For instance, the trace (12.40), obtained in any orthonormal basis $\{|u_i\rangle\}$, can be derived from the normalization of the $|\Psi_n\rangle$ and p_n:

$$Tr\{\rho\} = \sum_i \sum_n p_n \, |\langle\Psi_n|u_i\rangle|^2 = \sum_n p_n = 1 \tag{12.46}$$

12.1.6.b Pure states and statistical mixtures

The density operator is a Hermitian, positive (or, more precisely, nonnegative) operator; every diagonal element of ρ obeys:

$$0 \le \langle\Phi|\rho|\Phi\rangle \le 1 \tag{12.47}$$

for any normalized ket $|\Phi\rangle$. To check this property, it is sufficient to replace ρ by its definition (12.44), to use the Schwarz inequality, and finally to take (12.45) into account. The density operator can always be diagonalized and, if we write $|\theta_m\rangle$ its eigenvectors associated with eigenvalues q_m, we can write ρ in the form:

$$\rho = \sum_m q_m \, |\theta_m\rangle\langle\theta_m| \tag{12.48}$$

Since $\langle\theta_m|\rho|\theta_m\rangle = q_m$ is a real number between 0 and 1 – relation (12.47) – we have:

$$0 \le q_m \le 1 \tag{12.49}$$

Every eigenvalue q_m of ρ is therefore bounded between 0 and 1, and their sum gives the trace of ρ, equal to 1:

$$\sum_m q_m = \sum \langle\theta_m|\rho|\theta_m\rangle = \sum_{m,n} p_n \, |\langle\Psi_n|\theta_m\rangle|^2 = \sum_n p_n = 1 \tag{12.50}$$

We can therefore interpret the eigenvalues of ρ as the occupation probabilities of states $|\theta_m\rangle$; we find a form that is similar to (12.44), but now with projectors onto eigenstates that are necessarily orthogonal. If we square (12.48), we obtain:

$$\rho^2 = \sum_m (q_m)^2 \, |\theta_m\rangle\langle\theta_m| \tag{12.51}$$

One then distinguishes two cases for the density operator:

Pure state If only one of its eigenvalues q_m is equal to 1 (all the others being equal to zero), a single term plays a role in the m summation of (12.48); ρ is then the projector onto a single quantum states, its first eigenstate, which we can rename $|\Psi\rangle$. We are then in the case where the system is described by a "pure state" $|\Psi\rangle$, as in (12.39); this description corresponds to the maximal amount of information that can be given on a physical system within quantum mechanics. With (12.51), and since $1^2 = 1$ and $0^2 = 0$, we check that $\rho^2 = \rho$; in terms of density operators, we can characterize a pure state, either by this equality between ρ and ρ^2, or by relation:

$$Tr\left\{\rho^2\right\} = Tr\left\{\rho\right\} = 1 \qquad (12.52)$$

Statistical mixture If several eigenvalues q_m are nonzero, it is no longer possible to assign a single state $|\Psi\rangle$ to the system; several states are necessary, with weights given by probabilities that are neither 0, nor 1. One then says that the density operator ρ describes a "statistical mixture". In contrast with the preceding case, the quantum description provided by ρ is not the most accurate possible within quantum mechanics. In fact, the description may even be very inaccurate, if many states $|\theta_m\rangle$ are associated with equal (or similar) probabilities q_m. Because, for any number q between 0 and 1, we have $q^2 < q$, we see in (12.51) that the density operator is not equal to its square ($\rho^2 \neq \rho$). Another way to characterize a statistical mixture is the relation:

$$Tr\left\{\rho^2\right\} < Tr\left\{\rho\right\} = 1 \qquad (12.53)$$

12.1.6.c Time evolution

When the $|\Psi_n\rangle$ in (12.44) become functions of time $|\Psi_n(t)\rangle$ evolving according to the Schrödinger equation, the density operator ρ becomes time dependent (the kets change in time but the probabilities p_n, which define the initial random choice of the system, remain constant). By inserting (12.28) as well as the associated time evolution for the bra $\langle\Psi_n(t)|$ into (12.44), one obtains the equation of evolution of the density operator:

$$i\hbar\frac{d\rho(t)}{dt} = [H(t), \rho(t)] \qquad (12.54)$$

where $[H(t), \rho(t)]$ is the commutator of $H(t)$ and $\rho(t)$. This equation is often called the "von Neumann equation".

The same reasoning applies if one starts from expansion (12.48) of ρ onto its eigenvectors; one then notices that the eigenvalues of ρ remain constant during time evolution, as well as the trace and the trace of ρ^2.

12.1.6.d Statistical entropy

The statistical entropy (or von Neumann entropy) S associated with any density operator ρ is defined by:

$$S = -k_B Tr\left\{\rho\,\ln\rho\right\} \qquad (12.55)$$

where k_B is the Boltzmann constant of statistical mechanics and thermodynamics, and where ln denotes the natural logarithm (in quantum information, one often prefers to take $k_B = 1$ and the base 2 logarithm, but this changes nothing essential). Equation (12.48) provides:

$$S = -k_B \sum_m q_m \ln q_m \tag{12.56}$$

For a density operator describing a pure state, all eigenvalues are 0, except one that is equal to 1; one then has $S = 0$. For a statistical mixture, relations (12.49) imply[4] that $S > 0$; for instance, if the system is with equal probabilities in two orthogonal states, two eigenvalues of ρ are equal to $1/2$ and $S = 2 \ln 2$ (or $S = 2$ if the base 2 logarithm has been chosen). The value of S characterizes the distance between the quantum description provided by ρ and the optimal description with a pure state, the most accurate possible within quantum mechanics.

We have seen that the probabilities q_m remain constant during the time evolution of ρ according to (12.54); the Hamiltonian evolution of a density operator therefore conserves its entropy: $dS/dt = 0$.

12.1.7 Simple case, spin 1/2

The simplest space of states has two dimensions (except the trivial case with one dimension, which is of little interest since the system can never change state). This case occurs in the study of the properties of a spin $1/2$ particle with a single orbital wave function (§12.3.2); one then usually denotes $|+\rangle$ and $|-\rangle$ the two eigenstates of the Oz component of the spin, which provide a basis. The most general state then reads:

$$|\Psi\rangle = \alpha|+\rangle + \beta|-\rangle \tag{12.57}$$

where α and β are any two complex numbers, obeying condition $|\alpha|^2 + |\beta|^2 = 1$ if $|\Psi\rangle$ is normalized; this ket is associated with the column vector:

$$\begin{pmatrix} \alpha \\ \beta \end{pmatrix} \tag{12.58}$$

The three components $S_{x,y,z}$ of the spin along the three directions Ox, Oy, and Oz correspond to three Hermitian operators; their action onto the column vector (12.58) is given by the three "Pauli matrices" σ_x, σ_y, and σ_z defined by[5]:

$$\sigma_x = \begin{pmatrix} 0 & 1 \\ 1 & 0 \end{pmatrix} \quad \sigma_y = \begin{pmatrix} 0 & -i \\ i & 0 \end{pmatrix} \quad \sigma_z = \begin{pmatrix} 1 & 0 \\ 0 & -1 \end{pmatrix} \tag{12.59}$$

---

[4] Function $-x \ln x$ is positive when $0 < x < 1$.
[5] More precisely, the three components of the spin correspond to the three Pauli matrices multiplied by constant $\hbar/2$.; this is why we use the word "associated" and not "equal".

A simple calculation shows that the squares of the Pauli matrices are all the unit matrix:

$$[\sigma_x]^2 = [\sigma_y]^2 = [\sigma_z]^2 = \begin{pmatrix} 1 & 0 \\ 0 & 1 \end{pmatrix} \tag{12.60}$$

The commutation relations of these Pauli matrices are:

$$\left[\sigma_x, \sigma_y\right] \equiv \sigma_x \sigma_y - \sigma_y \sigma_x = 2i\sigma_z \tag{12.61}$$

but the Pauli matrices anticommute:

$$\left[\sigma_x, \sigma_y\right]_+ \equiv \sigma_x \sigma_y + \sigma_y \sigma_x = 0 \tag{12.62}$$

(in both cases, two other relations may be obtained by circular permutations of σ_x, σ_y, and σ_z).

The density operator ρ associated with state $|\Psi\rangle$ has the following matrix:

$$\rho = \begin{pmatrix} \alpha^*\alpha & \beta^*\alpha \\ \alpha^*\beta & \beta^*\beta \end{pmatrix} \tag{12.63}$$

which therefore corresponds to a pure state (one can easily check that $\rho^2 = \rho$ if $|\Psi\rangle$ is normalized). The most general density operator (pure state or statistical mixture) is described by the matrix:

$$\rho = \frac{1}{2}\left[1 + a\sigma_x + b\sigma_y + c\sigma_z\right] = \frac{1}{2}\left[1 + \mathbf{M}\cdot\sigma\right] \tag{12.64}$$

where vector $\mathbf{M}$, with components a, b, and c on three axes Ox, Oy, and Oz, is called the "Bloch vector". If its length is equal to 1, the state is pure as in (12.63); if its length is smaller than 1, one has a statistical mixture ($\rho^2 \neq \rho$). Chapter IV of $[12-1]$ gives more examples of quantum calculations performed in a space of states with dimension 2.

12.2 Grouping several physical systems

Consider two physical systems 1 and 2, the first with space of states $\mathcal{F}$ containing state vectors $|\Phi\rangle$, the second with space of states $\mathcal{G}$ containing state vectors $|\Xi\rangle$. Each of them may be considered as a subsystem of a larger physical system containing both, and having a space of states $\mathcal{E}$.

12.2.1 Tensor product

Assume that the ensemble of kets $\{|u_i\rangle\}$ provides a basis of $\mathcal{F}$; any vector $|\Phi\rangle$ can then be written as a linear combination:

$$|\Phi\rangle = \sum_i x_i |u_i\rangle \tag{12.65}$$

422 *Basic mathematical tools*

where the x_i are the (complex) components of $|\Phi\rangle$ in this basis. Similarly, the vectors $\{|v_j\rangle\}$ provide a basis of G and we can write:

$$|\Xi\rangle = \sum_j y_j |v_j\rangle \tag{12.66}$$

The space of states $\mathcal{E}$ of the physical system made by grouping the two preceding subsystems is spanned by vectors defining the individual state of each subsystem. These vectors are the "tensor products", which we write:

$$|1 : u_i\rangle \otimes |2 : v_j\rangle \tag{12.67}$$

We will often use a simplified notation for these kets, assuming that the first state always refer to subsystem 1:

$$|1 : u_i\rangle \otimes |2 : v_j\rangle \equiv |1 : u_i; 2 : v_j\rangle \equiv |u_i, v_j\rangle \tag{12.68}$$

If K is the dimension of $\mathcal{F}$ (number of distinct values for index i) and Q that of G (number of distinct values of j), the number of distinct product vectors is $R = KQ$, which determines the dimension of $\mathcal{E}$. Any state $|\Psi\rangle$ belonging to $\mathcal{E}$ can then be written:

$$|\Psi\rangle = \sum_{i=1}^{K} \sum_{j=1}^{Q} z_{i,j} |u_i, v_j\rangle \tag{12.69}$$

where the complex numbers $z_{i,j}$ are its KQ components. Space $\mathcal{E}$ is called the tensor product of $\mathcal{F}$ and G:

$$\mathcal{E} = \mathcal{F} \otimes G \tag{12.70}$$

In the simple case where all components $z_{i,j}$ of $|\Psi\rangle$ in (12.69) can be written as products:

$$z_{i,j} = x_i y_j \tag{12.71}$$

the state vector $|\Psi\rangle$ is a tensor product:

$$|\Psi\rangle = |\Phi\rangle \otimes |\Xi\rangle \tag{12.72}$$

This corresponds to two quantum systems without any correlation[6]. But this is only a special case: in general, one has to use relation (12.69), which does not factorize. We discuss the relation between this general expression and the notion of quantum entanglement in Chapter 7.

We have explicitly studied two physical systems grouped into a single system, as for instance a system of two particles (cf. §12.3.3). Needless to say, this operation

---

[6] Similarly, with classical probabilities, a distribution of two variables that is a product corresponds to noncorrelated variables.

can be generalized to any number of particles: the space of state $\mathcal{E}$ of a system made of N particles with individual spaces of states $\mathcal{F}_i$ ($i = 1, 2, ... N$) is the tensor product of all these spaces:

$$\mathcal{E} = \mathcal{F}_1 \otimes \mathcal{F}_2 \otimes ... \otimes \mathcal{F}_N \tag{12.73}$$

and (12.69) may be generalized in the same way.

Remark: to simplify the discussion, we have assumed that the individual spaces of states $\mathcal{F}$ and $\mathcal{G}$ have finite dimensions P and Q, but the reasoning can be generalized to cases where one of these dimensions is infinite, or both. For instance, in the case of a particle with spin 1/2 (§12.3.2.b), the orbital space of states (associated to the position of the particle) has an infinite dimension, while that of spin has dimension 2; the result is then that the dimension of the whole space of states (tensor product) is also infinite.

12.2.2 Ensemble of spins 1/2

The simplest case occurs when both $\mathcal{F}$ and $\mathcal{G}$ are state spaces of spins 1/2, with dimension 2 each. Space of states $\mathcal{E}$ then has dimension 4, with a basis provided by vectors:

$$|1 : +; 2 : +\rangle, \quad |1 : +; 2 : -\rangle, \quad |1 : -; 2 : +\rangle , \text{ and } \quad |1 : -; 2 : -\rangle \tag{12.74}$$

where, for instance, $|1 : +; 2 : -\rangle$ denotes the state where the Oz component of the first spin is equal to $+\hbar/2$, and that of the second spin equal to $-\hbar/2$. We will write these four vectors in a simpler way, without making the numbering of the spins explicit, as:

$$|+, +\rangle \qquad |+, -\rangle \qquad |-, +\rangle \qquad |-, -\rangle \tag{12.75}$$

The most general state vector belonging to $\mathcal{E}$ is any linear combination of these four vectors.

A particular state in this space, used in many examples, is the "singlet" state:

$$|\Psi\rangle = \frac{1}{\sqrt{2}} \left[|+, -\rangle - |-, +\rangle \right] \tag{12.76}$$

One special property of this state is that it is rotation invariant: it keeps exactly the same properties if, instead of a reference axis Oz to characterize both spins, we choose any other arbitrary direction Ou. Moreover, the two spins are anticorrelated: if the component of the first spin along any direction is found positive, the component of the second spin on a parallel axis is found negative and opposite. This property is essential for the discussion of §4.1.1.

For N spins 1/2, one proceeds in the same way. Their space of states is the tensor product of all individual spin states, with dimension 2^N and spanned by the 2^N vectors:

$$|\pm, \pm, \pm, ..., \pm\rangle \tag{12.77}$$

A very special state that generalizes (12.76) is the state often called the "GHZ state", or state "by all or nothing":

$$|\Psi\rangle = \left[\alpha|+, +, +, ..., +\rangle + \beta|-, -, -, ..., -\rangle\right] \tag{12.78}$$

(where α and β are two complex numbers with the sum of their squared modulus equal to 1). This state is remarkable because it contains only two components where all spins change from one individual state to an orthogonal state. It exhibits marked quantum properties, which were discussed in §§6.1. It should not be confused with a product state, where all spins are uncorrelated, and which can be written:

$$|\Psi\rangle = \left[\alpha|+\rangle + \beta|-\rangle\right]\left[\alpha|+\rangle + \beta|-\rangle\right]...\left[\alpha|+\rangle + \beta|-\rangle\right] \tag{12.79}$$

This state exhibits properties that are more similar to those of a classical state.

12.2.3 Partial traces

The density operator may be used to introduce the convenient notion of "partial traces". Consider a system made of two subsystems, 1 and 2. If they are uncorrelated and each in a pure state $|\Phi(1)\rangle$ and $|\Xi(2)\rangle$, the state of the whole system is given by:

$$|\Psi(1, 2)\rangle = |\Phi(1)\rangle \otimes |\Xi(2)\rangle \tag{12.80}$$

The density operator of this system is then simply the product of the projectors $\rho_1(1)$ and $\rho_1(2)$ over states $|\Phi(1)\rangle$ and $|\Xi(2)\rangle$:

$$\rho_{12}(1, 2) = |\Psi(1, 2)\rangle \langle\Psi(1, 2)| = |\Phi(1)\rangle \langle\Phi(1)|\otimes|\Xi(2)\rangle \langle\Xi(2)| = \rho_1(1)\,\rho_1(2) \tag{12.81}$$

All three systems, the whole system and the subsystems, are in pure states.

But, if $|\Psi(1, 2)\rangle$ is not a product (if the two subsystems are entangled), the situation is more complex. One can start from the density operator of the whole system:

$$\rho_{12} = |\Psi(1, 2)\rangle \langle\Psi(1, 2)| \tag{12.82}$$

and perform an operation called "partial trace", which transforms ρ into an operator ρ_1 that acts only in the space of states of the first subsystem:

$$\rho_1 = Tr_2 \{\rho_{12}\} \tag{12.83}$$

The matrix elements of ρ_1 are defined[7] by:

$$\langle u_i | \rho_1 | u_j \rangle = \sum_k \langle u_i, v_k | \rho_{12} | u_j, v_k \rangle \qquad (12.84)$$

From this partial trace one may obtain all probabilities and average values associated with measurements performed only on the first subsystem, since:

$$Tr\{A(1)\,\rho\} = Tr_1\{A(1)\,\rho_1\} \qquad (12.85)$$

where $A(1) \equiv A(1) \otimes I(2)$ is an operator acting only in the space of states of the first subsystem, while the identity operator in the space of states of the subsystem is denoted $I(2)$; the right-hand part of the equation is a trace taken only in the space of states of subsystem 1. Similarly, one defines another partial trace ρ_2, which relates only to the properties of the other subsystem 2.

A specific property of quantum mechanics is that, even if the whole system is known with the best possible accuracy (it is in a pure state), in general its subsystems are not in this case: they are described by statistical mixtures, and therefore with an accuracy that is not maximal. A classical example is given by two spins in a singlet state (12.76), for which the whole system is indeed in a pure state; if one calculates the partial trace over any of the two spins, one obtains the matrix associated with individual spins in the form:

$$\rho_1(1) = \rho_2(2) = \begin{pmatrix} 1/2 & 0 \\ 0 & 1/2 \end{pmatrix} \qquad (12.86)$$

Under these conditions, the state of each spin is completely unknown: it has the same probability to be in state $|+\rangle$ or $|-\rangle$ or, in fact, in any linear combination of these states[8]. Therefore, even if the whole system is known with the best possible accuracy in quantum mechanics, in this case no information whatsoever is available concerning the two subsystems, a situation that has no equivalent in classical physics (for a more detailed discussion of this unusual situation, in particular by Schrödinger, see §7.1).

12.3 Particles in a potential

We now apply the general formalism to a few simple cases.

---

[7] The partial density operator ρ_1 is independent of the basis $\{|u_i, v_j\rangle\}$ that is used to define it. One can show from (12.84) that, in any other basis $\{|u'_q, v'_l\rangle\}$ one obtains $\langle u'_q | \rho_1 | u'_r \rangle = \sum_l \langle u'_q, v'_l | \rho_{12} | u'_r, v'_l \rangle$.

[8] The matrix is propotional to the unit matrix, which keeps the same form in any basis.

12.3.1 Single particle

In classical mechanics, the position of a point particle is defined by the three components of its position $\mathbf{r}$ at time t. The momentum $\mathbf{p}$ of the particle is a vector with three components given by the conjugate momenta of the components of $\mathbf{r}$:

$$\mathbf{p} = m\frac{d}{dt}\mathbf{r} \tag{12.87}$$

where m is the mass of the particle. The dynamical state of the particle at any time is defined by the simultaneous specification of its position $\mathbf{r}$ and of its momentum $\mathbf{p}$, which corresponds to six scalar variables if the particle moves in ordinary three-dimensional space.

When a potential $V(\mathbf{r})$ acts on the particle, the evolution of the state of the particle may be obtained from an Hamiltonian $\mathcal{H}$, which is nothing but the sum of its kinetic energy and potential energy $\mathcal{V}$:

$$\mathcal{H}(\mathbf{r}, \mathbf{p}; t) = \frac{\mathbf{p}^2}{2m} + \mathcal{V}(\mathbf{r}; t) \tag{12.88}$$

12.3.1.a Wave function

In quantum mechanics, the state of the particle at time t is no longer defined by its position and momentum, but by a state vector $|\Psi(t)\rangle$ belonging to a space of states $\mathcal{E}_r$; in this space, a continuous "basis" (cf. note [1]) is given by the ensemble of kets $|\mathbf{r}\rangle$ where the particle is perfectly localized at point $\mathbf{r}$. According to (12.5), the components of $|\Psi\rangle$ in this basis are given by a $\mathbf{r}$-dependent function:

$$\Psi(\mathbf{r}, t) = \langle \mathbf{r} \,|\Psi(t)\rangle \tag{12.89}$$

with, according to (12.6):

$$|\Psi(t)\rangle = \int d^3r\, \Psi(\mathbf{r}, t)\, |\mathbf{r}\rangle \tag{12.90}$$

The complex function $\Psi(\mathbf{r}, t)$ is called the "wave function" of the particle. Since this function may extend over a whole domain of space, the position of the particle is not perfectly defined. Quantum mechanics only specifies that the probability to find at time t the particle in any volume D is given by:

$$\mathcal{P} = \int_D d^3r\, |\Psi(\mathbf{r}, t)|^2 \tag{12.91}$$

In other words, $|\Psi(\mathbf{r})|^2$ gives the density of probability $n(\mathbf{r})$ associated with the position variable of the particle:

$$n(\mathbf{r}, t) = |\Psi(\mathbf{r}, t)|^2 \tag{12.92}$$

The normalization condition of the total probability implies that the integral should be equal to 1 when D extends to all space:

$$\int d^3r \; |\Psi(\mathbf{r}, t)|^2 = \int d^3r \; n(\mathbf{r}, t) = 1 \tag{12.93}$$

Of course, this condition is crucial to interpret $n(\mathbf{r})$ as a density of probability. If a wave function does not provide a unit value for the integral of its square modulus, one says that it is not normalized; but it is then sufficient to divide it by the square root of this integral to normalize it. This operation is possible only if the integral in all space is finite; one can normalize only functions that obey this condition, an essential condition for a function to be acceptable as a wave function.

The probability to find the momentum within any domain is given by a formula that is similar to (12.91), but with the probability density $\bar{n}(\mathbf{p})$ replaced by $|\overline{\Psi}(\mathbf{p})|^2$, where $\overline{\Psi}(\mathbf{p})$ is the Fourier transform of $\Psi(\mathbf{r})$.

12.3.1.b Schrödinger equation, probability current

The time evolution of the wave function $\Psi(\mathbf{r},t)$ is given by the Schrödinger equation:

$$i\hbar \frac{\partial}{\partial t} \Psi(\mathbf{r},t) = -\frac{\hbar^2}{2m} \Delta \Psi(\mathbf{r},t) + V(\mathbf{r}; t) \; \Psi(\mathbf{r},t) \tag{12.94}$$

From this wave function, one can define a probability current:

$$\mathbf{J}(\mathbf{r}, t) = \frac{\hbar}{2im} \left\{ \Psi^*(\mathbf{r},t)\nabla\Psi(\mathbf{r},t) - \Psi(\mathbf{r},t)\nabla\Psi^*(\mathbf{r},t) \right\} \tag{12.95}$$

and, by using (12.94), derive the equation for local conservation of the probability:

$$\frac{\partial}{\partial t} n(\mathbf{r}, t) + \nabla \cdot \mathbf{J}(\mathbf{r}, t) = 0 \tag{12.96}$$

Integrated over all space, this relation shows that the norm of any wave function remains constant as a function of time; if it is initially normalized as in (12.93), the Schrödinger equation conserves this normalization during the time evolution.

12.3.2 Spin, Stern–Gerlach experiment

The formalism we have described applies only to spinless particles, also called spin zero particles. We now study how the formalism can be adapted to particles with nonzero spin, which will allow us to better understand the origin of the two-dimensional space of states that was introduced in §12.1.7.

12.3.2.a Introduction of spin

At the time quantum mechanics was invented, the experimental study of atomic spectral lines made physicists realize that a particle such as an electron must possess an additional degree of freedom, in addition to those associated with its position in space – this idea was proposed by Uhlenbeck and Goudsmit in 1925 [12 – 5]. The additional degree of freedom is called "spin"; it corresponds to an internal rotation of the particle (rotation around itself). The existence of spin can not be explained with a classical image: a classical object can turn around itself only if it extends over some domain of space, while in quantum mechanics even a point particle (the electron for instance) can have a spin.

Shortly before (1922), Stern and Gerlach [12 – 6] had performed an experiment that provided a direct evidence of an internal rotation of quantum particles such as atoms; the particles they had used were Silver atoms, which have a spin arising in particular from the electrons they contain. The experiment is shown very schematically in Figure 12.1. A beam of particles (atomic beam) originates from source S and propagates to region B, where a magnet creates a magnetic field with a strong gradient along direction Oz. The particles carry a magnetic moment that is proportional to their spin, and is therefore colinear with it. The local magnetic gradient creates a force acting on this magnetic moment, so that the trajectory of the particles is bent in a direction that depends on the Oz component of their spin. The impact of the particles is eventually registered on a screen E.

Within classical mechanics one would expect that, initially, the spins should have completely random orientations, uniformly spread in all directions; this component should then vary continuously between two extreme values. In other words, one should observe a continuum of possible values for the deviations of the particles, resulting in a broad spot on the screen. But the experimental surprise was to observe, instead, two well-separated spots: only two kinds of deviations were obtained, one along Oz, the other in the opposite direction. This result is interpreted by assuming that the component of each spin along Oz can only take two discrete values (which turn out to be equal to $\pm\hbar/2$): this is spin quantization, directly observed experimentally with this device.

In this experiment, direction Oz does not correspond to any particular direction for source S. This means that the spin component can take only one out of two opposite values, whatever measurement direction is chosen. Of course, in classical physics, such a situation is totally impossible: no vector has its component on any direction in space with constant modulus. This provides one more illustration of the completely quantum nature of spin, without any classical equivalent. For a more detailed discussion of the Stern–Gerlach experiment, and in particular of

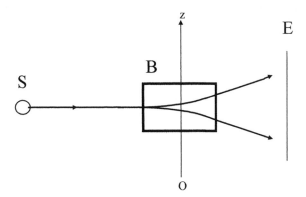

Figure 12.1 Schematical representation of a Stern–Gerlach experiment. A source S emits a beam of particles (Silver atoms), which propagate to region B where they undergo the influence of a strong magnetic gradient along direction Oz created by a magnet (not shown in the figure). The force exerted by this gradient on the magnetic moment of the particle, which is parallel to its spin, then depends on the direction of the spin; the trajectory is therefore bent in a way that depends on the Oz component of the spin.

In classical mechanics, one would expect a continuum of possible deviations for the particles (since this component can take continuous values within a whole interval). But two groups of completely different impacts are observed on screen E, corresponding to two possible deviations. This indicates that the Oz component of the spin can only take two discrete values – one says that it is quantized. In classical physics, no vector can have components on any axis that are equal to constant values. The experiment provides evidence, at the same time, for the existence of spin, and for its completely nonclassical properties.

measurements of spin components along various directions, see for instance §A in Chapter IV of [12 – 1].

In fact, quantum mechanics does not necessarily imply that the number of discrete components of a spin should be two: in general, their number is given by $2s + 1$, where s is any half integer number[9]. The case we have described corre-

---

[9] A half integer number is by definition a number that is, or becomes, integer when multiplied by 2.

sponds to $s = 1/2$, and this explains why one speaks of "spin 1/2 particle"; this applies to the electron for instance, as well as to other particles such as the proton, the neutron, etc. (but not the photon).

12.3.2.b Space of states

For a spin 1/2 particle, the formalism of nonrelativistic quantum mechanics introduces two wave functions instead of one: each spin component has a wave function. One first chooses a fixed reference axis, the "quantization axis" Oz, and one defines two wave functions $\Psi_{\pm}(\mathbf{r})$; index $\pm$ specifies the sign of the Oz component of the spin. For a particle with any spin s, one would introduce $2s + 1$ wave functions, but for the sake of simplicity here we limit ourselves to the case $s = 1/2$.

Introducing two functions $\Psi_{\pm}(\mathbf{r})$, instead of one, doubles the number of components of the state vector $|\Psi\rangle$, as well as the number of vectors in a basis of the space of states. We therefore now replace (12.90) by:

$$|\Psi(t)\rangle = \int d^3 r \left[\Psi_{+}(\mathbf{r};t) \, |\mathbf{r}, +\rangle + \Psi_{-}(\mathbf{r};t) \, |\mathbf{r}, -\rangle \right] \qquad (12.97)$$

where $|\mathbf{r}, +\rangle$ denotes the ket where the particle is localized at point $\mathbf{r}$ with a positive component of its spin along Oz, while $|\mathbf{r}, -\rangle$ is the corresponding state with a negative component.

One sometimes call "spinor" the ensemble of the two components of $|\Psi\rangle$, which are grouped in a column matrix as the two components of a vector in a two-dimensional space:

$$\begin{pmatrix} \Psi_{+}(\mathbf{r}) \\ \Psi_{-}(\mathbf{r}) \end{pmatrix} \qquad (12.98)$$

Applying the three Pauli matrices (12.59) to the column vector (12.98) defines (after multiplication by a constant factor $\hbar/2$) the action of the operators $S_{x,y,z}$ associated with the three components of the spin. We now see the emergence of a structure of a space of states that combines the properties of spin (space of states with dimension 2, studied in §12.1.7) and those of the orbital variable $\mathbf{r}$ (space of states with infinite dimension), and provides a good illustration of the notion of tensor product (§12.2.1).

12.3.3 Several particles

Assume now that the system under study is made of several particles. In classical mechanics, the evolution of a system containing N particles with masses m_1, m_2,

..., m_N and positions $\mathbf{r}_1, \mathbf{r}_2, ..., \mathbf{r}_N$ involves an Hamiltonian:

$$\mathcal{H}(\mathbf{r}, \mathbf{p}; t) = \sum_{i=1}^{N} \frac{\mathbf{p}_i^2}{2m_i} + \mathcal{V}(\mathbf{r}_1, \mathbf{r}_2, ..., \mathbf{r}_N; t) \qquad (12.99)$$

where $\mathcal{V}(\mathbf{r}_1, \mathbf{r}_2, ..., \mathbf{r}_N; t)$ is the sum of the external potential acting on the particles and of their mutual interaction potential.

In quantum mechanics, for an ensemble of N spinless particles, the single particle wave function $\Psi(\mathbf{r}, t)$ is replaced by a wave function $\Psi(\mathbf{r}_1, \mathbf{r}_2, ..., \mathbf{r}_N; t)$ depending of the positions $\mathbf{r}_1, \mathbf{r}_2, ..., \mathbf{r}_N$ of all particles:

$$\Psi(\mathbf{r}_1, \mathbf{r}_2, ..., \mathbf{r}_N; t) = \langle \mathbf{r}_1, \mathbf{r}_2, ..., \mathbf{r}_N | \Psi(t) \rangle$$

The function:

$$n(\mathbf{r}_1, \mathbf{r}_2, ..., \mathbf{r}_N; t) = |\Psi(\mathbf{r}_1, \mathbf{r}_2, ..., \mathbf{r}_N; t)|^2 \qquad (12.100)$$

generalizes (12.92) and provides the probability to find the first particle at point $\mathbf{r}_1$, the second at point $\mathbf{r}_2$, ... the last at point $\mathbf{r}_N$. The wave function Ψ is physically acceptable only if the integral of n in all configuration space (with $3N$ dimensions) is finite, which makes the normalization of Ψ possible.

The introduction of the notion of tensor space is natural in terms of wave functions. For two particles, we know that a function $\Psi(\mathbf{r}_1, \mathbf{r}_2)$ can be expanded as a sum of products of $\mathbf{r}_1$ functions by $\mathbf{r}_2$ functions:

$$\Psi(\mathbf{r}_1, \mathbf{r}_2) = \sum_\mu \sum_\nu \phi_\mu(\mathbf{r}_1)\, \phi_\nu(\mathbf{r}_2) \qquad (12.101)$$

(where the sums may possibly be infinite, or even continuous so that they become integrals over μ and ν): the space of two-particle wave functions $\Psi(\mathbf{r}_1, \mathbf{r}_2)$ is the tensor product of the spaces of single-particle wave functions. This notion can be generalized to N particles, for which the wave functions can be written:

$$\Psi(\mathbf{r}_1, \mathbf{r}_2, ..., \mathbf{r}_N) = \sum_\mu \sum_\nu ... \sum_\xi \phi_\mu(\mathbf{r}_1)\, \phi_\nu(\mathbf{r}_2) ... \phi_\xi(\mathbf{r}_N) \qquad (12.102)$$

and belong to the tensor product space of N spaces of single particle wave functions.

The Schrödinger equation is a generalization of (12.94), which reads:

$$i\hbar \frac{\partial}{\partial t} \Psi(\mathbf{r}_1, ..., \mathbf{r}_N; t) = -\frac{\hbar^2}{2m} \Delta \Psi(\mathbf{r}_1, ..., \mathbf{r}_N; t) + V(\mathbf{r}_1, ..., \mathbf{r}_N; t)\, \Psi(\mathbf{r}_1, ..., \mathbf{r}_N; t)$$

$$(12.103)$$

where, as in classical mechanics, the potential V may include a part due to an external potential acting separately on each particle as well as a mutual interaction part (generally the sum of binary interactions). One can define a probability current

J in a space with $3N$ dimensions (configuration space); this generalizes relations (12.95) and (12.96) and provides a multidimensional conservation relation.

If the N particles are spin 1/2 particles, the wave functions splits into 2^N components, labeled with N equal to $\pm$:

$$\Psi_{\pm,\pm,\,...,\pm}(\mathbf{r}_1, \mathbf{r}_2, ..., \mathbf{r}_N; t) \tag{12.104}$$

Except for this change, the general idea remains the same: the space of states of the whole system of particles is the tensor product of the single-particle spaces of states.

Remark: one can also quantize other physical systems than particles, for instance fields. For an introduction, see for instance $[12-7]$ and $[12-8]$.

References

$[12-1]$ C. Cohen-Tannoudji, B. Diu, and F. Laloë, *Quantum Mechanics*, Wiley (1977).

$[12-2]$ A. Messiah, *Mécanique Quantique*, Dunod (1962); *Quantum Mechanics*, Wiley (1961) and Dover (1999).

$[12-3]$ A. Peres, *Quantum theory, Concepts and Methods*, Kluwer Academic Press (1995).

$[12-4]$ M. Le Bellac, *Quantum Physics*, Cambridge Universty Press (2011).

$[12-5]$ G.E. Uhlenbeck and S. Goudsmit, *Naturwissenschaften* **47,** 953 (1925); *Nature* **117**, 264 (1926). For an interesting historical description of the way spin was discovered, see:

http://www.ilorentz.org/history/spin/goudsmit.html.

$[12-6]$ W. Gerlach and O. Stern, "Der experimentelle Nachweiss der Richtungsquantelung im Magnetfeld", *Zeit. Phys.* **9**, 349–355 (1922).

$[12-7]$ L. Schiff, *Quantum Mechanics*, McGraw Hill (1955).

$[12-8]$ C. Cohen-Tannoudji, J. Dupont-Roc, and G. Gryndberg, *Atom–Photon Interactions*, Wiley (1989).

Appendix A

Mental Content of the State Vector

(i) A relatively recent (1999) article by Englert, Scully, and Walther [109] provides an interesting illustration of the debate about the content of the state vector, in particular because of the wording chosen by the authors. Speaking of standard theory, they write: "(One) is led astray by regarding state reductions as physical processes, rather than accepting that they are nothing but mental processes". They then advocate a "minimalistic interpretation of state vectors" and actually even give a general warning that it is dangerous to go beyond it ("van Kampen's caveat[1]"), but do not expand very much on these dangers. It is then interesting to extend this line of thought further: if the state vector can evolve under the effect of a "purely mental process", it then necessarily acquires elements that are mental (subjective), instead of being related to external reality only (objective). The Schrödinger evolution, on the other hand, is determined by external macroscopic parameters, and therefore has a reality content that is similar with these parameters. Should we consider that the state vector is hybrid, and combines elements describing external reality, at least partially, and others that are purely mental, all contained in a single mathematical object[2]? In this view, the process of state vector reduction would correspond to times when the state vector suddenly acquires more mental elements (since the reduction process is considered as purely mental); the Schrödinger evolution would correspond to periods of time where the mental content remains constant.

Such a hybrid view concerning the reality content of the state vector is hardly compatible with the standard view where it defines a physical preparation procedure. But a preparation can be a measurement process involving state reduction (selection of atoms in one beam at the output of a Stern–Gerlach filter, for instance). If the effect of the measurement is a purely mental projection of the state

---

[1] "Whoever endows the state vector with more meaning than is needed for computing observable phenomena is responsible for the consequences" [108].

[2] This in itself would not necessarily be a problem. A classical statistical distribution function for an ensemble of time-dependent systems has similar properties: it does combine real evolution with elements related to imperfect initial knowledge on the systems.

vector, the resulting state vector should also be purely mental. With this definition, a state vector should then have no real content at all.

But we have already noted (§11.3.2) that most physicists would probably agree that, at least sometimes, the wave function does contain elements of reality, and illustrated this with a few examples (BCS state of electrons, etc.). One usually considers that Schrödinger evolution contains at the same time a physical evolution for those properties that quantum mechanics does attribute to the system (observables that have $|\Psi\rangle$ as an eigenvector), but also an evolution of probabilities (for all other observables) that represent only our knowledge of the system, and therefore can indeed be seen as mental processes. In fact, the notion of a "purely mental process" is not often mentioned by supporters of the Copenhagen interpretation (maybe with the exception of Wigner – see §11.1.1.b – but the Wigner interpretation is not really standard). Peres for instance, in the quotation of [44] given in §1.2.3.b, while also orthodox, never invokes human minds, but just preparations and tests on physical systems.

(ii) Another interesting illustration is provided by a note published by Fuchs and Peres [110] entitled "Quantum theory needs no interpretation" – especially since this note has stimulated many reactions from authors expressing various points of view. It goes even further than [109] since these authors seem to take explicitly a point of view where the wave function is not absolute, but observer dependent[3]. After stating that: "Quantum theory does *not* describe physical reality. What it does is provide an algorithm for computing *probabilities* for the macroscopic events that are the consequences of our experimental interventions", they add that "a wave function is only a mathematical expression for evaluating probabilities and depends on the knowledge of whoever is doing the computing". The wave function then becomes really similar to a classical probability distribution that, obviously, depends on the knowledge of the experimenter; several different distributions can be associated with the same physical system if there are several observers. As mentioned in 1.2.3.a, associating several different wave functions with one single system is not part of what is usually called the orthodox interpretation (except, of course, for a trivial phase factor); on the other hand, this idea is part of the relational interpretation (§11.3.1).

---

[3] As in §1.2.3.a, we assume that all observers use the same Galilean reference frame. Otherwise, they need different state vectors for describing the system, for trivial reasons (as in classical mechanics).

Appendix B

Bell Inequalities in Nondeterministic Local Theories

In the derivations of the Bell inequalities resulting from local realism (§4) given in §4.1, we have assumed that the results of experiments are known functions $A(a, \lambda)$ and $B(b, \lambda)$, which depend on the settings a and b and of the (possibly multidimensional) additional variable λ. This is a natural continuation of the EPR theorem, which derives the existence of these functions from the assumption of local realism (determinism is a consequence of local realism). Here we show that the Bell inequalities are actually more general: they can still be established when $A(a, \lambda)$ and $B(b, \lambda)$ are replaced by probabilities, that is, within a nondeterministic context, but provided locality is preserved. Figure 4.7 schematizes the scheme, where the correlations occur from fluctuating properties of the pairs of particles, and the results are modified by independent random local processes occurring in remote regions of space .

B.1 Assuming the factorization of probabilities

The basic idea behind this first generalization is straightforward: probabilities can always be considered as resulting from a deterministic process controlled by one more additional random variable, which we call μ here. Adding this variable does not change anything to the reasoning that leads to the inequalities; actually, λ can be seen as one single variable with several dimensions, one of its components being μ.

Consider a given value of λ. Within a nondeterministic theory, we replace $A(a, \lambda)$ by two probabilities $\mathcal{P}_+^A(a, \lambda)$ and $\mathcal{P}_-^A(a, \lambda)$, and $B(b, \lambda)$ by two probabilities $\mathcal{P}_+^B(b, \lambda)$ and $\mathcal{P}_-^B(b, \lambda)$. For any a and λ, we have:

$$\mathcal{P}_+^A(a, \lambda) + \mathcal{P}_-^A(a, \lambda) = 1 \tag{B.1}$$

with a similar condition for the probabilities $\mathcal{P}_\pm^B$. We can then introduce one additional variable μ having uniform distribution within interval $[0, 1]$, and define a function $\overline{A}(a, \lambda, \mu)$ by[1]:

$$\overline{A}(a, \lambda, \mu) = \begin{cases} +1 & \text{if} \quad 0 \leq \mu \leq \mathcal{P}_+^A(a, \lambda) \\ -1 & \text{if} \quad \mathcal{P}_+^A(a, \lambda) < \mu \leq 1 \end{cases} \tag{B.2}$$

We then have:

$$\int_0^1 d\mu \, \overline{A}(a, \lambda, \mu) = \mathcal{P}_+^A(a, \lambda) - \mathcal{P}_-^A(a, \lambda) \tag{B.3}$$

and, of course, a similar equation for the difference $\mathcal{P}_+^B(b, \lambda) - \mathcal{P}_-^B(b, \lambda)$.

Consider now two measurements performed in very remote regions of space with settings a and b and providing random results. The randomness arises from fluctuations of the additional variables λ, but also from some intrinsic random process occurring locally, one in the first region, the other in the second. For a given value of λ, randomness is only a consequence of these local processes, which can be considered as independent; the probability of occurrence of results $+$ and $+$ (for instance) is therefore the product:

$$\mathcal{P}_+^A(a, \lambda)\mathcal{P}_+^B(b, \lambda) \tag{B.4}$$

and similarly for the three other possibilities $(+, -)$, $(-, +)$, and $(-, -)$. Suppose now that we wish to calculate the average of the product of the two results observed. For each value of λ, we have to sum the probabilities associated with the $(+, +)$ and $(-, -)$ events, and to subtract the probabilities of the $(+, -)$ and $(-, +)$ events, which amounts to introducing the term:

$$\mathcal{P}_+^A(a, \lambda)\mathcal{P}_+^B(b, \lambda) + \mathcal{P}_-^A(a, \lambda)\mathcal{P}_-^B(b, \lambda) - \mathcal{P}_+^A(a, \lambda)\mathcal{P}_-^B(b, \lambda) - \mathcal{P}_-^A(a, \lambda)\mathcal{P}_+^B(b, \lambda)$$
$$= \left[\mathcal{P}_+^A(a, \lambda) - \mathcal{P}_-^A(a, \lambda)\right] \times \left[\mathcal{P}_+^B(b, \lambda) - \mathcal{P}_-^B(b, \lambda)\right] \tag{B.5}$$

The average value of this expression is obtained by a sum over λ, which, according to (B.3), gives the integral:

$$\int d\lambda \int_0^1 d\mu \, \overline{A}(a, \lambda, \mu) \int_0^1 d\mu' \, \overline{B}(b, \lambda, \mu') \tag{B.6}$$

At this point, we see that we obtain the same expression as with a deterministic theory, with an integral of the product of two functions that are equal to ± 1. The only difference is the presence of two additional integration variables μ and μ': the local stochastic theory is therefore equivalent to a deterministic theory with more additional variables. The rest of the reasoning leading to Bell inequalities remains unchanged.

---

[1] We use the notation $\overline{A}$ to make a distinction from A, which was defined in the main text as the result of the experiment. Here, $\overline{A}$ is just a mathematical variable introduced to express probabilities in a convenient way.

The conclusion is that these inequalities are also valid for nondeterministic theories, provided the dependences of the probabilities are local – if we had assumed that $\mathcal{P}_+^A(a, \lambda)$ also depends on b, the proof of the inequalities would no longer have been possible.

B.2 Deriving the factorization

In §4.2.2.b, we have discussed a derivation of the BCHSH inequality in the line of Bell's "La nouvelle cuisine" [169]. We have also mentioned that slightly other derivations are possible and have been proposed by Jarrett, Ballentine, Shimony and others [168, 174–178]. The common starting point of these derivation is the use of the Bayes law of conditional probabilities. With the same notation as in §4.2.2.b, the probability to observe results A and B with two remote experiments and choices a and b for their respective settings is:

$$\mathcal{P}(A, B \mid a, b, \lambda_1, \lambda_2) = \mathcal{P}(A \mid B, a, b, \lambda_1, \lambda_2)\ \mathcal{P}(B \mid a, b, \lambda_1, \lambda_2) \qquad (B.7)$$

(the probability of obtaining result B if values $a, b, \lambda_1, \lambda_2$ occur is multiplied by the probability of obtaining result A if, in addition, result B occurs). This form is completely general, but not sufficient to derive a Bell inequality (it is valid in quantum mechanics, which violates the BCHSH inequality). Two assumptions are then introduced:

- The result A observed in a region of space cannot depend on the setting arbitrarily chosen in another very remote region of space, outside of its past cone. Jarrett [174] calls this condition "locality", but it often referred to as "parameter independence"; Shimony [176] prefers the words "remote context independence". With this assumption, in the product of probabilities we can then cross out the b dependence of the first[2], as well as the a dependence of the second. We then obtain:

$$\mathcal{P}(A, B \mid a, b, \lambda_1, \lambda_2) = \mathcal{P}(A \mid B, a, \lambda_1, \lambda_2)\ \mathcal{P}(B \mid b, \lambda_1, \lambda_2) \qquad (B.8)$$

This form is still not sufficient to obtain a Bell inequality, because of the B dependence of the first term in the product.
- Each result may depend on all possible causes taking place in the past light cone, but not on the random result obtained in a very remote regions of space. In other words, the results A and B may be correlated in many ways through the dependences on λ_1 and λ_2 of their probabilities (λ_1 and λ_2 may have arbitrary correlations); but then the final random event is a local space-time event, which cannot be influenced by another similar random event taking place very far away.

---

[2] This independence is a relativistic necessity corresponding to the "no signaling condition" (§5.4.1).

Ballentine and Jarrett [175] call this condition "predictive completeness", but the words "outcome independence" are also often used; Shimony speaks[3] of "remote outcome independence". We can now cross out the B dependence in the first term of the product, and write:

$$\mathcal{P}(A, B \mid a, b, \lambda_1, \lambda_2) = \mathcal{P}(A \mid a, \lambda_1, \lambda_2) \, \mathcal{P}(B \mid b, \lambda_1, \lambda_2) \qquad \text{(B.9)}$$

This form is sufficient to obtain the BCHSH inequality, by the same reasoning as in §4.2.2.b: probability $\mathcal{P}(A, B \mid a, b)$ is given by (4.25), averages of products are calculated in the same way, and finally lead to inequality (4.32).

It is interesting to compare the steps of this derivation with the predictions of quantum mechanics. Parameter independence applies to quantum mechanics to some extent, since this theory provides a value of $\mathcal{P}(B \mid a, b, \lambda_1, \lambda_2)$ that is independent[4] of a (otherwise, instantaneous communication would be possible, in contradiction with relativity – see §5.4). Nevertheless, $\mathcal{P}(A \mid B, a, b, \lambda_1, \lambda_2)$ is not independent of B and b in general. For instance, equations (4.3) and (4.4) express that, for two spins in a singlet state, the conditional probability for a second measurement depends on both the direction chosen for the first measurement and its result. For this conditional probability, neither parameter independence nor outcome independence are obeyed; nevertheless, quantum mechanics remains compatible with relativity [175].

The conclusion is that the Bayes law of conditional probabilities, combined with parameter independence and outcome independence, is sufficient to lead to the BCHSH inequality. A violation of this inequality therefore means that at least one of these independences does not apply, as is the case for quantum mechanics.

---

[3] What he calls "locality" is the conjunction of this condition and of "remote context independence".

[4] In quantum mechanics, λ_1 and λ_2 do not occur if the measured system is prepared in a pure state. If it is prepared in a statistical mixture of states, λ_1 and λ_2 can be introduced to describe this mixture.

Appendix C

An Attempt to Construct a "Separable" Quantum Theory (Nondeterministic but Local)

We give an example of such a nondeterministic local theory that looks similar to quantum mechanics, and actually even makes use of its formalism, while it is in fact significantly different. This theory includes the nondeterminism of quantum mechanics, but gives to the state vector a role that is more local than in standard quantum mechanics. For this, we consider a physicist who has well understood the basic rules of quantum mechanics concerning nondeterminism, but who remains skeptical about nonlocality (or nonseparability; for a detailed discussion of the meaning of these terms, see §3.3.3.c as well as, for instance, [24, 55]). This physicist thinks that, if measurements are performed in remote regions of space, it is more natural to apply the rules of quantum mechanics separately in these two regions. In order to calculate the probability of any measurement result, he/she will then apply the rules of quantum mechanics, in a way that is locally perfectly correct, but that also assumes that it is possible to reason separately in the two regions of space. If for instance the two measurements take place in two different galaxies, our skeptical physicist is prepared to apply quantum mechanics at the scale of a galaxy, but not at an intergalactic scale!

How can one then treat the measurement process that takes place in the first galaxy? It is very natural to assume that the spin it contains is described by a state vector (or by a density operator; it makes no difference here) that may be used to apply the orthodox formula for obtaining the probabilities of each possible result. Of course, it would not be a good idea to assume that each spin is described by a single, fixed, density operator: obviously, the model would then exclude correlations between the results of measurements performed in the two galaxies. A better idea is to assume that the density operators in question are random mathematical objects $\rho_1(\lambda)$ and $\rho_2(\lambda)$ that fluctuate because they are functions of a random variable λ, corresponding for instance to fluctuating conditions of emission of the particles. The method is then clear: for any possible condition of the emission, one performs an orthodox quantum calculation in each region of space, and then takes an average value over the conditions in question. After all, this is nothing but the universal method for calculating correlations in all the rest of physics! This approach takes into account the indeterministic character of quantum mechanics[1], but introduces a notion of space separability that is directly in the line of the EPR reasoning. Our physicist may for instance assume that the two measurement events are separated by a space-like interval in the sense of relativity, so that no causal relation can relate them in any circumstance; this seems to fully justify an independent calculation of both phenomena.

If we note $|+(a)\rangle$ the eigenstate of the measurement corresponding to result $+1$ and $\rho_1(\lambda)$ the density operator of s, the probability of obtaining result $+1$ if the first measurement is made along direction a is then:

$$\mathcal{P}_+^A(a, \lambda) = \langle +(a)|\rho_1(\lambda)|+(a)\rangle \tag{C.1}$$

In the same way, we write the probability for the result -1 in the form:

$$\mathcal{P}_-^A(a, \lambda) = \langle -(a)|\rho_1(\lambda)|-(a)\rangle \tag{C.2}$$

If, instead of direction a, another different direction a' is chosen, the calculations remain the same and lead to two functions $\mathcal{P}_\pm^A(a', \lambda)$. As for measurements performed in the second region of space, they provide two functions $\mathcal{P}_+^B(b, \lambda)$ and $\mathcal{P}_-^B(b, \lambda)$.

We now calculate the average of the product of the two results, which is nothing but the average over λ of the expression already written (B.5). If now we define $\widehat{A}(\lambda)$ and $\widehat{B}(\lambda)$ by:

$$\begin{aligned}\widehat{A}(\lambda) &= \mathcal{P}_+^A(a, \lambda) - \mathcal{P}_-^A(a, \lambda)\\ \widehat{B}(\lambda) &= \mathcal{P}_+^B(b, \lambda) - \mathcal{P}_-^B(b, \lambda)\end{aligned} \tag{C.3}$$

we can write the average of the product of results as:

$$\int d\lambda\, n(\lambda)\, \widehat{A}(\lambda)\, \widehat{B}(\lambda) \tag{C.4}$$

where $n(\lambda)$ is the distribution density of variable λ.

The difference with the usual result is that, here, $\widehat{A}(\lambda)$ and $\widehat{B}(\lambda)$ are not defined as functions that are always equal to ± 1. To complete the proof, we have the choice between two different methods:

(i) either we proceed as in Appendix B and introduce an additional variable μ to express the probability differences as in (B.3), in terms of new quantities $\overline{A}(a, \lambda, \mu)$ and $\overline{B}(b, \lambda, \mu')$ that are always equal to ± 1. This leads to the same mathematical expression and proof of the BCHSH inequalities as before.

---

[1] In §8.2.2.b, we remarked that this is an appropriate method, within standard quantum mechanics, to study situations where Alice and Bob observe correlations when Eve interferes and measures polarizations before the particles reach the two partners.

(ii) or we introduce the couples of orientations (a, b), (a, b'), (a', b), (a', b') to calculate the average over λ of the expression:

$$\widehat{A}(\lambda)\widehat{B}(\lambda) - \widehat{A}(\lambda)\widehat{B'}(\lambda) + \widehat{A'}(\lambda)\widehat{B}(\lambda) + \widehat{A'}(\lambda)\widehat{B'}(\lambda) \tag{C.5}$$

The $\widehat{A}$ and $\widehat{B}$, which are now defined as probability differences, are no longer necessarily ± 1; but it is nonetheless easy to see that they have values between $+1$ and -1, whatever the value of λ is. This implies[2] that expression (C.5) is necessarily between ± 2, which brings us back to the calculation of Section 4.1.2.

Once more, we find that the Bell theorem holds in a large variety of situations! One may wonder what exactly went wrong in the approach of our skeptical physicist, and why his results, because they satisfy the Bell inequalities, are necessarily incompatible with standard mechanics (not necessarily always, but at least sometimes). After all, the physicist's reasoning was based on the use of the usual formalism of quantum mechanics. In fact, what caused the error was the insistence of treating the quantum measurements as separable events, while orthodox quantum mechanics requires us to consider the whole two-spin system as a single, nonseparable, system; in this system, no attempt should be made to distinguish subsystems. The correct reasoning uses only state vectors/density operators that describe this whole system with one single mathematical object. This example illustrates how it is really separability and/or locality that are at stake in a violation of the Bell inequalities, not determinism.

---

[2] To see why, let us for a moment consider λ, $\widehat{A}$ and $\widehat{A'}$ as fixed, keeping only $\widehat{B}$ and $\widehat{B'}$ as variables; in the space of these variables, expression (C.5) corresponds to a plane surface which, at the four corners of the square $\widehat{B} = \pm 1$, $\widehat{B'} = \pm 1$, takes values $\pm 2\widehat{A}$ or $\pm 2\widehat{A'}$, which are between ± 2; at the center of the square, the plane goes through the origin. By linear interpolation, it is clear that, within the inside of the square, the function given by (C.5) also remains bounded between ± 2; finally, its average value has the same property.

Appendix D
Maximal Probability for a State

In this appendix, we give more details on the calculations of §6.3; the two-particle state corresponding to the measurement considered in (i) is the tensor product of ket (6.40) by its correspondent for the second spin:

$$\cos^2 \theta \, |+, +\rangle + \sin \theta \cos \theta \, [|+, -\rangle + |-, +\rangle] + \sin^2 \theta \, |-, -\rangle \qquad \text{(D.1)}$$

which has the following scalar product with ket (6.45):

$$\cos^2 \theta \sin \theta - 2 \sin \theta \cos^2 \theta = -\sin \theta \cos^2 \theta \qquad \text{(D.2)}$$

The requested probability is obtained by dividing the square of this expression by the square of the norm of the state vector 6.45:

$$\mathcal{P} = \frac{\sin^2 \theta \cos^4 \theta}{2 \cos^2 \theta + \sin^2 \theta} = \frac{\sin^2 \theta \left(1 - \sin^2 \theta\right)^2}{2 - \sin^2 \theta} \qquad \text{(D.3)}$$

A plot of this function shows that it has a maximum of about 0.09.

Appendix E

The Influence of Pair Selection

In the proof of the Bell theorem, we have assumed that all pairs of particles emitted by the source are actually detected, whatever choice is made for the measurement settings a and b; within local realism, the Bell inequalities are then obeyed, which means that it is impossible to reproduce the a and b dependence (4.10) of the correlation function predicted by quantum mechanics (since it allows violations of the Bell inequalities). In this appendix, we examine what happens when the detection process introduces a selection in the ensemble of emitted pairs; in a first step (§E.1), we assume that this selection is independent of a and b, and in a second step (§E.2), we generalize to include a possible dependence. In the latter case, we will see that it then becomes possible to reproduce any variation of the correlation rate as a function of a and b, including the prediction (4.10) in $\cos(a - b)$ of quantum mechanics, while remaining within local realism; this is the origin of the "loophole" discussed in §4.3.1.a.

E.1 Setting independent selection

Our first model is as follows:

(i) We assume that some process selects particles in a way that is independent of the choice of the experimental settings a and b. We can for instance assume that the particles are emitted randomly by the source in many directions, while only the particles that are emitted into a small solid angle can reach the analyzers and detectors. For each particle, we characterize this emission direction with a random variable ω. The first particle is detected only if its random variable ω_1 falls inside a range $D_\lambda(\Omega_1)$ corresponding to a cone originating at the source and going to the periphery of the input diaphragm of the detector:

$$\omega_1 \in D_1(\Omega_1) \tag{E.1}$$

(Otherwise it is lost). Similarly, the second particle is detected only if its random variable ω_2 falls inside domain $D_\lambda(\Omega_2)$, and is lost otherwise. This sort of angular selection occurs in practice in all experiments where the small size of detectors puts a strong angular limit; the two domains are then small ω domains with sizes $d\Omega_{1,2}$ centered around values $\omega_{\text{det }1,2}$.

(ii) The particles that have survived the preceding selection arrive on the analyzers and detectors, and provide results that depend on a random variable λ_1 for the first, λ_2 for the second. As in §4.1.2, these results depend also of the local choice of the measurement setting; they are noted $A(a, \lambda_1) = \pm 1$ for the first particle, $B(b, \lambda_2)$ for the second (we assume that two channel detectors are used, as in an EPRB experiment). For the particles that have not been selected in process (i), we choose by convention to attribute to them result 0.

The ensemble of variables λ_1, λ_2, ω_1, ω_2, can be grouped formally into a single parameter Λ with several components (a vector in a space with many dimensions). Every pair is characterized by a given value of Λ, which determines if the pair will be detected or not by the detection apparatuses and the results that they will provide. Assume now that the source emits the pairs of particles randomly; the values of Λ are then associated with a probability distribution $\rho(\Lambda)$, with the normalization condition:

$$\int d\Lambda \, \rho(\Lambda) = \int d\lambda_1 \int d\lambda_2 \int d\omega_1 \int d\omega_2 \, \rho(\lambda_1, \lambda_2, \omega_1, \omega_2) = 1 \quad (E.2)$$

The average value of the product of results is then:

$$\langle AB \rangle = \int d\lambda_1 \int d\lambda_2 \int_{D_1(\Omega_1)} d\omega_1 \int_{D_2(\Omega_2)} d\omega_2 \, \rho(\Lambda) \, A(a, \lambda_1) \, B(b, \lambda_2) \quad (E.3)$$

which, if the sizes of the domains $d\Omega_{1,2}$ centered around values $\omega_{\text{det }1,2}$ are small, is also given by:

$$\langle AB \rangle \simeq d\Omega_1 d\Omega_2 \int d\lambda_1 \int d\lambda_2 \, \rho(\lambda_1, \lambda_2, \omega_{\text{det }1}, \omega_{\text{det }2}) \, A(a, \lambda_1) \, B(b, \lambda_2) \quad (E.4)$$

The two formulas are exactly of the type that can be handled by the Bell theorem; they allow to derive inequalities such as, for instance, the BCHSH inequality (4.9). We note in passing that, if the selection is very efficient, most particles give result zero and do not contribute to the average value; as a consequence, the value of expression (E.4) is necessarily very small (it contains the very small product $d\Omega_1 d\Omega_2$); it is then not surprising that the sum of four average values can never reach values ± 2. In quantum mechanics, the probability of detecting pairs is also proportional to $d\Omega_1 d\Omega_2$, and therefore low; there is no hope either to obtain violations of the Bell inequalities beyond ± 2, and the situation is not extremely interesting.

Nevertheless, one can also take a different point of view: instead of normalizing the average values over the ensemble of emitted pairs, we can normalize over the smaller ensemble of detected pairs. The normalization condition (E.2) then becomes:

$$\int d\lambda_1 \int d\lambda_2 \int_{D_1(\Omega_1)} d\omega_1 \int_{D_2(\Omega_2)} d\omega_2 \, \rho(\lambda_1, \lambda_2, \omega_1, \omega_2) = 1 \qquad \text{(E.5)}$$

or:

$$d\Omega_1 d\Omega_2 \int d\lambda_1 \int d\lambda_2 \, \rho(\lambda_1, \lambda_2, \omega_{\det 1}, \omega_{\det 2}) \simeq 1 \qquad \text{(E.6)}$$

This change of normalization introduces a factor $1/d\Omega_1 d\Omega_2$ into ρ, which cancels that of the average value (E.4). As a consequence, the strong reduction of the average values does not occur anymore, and we merely come back to a case that is exactly equivalent to what happens in the absence of selection. The BCHSH sum of four average values can then reach the values ± 2. In quantum mechanics, since the small factor has now disappeared, one recovers the $\cos(a - b)$ dependence and the violation of the Bell inequality.

E.2 Setting dependent selection

We now replace assumption (i) by a more general assumption, which no longer excludes a setting dependence in the selection process:

(iii) An *a*- and *b*- dependent selection process occurs; for instance, after being emitted, when the particles fly towards the detectors, they are part of a physical process (absorption, for instance) that destroys some fraction of them in a way that depends on *a* for the first particle, *b* for the second. The no-absorption condition for the first particle then becomes:

$$\omega_1 \in D_1(a) \qquad \text{(E.7)}$$

and a similar condition applies to the second particle. If *a* and *b* are angles, we can write the no-absorption condition in the form:

$$a - \Delta a < \omega_1 < a + \Delta a \quad \text{and} \quad b - \Delta b < \omega_2 < a + \Delta b \qquad \text{(E.8)}$$

where Δa and Δb are some fixed small angles. We can then replace the distribution $\rho(\lambda_1, \lambda_2, \omega_1, \omega_2)$ by $\rho(\lambda_1, \lambda_2, a, b)$, and relation (E.4) by:

$$\langle AB \rangle \simeq \Delta a \, \Delta b \int d\lambda_1 \int d\lambda_2 \, \rho(\lambda_1, \lambda_2, a, b) \, A(a, \lambda_1) \, B(b, \lambda_2) \qquad \text{(E.9)}$$

At this stage, it becomes clear that the model gives much more flexibility than that of §E.1 to reproduce arbitrary *a* and *b* dependences; this is because we are indeed free to choose the positive distribution $\rho(\Lambda)$ as we wish, in particular its ω_1

and ω_2 dependences, which, in turn, introduces an a and b dependence of ρ into (E.9).

To see mathematically why, we can simplify the model even more, while preserving the possibility of reproducing an arbitrary dependence on the settings. Assume for instance that the domains $D_1^{\pm}$ of variable λ_1 where $A(a, \lambda_1) = \pm 1$ are independent of a, and, similarly, for the other particle. Then:

$$\langle AB \rangle \simeq \Delta a \, \Delta b \, [I_{+,+} + I_{-,-} - I_{+,-} - I_{-,+}] \qquad (E.10)$$

with:

$$I_{\pm,\pm} = \int_{D_1^{\pm}} d\lambda_1 \int_{D_2^{\pm}} d\lambda_2 \, \rho(\lambda_1, \lambda_2, a, b) \qquad (E.11)$$

The four integrals $I_{\pm,\pm}$ correspond to the probabilities of the four possible measurement results $(\pm, \pm)$, which in quantum mechanics have the expressions given in (4.3) and (4.4). By arbitrarily choosing the ω_1 and ω_2 dependence of ρ, and therefore the a and b dependence of $\rho(\lambda_1, \lambda_2, a, b)$, we can obtain any dependence of the four probabilities on the experimental settings, while remaining within local realism. For instance, we can choose a distribution ρ that is constant (independent of λ_1 and λ_2) within the four domains $D_{1,2}^{\pm}$, and select for ρ a value that is proportional to the target probability, to obtain the desired dependence on a and b. In this way, we reproduce the predictions of any arbitrary theory[1], provided of course it gives positive probabilities for all values of a and b.

Physically, what we have done is merely to assume that, for each value of the settings a and b, a narrow selection occurs for the detected particles; each time the settings are changed, a different class of pairs is detected; it is then sufficient to attribute arbitrary properties to these classes in order to reproduce all possible variations as functions of a and b, including those of quantum mechanics. Nevertheless, we should remark that what is reproduced is only these a and b variations, and not the absolute value of coincidence rates. There is a compromise between accuracy and detection efficiency: the smaller the values of Δa and Δb, the better (E.9) can approximate any variation, but the price to pay is that more and more pairs go undetected.

The conclusion is that the validity of the Bell inequalities is strongly related to assuming that all pairs of a well-defined sample that is independent of a and b are detected. If, each time one changes the measurement settings one changes the category of detected pairs, no Bell-type limit may exist for the dependence of the correlation rates with respect to the settings a and b.

---

[1] A few examples of setting dependent pair selection processes that lead to artificial violations of the BCHSH inequalities are given in [209]; in its §3-3 it shows that the selection can lead to a violation of the Cirelson bound and, in §3-4, to the maximal mathematically possible violation, even with a pair of uncorrelated spins.

Appendix F

Impossibility of Superluminal Communication

F.1 Introduction

In EPR schemes, applying the reduction postulate projects the second particle instantaneously onto an eigenstate corresponding to the same quantization axis as the first measurement. If it were possible to determine this state completely, superluminal communication would become accessible: from this state, the second experimenter could calculate the direction of the quantization axis to which it corresponds, and rapidly know what direction was chosen by the first experimenter[1], even if the experimenters are in two different and remote galaxies. This, obviously, could be used as a sort of telegraph, completely free of any relativistic minimum delay (proportional to the distance covered) for the transmission of information. Nevertheless, we have seen in §8.2.1 that it is impossible to obtain a complete determination of a quantum state from a single realization of this state. Such a realization allows only one single measurement, which (almost always) perturbs the state, so that a second measurement on the same state is not feasible; there is not, and by far, sufficient information in the first measurement for a full determination of the quantum state – see discussion in §8.2. This telegraph would therefore not function.

If a single particle is not sufficient for Bob to get a message, could he use more particles? Suppose for a moment that a perfect "cloning" of quantum states could be performed – more precisely, the reproduction with many particles of the unknown state of a single particle[2]. Applying the cloning process to the second particle of an EPR pair, one could then make a large number of perfect copies of its

---

[1] What is envisaged here is communication through the choice of the settings of the measurement apparatuses; this makes sense since the settings are chosen at will by the experimenters. On the other hand, the results of the experiments are not controlled, but random, so that they cannot be directly used as signals.

[2] The "cloning" operation is not to be confused with the preparation of a series of particles into a same quantum state that is chosen by the experimentalist. The latter operation can be realized by sending many spin 1/2 particles through the same Stern–Gerlach magnet, or many photons through the same polarizing filter. What is theoretically impossible is to perfectly duplicate an initially unknown (and arbitrary) state.

state; one could then perform a series of measurements on each of these copies, and progressively determine the state in question with arbitrary accuracy. In this way, the possibility for superluminal communication would be restored! But we have also seen in §8.1 that quantum mechanics does not allow for such a perfect reproduction of quantum states [376, 377]; for instance, if one considers using stimulated emission in order to clone the state of polarization of one single photon into many copies, the presence of spontaneous emission introduces noise in the process and prevents perfect copying.

Even if it is impossible to clone quantum states and to measure single states accurately, this does not provide an obvious answer to the general question: by using only the information that is available in one single measurement in each region of space, is it possible to make use of the instantaneous reduction of the state vector for superluminal communication? After all, it is possible to repeat the experiment many times with many independent pairs of correlated particles, and to try to extract some information from the statistical properties of the results of all measurements. The EPR correlations are very special and exhibit such completely unexpected properties (e.g., violations of the Bell inequalities)! Why not imagine that, by using or generalizing EPR schemes (more than two systems, delocalized systems, etc.), one could invent schemes where superluminal communication becomes possible? Here we show why such schemes do not exist; we will give the general impossibility proof in the case of two particles (or two regions of space), but the generalization to more systems in several different regions of space is straightforward.

F.2 A first scheme

Suppose that, initially, the two remote observers already possess a collection of pairs of correlated particles, which have propagated from their common source to two remote regions of space before the experiment starts; the first has propagated to region A where Alice does experiments, the second to region B containing Bob's laboratory. Each pair is in an arbitrary state of quantum entanglement; we describe its state before measurement with a density operator $\rho(t_0)$ in a completely general way. When the two particles are very distant, they do not interact, and their Hamiltonian is merely the sum $H_1 + H_2$. Alice then chooses a setting a or, more generally, any local observable $O_A(1)$ to measure. Bob is equally free to choose any local observable $O_B(2)$, and may use as many particles as necessary to measure the frequency of occurrence of each result (i.e., probabilities). The question is whether the second observer can extract some information on the choice of O_A from any statistical property of the observed results.

Equation (11.9) provides the probability of an event where Alice observes result m at time t_1 and Bob result n at time t_2:

$$\mathcal{P}(m, t_1; n, t_2) = Tr\left\{\widehat{P}_{O_B}(n, t_2)\,\widehat{P}_{O_A}(m, t_1)\,\rho(t_0)\,\widehat{P}_{O_A}(m, t_1)\,\widehat{P}_{O_B}(n, t_2)\right\} \tag{F.1}$$

where $\widehat{P}_{O_A}(m, t_1)$ is the projector onto the eigenstates corresponding to Alice's measurement in the Heisenberg picture, and $\widehat{P}_{O_B}(n_2, t_2)$ the corresponding projector for Bob. But Bob does not have access to the results obtained by Alice, so that the probability of the events that he observes is the sum of (F.1) over m (§5.4.1):

$$\mathcal{P}_B(n, t_2) = \sum_m \mathcal{P}(m, t_1; n, t_2) \tag{F.2}$$

To calculate this sum, we first remark that the two projectors $\widehat{P}_{O_A}(m, t_1)$ and $\widehat{P}_{O_B}(n, t_2)$ commute: they correspond to operators acting on different particles, and evolve in the Heisenberg picture under the effect of independent Hamiltonians H_1 and H_2. On the right-hand side of (F.1), consider the operator $\widehat{P}_{O_A}(m, t_1)$ appearing just before last in the trace; we can bring it to the last position and then, by circular permutation of the operators inside the trace, put it first. It then sits just before the first $\widehat{P}_{O_B}(n, t_2)$, and can be put just after since it commutes with that operator; but, since the square of projector $\widehat{P}_{O_A}(m, t_1)$ is equal to the projector itself, eventually the result of our operation is just to suppress one of the operators $\widehat{P}_{O_A}(m, t_1)$ in (F.1). We now have to perform the summation over m. Since:

$$\sum_m \widehat{P}_{O_A}(m, t_1) = 1 \tag{F.3}$$

we obtain at the end:

$$\mathcal{P}_B(n, t_2) = Tr\left\{\widehat{P}_{O_B}(n, t_2)\,\rho(t_0)\,\widehat{P}_{O_B}(n, t_2)\right\} \tag{F.4}$$

We then see that, in this probability, any dependence on the choice of operator O_A made by Alice has disappeared. By measuring probabilities locally, Bob obtains no information on the choice of operator made by Alice.

The proof can easily be generalized to a situation where Alice and Bob perform several measurements at different times[3]. We have therefore shown that, in a very general way, the second observer receives exactly the same information, in a way

---

[3] The proof proceeds by the same method just described. One takes the last projector corresponding to Alice's measurements, pushes it to the end of the product in the trace, then at the beginning, and finally groups it with the other occurrence of the same operator. The summation of this residual projector over the last result obtained by Alice then gives 1. Then one proceeds in the same way for the second projector associated to Alice's measurements, etc. When all the m summations are performed, one obtains a probability that is completely independent of the choice of Alice's operators.

Of course, the proof is meaningful only if the time intervals between Alice's measurements remain shorter than the time of propagation to Bob's site at the speed of light. Otherwise the particles would have the time to propagate from one region to the other, and the comparison with relativity would be irrelevant.

that is completely independent of the decisions made by the first observer; even the fact that Alice makes experiments or no measurement at all is impossible to detect in Bob's laboratory. No faster-than-light-communication is therefore possible by this method.

F.3 Generalization

One could object that it is not necessary that particle 1 should be located in region *A* of space and particle 2 in region *B*: if each of the particles is at the same time in both regions, can we not imagine cases where the probabilities observed by Bob depend on the choices made by Alice? In fact, this situation is not so different from the previous case since, again, all operators corresponding to the measurements performed in region *A* commute with all those associated with measurements performed in region *B*. In field theory, this is a consequence of the commutation of field operators at points of space-time that correspond to space like intervals in relativity. In a more elementary theory, if Alice can detect both particles, we can write the corresponding measurement operator as a sum of operators acting on the two particles:

$$O_{Alice} = P_A(1)O_A(1)P_A(1) + P_A(2)O_A(2)P_A(2) \qquad (F.5)$$

where $P_A(1,2)$ is the projector over all states of particle 1, 2 localized in region *A*; the first term corresponds to the case where Alice detects particle 1, the second to the case where she detects particle 2. Similarly, the measurement performed by Bob corresponds to the operator:

$$O_{Bob} = P_B(1)O_B(1)P_B(1) + P_B(2)O_B(2)P_B(2) \qquad (F.6)$$

It is then easy to check that O_{Alice} and O_{Bob} commute. The reason is that, in the products between these two operators, the terms in $P_A P_B$ vanish when the two projectors act on the same particle, and the two remaining terms have the form:

$$P_A(1)O_A(1)P_A(1) \times P_B(2)O_B(2)P_B(2) \qquad (F.7)$$

added to the similar term where the particle numbers are exchanged; this is because, if each of the two operators performs a measurement on a particle, either particle 1 is in region *A* and particle 2 in region *B*, or the reverse, but the two particles are never in the same region. One then immediately sees that it is possible to reverse the order of the two factors in (F.7) without changing the result, so that the two operators O_{Alice} and O_{Bob} commute. This result allows us to use the same proof as before and to obtain the same conclusion: the information available in one region of space is completely independent of the sort of measurements that are performed in the other region. No message can then be sent, and quantum mechanics is not contradictory with relativity!

Appendix G

Quantum Measurements at Different Times

In this appendix, we discuss the probabilities associated with several measurements of the same quantum system at different times. We first give a proof of relation (11.7), which we have used in §11.1.2 without justification, in order to derive the Wigner rule (11.9) from the postulate of state vector reduction (§1.2.2). Conversely, this postulate can also be considered as a consequence of a generalized Born rule, dealing with several measurements performed at different times. In a second part of the appendix, we discuss this derivation, with a reasoning that involves the coupling of the system with the environment provided by the measurement apparatuses.

G.1 Wigner formula

To understand how (11.7) can be obtained, let us first calculate the probability that the first measurement provides result m at time t_1. The usual Born rule (1.6) indicates that this probability is given by the square norm of $|\Psi_m(t_1)\rangle$ defined in (11.3):

$$\mathcal{P}_1(m, t_1) = \langle \Psi_m(t_1) | \Psi_m(t_1) \rangle \tag{G.1}$$

Equivalently, we can express this probability as a trace:

$$\mathcal{P}_1(m, t_1) = Tr\{|\Psi_m(t_1)\rangle \langle \Psi_m(t_1)|\} \tag{G.2}$$

(The definition of the trace in any orthonormal basis immediately shows that these two expressions are equal). We now calculate the conditional probability $Q_{N/M}$ that, if result m has been obtained, the second measurement will provide result n. After the first measurement, the state vector reduction postulate – relation (1.8) – indicates that the normalized state vector is:

$$|\Psi'_m(t_1)\rangle = \frac{1}{\sqrt{\langle \Psi_m(t_1) | \Psi_m(t_1) \rangle}} |\Psi_m(t_1)\rangle \tag{G.3}$$

which, after evolution between time t_1 and time t_2, is nothing but the ket $|\Psi_m(t_2)\rangle$ given by (11.4) and (11.5), divided by $\sqrt{\langle\Psi_m(t_1)|\Psi_m(t_1)\rangle}$. The same reasoning as in the preceding then gives the conditional probability as:

$$Q_{N/M}(m, t_1; n, t_2) = \frac{1}{\langle\Psi_m(t_1)|\Psi_m(t_1)\rangle} \langle\Psi_{m,n}(t_2)|\Psi_{m,n}(t_2)\rangle \qquad \text{(G.4)}$$

where $|\Psi_{m,n}(t_2)\rangle$ has been defined in (11.6). From this, we obtain the probability of obtaining the sequence m and n:

$$\mathcal{P}_1(m, t_1; n, t_2) = \mathcal{P}_1(m, t_1) \times Q_{N/M}(m, t_1; n, t_2) = \langle\Psi_{m,n}(t_2)|\Psi_{m,n}(t_2)\rangle \qquad \text{(G.5)}$$

(Two factors $\langle\Psi_m(t_1)|\Psi_m(t_1)\rangle$ cancel each other in the numerator and the denominator). Equivalently, one can express the right-hand side of (G.5) as a trace:

$$\mathcal{P}_1(m, t_1; n, t_2) = Tr\left\{|\Psi_{m,n}(t_2)\rangle\langle\Psi_{m,n}(t_2)|\right\} \qquad \text{(G.6)}$$

By recurrence, (G.5) leads to (11.7).

To prove the Wigner formula (11.9), we introduce the projection operator in the Heisenberg picture:

$$\widehat{P}_M(m, t_1) = U^\dagger(t_1, t_0)P_M(m)U(t_1, t_0) \qquad \text{(G.7)}$$

where $U(t_1, t_0)$ is the unitary evolution operator between times t_0 and t_1 (§12.1.4). From (11.3) and (11.1), we obtain, since $U(t_1, t_0)U^\dagger(t_1, t_0) = 1$:

$$|\Psi_m(t_1)\rangle = P_M(m)U(t_1, t_0)|\Psi(t_0)\rangle = U(t_1, t_0)\widehat{P}_M(m, t_1)|\Psi(t_0)\rangle$$

and relation (G.2) gives, since $P_M(m)$ is Hermitian:

$$\begin{aligned}\mathcal{P}_1(m, t_1) &= Tr\left\{U(t_1, t_0)\widehat{P}_M(m, t_1)|\Psi(t_0)\rangle\langle\Psi(t_0)|\widehat{P}_M(m, t_1)U^\dagger(t_1, t_0)\right\}\\ &= Tr\left\{\widehat{P}_M(m, t_1)|\Psi(t_0)\rangle\langle\Psi(t_0)|\widehat{P}_M(m, t_1)\right\} \qquad \text{(G.8)}\\ &= Tr\left\{\widehat{P}_M(m, t_1)\rho(t_0)\widehat{P}_M(m, t_1)\right\}\end{aligned}$$

(We have used the circular permutation of operators under the trace, which makes the product of two operators disappear).

A similar calculation can also be made from (G.5). The ket $|\Psi_{m,n}(t_2)\rangle$ is given by:

$$\begin{aligned}|\Psi_{m,n}(t_2)\rangle &= P_N(n)\, U(t_2, t_1)\, |\Psi_m(t_1)\rangle\\ &= P_N(n)\, U(t_2, t_1)U(t_1, t_0)U^\dagger(t_1, t_0)\, P_M(m)\, U(t_1, t_0)|\Psi(t_0)\rangle\end{aligned} \qquad \text{(G.9)}$$

where we have inserted the relation $U(t_1, t_0)U^\dagger(t_1, t_0) = 1$ in order to make the Heisenberg projector $\widehat{P}_M(m, t_1)$ appear again; we then simplify the product of operators $U(t_2, t_1)U(t_1, t_0)$ into $U(t_2, t_0)$ and use the unitarity of U to write:

$$|\Psi_{m,n}(t_2)\rangle = U(t_2, t_0)\widehat{P}_N(n, t_2)\widehat{P}_M(m, t_1)|\Psi(t_0)\rangle \qquad \text{(G.10)}$$

Finally, using (G.6) and after circular permutation of operators under the trace (which makes two U operators disappear), we obtain:

$$\mathcal{P}_1(m, t_1; n, t_2) = Tr\left\{\widehat{P_N}(n, t_2)\widehat{P_M}(m, t_1)\,|\Psi(t_0)\rangle\,\langle\Psi(t_0)|\,\widehat{P_M}(m, t_1)\widehat{P_N}(n, t_2)\right\}$$
(G.11)

which leads to (11.9). By the same method, it is straightforward to generalize this formula to more than two measurements, with additional projectors in the reverse order on both sides of $\rho(t_0)$. By linearity, this result also applies to situations where the density operator $\rho(t_0)$ is not a projector (a pure state) as in (11.8), but a statistical mixture.

G.2 Generalized Born rule

We now take a different approach by taking into account the entanglement of the measured system S with the measurement apparatuses[1]. A measurement associated with operator M is performed at time t_1, another measurement associated with operator N at time t_2, etc. (in the calculation, we assume that two measurements are performed, but the generalization to an arbitrary number of measurements is straightforward).

Initially, at time t_0, the system S is in state $|\Psi^S(t_0)\rangle$; the two measurement apparatuses M and N, which have not yet operated, are in states $|\Psi^M(t_0)\rangle$ and $|\Psi^N(t_0)\rangle$, and the state vector $|\overline{\Psi}(t_0)\rangle$ of the whole system including these apparatuses is the (tensor) product:

$$|\overline{\Psi}(t_0)\rangle = |\Psi^S(t_0)\rangle \otimes |\Psi^M(t_0)\rangle \otimes |\Psi^N(t_0)\rangle$$
(G.12)

(All these states are normalized). We assume that the three systems evolve independently (without any interaction), except between times t_1 and t_1', where S interacts with the first measurement apparatus M, and between times t_2 and t_2', where it interacts with the second measurement apparatus N. The two measurement apparatuses are macroscopic; each includes a pointer that indicates, after the measurement, the result that has been obtained. They never interact with each other, but only with system S.

Between time t_0 and time t_1, the state of S evolves from $|\Psi^S(t_0)\rangle$ to $|\Psi^S(t_1)\rangle$ according to the Schrödinger equation (11.1) and, similarly, the state vectors of the measurement apparatuses become $|\Psi^M(t_1)\rangle$ and $|\Psi^N(t_1)\rangle$.

We now consider the effect of the first measurement. As in §§1.2.2.a and 11.1.2, we call $P_M(m)$ the projectors over the possible eigenvectors of operator M, with eigenvalues $m = m_1, m_2, ..., m_i, ...$ and expand $|\Psi^S(t_1)\rangle$ over these eigenvectors;

---

[1] This approach is necessary if one chooses the Everett interpretation (§11.12)

with our notation, relations (11.2) and (11.3) become:

$$\left|\Psi^S(t_1)\right\rangle = \sum_m \left|\Psi_m^S(t_1)\right\rangle \tag{G.13}$$

and:

$$\left|\Psi_m^S(t_1)\right\rangle = P_M(m)\left|\Psi^S(t_1)\right\rangle \tag{G.14}$$

If it turns out that $\left|\Psi^S(t_1)\right\rangle$ is an eigenvector of M with eigenvalue m_i, then only one term $m = m_i$ occurs in the summation; the result of the first measurement is certain. At a time t_1' just after this measurement, the first apparatus will reach a well-defined normalized state $\left|\Psi_{m_i}^M(t_1')\right\rangle$ where its pointer indicates the result. Therefore, in this particular case, just after the first measurement the state of the whole system will be:

$$\left|\overline{\Psi}(t_1')\right\rangle = \left|\Psi_{m_i}^S(t_1')\right\rangle \otimes \left|\Psi_{m_i}^M(t_1')\right\rangle \otimes \left|\Psi^N(t_1')\right\rangle \tag{G.15}$$

In general, for any $\left|\Psi^S(t_1)\right\rangle$, the linearity of the Schrödinger equation implies that the state of the whole system just after the first measurement is:

$$\left|\overline{\Psi}(t_1')\right\rangle = \sum_m \left|\Psi_m^S(t_1')\right\rangle \otimes \left|\Psi_m^M(t_1')\right\rangle \otimes \left|\Psi^N(t_1')\right\rangle \tag{G.16}$$

The first measurement apparatus has then reached a state that depends on the eigenvalue m, but not the second, which has not yet interacted with S. All the evolutions of the individual state vectors correspond to unitary time evolutions, which conserve the norm; nevertheless, expansion (G.13) contains kets $\left|\Psi_m^S(t_1)\right\rangle$ having a norm that is in general smaller than that of $\left|\Psi^S(t_1)\right\rangle$.

For the second measurement, we repeat the same calculation. Between time t_1' and t_2, for each value of m in (G.16) every term in the product evolves independently (unitary evolution), and $\left|\overline{\Psi}(t_2)\right\rangle$ is obtained by replacing t_1' by t_2 in (G.16). The analogous expressions of (G.13) and (G.14) are now:

$$\left|\Psi_m^S(t_2)\right\rangle = \sum_n \left|\Psi_{m,n}^S(t_2)\right\rangle \tag{G.17}$$

with:

$$\left|\Psi_{m,n}^S(t_2)\right\rangle = P_N(n)\left|\Psi_m^S(t_2)\right\rangle \tag{G.18}$$

The second measurement correlates the state of the second apparatus to that of system S. Finally, the state of the whole system at time t_2' after the second measurements is:

$$\left|\overline{\Psi}(t_2')\right\rangle = \sum_m \sum_n \left|\Psi_{m,n}^S(t_2')\right\rangle \otimes \left|\Psi_m^M(t_2')\right\rangle \otimes \left|\Psi_n^N(t_2')\right\rangle \tag{G.19}$$

This state vector contains a coherent superposition of various components associated with all possible pairs (m, n) of results of measurements; in each of these components, the state of the measurement apparatuses have registered the results.

At this stage, we can introduce a "generalized Born rule" and postulate that the probability of obtaining results (m, n) is given by the square of the norm of each component. We have already remarked that, as long as the three subsystems do not interact, the norm of each factor in one component remains constant. Actually, the changes of norm occur only during the interactions (the measurements) and are contained in expansions (G.13) and (G.17). Since the states of the measurement apparatuses in (G.19) are normalized, the probability $\mathcal{P}_{m,n}$ of any result (m, n) is then:

$$\mathcal{P}_{m,n} = \left\langle \Psi_{m,n}^{S}(t_2') \,\middle|\, \Psi_{m,n}^{S}(t_2') \right\rangle \tag{G.20}$$

which is nothing but the norm of the ket:

$$
\begin{aligned}
P_N(n)U(t_2, t_1')\,\middle|\Psi_m^S(t_1')\rangle &= \\
&= P_N(n)U(t_2, t_1')P_M(m)U(t_1, t_0)\,\middle|\Psi^S(t_0)\rangle
\end{aligned}
\tag{G.21}
$$

The state vectors associated to the measurement apparatuses have completely disappeared from this result. It has a simple interpretation in terms of system S only, if one reads the second line from the left to the right: system S was initially in state $\left|\Psi^S(t_0)\right\rangle$, evolved freely from time t_0 to time t_1, then underwent a projection over the eigenstates corresponding to the result of measurement m, then evolved freely again from time t_1' to time t_2, to be projected a second time over the eigenstates corresponding to the result of measurement n. Moreover, (G.20) is equivalent to relation (G.5), which was obtained by applying the postulate of state vector reduction. From the probabilities of combined (m, n) events, one can derive the conditional probabilities of other events, using the usual rules of statistics. The final conclusion is that both methods provide exactly the same results for all probabilities: one may either involve the measurement apparatuses and use a generalized Born rule, or consider the measured system S only and apply the state vector reduction postulate (§1.2.2.a). It is a matter of taste to choose one method as a postulate and derive the other as a consequence, or the reverse.

Appendix H

Manipulating and Preparing Additional Variables

Introducing the hydrodynamic equations (Madelung [593]) associated with the evolution of the wave function, and using them to guide the evolution of the additional variables (positions of particles), is a natural idea. In the dynamics of fluids, hydrodynamic equations can be obtained by taking averages of microscopic quantities over positions and velocities of point-like particles; for instance, the Navier–Stokes macroscopic equations can be derived from the Boltzmann transport equation by appropriate microscopic averages (the Chapman–Enskog method); conversely, the hydrodynamic variables will influence the motion of individual particles. Moreover, there is some analogy between the guiding term and the force term in a Landau-type kinetic equation, where each particle is subject to an average force proportional to the gradient of the density of the others. Nevertheless, here we are dealing with a single particle, so that the guiding term cannot be associated with interactions between particles. Moreover, we also know from the beginning that rather unusual properties must be contained in the guiding equations, at least if we wish to exactly reproduce the predictions of usual quantum mechanics: the Bell theorem states that the additional variables have to evolve nonlocally in ordinary three-dimensional space (they evolve locally only in the configuration space of the system, exactly as for the state vector). In other words, in real space the additional variables must be able to influence each other at an arbitrary distance. Indeed, in the Bohmian equation of motion of the additional variables, the velocity of a particle contains an explicit dependence on its own position, as expected, but also a dependence on the positions of all the other particles (assuming that the particles are entangled). This is not a problem in itself: as mentioned in the main text, one can consider that making nonlocality completely explicit in the equations is actually an advantage of Bohmian mechanics.

But one also has to be careful when this nonlocal term is included in the equations of motion: since relativity is based on the idea that it is totally impossible to send a message at a velocity exceeding the velocity of light, one must avoid features in the theory that would create conflicts with this principle. We must distinguish two cases, depending whether we consider influences on the additional variables that are direct (one modifies them "by hand", in a completely arbitrary way, as for instance the position of a billiard ball), or indirect (applying external fields changes the Hamiltonian of the system, and therefore modifies the evolution of the wave function so that, in turn, the evolution of the additional variables is affected). In the latter case, one can check that the nonlocal Bohmian term creates no problem: it cannot be used to transmit instantaneous information through the additional variables. This is a general result, which holds simply because the statistical predictions of Bohmian theory are equivalent to usual quantum mechanics, which itself does not allow superluminal communication (§5.4 and Appendix F). But assume for instance that we could manipulate directly the additional variable attached to a particle belonging to an EPR correlated pair, in a completely arbitrary way (even at a microscopic scale) and without changing the wave function. Then, the "quantum velocity term" acting on the additional variables of the other particle would instantaneously be affected, and so would be its subsequent position in space. Since that particle may be in principle at an arbitrary distance, one could use this property to send messages at a velocity exceeding the velocity of light. The conclusion is that such manipulations should be considered as impossible: the only possible source of evolution of the additional variables has to be the wave function-dependent term, without any possibility for direct human action.

If the additional variables cannot be directly manipulated at a microscopic scale, can we then assume that it is possible somehow to filter them in a range of values, as one does for the state vector when the Oz component is filtered in a Stern–Gerlach apparatus? Suppose for instance that we could, for a particle in an eigenstate of the Oz component of its spin, select the values of the additional variable that will correspond to a result $+1$ in a future measurement of the Ox component; were such a selection possible with the help of any physical device, the theory with additional variables would obviously no longer be completely equivalent to standard quantum mechanics, introducing determinism where this theory forbids it[1]. Moreover, Valentini has shown [595] that, if the initial distribution of the Bohmian positions differs from the usual "quantum equilibrium" distribution (§11.8.1.b), faster-than-light communication becomes possible. If one could somehow prepare by hand a distribution of the Bohmian positions that differs from $|\Psi(\mathbf{Q}_1, \mathbf{Q}_2, ...)|^2$, a narrower distribution, for instance, then contradictions with relativity would occur. This is the reason why one generally considers that such preparations are impossible[2].

---

[1] Within orthodox theory, if a spin $1/2$ particle is initially selected into the $+1$ spin state by an Oz oriented Stern–Gerlach apparatus, it becomes completely impossible to make any prediction on the deviation observed later in an Ox oriented Stern–Gerlach apparatus.

[2] We should mention that, historically, Bohm and Bud have developed a theory where this possibility is not excluded [7] – see §11.10.1.a. Conceptual revolutions are of course always possible, but for the moment it

To summarize, if one provides the additional variable theories with features that make them equivalent to orthodox theory, it is necessary to assume that the additional variables can neither be manipulated directly nor filtered, as opposed to the state vector. The additional variables describe an objective reality, but at a different level from the reality of the field of the wave function, since only the latter can be influenced directly by human decisions. Additional variables are indeed easily visible (the results of the experiments) but not controllable, while wave functions have the complementary properties. At the end, we have two levels of reality, one for experimentally controllable classical fields (the wave functions), and one for positions that can be observed but not controlled.

Moreover, when empty waves appear during measurement processes, classical wave functions in turn split into two sublevels of reality: part of them continues to play an effective role as a pilot of the position of particles, while another part is made of empty waves that play no role at all in determining the result of future measurements. This second part becomes, so to say, virtual. If for instance one considers the wave function of the universe, as in the Everett theory it is split into a fantastic number of orthogonal components (although, here, these components have a different status, with no special relation to memory registers of observers). Among all these components, only one plays a role to pilot the position of the point representing the universe, in a configuration space with enormous dimension. All the others are too far from the real position in this space, and will remain in limbo forever.

In conclusion, the description of physical reality within the de Broglie–Bohm theory is certainly not as direct as in classical theory of particles and fields. It is not entirely free of conceptual difficulties, which have some similarity with those encountered in the standard interpretation of quantum mechanics.

---

may seem safer to provide the additional variable theories with features that make them equivalent to orthodox theory.

Appendix I

Correlations and Trajectories in Bohmian Theory

In this appendix, we first study two-time correlation functions and two-particle correlation functions within dBB theory (§I.1). We then proceed to show that the Bohmian trajectories are consistent with the results of successive position measurements performed at various places (§I.2), including the case when the macroscopic pointer is slow (I.3).

I.1 Two-particle correlations

I.1.1 Harmonic oscillators

In standard quantum mechanics, the calculation of any two-time correlation function has to include the evolution of the system between the two times considered; this evolution is described by the unitary evolution operator $U(t', t)$, as for instance in relation (11.9), and takes place from a quantum state that depends on the result of the first measurement (projection postulate). In Bohmian theory, it is equally important to take into account the effect of the first measurement, which entangles the system under study with a measurement apparatus; the Bohmian variables of the apparatus then play an important role, determining which wave is "full" or "empty" (§11.8.1.d). Otherwise, contradictions with the standard predictions are obtained.

For instance, the author of [648] considers a one-dimensional harmonic oscillator that is initially in a stationary state, and studies the correlation function of the position at times t and time t', in the particular case where $t' - t$ is equal to half the period $2\pi/\omega$ of the oscillator. In standard quantum mechanics, it is easy to show that the corresponding position operators $X(t)$ and $X(t')$ are then opposite, so that the correlation function $\langle X(t)X(t')\rangle$ is equal to $-\langle [X(t)]^2\rangle$, and therefore negative. In Bohmian mechanics, the particle is initially static since the wave function is real. If one ignores the effect of the first measurement, the position of the particle will remain at the same place, which corresponds to a correlation function equal to

$\left\langle [X(t)]^2 \right\rangle$, and therefore positive; one reaches an apparent complete contradiction. But, if one takes into account the effect of the first measurement on the particle, one finds that, just after this measurement, each position of the oscillator becomes correlated with a different Bohmian position of the pointer. Since the wave function is no longer a product, the motions of both systems become correlated: for each position of the pointer, the measured particle takes a different velocity. In practice, just after the measurement, the position distribution of the oscillator becomes a narrow function that starts to oscillate in the potential well, giving an oscillating character to two-time position distribution. The sign difference with the standard correlation function then disappears and a perfect agreement is obtained.

A similar case is studied in [647], with two independent one-dimensional harmonic oscillators, initially in the state:

$$|\Psi\rangle = \frac{1}{\sqrt{2}} \left[|1, 0\rangle + |0, 1\rangle \right] \tag{I.1}$$

where $|n, p\rangle$ is the state where the first oscillator has quantum number n and the second quantum number p; for the sake of simplicity, we assume that the frequencies of the two oscillators are the same, equal to $\omega/2\pi$. Since the stationary wave functions of the harmonic oscillator can be chosen as real functions, the associated wave functions are also real, meaning that none of the Bohmian positions of the two particles moves at all.

The position operators of the two oscillators commute and can be measured, either simultaneously, or with any delay between the measurements (one can assume for instance that the two oscillators are centered at different points of space, so that differentiating between the two positions is easy). Since the two operators commute, the standard quantum calculation of the correlation coefficient of the two positions is simple. After the measurement of the position of the first oscillator at position x_1^0, the state of the two particles is proportional to the product of a state of the first oscillator localized around x_1^0 by a state of the second oscillator in a coherent superposition:

$$\varphi_1(x_1^0)\, \varphi_0(x_2) + \varphi_0(x_1^0)\, \varphi_1(x_2) \tag{I.2}$$

At later times, the evolution of this superposition introduces a factor $e^{-i\omega t}$ in the second term; the wave function is no longer real. Since the two oscillators are independent, and since the wave function is now a product, the calculation of the probability of finding a value x_2^0 can be obtained from this single particle time-dependent wave function; one easily finds that it contains in general a component oscillating at frequency $\omega/2\pi$ (except if x_1^0 is by chance a node of function φ_1). On the other hand, we have seen that the Bohmian positions are static, so that the average of their product over all possible trajectories gives a constant result. It therefore seems again that one reaches a contradiction between the predictions of standard mechanics and those of Bohmian mechanics.

But actually this is not the way correlation functions should be calculated within Bohmian mechanics: again, the effect of the first measurement must be properly taken into account, even if the two measured observables commute, and even if they correspond to independent (but entangled) systems. The first measurement correlates the Bohmian position of the particle with that of a pointer, resulting in "empty waves" and modified dynamics of the system, which then affects the correlation function. When particle 1 is measured, the system becomes a three (or more) body problem including the positions of the two particles and that of the pointer. Each Bohmian trajectory selects a pointer position (for instance that corresponding to a positive measurement in the detection volume), and for each such trajectory the two particles move guided by a wave function that is now a product. The entanglement has disappeared, so that one does not have to worry about possible nonlocal effects; both particles then follow a local Schrödinger evolution. After the measurement, both particles oscillate in their potential wells, in a way that exactly reproduces the standard quantum result for the correlation function.

I.1.2 EPRB experiment

Consider an EPRB experiment (§3.3.1) with two remote particles, the first with wave function $\varphi(\mathbf{r}_1)$ and the second with wave function $\chi(\mathbf{r}_2)$. We assume that each of these particles has a spin $1/2$ and that their spins are entangled in a singlet spin state. In Bohmian theory, it is convenient to write explicit wave functions, so that here we will use a mixed notation with wave functions for orbital variables and spins treated as kets in the Dirac notation. We can then write the wave function/state vector of the two-particle system as:

$$|\Psi\rangle = \varphi(\mathbf{r}_1)\,\chi(\mathbf{r}_2)\frac{1}{\sqrt{2}}\Big[\,|1:+,2:-\rangle - |1:-,2:+\rangle\Big] \tag{I.3}$$

Assume now that the first particle enters a magnetic field gradient such as that of a Stern–Gerlach magnet oriented along direction a, which splits the wave function $\varphi(\mathbf{r}_1)$ into two spatially separated components $\varphi'_+(\mathbf{r}_1)$ and $\varphi'_-(\mathbf{r}_1)$. Then $|\Psi\rangle$ becomes:

$$|\Psi'\rangle = \frac{1}{\sqrt{2}}\Big[\varphi'_+(\mathbf{r}_1)\,|1:+\rangle_a\,|2:-\rangle_a - \varphi'_-(\mathbf{r}_1)\,|1:-\rangle_a\,|2:+\rangle_a\Big]\chi(\mathbf{r}_2) \tag{I.4}$$

where the indices a in the spin states refer to a quantization direction that has been chosen parallel to a. In such a situation, the Bohmian position of the first particle must be either in wave packet $\varphi'_+(\mathbf{r}_1)$, or in wave packet $\varphi'_-(\mathbf{r}_1)$. The other wave is necessarily an "empty wave", which does not play any role and may be ignored. One of the components of (I.4) then disappears, so that the spin state of the second particle has been projected onto the opposite state of that of the first particle. We see that a spatial separation of the wave packet of one of the particles is already sufficient to project both spin states, even before the second particle has entered any magnetic gradient. One therefore obtains in this way a very efficient mechanism for reproducing the standard state vector projection.

I.1.3 Two-photon interferences

An attempt to experimentally invalidate Bohmian mechanics was published in [649, 650], involving correlations between two photons, each going through different slits of a screen; the refutation was based on a previous theoretical claim of a discrepancy between Bohmian theory and standard quantum mechanics [651]. But, here again, what is disproved is not the full Bohmian theory, but a modified version. The full Bohmian theory is built to reproduce exactly the same results as standard quantum mechanics, provided the same wave function is used in both cases (symmetrical with respect to exchange between the two bosons in this case); such discrepancies in predictions concerning position measurements cannot occur. In the present case, the extra assumption is that the trajectories of the two bosons are always symmetric with respect to the symmetry plane of the experiment. But, in Bohmian theory, the positions of the particles fill the whole volume available in six-dimensional configuration space, and within this volume nonsymmetric pairs of positions are perfectly accessible. Provided the state vector has been properly symmetrized, there is no reason to add a second postulate of symmetrization for the Bohmian positions themselves.

I.2 Bohmian trajectories are consistent with position measurements

We come back to a question discussed in §11.8.1.d, and study the relation between quantum position measurements and Bohmian trajectories. We will see that, if a particle has been detected by a series of apparatuses located in different regions of space, the Bohmian trajectory of the particle necessarily crossed these regions; in other words, the trajectory always remains consistent with the results of successive quantum position measurements. Inconsistent trajectories, or "surrealistic trajectories" as they are sometimes called in the literature, do not exist (see §11.8.1.h). To show this, we consider a situation where a particle is initially described by a wave packet $|\varphi_0\rangle$ propagating in direction Oz, and then crosses two successive planes P_D and P_E, where (nonabsorbing) particle detectors can register its presence (Figure I.1). Two detectors D_1 and D_2 have disjoint detection ranges in plane P_D; similarly, P_E contains the disjoint ranges of two other detectors E_1 and E_2. We assume that

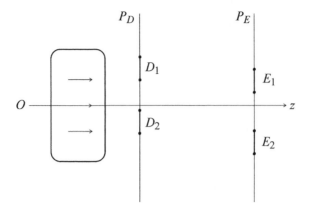

Figure I.1 A broad wave packet propagates along axis Oz from the left. It crosses two planes P_D and P_E containing position detectors, D_1 and D_2 in the first plane, E_1 and E_2 in the second. The range of all detectors is disjoint and much larger than the de Broglie wavelength of the particle; the detectors are assumed to be perfect (100% efficiency). We show in the text that, if detectors D_i and E_j registered the presence of a particle ($i, j = 1, 2$), the Bohmian trajectory of the particle necessarily crossed these detectors; if none of the detectors registers a particle, the trajectory avoided all detectors. The detection of positions is therefore consistent with Bohmian trajectories.

the detection ranges of all detectors are much larger than the de Broglie wavelength of the particle, and that they are ideal detectors (100% efficiency). The initial wave packet is sufficiently broad to cover all detectors during its propagation.

The initial state vector of the whole system (particle + detectors) at the initial time t_0 is:

$$|\Psi(t_0)\rangle = |\varphi(t_0)\rangle \left|D_1^0\right\rangle \left|D_2^0\right\rangle \left|E_1^0\right\rangle \left|E_2^0\right\rangle \tag{I.5}$$

where $\left|D_i^0\right\rangle$ is the state of detector D_i before it has detected a particle ($i = 1, 2$), and similarly $\left|E_j^0\right\rangle$ is the initial state of detector E_j in the second plane ($j = 1, 2$). Each of these state vectors actually characterizes the properties of many particles that are contained in each detector; nevertheless, for simplicity, we take into account only one variable per detector, the position of its macroscopic "pointer". We now propagate this state vector in time, as summarized in Figure I.2.

Consider a time t_1 just before the wave packet crosses plane P_D. The ket describing the particle is then:

$$|\varphi(t_1)\rangle = U(t_0, t_1)|\varphi_0\rangle \tag{I.6}$$

where $U(t, t')$ is the unitary evolution operator of the particle between times t and t'. We can expand this ket as the sum:

$$|\varphi(t_1)\rangle = |\overline{\varphi}_1\rangle + |\overline{\varphi}_2\rangle + |\overline{\varphi}_3\rangle \tag{I.7}$$

of three components of the incoming wave packet: the component $|\overline{\varphi}_1\rangle$ that will interact with D_1, the component $|\overline{\varphi}_2\rangle$ that will interact with D_2, and $|\overline{\varphi}_3\rangle$ that will miss both detectors in P_D because its wave function crosses the plane outside of their detection ranges. To obtain the first component, one may consider a cylinder that is parallel to the direction of propagation Oz and intersects the plane P_D at the outside limit of the detection range of D_1; one then projects the wave function of $|\varphi(t_1)\rangle$ into the inside of this cylinder (its values outside of the cylinder are cancelled), which provides the wave function $\overline{\varphi}_1(\mathbf{r})$ of $|\overline{\varphi}_1\rangle$. The same construction from the range of detection of D_2 is applied to obtain $|\overline{\varphi}_2\rangle$. Finally, $|\overline{\varphi}_3\rangle$ is merely defined as the difference between $|\varphi(t_1)\rangle$ and the sum of $|\overline{\varphi}_1\rangle$ and $|\overline{\varphi}_2\rangle$; the corresponding wave function is $\overline{\varphi}_3(\mathbf{r})$. Since we have assumed that the size of the detectors is much larger than the de Broglie wavelength, diffraction effects remain negligible during the propagation of the three components, which keep a negligible overlap. Therefore, to a very good approximation, a wave function starting from $\overline{\varphi}_1(\mathbf{r})$ at time t_1 crosses D_1 only during its time propagation, a wave function starting from $\overline{\varphi}_2(\mathbf{r})$ crosses D_2 only, while a wave function starting from $\overline{\varphi}_3(\mathbf{r})$ crosses none of the detectors. The state of the whole system just before measurement can then be written as:

$$|\Psi(t_1)\rangle = \left[|\overline{\varphi}_1\rangle + |\overline{\varphi}_2\rangle + |\overline{\varphi}_3\rangle\right]|D_1^0\rangle|D_2^0\rangle|E_1^0\rangle|E_2^0\rangle \tag{I.8}$$

At a later time t_1', after the wave packet has crossed plane P_D, this state becomes:

$$\left|\Psi\left(t_1'\right)\right\rangle = \left[|\overline{\chi}_1\rangle|D_1^*\rangle|D_2^0\rangle + |\overline{\chi}_2\rangle|D_1^0\rangle|D_2^*\rangle + |\overline{\chi}_3\rangle|D_1^0\rangle|D_2^0\rangle\right]|E_1^0\rangle|E_2^0\rangle \tag{I.9}$$

where $|D_1^*\rangle$ is the state of detector D_1 when it has recorded a particle (its pointer has moved to a different position to indicate the result of measurement), and $|\overline{\chi}_1\rangle$ the state of the particle at time t_1' after interaction with D_1. Similar notations are used for $|D_2^*\rangle$ and $|\overline{\chi}_2\rangle$, and $|\overline{\chi}_3\rangle$ is the state of the particle at time t_1' if it has not interacted with any measurement apparatus.

The wave packet then propagates towards plane P_E. Consider a time t_2 just before it crosses this plane and call $|\overline{\varphi}_i(t_2)\rangle$ (with $i = 1, 2, 3$) the ket obtained by applying to $|\overline{\chi}_i\rangle$ the unitary time evolution from t_1' to t_2:

$$|\overline{\varphi}_i(t_2)\rangle = U\left(t_1', t_2\right)|\overline{\chi}_i\rangle \tag{I.10}$$

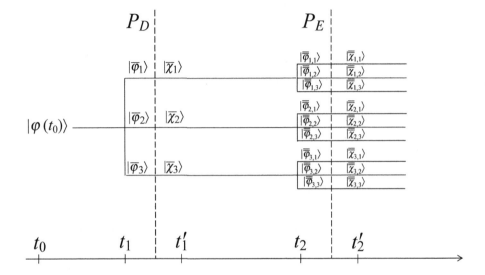

Figure I.2 This figure summarizes the successive expansions of the state vector or the particle when it crosses planes P_D and P_E. In fact, this state vector is entangled with states vectors describing the states of the position measurement apparatuses, but for clarity they are not shown in the figure. At time t_1, just before the first series of measurements, the state vector of the particle is expanded as the sum of three kets: $|\overline{\varphi}_1\rangle$, which will interact with the first measurement apparatus; $|\overline{\varphi}_2\rangle$, which will interact with the second measurement apparatus; and $|\overline{\varphi}\rangle$, which interacts with none. After the measurement, these kets have evolved and become $|\overline{\chi}_1\rangle$, $|\overline{\chi}_2\rangle$, and $|\overline{\chi}_3\rangle$ respectively, while each of these kets is associated with a different quantum state of the measurement apparatuses. A second similar expansion is made when the particle crosses the second plane containing the second series of measurement apparatuses.

By the same method as the preceding, we expand each of the $|\overline{\varphi}_i(t_2)\rangle$ (with $i = 1, 2, 3$) onto a component $|\overline{\overline{\varphi}}_{i,1}\rangle$ that will interact with detector E_1, another component $|\overline{\overline{\varphi}}_{i,2}\rangle$ that will interact with detector E_2, and a component $|\overline{\overline{\varphi}}_{i,3}\rangle$ that will miss both detectors E_1 and E_2:

$$\left|\overline{\varphi}_i(t_2)\right\rangle = \left|\overline{\overline{\varphi}}_{i,1}\right\rangle + \left|\overline{\overline{\varphi}}_{i,2}\right\rangle + \left|\overline{\overline{\varphi}}_{i,3}\right\rangle \tag{I.11}$$

We then have:

$$\begin{aligned}
\left|\Psi(t_2)\right\rangle = & \left[\left|\overline{\overline{\varphi}}_{1,1}\right\rangle + \left|\overline{\overline{\varphi}}_{1,2}\right\rangle + \left|\overline{\overline{\varphi}}_{1,3}\right\rangle\right]\left|D_1^*\right\rangle\left|D_2^0\right\rangle\left|E_1^0\right\rangle\left|E_2^0\right\rangle \\
& + \left[\left|\overline{\overline{\varphi}}_{2,1}\right\rangle + \left|\overline{\overline{\varphi}}_{2,2}\right\rangle + \left|\overline{\overline{\varphi}}_{2,3}\right\rangle\right]\left|D_1^0\right\rangle\left|D_2^*\right\rangle\left|E_1^0\right\rangle\left|E_2^0\right\rangle \\
& + \left[\left|\overline{\overline{\varphi}}_{3,1}\right\rangle + \left|\overline{\overline{\varphi}}_{3,2}\right\rangle + \left|\overline{\overline{\varphi}}_{3,3}\right\rangle\right]\left|D_1^0\right\rangle\left|D_2^0\right\rangle\left|E_1^0\right\rangle\left|E_2^0\right\rangle
\end{aligned} \tag{I.12}$$

Finally, at time t_2', after the wave packet has crossed plane P_E, the state vector of the system becomes:

$$\begin{aligned}
\left|\Psi\left(t_2'\right)\right\rangle = & \left|\overline{\overline{\chi}}_{1,1}\right\rangle\left|D_1^*\right\rangle\left|D_2^0\right\rangle\left|E_1^*\right\rangle\left|E_2^0\right\rangle + \left|\overline{\overline{\chi}}_{1,2}\right\rangle\left|D_1^*\right\rangle\left|D_2^0\right\rangle\left|E_1^0\right\rangle\left|E_2^*\right\rangle \\
& + \left|\overline{\overline{\chi}}_{1,3}\right\rangle\left|D_1^*\right\rangle\left|D_2^0\right\rangle\left|E_1^0\right\rangle\left|E_2^0\right\rangle \\
& + \left|\overline{\overline{\chi}}_{2,1}\right\rangle\left|D_1^0\right\rangle\left|D_2^*\right\rangle\left|E_1^*\right\rangle\left|E_2^0\right\rangle + \left|\overline{\overline{\chi}}_{2,2}\right\rangle\left|D_1^0\right\rangle\left|D_2^*\right\rangle\left|E_1^0\right\rangle\left|E_2^*\right\rangle \\
& + \left|\overline{\overline{\chi}}_{2,3}\right\rangle\left|D_1^0\right\rangle\left|D_2^*\right\rangle\left|E_1^0\right\rangle\left|E_2^0\right\rangle \\
& + \left|\overline{\overline{\chi}}_{3,1}\right\rangle\left|D_1^0\right\rangle\left|D_2^0\right\rangle\left|E_1^*\right\rangle\left|E_2^0\right\rangle + \left|\overline{\overline{\chi}}_{3,2}\right\rangle\left|D_1^0\right\rangle\left|D_2^0\right\rangle\left|E_1^0\right\rangle\left|E_2^*\right\rangle \\
& + \left|\overline{\overline{\chi}}_{3,3}\right\rangle\left|D_1^0\right\rangle\left|D_2^0\right\rangle\left|E_1^0\right\rangle\left|E_2^0\right\rangle
\end{aligned} \tag{I.13}$$

where the $\left|\overline{\overline{\chi}}_{i,j}\right\rangle$ denote the states of the particle in each component of the state vector at time t_2' and after interaction with E_1 if $j = 1$, with E_2 if $j = 2$, and none of these detectors if $j = 3$. We note that the positions of the pointers are different in each of the nine components of this state vector: for instance, the first component is the only one for which the pointers of D_1 and E_1 have moved; the last component is the only one that corresponds to a state where none of these pointers has moved.

We now investigate the evolution of the Bohmian positions. We call $\mathbf{Q}$ the Bohmian position of the incident particle, $\mathbf{R}_1$ and $\mathbf{R}_2$ those associated with the pointers of D_1 and D_2 respectively, and finally $\mathbf{T}_1$ and $\mathbf{T}_2$ those associated with the pointers of E_1 and E_2 respectively. The full Bohmian position is then a 15-dimensional vector:

$$\mathbf{X} = \{\mathbf{Q}, \mathbf{R}_1, \mathbf{R}_2, \mathbf{T}_1, \mathbf{T}_2\} \tag{I.14}$$

propagating in a large configuration space. The initial value of this vector is:

$$\mathbf{X}^0 = \left\{\mathbf{Q}^0, \mathbf{R}_1^0, \mathbf{R}_2^0, \mathbf{T}_1^0, \mathbf{T}_2^0\right\} \tag{I.15}$$

The initial value $\mathbf{Q}^0$ of the Bohmian position of the particle can be anywhere inside the incident wave packet of $|\varphi_0\rangle$, but it necessarily falls into the spatial range of one of the three wave functions associated with $|\bar{\varphi}_1\rangle$, $|\bar{\varphi}_2\rangle$, or $|\bar{\varphi}_3\rangle$; this position is therefore guided by one of the three wave functions $\bar{\varphi}_1(\mathbf{r}, t)$, $\bar{\varphi}_2(\mathbf{r}, t)$, or $\bar{\varphi}_3(\mathbf{r}, t)$. But we have remarked that a Bohmian position moves continuously and can never leave the regions of space where its guiding wave function is nonzero. Moreover, we have seen that the three regions of space of $\bar{\varphi}_1(\mathbf{r}, t)$, $\bar{\varphi}_2(\mathbf{r}, t)$ and $\bar{\varphi}_3(\mathbf{r}, t)$ remain disjoint until the particle crosses plane P_D (except negligible overlaps that may occur at their borders, due to small diffraction effects); these waves do not interfere. This means that, whatever the initial value $\mathbf{Q}^0$ is, until the wave packet crosses plane P_D the position $\mathbf{Q}$ is guided by one of the three waves only. Therefore, if the position follows the first wave, it reaches the first detector D_1; if it follows the second wave, it reaches the second detector; if it follows the third, it reaches no detector. During the crossing of P_D, for the first wave the pointer position $\mathbf{R}_1$ moves from $\mathbf{R}_1^0$ to a new value $\mathbf{R}_1^*$ indicating the registration of a particle by D_1 (while the other positions of pointers remain unchanged); for the second wave, the position $\mathbf{R}_2$ moves from $\mathbf{R}_2^0$ to a new value $\mathbf{R}_1^*$; and for the third wave no pointer position changes. So, until this time, we see that there is a perfect match between the Bohmian positions of the pointers of the detectors (which indicate the results of quantum measurements) and the Bohmian trajectory followed by the particle.

After the crossing of P_D, the three waves are associated with three different combinations of positions of the pointers, which means that $\mathbf{X}$ propagates in three nonoverlapping regions of configuration space: as soon as the first series of measurement apparatuses provide their results registered and as long as they keep them registered, only one wave remains active and the two others play no role (empty waves). The two first pointer variables now select one of the three branches of the state vector, $i = 1$, or 2 or 3, the two others are empty waves playing no role in the future[1]. The only remaining wave that is effective is then a product. Therefore, the situation for the second series of measurements in plane P_E is very similar to that for the first series in plane P_D. The wave function $\bar{\bar{\varphi}}_i'(\mathbf{r}, t)$ divides into three branches $\bar{\bar{\varphi}}_{i,j}''(\mathbf{r}, t)$, which guide the Bohmian position towards E_1 if $j = 1$, towards E_2 if $j = 2$, or outside of their range if $j = 3$. In each of these branches, the Bohmian positions of the pointers reach different values: $\mathbf{R}_1^*, \mathbf{R}_2^0, \mathbf{T}_1^*, \mathbf{T}_2^0$ if $i = j = 1$, or $\mathbf{R}_1^*, \mathbf{R}_2^0, \mathbf{T}_1^0, \mathbf{T}_2^*$ if $i = 1$ and $j = 2$, or $\mathbf{R}_1^0, \mathbf{R}_2^*, \mathbf{T}_1^*, \mathbf{T}_2^0$ if $i = 2$ and $j = 1$, etc. Again, we obtain an complete agreement between the results of quantum measurements at two different times, the Bohmian positions of the detectors, and the Bohmian trajectory followed by the particle. Indeed, the quantum measurements of positions reveal through which detector the trajectory of the particle went. Note, nevertheless, that, since we have assumed that the range of the detectors are

---

[1] We assume that, once a measurement result has been recorded, it remains safely stored and cannot change any longer.

sufficiently large to avoid strong diffraction effects, this determination of the trajectory is obtained only with a limited precision; a general property of the dBB theory is that the Bohmian trajectory can never be determined with perfect accuracy – see for instance the discussion of the experiment illustrated in Figure 11.3.

Remarks:

(i) We have limited our analysis to two planes of detection and two detectors in each plane, but the discussion can be generalized to arbitrary numbers of position detections at an arbitrary number of times. Also, we have taken into account only one variable per detector, which is unrealistic since they are macroscopic objects made of a large number of particles. The variable we have considered symbolizes a collective variable resulting from the position of many particles; we discuss in §*I.3.2* the limits of such a treatment. Nevertheless, the preceding reasoning can be transposed to this case, where each of the variables **R** or **T** contains a large number of Bohmian variables.

(ii) Our reasoning shows that the position correlation functions obtained from quantum measurements and that inferred from the Bohmian trajectory are in agreement. But it is important to note that, in our reasoning, we have assumed that two quantum measurements are actually performed, and taken into account their effect on the state vector and the Bohmian variables. In an experiment where only one quantum measurement is performed, it would be incorrect to use trajectories to infer from the past value of positions the expression of correlation functions; one cannot guess what would have been observed if a previous measurement apparatus had been inserted (counterfactual reasoning, §§4.2.1 and 4.3.2). We have now seen on several occasions that, as in standard quantum mechanics, correlation functions of the results of quantum measurements must always be evaluated by taking into account the coupling with the two measurement apparatus; otherwise incorrect results may be obtained (see also §11.8.1.h).

I.3 Slow and fast pointers

We now come back the the problem discussed in §11.8.1.h: if the pointer of the measurement apparatus is slow (it provides the Welcher-Weg information only long after the test particle has crossed the interference region), can it provide an information on the hole through which the particle went that is contradictory with the Bohmian trajectory? We summarize the discussion given in Reference [294], but more details can be found in this reference.

I.3.1 Microscopic entanglement with a slow pointer

To study the entanglement of the test particle with another slow particle, we replace relation (11.35) of Chapter 11 (we keep the same notation) with:

$$\Psi(\mathbf{r}, \mathbf{r}_1; t = 0) = \frac{1}{\sqrt{2}} \left[\varphi_{\text{upper}}(\mathbf{r}) \chi_{\text{upper}}(\mathbf{r}_1) + \varphi_{\text{lower}}(\mathbf{r}) \chi_{\text{lower}}(\mathbf{r}_1) \right] \qquad (\text{I.16})$$

where $\chi_{\text{upper}}(\mathbf{r}_1)$ and $\chi_{\text{lower}}(\mathbf{r}_1)$ are the two initial wave functions of the pointer particle, with the same modulus:

$$\left| \chi_{\text{upper}}(\mathbf{r}_1) \right| = \left| \chi_{\text{lower}}(\mathbf{r}_1) \right| \qquad (\text{I.17})$$

but with a different phase:

$$\chi_{\text{upper}}(\mathbf{r}_1) = e^{i\mathbf{K} \cdot \mathbf{r}_1} \left| \chi_{\text{upper}}(\mathbf{r}_1) \right| \quad ; \quad \chi_{\text{lower}}(\mathbf{r}_1) = e^{-i\mathbf{K} \cdot \mathbf{r}_1} \left| \chi_{\text{lower}}(\mathbf{r}_1) \right| \qquad (\text{I.18})$$

(The average momentum in the two states has opposite directions). At time $t = 0$, the two wave functions of the pointer therefore overlap perfectly well. This is no longer the case when time increases: when t increases, the overlap decreases progressively, with a time constant of the order of a time T_{PO} (PO is for "pointer ovelap"). For times $t \lesssim T_{\text{PO}}$, the evolution of the two Bohmian positions $\mathbf{Q}$ and $\mathbf{Q}_1$ associated with $\mathbf{r}$ and $\mathbf{r}_1$ obeys a coupled dynamics, because of the entanglement of the wave function, and their trajectories may be complicated. But, for times $t \gtrsim T_{\text{PO}}$, the situation is simple again, since Bohmian position $\mathbf{Q}_1$ can follow only one of the components (the other is then empty); the entanglement has no role any longer on the evolution of the Bohmian positions, which propagate as if the particle were independent.

If the time at which the wave packets cross in the interference region is longer that T_{PO}, during the crossing the Bohmian positions are sensitive to one component of the wave function only, and their trajectories are practically straight (left part of Figure I.3); the no-crossing rule does not apply, and no paradoxical trajectory appears. This is what happens if the marker particle is sufficiently fast. But if it is very slow, and if the test particle crosses the interference region before the wave packets of the marker have separated, the dynamics of the two Bohmian positions is more complex. Numerical simulations [294, 641] show that, in the limit where the marker particle is very slow, the trajectories of the test particle depend on the initial Bohmian position of the particle playing the role of the pointer. The right part of Figure I.3 shows a case where this position is neutral (it is centered on its wave packet), and where all the trajectories obey the no-crossing rule again. The middle part of Figure I.3 shows an intermediate situation where the marker is neither very slow or very fast. The conclusion is that nonlocal effects between the trajectories may indeed appear with two microscopic particles.

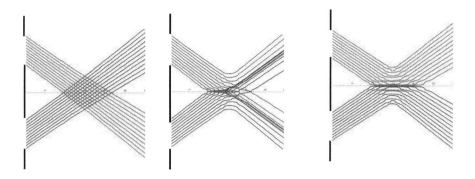

Figure I.3 This figure shows Bohmian trajectories of the test particle starting from various initial positions inside one of the two holes in the screen. The particle flies towards the right, reaches the interference region, and finally goes away. A single particle plays the role of the pointer of the measurement apparatus; its trajectory is not shown (for the sake of simplicity, it is assumed that the initial Bohmian position of the pointer is at the center of its wave packet).

The left part of the figure shows what happens when the pointer is fast (it indicates a result before the particle reaches the interference region): the trajectories are (almost) straight lines and merely cross; no "surrealistic trajectory" then occurs. The right part of the figure shows the trajectories when the pointer is slow, so that the which-way information is obtained only long after the wave packets have crossed in the interference region. The shape of the trajectories then depend on the initial Bohmian position of the pointer; what is shown is a situation where this initial position is centered and where, instead of crossing, all trajectories bounce on the symmetry plane of the experiment (they are all "surrealistic"). The middle part shows an intermediate situation where the pointer is neither very fast nor very slow; some trajectories cross, some others bounce (figure kindly provided by Geneviève Tastevin).

I.3.2 Macroscopic measurement apparatus and entanglement

A macroscopic measurement apparatus is made of a huge number of particles. We now assume that such an apparatus has a slow pointer, i.e., a pointer that will indicate which hole the particle has passed only long after it has crossed the interference region. The question is whether or not the preceding nonlocal effects, leading to surrealistic trajectories, continue to occur at a macroscopic level. As we have seen, the crucial element to answer the question is whether or not two components of the wave function are still simultaneously active to pilot the test particle.

We now have to replace (I.16) by:

$$\Psi(\mathbf{r}, \mathbf{r}_1, .., \mathbf{r}_k, ; .., \mathbf{r}_N; t = 0) = \frac{1}{\sqrt{2}} \left[\varphi_{\text{upper}}(\mathbf{r}) \prod_{k=1}^{N} \chi_{\text{upper}}^k(\mathbf{r}_k) + \varphi_{\text{lower}}(\mathbf{r}) \prod_{k=1}^{N} \chi_{\text{lower}}^k(\mathbf{r}_k) \right]$$

(I.19)

where the product over k corresponds to all N particles contained in the pointer. The two components of the wave function can be simultaneously active only if all the positions $\mathbf{r}_k$ have values for which neither $\chi_{\text{upper}}^k(\mathbf{r}_k)$ nor $\chi_{\text{lower}}^k(\mathbf{r}_k)$ vanishes, that is values belonging to the overlap between the two functions. This is automatically the case at $t = 0$ since relation (I.17) is obeyed for all particles. But, at later times, the situation is different, since, when time increases while remaining sufficiently small, the spatial overlap between two wave functions of the pointer varies roughly as $\left[1 - (t/T_{\text{PO}})^2 \right]$; it can remain large if $t \ll T_{\text{PO}}$. The overlap between the N wave functions of all the particles of the pointer varies as:

$$\left[1 - \left(\frac{t}{T_{\text{PO}}} \right)^2 \right]^N = 1 - N \left(\frac{t}{T_{\text{PO}}} \right)^2$$

(I.20)

which decreases much faster in time, 10^{20} times faster if the pointer contains 10^{20} particles! The time at which this overlap has significantly decreased is roughly $\tau = T_{\text{po}}/\sqrt{N}$, which is 10^{10} shorter than T_{po}. The consequence is that, in practice, a macroscopic pointer always acts as a a fast pointer so that the no-crossing rule does not apply. In this case, one of the waves becomes empty almost immediately, and does not play any role in the further evolution of the Bohmian positions. The nonlocal effects discussed earlier may therefore exist with few degrees of freedom, but not with the macroscopic pointer of a measurement apparatus (which, moreover, rapidly gets entangled with its environment, involving even more Bohmian positions in its dynamics).

In conclusion, it is perfectly appropriate to study the entanglement between microscopic quantum systems with a small number (often two) of quantum and Bohmian variables. Interesting nonlocal effects then may appear. Nevertheless, when one of the two entangled quantum systems is macroscopic, a measurement apparatus for instance, it becomes necessary to include a macroscopic number of Bohmian variables. The surrealistic trajectories then disappear. The same necessity of including all degrees of freedom is not specific to the dBB theory, but also occurs in standard quantum mechanics: even when the quantum dynamics of a collective variable (center of mass, for instance) can be treated as a single particle of large mass, all other variables cannot be ignored when a partial trace is taken.

Appendix J

Models for Spontaneous Reduction of the State Vector

In this appendix, we introduce a few simple models involving modified Schrödinger dynamics with stochasticity, in order to illustrate how such models may lead to an evolution that reproduces the reduction of the state vector during a measurement (emergence of a single eigenvalue during a single realization, with a random value). We discuss the continuous process CLS theory (*cf.* §11.10.1.c); in Appendix L, we will study another modification of the Schrödinger dynamics (*cf.* §11.10.4), where spatially attractive terms are inserted. For the sake of simplicity, here we ignore the usual Hamiltonian evolution during the time of measurement, assuming for instance that this time is too short for this evolution to be significant; otherwise, it would be necessary to use the interaction representation with respect to the Hamiltonian, which does not change the calculations much, except that this introduces a time dependence of the operators.

J.1 Single operator

We consider the measurement of some quantum observable associated with an Hermitian operator A; we look for an equation of evolution containing a state vector reduction process associated with this particular measurement. Since the final eigenvector must vary randomly from one realization to the next, the evolution equation necessarily contains a random component. In our case, it will take the form of a random function of time (as opposed to the GRW theory where the stochasticity is introduced by the discontinuous "hitting processes"; see §11.10.1.b).

J.1.1 Equation of evolution

We assume that the state vector $|\Psi(t)\rangle$ evolves according to:

$$\frac{d}{dt}|\Psi(t)\rangle = -[w(t) - A]^2 \, |\Psi(t)\rangle \tag{J.1}$$

where $w(t)$ is a real random function of time. In order to simplify the model as much as possible, we may discretize time into small finite intervals Δt, during which we assume that $w(t)$ remains constant; moreover, we may also assume that the possible values of $w(t)$ belong to a finite discrete ensemble w_1, w_2, ..., w_N. One may then choose a rule to specify the time progression of $w(t)$, and possibly an interpolation rule to make the functions continuous. Another choice would be to assume that $w(t)$ corresponds to a white noise (Wiener process), with no time memory. For the moment, we do not specify the properties of this random function any further.

Equation (J.1) clearly does not conserve the norm of $|\Psi(t)\rangle$, but it is possible to define a normalized state vector $|\Phi(t)\rangle$ by:

$$|\Phi(t)\rangle = \frac{|\Psi(t)\rangle}{\langle\Psi(t)\,|\Psi(t)\rangle^{1/2}} \tag{J.2}$$

We then have:

$$\begin{aligned}
\frac{d}{dt}|\Phi(t)\rangle &= -[w(t) - A]^2 \, |\Phi(t)\rangle \\
&\quad - \frac{1}{2\,\langle\Psi(t)\,|\Psi(t)\rangle^{3/2}}(-2)\,\langle\Psi(t)|\,[w(t) - A]^2\,|\Psi(t)\rangle \times |\Psi(t)\rangle
\end{aligned} \tag{J.3}$$

or:

$$\frac{d}{dt}|\Phi(t)\rangle = \left\{-[w(t) - A]^2 + \langle\Phi(t)|\,[w(t) - A]^2\,|\Phi(t)\rangle\right\}|\Phi(t)\rangle \tag{J.4}$$

With this nonlinear equation, the norm of $|\Phi(t)\rangle$ does not vary in time, whatever choice is made for the random function $w(t)$.

J.1.2 Solution of the equation

We denote $|a_n\rangle$ the eigenvectors of A with eigenvalues[1] a_n; we may then expand $|\Psi(t)\rangle$ as:

$$|\Psi(t)\rangle = \sum_n x_n(t)\,|a_n\rangle \tag{J.5}$$

We then have:

$$\frac{d}{dt}x_n(t) = -[w(t) - A]^2\,x_n(t) \tag{J.6}$$

---

[1] In case of degeneracy, several consecutive values of a_n are equal, but they correspond to different (orthogonal) eigenstates.

so that:

$$x_n(t) = e^{-\int_0^t dt' \left[w(t') - a_n \right]^2} \, x_n^0 \tag{J.7}$$

where x_n^0 is the initial value of $x_n(t)$ at time $t = 0$. Equivalently, we may also write:

$$|\Psi(t)\rangle = e^{-\int_0^t dt' \left[w(t') - A \right]^2} \, |\Psi(0)\rangle \tag{J.8}$$

According to (J.7), all components of $|\Psi(t)\rangle$ constantly decrease in time, except if $w(t) = a_n$; in this particular case, the components associated with one eigenvalue of A remain constant, as long as the equality is obeyed.

J.1.3 CSL probability rule

We now assume that the probability of each realization of $w(t)$ (the number of these realizations is finite if we take the simple aforementioned assumptions) is proportional to the square of the norm of the value of $|\Psi(t)\rangle$ obtained from (J.1):

$$\mathcal{P}(w_1, w_2, ..., w_N) = c_N \langle \Psi(t) | \Psi(t) \rangle \tag{J.9}$$

where $t = N\Delta t$, and where $w_1, w_2, ..., w_N$ are the selected values for $w(t)$ during the successive time intervals; the normalization coefficient c_N is obtained by writing that the sum of probabilities associated with all realizations is 1. Relation (J.9) may be called the CSL probability rule. From this condition, the Bayes theorem provides the probability that, if $w(t)$ has a given value at some time, the function will jump to any other value in the next discrete time interval.

Among all possible realizations of the random functions $w(t)$, the CSL probability rule strongly favors a small subensemble, that containing functions conserving a large norm for $|\Psi(t)\rangle$, in other words functions that remain constantly equal (or almost equal) to one of the eigenvalues a_n. All other possibilities, even if there are very many of them, are assumed to be very unlikely. For each realization, the mechanism obtained in this way breaks the symmetry between all eigenvalues: the same random function cannot remain very close all the time to more than one eigenvalue a_n; it has to make a choice among all of them. The net result is that, after some time, one always ends up with a state vector that differs very little from an eigenvector of A. Nevertheless, depending on the random function $w(t)$, which may be different for each realization of the experiment, a different eigenvalue is obtained each time. This is precisely the behavior that is needed to reproduce state vector reduction.

Remarks:

(i) This model does not correspond to a hidden variable theory stricto sensu, but is relatively similar; what is added to standard mechanics is a random function playing a role in the dynamics of the state vector, not a hidden variable giving directly the measured observable, the position for instance. One may call it a "hidden function model".

(ii) The CSL probability rule gives the probability of a particular realization of a random function, but not of a particular value of $|\Psi(t)\rangle$, since several different random functions may lead to the same state vector.

J.2 Several operators

To measure the position of a particle, one may imagine that many detectors are located at different places in space, each signalling the possible presence of the particle within its spatial range. This amounts to the simultaneous measurement of a large number of observables, all commuting with each other. We therefore generalize the preceding model to a case where several commuting operators A_k play a role in the dynamics of a state vector.

J.2.1 Equation of evolution

We now postulate the equation of evolution:

$$\frac{d}{dt}|\Psi(t)\rangle = -\sum_k [w_k(t) - A_k]^2 \, |\Psi(t)\rangle \tag{J.10}$$

which contains a series of random real functions $w_k(t)$ and a series of commuting operators A_k. For instance, these operators are assumed to be diagonal in the position representation, and their effect is to multiply the wave function by a given function $\varphi_k(\mathbf{r})$. These functions are, for instance, Gaussian functions with width $\alpha^{-1/2}$, each centered on a point of space that depends on k; when k varies, these points form a regular lattice in all space. The A_k then all commute, but their product for two different values of k is not necessarily 0 since the $\varphi_k(\mathbf{r})$ have mutual overlap.

We now introduce another, much finer, lattice, made of cubic "cells" having dimensions significantly smaller than $\alpha^{-1/2}$, which are labeled with index q. Within each of these cells, each function $\varphi_k(\mathbf{r})$ remains practically constant, so that in this volume the action of A_k may be approximated by a multiplication by a constant, the value φ_k^q of $\varphi_k(\mathbf{r})$ at the center of the cell. We then expand the state vector onto its components in the cells[2]:

$$|\Psi(0)\rangle = \sum_q \left|\overline{\Psi}_q^0\right\rangle \tag{J.11}$$

We then have:

$$A_k \left|\overline{\Psi}_q^0\right\rangle \simeq \varphi_k^q \left|\overline{\Psi}_q^0\right\rangle \tag{J.12}$$

which means that $\left|\overline{\Psi}_q^0\right\rangle$ is almost an eigenvector of A_k. Under these conditions:

$$|\Psi(t)\rangle \simeq \sum_q \left[e^{-\int_0^t dt' \, \Sigma_k \left[w_k(t') - \varphi_k^q\right]^2} \right] \left|\overline{\Psi}_q^0\right\rangle \tag{J.13}$$

We can also write the evolution equation of the normalized state vector $|\Phi(t)\rangle$ defined in (J.2); we then obtain:

$$\frac{d}{dt}|\Phi(t)\rangle = \left\{ -\sum_k [w_k(t) - A]^2 + \langle\Phi(t)| \sum_k [w_k(t) - A]^2 |\Phi(t)\rangle \right\} |\Phi(t)\rangle \tag{J.14}$$

J.2.2 Spontaneous localization of the state vector

The situation is then rather similar to that occurring for one operator A. Equation (J.13) shows that the component associated with a particular cell q can keep a large norm only if each random function $w_k(t)$ remains very close to value φ_k^q during all the time interval $[0, t]$. This means that the function $w_k(t)$ corresponding to the point k that is closest to cell q must take a significant value; as for the other random functions $w_{k'}(t)$ with $k' \neq k$, they must be much smaller, since $\varphi_{k'}^q$ takes an exponentially smaller and smaller value when point k' is far from the considered cell. As a consequence, the random functions $w_k(t)$ may select a particular cell q, or even an ensemble of close cells if they are small and if the time is not too large, but certainly not remote cells at the same time. Finally, if one postulates the CSL probability rule as stated earlier, one favors a very particular ensemble of functions, those for which the wave function becomes localized in close cells, or even a single cell after some time. This realizes the equivalent of a state vector reduction in an arbitrary and random small region of space.

---

[2] By definition, the component $\left|\overline{\Psi}_q^0\right\rangle$ has the same wave function as $|\Psi(0)\rangle$ within cell q, but is 0 outside of this cell.

Appendix K

Consistent Families of Histories

This appendix provides a discussion of the consistency condition (11.22) and of the construction of consistent families of histories. First, we should mention that other conditions have been proposed and used in the literature; in the initial article on histories [17], a weaker condition involving only the cancellation of the real part of (11.22) was introduced. For simplicity, here we limit ourselves to the stronger condition (11.22), which is a sufficient but not necessary condition to the weaker form; it turns out that, as noted in [578], it seems more useful in this context to introduce selectivity than generality in the definition of consistent histories.

At first sight, a natural question that comes to mind is whether or not it is easy, or even possible at all, to fulfill exactly the large number of conditions contained in (11.22). Actually, it has been proposed by Gell-Mann and Hartle to give a fundamental role to families that satisfy consistency conditions in only an approximate way [569], but here we leave aside this possibility and consider only exact consistency conditions. Let us assume, for instance, that the system under study is a particle propagating in free space; the various projectors may then define ranges of positions for the particle, playing a role similar to diaphragms or spatial filters in optics that confine an optical beam in the transverse direction. Then the consistency condition will appear as similar to a non-interference condition for the Huyghens wavelets that are radiated by the inner surface of each diaphragm. But we know that diffraction is unavoidable in the propagation of light; even if it can be a very small effect when the wavelength is sufficiently short and the diaphragms sufficiently broad, it is never strictly 0. Can we then satisfy the non-interference conditions exactly? The answer is not obvious. It turns out to be yes, but it is necessary to exploit the enormous flexibility that we have in the choice of subspaces and projectors in a large space of states, and not to limit ourselves to projectors over well-defined positions only. To understand why, we now briefly sketch one possible systematic method to construct consistent families of histories.

A simple method is to choose any basis $\{|u_k\rangle\}$ in the space of states of the system, and to introduce the projectors:

$$P(u_k) = |u_k\rangle\langle u_k| \tag{K.1}$$

The first time t_1 for constructing histories is taken as the origin of times. We assume that the projectors P_{j_1} defining the family at this time are given by :

$$P_{j_1}(t_1) = P(u_{j_1}) \tag{K.2}$$

If, in Equation (11.22), we insert these projectors on both sides of the density operator, we obtain:

$$Tr\left\{P_{j_1}\rho(t_0)P_{j_1'}\right\} \tag{K.3}$$

which, by circular permutation under the trace, is equal to:

$$Tr\left\{P_{j_1'}P_{j_1}\rho(t_0)\right\} = Tr\left\{|u_{j_1'}\rangle\langle u_{j_1'}|u_{j_1}\rangle\langle u_{j_1}|\rho(t_0)\right\}$$
$$= \delta_{j_1',j_1}Tr\left\{|u_{j_1'}\rangle\langle u_{j_1}|\rho(t_0)\right\} \tag{K.4}$$

Since this expression contains the delta function required by (11.22), the one-time consistency conditions at time t_1 are obeyed.

Now we choose a second (later) time t_2 and call $U(t_2, t_1)$ the evolution operator between t_1 and t_2. We then propagate in time the $|u_k\rangle$ and the P_k, to obtain new projectors:

$$P_k^{(2)} = U(t_2, t_1)P(u_k)U^\dagger(t_2, t_1) \tag{K.5}$$

We choose the $P_{k=j}^2(t_2)$ as the projectors (in the Schrödinger picture) defining the family of histories at time t_2. In the Heisenberg picture, these operators become:

$$\widehat{P}_{j_2}^{(2)}(t_2) = U^\dagger(t_2, t_1)P_{j_2}^{(2)}U(t_2, t_1) = P(u_{j_2}) \tag{K.6}$$

since the evolution operator U is unitary ($U^\dagger U = 1$); they are independent of t_2. The two-time consistency condition (11.22) then involves the trace:

$$Tr\left\{\widehat{P}_{j_2}(t_2)\widehat{P}_{j_1}(0)\rho(t_0)\widehat{P}_{j_1'}(0)\widehat{P}_{j_2'}(t_2)\right\} = Tr\left\{P(u_{j_2})P(u_{j_1})\rho(t_0)P(u_{j_1'})P(u_{j_2'}))\right\} \tag{K.7}$$

inside which all projectors commute, so that it can be written as:

$$Tr\left\{P(u_{j_2'})P(u_{j_2})P(u_{j_1'})P(u_{j_1})\rho(t_0)\right\} \tag{K.8}$$

But we have:

$$P(u_{j_2'})P(u_{j_2}) = \delta_{j_2',j_2}P(u_{j_2}) \quad \text{and} \quad P(u_{j_1'})P(u_{j_1}) = \delta_{j_1',j_1}P(u_{j_1}) \tag{K.9}$$

which shows the occurrence of the delta functions ensuring the two-time consistency condition.

We can now add a third time t_3 and propagate the basis vectors and their projectors between t_2 and t_3; we then define, as in (K.5):

$$P^{(3)}_k = U(t_3,t_1)P(u_k)U^\dagger(t_3,t_1) \qquad (\text{K.10})$$

and choose the operators obtained in this way to define the family of histories at time t_3. Again, the consistency condition (11.22) contains only time-independent orthogonal and commuting projectors P_i, so that the condition is obeyed. An arbitrary number of times can be added in this way to define a family of histories.

Every basis $\{|u_k\rangle\}$ will generate by this construction a different family of histories, and a single one. Each of them is nevertheless a very special family, since each projector corresponds to a subspace of dimension 1 only, which defines histories that are "maximally accurate". But it is possible to make the family more general, and less singular, by grouping together, for each time t_i, several projectors into one single projector over a subspace:

$$P_j(t_i) = \sum_k P^{(i)}_k \qquad (\text{K.11})$$

where the $P^{(i)}_k$ are defined as in (K.5) and (K.10), and where the sum contains any domain of k values. These domains can even be different at the various times t_i, but one important condition should be obeyed: for every time t_i, each value of k should always appear in one $P_j(t_i)$, never in two (or more) different $P_j(t_i)$. Then, for each t_i, the sum of the projectors $P_j(t_i)$ over all values of j gives the unity operator, and the product of two projectors for different values of j is always zero. In the Heisenberg picture, for the same reason as before, the projectors become time independent; moreover, all these projectors commute, even if they correspond to different times, since they are built from the same basis. One can then see by the same reasoning as before that the consistency condition is then still obeyed, so that the family obtained by grouping projectors remains consistent. In this way, we can generate many histories from each basis $\{|u_k\rangle\}$. Nevertheless, other methods for constructing consistent families are possible.

Finally we note that, in our construction of consistent families, the initial density operator $\rho(t_0)$ played no role: the method is universal, whatever the initial state of the system is. But it is also possible to choose a basis $\{|u_k\rangle\}$ made of eigenvectors of the density operator. The history description then obtained is, in a sense, trivial: initially, the system can be seen as being in one of the eigenstates of $\rho(t_0)$, with a probability equal to the corresponding eigenvalue, and then evolves deterministically under the effect of the evolution operator. One interest of this method is that it shows that, in all eigen-subspaces of the density operator that have zero eigenvalues, any projectors can be used since the consistency condition is automatically satisfied for this initial state.

Appendix L

Attractive Schrödinger Dynamics

In this appendix, we give more details on the model of spontaneous collapse with attractive densities introduced in §11.10.4. In the first part, we introduce the localization operator L, the modified Schrödinger equation obtained from this operator, and then study the coupled dynamics of the wave function and the Bohmian density. In the second part, we investigate the predictions of the model concerning small and big quantum systems; we discuss why a quick collapse takes place at the "branching points" where superpositions of macroscopically distinct states start to appear.

L.1 Modified Schrödinger dynamics

An additional term can be introduced into the Schrödinger equation in order to attract the wave functions in regions of space where many Bohmian positions are localized.

L.1.1 Localization operator L

A quantum field operator $\Psi(\mathbf{r})$ can be associated with a system of N identical particles. If the system is described by a quantum state $|\Phi\rangle$, the local (number) density $D_\Phi(\mathbf{r})$ of particles at $\mathbf{r}$ is:

$$D_\Phi(\mathbf{r}) = \frac{\langle\Phi|\Psi^\dagger(\mathbf{r})\Psi(\mathbf{r})|\Phi\rangle}{\langle\Phi|\Phi\rangle} \tag{L.1}$$

In dBB theory, the N particles have Bohmian positions $\mathbf{q}_n(t)$, with $n = 1, 2, .., N$; the local density $D_B(\mathbf{r})$ of Bohmian positions is a sum of delta functions:

$$D_B(\mathbf{r},t) = \sum_{n=1}^{N} \delta(\mathbf{r} - \mathbf{q}_n) \tag{L.2}$$

where the sum runs over all particles. We introduce the following more regular space integral of $D_B(\mathbf{r})$ over a volume of size a_L:

$$N_B(\mathbf{r},t) = \int d^3 r'\ e^{-(\mathbf{r}-\mathbf{r}')^2/(a_L)^2} D_B(\mathbf{r}',t) = \sum_{n=1}^{N} e^{-(\mathbf{r}-\mathbf{q}_n)^2/(a_L)^2} \qquad \text{(L.3)}$$

The order of magnitude of $N_B(\mathbf{r},t)$ is the number of Bohmian positions within a sphere or radius a_L centered at point $\mathbf{r}$. A Bohmian space density $n_B(\mathbf{r},t) = N_B(\mathbf{r},t)/(a_0)^3$ can also be introduced.

We then define the localization operator $L(t)$ by:

$$L(t) = \int d^3 r\ N_B(\mathbf{r},t)\ \Psi^\dagger(\mathbf{r})\Psi(\mathbf{r}) = \sum_{n=1}^{N} N_B(\mathbf{R}_n,t) \qquad \text{(L.4)}$$

($\mathbf{R}_n$ is the position operator of particle n). This operator combines the quantum density operator $\Psi^\dagger(\mathbf{r})\Psi(\mathbf{r})$ with the averaged Bohmian number $N_B(\mathbf{r},t)$. It has the form of a single-particle potential energy operator; $L(t)$ multiplies any wave function $\Phi(\mathbf{r}_1,\mathbf{r}_2,..,\mathbf{r}_N)$ by the the sum over n of the individual potentials $N_B(\mathbf{r}_n,t)$. It therefore takes large positive values at points of space where the Bohmian density is high, and zero values where this density vanishes.

L.1.2 Attractive quantum dynamics

We require a dynamics that favors evolutions of the state vector such that $D_\Phi(\mathbf{r})$ is attracted towards regions where $N_B(\mathbf{r})$ is high. This can be obtained by adding to the usual Hamiltonian $H(t)$ a localization term proportional to $L(t)$, and leads to the modified Schrödinger equation:

$$i\hbar\frac{d}{dt}|\Phi(t)\rangle = [H(t) + i\hbar\gamma_L\,L(t)]\,|\Phi(t)\rangle \qquad \text{(L.5)}$$

where γ_L is a constant localization rate and a_L a localization length. The new term in the right-hand side of the equation increases the modulus of the wave function in regions where the Bohmian density is large. Because of the presence of the imaginary i coefficient, it is not Hermitian, and therefore no longer conserves the norm of $|\Phi\rangle$. In fact, we can consider that $|\Phi\rangle$ defines a direction in the space of states (a one-dimension subspace of this space, what von Neumann calls a "ray"), so that its norm is irrelevant.

But one can also easily obtain an evolution equation for the normalized state vector $|\overline{\Phi}\rangle$, which is:

$$i\hbar\frac{d}{dt}|\overline{\Phi}(t)\rangle = \left[H(t) + \overline{H}_L(t)\right]|\overline{\Phi}(t)\rangle \qquad \text{(L.6)}$$

with:

$$\overline{H}_L(t) = i\hbar\gamma_L \int d^3r \left[\Psi^\dagger(\mathbf{r})\Psi(\mathbf{r}) - D_\Phi(\mathbf{r})\right] N_B(\mathbf{r},t) \tag{L.7}$$

The only difference with the nonnormalized equation (L.5) is that the operator $\Psi^\dagger(\mathbf{r})\Psi(\mathbf{r})$ has been replaced by the difference appearing inside the bracket in the integral, which is actually nothing but the operator associated with the fluctuation of the local density.

Equation (L.5) is linear but time dependent (even if the Hamiltonian H is time independent), since the Bohmian positions and thus $N_B(\mathbf{r},t)$ depend on time. The norm-conserving version (L.7) is nonlinear because $D_\Phi(\mathbf{r}')$ depends on the state vector $|\Phi\rangle$.

L.1.3 Coupled evolutions

We write the normalized wave function $\overline{\Phi}(\mathbf{r}_1, \mathbf{r}_2, .., \mathbf{r}_N)$ as the product of its modulus R by a phase factor:

$$\overline{\Phi}(\mathbf{r}_1, \mathbf{r}_2, .., \mathbf{r}_N) = R(\mathbf{r}_1, \mathbf{r}_2, .., \mathbf{r}_N)\, e^{i\xi(\mathbf{r}_1, \mathbf{r}_2, .., \mathbf{r}_N)} \tag{L.8}$$

and we assume that the positions $\mathbf{q}_n$ evolve according to the usual dBB equation:

$$\frac{d\mathbf{q}_n(t)}{dt} = \hbar \frac{\nabla_n \xi}{m} \tag{L.9}$$

The operator $L(t)$ is diagonal and real in the position representation: it does not change the phase of the wave function, but only its modulus. The equation of evolution of the phase is therefore the standard guiding equation:

$$\hbar\frac{\partial}{\partial t}\xi(\mathbf{r}_1, \mathbf{r}_2, .., \mathbf{r}_N) = -\frac{\hbar^2}{2m}\sum_n \left[\nabla_n \xi \cdot \nabla_n \xi - \frac{\Delta_n R}{R}\right] - V(\mathbf{r}_1, \mathbf{r}_2, .., \mathbf{r}_N) \tag{L.10}$$

where ∇_n and Laplacian Δ_n contain derivatives with respect to the 3 coordinates of particle n; V is the usual potential operator. The term in $\Delta_n R/R$ is often called the "quantum potential". This equation shows that the localization process therefore does not change the Bohmian velocities directly. It nevertheless changes them indirectly, because the evolution of the modulus R is now given by:

$$\hbar\frac{\partial}{\partial t}R(\mathbf{r}_1, \mathbf{r}_2, .., \mathbf{r}_N) = -\frac{\hbar^2}{2m}\sum_n \left[2\nabla_n \xi \cdot \nabla_n R + R\Delta_n \xi\right]$$
$$+ \hbar\gamma_L\, R(\mathbf{r}_1, \mathbf{r}_2, .., \mathbf{r}_N)\left[\sum_n N_B(\mathbf{r}_n) - \int d^3r\, D_\Phi(\mathbf{r})\, N_B(\mathbf{r})\right] \tag{L.11}$$

We note that the variation of R due to the localization process contains a decay term with a minus sign that is proportional to the integral in all space of the product $D_\Phi(\mathbf{r})\, N_B(\mathbf{r})$. This term does not depend on the variables $\mathbf{r}_1, \mathbf{r}_2, .., \mathbf{r}_N$, but introduces a uniform decay of the wave function in all configuration space. The localization also introduces a source term, which is proportional to the sum over n of all Bohmian average densities $N_B(\mathbf{r}_n)$; this term is localized at all points of configuration space that are at a sufficiently small distance of one Bohmian position $\mathbf{q}_n$ (a distance smaller than α_L or comparable to it). Together, the two effects maintain a constant norm for the integral of R^2 in all configuration space.

Another remark that is useful for what follows is that, in the limit $a_L \to \infty$, the localization term has no effect: in (L.3), $N_B(\mathbf{r})$ then becomes equal to the number of particles N (a constant) and, in (L.4), $L(t)$ becomes the product $N\widehat{N}$ (where $\widehat{N}$ is the operator associated with the total number of particles). The right-hand side of (L.7) then becomes proportional to $N(\widehat{N} - N)$, which gives zero when acting on any ket $|\overline{\Phi}(t)\rangle$ with a fixed number of particles; nothing is then changed with respect to standard Schrödinger dynamics. Therefore, if either $a_L \to \infty$ or $\gamma_L \to 0$, one recovers the usual quantum evolution.

L.2 Small and large physical systems

We now discuss the effect of the additional localization term in various situations: an isolated microscopic system, a macroscopic system localized in one region of space, and a macroscopic system in a superposition of two quantum states localized in different regions of space. In the latter case, we will see that the superposition is rapidly reduced to one of its components. This ensures macroscopic uniqueness and the emergence of a single, well-defined (but random), result in a quantum measurement. For this discussion, we will use either the nonnormalized form of the Equation (L.5), or the normalized form (L.7), depending on which turns out to be more convenient.

We assume that the values of the two parameters γ_L and α_L are similar to those usually chosen in GRW and CSL theories, for instance, that:

$$\gamma_L = 10^{-16}\ \mathrm{s}^{-1}$$
$$a_L \simeq 10^{-6}\ m \tag{L.12}$$

Of course, our purpose here is not to define accurate values of these constants; we just wish to show that there is a wide range of values that are compatible with the enormous body of experimental data agreeing with standard quantum mechanics (sometimes with an incredible precision of 10^{-12}!), but also introduce a very rapid dynamics of state vector projection when a quantum measurement is made.

L.2.1 Microscopic system

Consider first a single particle; the corresponding function $N_B(\mathbf{r}_1)$ is a Gaussian function centered at a point $\mathbf{q}_1$, which is necessarily a point where the wave function does not vanish (the Bohmian position cannot leave the wave function). The localization process of Equation (L.5) increases the wave function at all points that are within the range of this Gaussian function, leaving it unchanged at points that fall outside of this range; if one prefers to use the normalized equation (L.5), this increase is compensated by a decrease of the wave function as a whole, which maintains its norm constant. The localization process therefore has a negligible effect on any wave function that is already contained in a domain that is smaller than a_L, but tends to reduce the tails of wave functions that extend over larger distances. Equivalently, one can say that the new term spreads the Fourier components of the wave function over a range $\Delta k \simeq 1/a_L$; therefore, only particles having de Broglie wavelengths λ that are of the order of (or larger than) a_L can undergo an appreciable change of their Bohmian velocity. With the mesoscopic value (L.12) chosen for a_L, this corresponds to very low velocities; they are transferred to the Bohmian velocities at a rate γ_L, for which we have chosen a very small value. Altogether, after time integration to obtain the Bohmian position, the localization term produces extremely small changes of the position.

Then consider an atom with a size of the order of a_0 (comparable to the Bohr radius). Since the $\mathbf{q}_n$'s can never reach regions of space where the wave function vanishes, they also remain localized in a region of space of dimension a_0. Then the sum over n of the functions $N_B(\mathbf{r}_n)$ is of the order of the total number of particles N when all positions $\mathbf{r}_n$ fall in a domain of size a_L centered on the atom, and tends rapidly to zero when all the positions leave this domain. In the limit $a_0/a_L \to 0$, we have seen that $L(t) \to N\widehat{N}$, so that in (L.5) the localization term has no effect on the wave function (except a multiplication by an overall factor without any physical consequence). If $a_0/a_L \ll 1$, the exponential in (L.3) can be approximated by $1 - c(a_0/a_L)^2$, where the term in 1 does not contribute (this is the limit $a_L \to \infty$ discussed previously), and where $c \simeq 1$ (the exact value of c depends on the Bohmian positions). So, retaining only the term in $(a_0/a_L)^2$, we see that the parts of the wave function at the periphery of the atom are reduced at a rate γ given by:

$$\gamma \lesssim \gamma_L \left(\frac{a_0}{a_L}\right)^2 N^2 \tag{L.13}$$

while the parts near the center of the atom remain unaffected.

If the atom is small (hydrogen or helium for instance), we have (L.12), $a_0/a_L \simeq 10^{-4}$ so that $\gamma \leq 10^{-24}N^2 \text{ s}^{-1}$, where N is a few units; this rate is clearly extremely low and undetectable. For a molecule, a size of 10 nm is already large, which corresponds to $a_0/a_L \simeq 10^{-2}$ and to $\gamma \leq 10^{-20} N^2$ s^{-1}; even with a number of constituents (protons, neutrons) of the order of 10^4, we still obtain an extremely small rate.

Now consider an interference experiment made with the same microscopic system. In the interferometer, its wave function is localized at the same time in two very different regions of space; but, as the different constituents of the atom remain bound together (the Hamiltonian of the microscopic system keeps its cohesion), the Bohmian positions are all clustered, either in one region, or in the other. Therefore, in one region $N_B(\mathbf{r})$ is equal to N as before, but in the other it is zero. In (L.5), this clearly introduces an imbalance between the "full" wave, which increases at a rate $\gamma_L N^2$, and the "empty" wave, which is not affected by the localization term. The rate of growth of this imbalance is:

$$\gamma \simeq \gamma_L \, N^2 \tag{L.14}$$

Therefore, even for a long experiment lasting one second, if $N < 10^7$, the localization rate remains negligible, and the interference takes place as predicted by standard Schrödinger dynamics; but, for larger values of N, the model predicts that the contrast of fringes should decrease and vanish in the limit $N \gg 1/\sqrt{\gamma_L t}$, where t is the time spent by the particles in the interferometer.

Remark: in an interference experiment with a Bose condensed gas, the situation is very different since no cohesive force maintains all the atoms together. The wave function then contains components with some atoms localized in one arm of the interferometer, some in the other arm. As a consequence, the Bohmian positions may now be spread over the two paths of the interferometer, so that the differential damping effect occurs at a rate that is much smaller than (L.14).

L.2.2 Macroscopic system

The situation is significantly different for macroscopic systems. Consider a macroscopic piece of matter, for instance, a cube of metal with a $10 \, \mu$ side. For the same reasons as before, the Bohmian positions of the ions and of the electrons are localized inside the range of the wave function; they are spread randomly inside this cube, at a mutual distance that is much smaller than α_L. As a consequence, the Bohmian function $N_B(\mathbf{r}_n)$ is practically constant inside the cube, and decays slowly outside over a distance a_L. In equation (L.5), the localization of each constituent of the metal (electron, ion) then occurs with a very smooth function, with a range that is of the order of $10 \, \mu$ (much larger than a_L). This is because each Bohmian position creates an attraction for the wave function of all other particles (the localization process does not act on the particles independently).

The rate of the localization process is given by the term in $\gamma_L \sum_n N_B(\mathbf{r}_n)$ in Equation (L.11). When all current positions $\mathbf{r}_1, \mathbf{r}_2, .., \mathbf{r}_N$ of the particles fall inside the piece of metal, the rate of this localization process is of the order of:

$$\gamma \simeq \gamma_L \, N_B \, N \tag{L.15}$$

Even if we choose conservatively small values $N_P = 10^{20}$, $N_B = 10^{11}$, we obtain a very fast rate $\gamma \simeq 10^{15}$; the dynamical equation leads to an extremely fast collapse of the wave function.

Now, two situations can occur:

(i) A "normal" situation if the N body wave function contains only components where the constituent particles are localized within the volume $\mathcal{V}$ of the cube. Then Equation (L.5) predicts a constant uniform increase of the wave function that has no physical effect; alternatively, one can use the normalized version (L.6) and note that the term containing the space integral of $D_\Phi \, (\mathbf{r})$ exactly compensates for this increase. In this case, the localization terms have no effect inside the volume of the cube.

(ii) A "Schrödinger cat situation" with an N body wave function containing two (or more) components where the constituent particles are localized in different regions of space. In a quantum measurement, this happens for instance if the pointer of the apparatus ends up in a quantum superposition of two states indicating different measurement results. It is then clear that the localization process affects the various components of the wave function in a completely different way. This is because all Bohmian positions have to follow the same component of the wave function (points of configuration space where the wave function vanishes are never reached). So, one component of the wave function (the "full" wave) benefits from a fast localization process, with a rate approximately given by (L.15), while the others (the "empty" waves) do not. The wave function then collapses around the Bohmian positions of the particles in the pointer, with a rate of the order of (L.15).

The model therefore predicts that any quantum superposition of states corresponding to macroscopic localizations in different regions of space collapses very quickly to one single component of the superposition. In particular, in a measurement apparatus, if the pointer reached a superposition of states where it indicates several different measurement results at the same time, then a rapid spontaneous reduction would occur to one of these components only. This explains why a single result is observed in a single realization of the experiment.

L.2.3 Branching points, collapse, return to quantum equilibrium

The preceding analysis shows that the new term introduced in the Schrödinger dynamics has a negligible effect in most cases. The exception is that of "branching points" (or bifurcations of the wave function, or "birth of a Schrödinger cat"), occurring during the appearance of a superposition of quantum states where a macroscopic object occupies different locations of space. The localization term then efficiently reduces all components of the superposition into a single one.

This is nevertheless still not sufficient to solve the measurement problem: we have to check that, if a second measurement is performed, the standard predictions of quantum mechanics (Born rule) are still obtained. This is not obvious, since the change of the dynamics that we have introduced modifies the state vector during the collapse process, and therefore also changes the Bohmian trajectories. The problem is not crucial for a first measurement, since the localization term plays very little role before the branching point. One then expects that the Bohmian positions will follow the usual trajectories until this point, and will not be able to change branch during the very short time taken by the collapse; they will follow one route or another with a probability that corresponds with the standard Born rule. But what happens if a second measurement is performed is not obvious, since nothing guarantees that the Bohmian positions still obey the quantum equilibrium condition after the first crossing point, while this condition is essential to lead to probabilities that correspond to the standard Born rule. Indeed, if the distribution of the $\mathbf{q}_n$ was very different from that of the dBB theory, the predictions concerning a second measurement could be completely modified.

Nevertheless, Towler, Russell and Valentini [637] have shown that a fast relaxation process tends to constantly bring back any distribution of the $\mathbf{q}_n$'s in configuration space towards that of quantum equilibrium[1] (the coarse graining considered in this reference is immediately provided here by the average over the localization length a_L); the distribution of the $\mathbf{q}_n$'s then has to closely follows the square of the reduced wave function. As a consequence, shortly after the first measurement, the usual distribution of the Bohmian positions is restored, and one can start from this new initial value to calculate the further evolution of the state vector in the usual way.

The conclusion is that the collapse term plays a significant role only at the branching points, preventing the simultaneous appearance of macroscopically distinct quantum states, but very little role otherwise. Between these branching points, a rapid relaxation process ensures a quick return to the results of the usual dBB theory.

---

[1] These authors study the evolution of the distribution of the $\mathbf{q}_n$'s when the wave function obeys the standard Schrödinger equation (without any localization term). Since they predict relatively short relaxation times, their conclusions for microscopic objects should not be changed significantly by the introduction of a localization term with a very small coupling constant.

References

[1] N. Bohr, *Atomic Physics and Human Knowledge*, Wiley (1958), and Dover (2011), see in particular chapter "Discussions with Einstein on epistemological problems in atomic physics" – or, in French, with bibliography and glossary by C. Chevalley: *Physique atomique et connaissance humaine*, Folio essais, Gallimard (1991); *Essays 1933 to 1957 on Atomic Physics and Human Knowledge*, Ox Bow Press (1987); *Essays 1958–62 on Atomic Physics and Human Knowledge*, Wiley (1963) and Ox Bow Press (1987); *Atomic Physics and the Description of Nature*, Cambridge University Press (1934 and 1961).

[2] *Albert Einstein: Philosopher-Scientist*, P.A. Schilpp editor, Open Court and Cambridge University Press (1949).

[3] G. Bacchiagaluppi and A. Valentini, *Quantum Theory at the Crossroads: Reconsidering the 1927 Solvay Conference*, Cambridge University Press (2009); https://arxiv.org/abs/quant-ph/0609184v2.

[4] J. von Neumann, *Mathematische Grundlagen der Quantenmechanik*, Springer, Berlin (1932); *Mathematical Foundations of Quantum Mechanics*, Princeton University Press (1955).

[5] J.S. Bell, "On the problem of hidden variables in quantum mechanics", *Rev. Mod. Phys.* **38**, 447–452 (1966); reprinted in *Quantum Theory and Measurement*, J.A. Wheeler and W.H. Zurek editors, Princeton University Press (1983), 396–402 and in chapter 1 of [6].

[6] J.S. Bell, *Speakable and Unspeakable in Quantum Mechanics*, Cambridge University Press (1987); second augmented edition (2004), which contains the complete set of J. Bell's articles on quantum mechanics.

[7] D. Bohm and J. Bub, "A proposed solution of the measurement problem in quantum mechanics by a hidden variable theory", *Rev. Mod. Phys.* **38**, 453–469 (1966).

[8] D. Bohm and J. Bub, "A refutation of the proof by Jauch and Piron that hidden variables can be excluded in quantum mechanics", *Rev. Mod. Phys.* **38**, 470–475 (1966).

[9] N.D. Mermin, "Hidden variables and the two theorems of John Bell", *Rev. Mod. Phys.* **65**, 803–815 (1993); in particular, see §III.

[10] A. Shimony, "Role of the observer in quantum theory", *Am. J. Phys.* **31**, 755–773 (1963).

[11] D. Bohm, "A suggested interpretation of the quantum theory in terms of 'hidden' variables", *Phys. Rev.* **85**, 166–179 and 180–193 (1952); by the same author, see also

Quantum Theory, Constable (1954), although this book does not discuss theories with additional variables.

[12] N. Wiener and A. Siegel, "A new form for the statistical postulate of quantum mechanics", *Phys. Rev.* **91**, 1551–1560 (1953); A. Siegel and N. Wiener, "Theory of measurement in differential space quantum theory", *Phys. Rev.* **101**, 429–432 (1956).

[13] P. Pearle, "Reduction of the state vector by a nonlinear Schrödinger equation", *Phys. Rev.* **D 13**, 857–868 (1976).

[14] P. Pearle, "Toward explaining why events occur", *Int. J. Theor. Phys.* **18**, 489–518 (1979).

[15] G.C. Ghirardi, A. Rimini, and T. Weber, "Unified dynamics for microscopic and macroscopic systems", *Phys. Rev.* **D 34**, 470–491 (1986); "Disentanglement of quantum wave functions", *Phys. Rev.* **D 36**, 3287–3289 (1987).

[16] B.S. DeWitt, "Quantum mechanics and reality", *Phys. Today* **23**, 30–35 (September 1970).

[17] R.B. Griffiths, "Consistent histories and the interpretation of quantum mechanics", *J. Stat. Phys.* **36**, 219–272 (1984); *Consistent Quantum Theory*, Cambridge University Press (2002).

[18] S. Goldstein, "Quantum theory without observers", *Phys. Today* **51**, 42–46 (March 1998) and 38–41 (April 1998).

[19] "Quantum mechanics debate", *Phys. Today* **24**, 36–44 (April 1971); "Still more quantum mechanics", *Phys. Today* **24**, 11–15 (Oct. 1971).

[20] B.S. DeWitt and R.N. Graham, "Resource letter IQM-1 on the interpretation of quantum mechanics", *Am. J. Phys.* **39**, 724–738 (1971).

[21] M. Jammer, *The Conceptual Development of Quantum Mechanics*, McGraw-Hill (1966), second edition (1989).

[22] J. Mehra and H. Rechenberg, *The Historical Development of Quantum Theory*, Springer (1982).

[23] O. Darrigol, *From c-Numbers to q-Numbers: The Classical Analogy in the History of Quantum Theory*, University of California Press (1992).

[24] B. d'Espagnat, *Conceptual Foundations of Quantum Mechanics*, Benjamin (1971).

[25] B. d'Espagnat, *Veiled Reality: An Analysis of Present Day Quantum Mechanics Concepts*, Addison Wesley (1995); *Le réel voilé, analyse des concepts quantiques*, Fayard (1994); *Une incertaine réalité, la connaissance et la durée*, Gauthier-Villars, (1985); *A la recherche du réel*, Gauthier-Villars Bordas (1979).

[26] M. Planck, "Über eine Verbesserung der Wienerschen Spektralgleichung", *Verhandlungen der Deutschen Physikalischen Gesellschaft* **2**, 202–204 (1900). *Physikalische Abhandlungen und Vorträge*, vol. 1, 493–600, Friedrich Vieweg und Sohn (1958).

[27] G.N. Lewis, "The conservation of photons", *Nature* **118**, 874–875 (1926).

[28] A. Pais, "Einstein and the quantum theory", *Rev. Mod. Phys.* **51**, 863–914 (1979).

[29] E.H. Lieb and R. Seiringer, *The Stability of Matter in Quantum Mechanics*, Cambridge University Press (2010).

[30] L. de Broglie, "Recherches sur la théorie des quanta", thesis Paris (1924).

[31] C.J. Davisson and L.H. Germer, "Reflection of electrons by a crystal of nickel", *Nature* **119**, 558–560 (1927).

[32] O. Darrigol, "Strangeness and soundness in Louis de Broglie's early works", *Physis* **30**, 303–372 (1993).

[33] E. Schrödinger, "Quantisierung als Eigenwert Problem", *Annalen der Physik*, 1st communication: **79**, 361–376 (1926); 2nd communication: **79**, 489–527 (1926); 3rd communication: **80**, 437–490 (1926); 4th communication: **81**, 109–139 (1926).

[34] M. Born, "Quantenmechanik der Stossvorgänge", *Zeitschrift für Physik* **38**, 803–827 (1926); "Zur Wellenmechanik der Stossvorgänge",*Nachrichten von der Gesellschaft der Wissenschaften zu Göttingen, Mathematisch-Physikalische Klasse* 146–160 (1926).

[35] E.A. Cornell and C.E. Wieman, "The Bose–Einstein condensate", *Scientific American* **278**, 26–31 (March 1998).

[36] W. Heisenberg, *The Physical Principles of the Quantum Theory*, University of Chicago Press (1930).

[37] P. Jordan, "Bemerkungen zur Theorie der Atomstruktur", *Zeitschrift für Physik* **33**, 563–570 (1925); "Über eine neue Begründung der Quantenmechanik I und II", *Zeitschrift für Physik* **40**, 809–838 (1926) and **44**, 1–25 (1927); "Austauschprobleme und zweite Quantelung", *Zeitschrift für Physik* **91**, 284–288 (1934).

[38] B. Schroer, "Pascual Jordan, his contibutions to quantum mechanics, and his legacy in contemporary local quantum physics", arXiv:hep-th/0303241v2 (2003).

[39] P.A.M. Dirac, *The Principles of Quantum Mechanics*, Oxford University Press (1930, 1958).

[40] D. Howard, "Who invented the Copenhagen interpretation? A study in mythology", *Philos. Sci.* **71**, 669–682 (2004).

[41] N. Bohr, "Can quantum-mechanical description of physical reality be considered complete?", *Phys. Rev.* **48**, 696–702 (1935).

[42] H.P. Stapp, "S-matrix interpretation of quantum theory", *Phys. Rev.* **D 3**, 1303–1320 (1971).

[43] H.P. Stapp, "The Copenhagen interpretation", *Am. J. Phys.* **40**, 1098–1116 (1972).

[44] A. Peres, "What is a state vector?", *Am. J. Phys.* **52**, 644–650 (1984).

[45] J.B. Hartle, "Quantum mechanics of individual systems", *Am. J. Phys.* **36**, 704–712 (1968).

[46] N. Bohr, "On the notions of causality and complementarity", *Dialectica* **2**, 312–319 (1948).

[47] W. Pauli, "Über den Zusammenhang des Abschlusses der Elektronengruppen im Atom mit der Komplexstruktur der Spektren", *Zeit. Phys.* **31**, 765–783 (1925).

[48] P.A.M. Dirac, "The quantum theory of the emission and absorption of radiation", *Proc. Roy. Soc A* **114**, 243–265 (1927).

[49] V. Fock, "Konfigurationraum und zweite Quantelung", *Zeit. Phys.* **75**, 622–647 (1932).

[50] P. Jordan and O. Klein, "Zum Mehrkörperproblem in der Quantentheorie", *Zeit. Phys.* **45**, 751–765 (1927); P. Jordan, "Über Wellen und Korpuskeln in der Quantenmechanik", *Zeit. Phys.* **45**, 766–775 (1927); P. Jordan and E. Wigner, "Über das Paulische Äquivalenzverbot", *Zeit. Phys.* **47**, 631–651 (1928).

[51] R.P. Feynman, "Space-time approach to non-relativistic quantum mechanics", *Rev. Mod. Physics* **20**, 367–387 (1948).

[52] R.P. Feynman and A.R. Hibbs, *Quantum Mechanics and Path Integrals*, McGraw-Hill (1965).

[53] J.S. Bell, "Six possible worlds for quantum mechanics", *Found. Phys.* **22**, 1201–1215 (1992).

[54] N.D. Mermin, "Quantum mechanics: fixing the shifty split", *Physics Today*, 8–10 (July 2012).

[55] J.S. Bell, "Quantum mechanics for cosmologists", in *Quantum Gravity*, C. Isham, R. Penrose, and D. Sciama editors, **2**, 611–637, Clarendon Press (1981); pp. 117–138 of [6].

[56] N.D. Mermin, "Is the moon there when nobody looks? Reality and the quantum theory", *Phys. Today* **38**, 38–47 (April 1985).

[57] F. London and E. Bauer, "La théorie de l'observation en mécanique quantique", n° 775 of *Actualités scientifiques et industrielles, exposés de physique générale*; Hermann (1939); translated into English as "The theory of observation in quantum mechanics" in *Quantum Theory of Measurement*, J.A. Wheeler and W.H. Zurek editors, Princeton University Press (1983), pp. 217–259; see in particular §11, but also 13 and 14.

[58] M. Jammer, *The Philosophy of Quantum Mechanics*, Wiley (1974).

[59] E.P. Wigner, "Remarks on the mind-body question" in *The Scientist Speculates*, I.J. Good editor, Heinemann (1961), pp. 284–302; reprinted in E.P. Wigner, *Symmetries and Reflections*, Indiana University Press (1967), pp. 171–184.

[60] E. Schrödinger, "Die gegenwärtige Situation in der Quantenmechanik", *Naturwissenschaften* **23**, 807–812, 823–828, 844–849 (1935).

[61] J.D. Trimmer, "The present situation in quantum mechanics: a translation of Schrödinger's cat paradox paper", *Proc. Amer. Phil. Soc.* **124**, 323–338 (1980). Also available in *Quantum Theory of Measurement*, J.A. Wheeler and W.H. Zurek editors, Princeton University Press (1983), pp. 152–167.

[62] A. Einstein, letter to Schrödinger dated 8 August 1935; available for instance (translated into French) p. 238 of [99].

[63] E. Schrödinger, *The Interpretation of Quantum Mechanics*, edited and with an introduction by M. Bitbol, Ox Bow Press (1995).

[64] K. Hornberger, S. Gerlich, P. Haslinger, S. Nimmrichter, and M. Arndt, "Colloquium: quantum interference of clusters and molecules", *Rev. Mod. Phys.* **84**, 157–173 (2012).

[65] E.P. Wigner, "The problem of measurement", *Am. J. Phys.* **31**, 6–15 (1963); reprinted in *Symmetries and Reflections*, Indiana University Press, pp. 153–170 (1967); or again in *Quantum Theory of Measurement*, J.A. Wheeler and W.H. Zurek editors, Princeton University Press (1983), pp. 324–341.

[66] M. Renninger, "Zum Wellen-Korpuskel Dualismus", *Zeit. Phys.* **136**, 251–261 (1953).

[67] M. Renninger, "Messungen ohne Störung des Messobjekts", *Zeit. Phys.* **158**, 417–421 (1960).

[68] R.H. Dicke, "Interaction-free quantum measurement: a paradox?", *Am. J. Phys.* **49**, 925–930 (1981).

[69] A.C. Elitzur and L. Vaidman, "Quantum mechanical interaction-free measurements", *Found. Phys.* **23**, 987–997 (1993).

[70] P. Kwiat, H. Weinfurter, T. Herzog, A. Zeilinger, and M.A. Kasevich, "Interaction-free measurement", *Phys. Rev. Lett.* **74**, 4763–4766 (1995).

[71] L. Hardy, "On the existence of empty waves in quantum theory", *Phys. Lett.* **A 167**, 11–16 (1992).

[72] L. Hardy, "Quantum mechanics, local realistic theories, and Lorentz invariant realistic theories", *Phys. Rev. Lett.* **68**, 2981–2984 (1992).

[73] A.J. DeWeerd, "Interaction-free measurement", *Am. J. Phys.* **70**, 272–275 (2001).

[74] A.G. White, J.R. Mitchell, O. Nairz, and P. Kwiat, "Interaction-free imaging", *Phys. Rev.* **A 58**, 605–613 (1998).

[75] W. Putnam and M. Yanik, "Noninvasive electron microscopy with interaction-free quantum measurements", *Phys. Rev.* **A 80**, 040902 (2009).

[76] S. Thomas, C. Kohstall, P. Kruit, and P. Hommelhoff, "Semitransparency in interaction-free measurements", *Phys. Rev.* **A 90**, 053840 (2014); arXiv:1409.0044 [quant-ph] (2014).

[77] P. Kruit, R.G. Hobbs, C-S. Kim, Y. Yang, V.R. Manfrinato, J. Hammer, S. Thomas, P. Weber, B. Klopfer, C. Kohstall, T. Juffmann, M.A. Kasevich, P. Hommelhoff, and K.K. Berggren, "Designs for a quantum electron microscope", *Ultramicroscopy* **164**, 31–45 (2015); arXiv:1510.05946.

[78] G. Ghirardi, *Sneaking a Look at God's Cards, Unraveling the Mysteries of Quantum Mechanics*, revised edition, Princeton University Press (2007).

[79] C. Robens, A. Alt, C. Emary, D. Meschede, and A. Alberti, "Atomic 'bomb testing': the Elitzur-Vaidman experiment violates the Leggett-Garg inequality", *Appl. Phys. B* 123:12 (2017).

[80] X.Y. Zou, L.J. Wang, and L. Mandel, "Induced coherence and indistinguishability in optical coherence", *Phys. Rev. Lett.* **67**, 318–321 (1991).

[81] L.J. Wang, X.Y. Zou, and L. Mandel, "Induced coherence without induced emission", *Phys. Rev.* **A 44**, 4614–4622 (1991).

[82] G. Baretta Lemos, V. Borish, G.D. Cole, S. Ramelow, R. Lapkiewicz, and A. Zeilinger, "Quantum imaging with undetected photons", *Nature* **512**, 409–412 (2014).

[83] Tae-Gon Noh, "Counterfactual quantum cryptography", *Phys. Rev. Lett.* **103**, 230501 (2009).

[84] A. Petersen, "The philosophy of Niels Bohr", in *Bulletin of the Atomic Scientists* **XIX**, 8–14 (September 1963).

[85] C. Chevalley, "Niels Bohr's words and the Atlantis of Kantianism", in *Niels Bohr and Contemporary Philosophy*, J. Faye and H. Folse editors, Dordrecht Kluwer (1994), pp. 33–57.

[86] N. Bohr, "The unity of human knowledge" (October 1960) in *Atomic Physics and Human Knowledge*, Wiley (1958 and 1963).

[87] C. Norris, *Quantum Theory and the Flight from Realism: Philosophical Responses to Quantum Mechanics*, Routledge (2000), p. 233.

[88] N. Bohr, "Quantum physics and philosophy: causality and complementarity", in *Philosophy in the Mid-Century: A Survey*; R. Klibansky editor, La Nuova Italia Editrice, Firenze (1958). See also "The quantum of action and the description of nature", in *Atomic Theory and the Description of Nature*, Cambridge University Press (1934), pp. 92–101.

[89] P. Bokulich and A. Bokulich, "Niels Bohr's generalization of classical mechanics", *Found. Phys.* **35**, 347–371 (2005).

[90] N. Bohr, "Atomic theory and mechanics", *Nature* **116**, 845–852 (1925).

[91] N. Bohr, *Collected Works*, edited by F. Aaserud, Elsevier (2008); see also *Collected Works, Complementarity beyond Physics (1928–1962)*.

[92] N. Bohr, "The quantum postulate and the recent development of atomic theory", *Nature*, 580–590 (Supplement April 14, 1928).

[93] M. Born, "Physical aspects of quantum mechanics", *Nature* **119**, 354–357 (1927).

[94] W. Heisenberg, *Physics and Philosophy*, Harper & Brothers (1958); Harper Perennial Modern Classics (2007).

[95] J.S. Bell, "Bertlmann's socks and the nature of reality", *J. Physique colloques* **C2**, 41–62 (1981). This article is reprinted in pp. 139–158 of [6].

[96] L.D. Landau and E.M. Lifshitz, *Quantum Mechanics, Non-Relativistic Theory*, Pergamon Press (1958), Butterworth-Heinemann Ltd (1996).

[97] E. Schrödinger, *What is Life? Mind and Matter*, Cambridge University Press (1944 and 1967), p. 137.

[98] A. Einstein, letter to Schrödinger dated May 31, 1928; available for instance (translated into French) p. 213 of [99].

[99] F. Balibar, O. Darrigol, and B. Jech, *Albert Einstein, œuvres choisies I, quanta*, Editions du Seuil et Editions du CNRS (1989).

[100] A. Einstein, "Physik und Realität", *Journal of the Franklin Institute* **221**, 313–347 (1936).

[101] L. de Broglie, "La physique quantique restera-t-elle indéterministe?", *Revue des sciences et de leurs applications*, **5**, 289–311 (1952). French Academy of Sciences, session of April 25, 1953; http://www.sofrphilo.fr/telecharger.php?id=74

[102] J.S. Bell, "Against measurement", pp. 17–31 of *Sixty Two Years of Uncertainty: Historical, Philosophical and Physical Enquiries into the Foundations of Quantum Mechanics*, Erice meeting in August 1989, A.I. Miller editor (Plenum Press); reprinted in pp. 213–231 of the 2004 edition of [6].

[103] L. Rosenfeld, "The measuring process in quantum mechanics", *Suppl. Prog. Theor. Phys.*, extra number 222 "Commemoration of the thirtieth anniversary of the meson theory by Dr. H. Yukawa" (1965).

[104] K. Gottfried, *Quantum Mechanics*, Benjamin (1966); second edition, K. Gottfried and Yan Tun-Mow, Springer (2003).

[105] A.J. Leggett, "Testing the limits of quantum mechanics: motivation, state of play, prospects", *J. Phys. Condens. Matter* **14**, R415–R451 (2002).

[106] A.J. Leggett, "Macroscopic quantum systems and the quantum theory of measurement", Supplement of the *Progr. Theor. Phys.* n^r 69, 80–100 (1980).

[107] A.J. Leggett, *The Problems of Physics*, Oxford University Press (1987).

[108] N.G. van Kampen, "Ten theorems about quantum mechanical measurements", *Physica* **A 153**, 97–113 (1988).

[109] B.G. Englert, M.O. Scully, and H. Walther, "Quantum erasure in double-slit interferometers with which-way detectors", *Am. J. Phys.* **67**, 325–329 (1999); see the first few lines of §IV.

[110] C.A. Fuchs and A. Peres, "Quantum theory needs no 'interpretation'", *Phys. Today* **53**, March 2000, 70–71; see also various reactions to this text in the letters of the September 2000 issue.

[111] C.F. von Weizsäcker, *Voraussetzungen des naturwissenschaftlichen Denkens*, Hanser Verlag (1971) and Herder (1972).

[112] A. Einstein, B. Podolsky, and N. Rosen, "Can quantum-mechanical description of physical reality be considered complete?", *Phys. Rev.* **47**, 777–780 (1935); or in *Quantum Theory of Measurement*, J.A. Wheeler and W.H. Zurek editors, Princeton University Press (1983), pp. 138–141.

[113] M. Born, editor, *The Einstein–Born letters (1916–1955)*, MacMillan, London (1971).

[114] A. Einstein, letter to Schrödinger dated June 19, 1935; available for instance (translated into French) p. 234 of [99].

[115] A. Einstein, "Quantenmechanik und Wirklichkeit", *Dialectica* **2**, 320–324 (1948).

[116] A. Einstein, "Autobiographical notes", pp. 5–94 (especially p. 85) and "Reply to criticism" pp. 663–688 (especially pp. 681–683) in *Albert Einstein: Philosopher-Scientist*, edited by P.A. Schilpp, Open Court and Cambridge University Press (1949).

[117] T. Sauer, "An Einstein manuscript on the EPR paradox for spin observables", *Studies in History and Philosophy of Modern Physics*, **38**, 879–887 (2007).

[118] A. Peres, "Einstein, Podolsky, Rosen, and Shannon", *Found. Phys.* **35**, 511–514 (2005).

[119] D. Home and F. Selleri, "Bell's theorem and the EPR paradox", *Rivista del Nuov. Cim.* **14**, 1–95 (1991).

[120] D. Bohm, *Quantum Theory*, Prentice Hall (1951).

[121] N. Bohr, "Quantum mechanics and physical reality", *Nature* **136**, 65 (1935).

[122] P. Pearle, "Alternative to the orthodox interpretation of quantum theory", *Am. J. Phys.* **35**, 742–753 (1967).

[123] J.F. Clauser and A. Shimony, "Bell's theorem: experimental tests and implications", *Rep. Progr. Phys.* **41**, 1881–1927 (1978).

[124] *The Born–Einstein Letters, with Commentaries by Max Born*, translated by Irene Born, Macmillan Press (1971).

[125] C.A. Kocher and E.D. Commins, "Polarization correlation of photons emitted in an atomic cascade", *Phys. Rev. Lett.* **18**, 575–577 (1967).

[126] M. Fadel, T. Zibold, B. Décamps, and P. Treutlein, "Spatial entanglement patterns and Einstein-Podolsky-Rosen steering in Bose-Einstein condensates", *Science* **360**, 409–413 (2018).

[127] P. Kunkel, M. Prüfer, H. Strobel, D. Linnemann, A. Frölian, T. Gasenzer, M. Gärttner, and M.K. Oberthaler, "Spatially distributed multipartite entanglement enables EPR steering of atomic clouds", *Science* **360**, 413–416 (2018).

[128] K. Lange, J. Peise, B. Lücke, I. Kruse, G. Vitagliano, I. Apellaniz, M. Kleinmann, G. Tóth, and Carsten Klepmt, "Entanglement between two spatially separated atomic modes", *Science* **360**, 416–418 (2018).

[129] D.J. Hemmick and A.M. Shakur, *Bell's Theorem and Quantum Realism; Reassessment in Light of the Schrödinger Paradox*, Springer (2012).

[130] J. Conway and S. Kochen, "The free will theorem", *Found. of Phys.* **36**, 1441–1473 (2006); "The strong free will theorem", *Notices of the AMS* **56**, 1441–1473 (2009).

[131] A. Bassi and G. Ghirardi, "The Conway–Kochen argument and relativistic GRW models", *Found. of Phys.* **37**, 169–185 (2007).

[132] F. Laloë, "The hidden phase of Fock states; quantum non-local effects", *Europ. Phys. J.* **33**, 87–97 (2005); "Bose–Einstein condensates and quantum non-locality", in *Beyond the Quantum*, T.M. Nieuwenhiuzen *et al.* editors, World Scientific (2007).

[133] P.W. Anderson, in *The Lesson of Quantum Theory*, editors J. de Boer, E. Dahl, and O. Ulfbeck, Elsevier, New York (1986).

[134] W.J. Mullin and F. Laloë, "Quantum non-local effects with Bose–Einstein condensates", *Phys. Rev. Lett.* **99**, 150401 (2007); "EPR argument and Bell inequalities for Bose–Einstein spin condensates", *Phys. Rev.* **A 77**, 022108 (2008).

[135] J.S. Bell, "On the Einstein–Podolsky–Rosen paradox", *Physics*, **I**, 195–200 (1964); reprinted in chapter 2 of [6].

[136] F. Laloë, "Les surprenantes prédictions de la mécanique quantique", *La Recherche* no. 182, 1358–1367 (November 1986).

[137] F. Laloë, "Cadre général de la mécanique quantique; les objections de Einstein, Podolsky et Rosen", *J. Physique Colloques* **C-2**, 1–40 (1981). See also the other articles following in this issue, especially that by J. Bell on Bertlmann socks, which is a classic!

[138] P. Eberhard, "Bell's theorem without hidden variables", *Nuo. Cim.* **38 B**, 75–79 (1977); "Bell's theorem and the different concepts of locality", *Nuo. Cim.* **46 B**, 392–419 (1978).

[139] J.F. Clauser, M.A. Horne, A. Shimony, and R.A. Holt, "Proposed experiment to test local hidden-variables theories", *Phys. Rev. Lett.* **23**, 880–884 (1969).

[140] A. Peres, "Unperformed experiments have no results", *Am. J. Phys.* **46**, 745–747 (1978).

[141] J.A. Wheeler, "Niels Bohr in today's words" in *Quantum Theory and Measurement*, J.A. Wheeler and W.H. Zurek editors, Princeton University Press (1983), pp. 182–213.

[142] E.P. Wigner, "On hidden variables and quantum mechanical probabilities", *Am. J. Phys.* **38**, 1005–1009 (1970).

[143] K. Hess and W. Philipp, "The Bell theorem as a special case of a theorem of Bass", *Found. Phys.* **35**, 1749–1767 (2005).

[144] J. Bass, "Sur la compatibilité des fonctions de répartition", *C.R. Académie des Sciences* **240**, 839–841 (1955).

[145] B. d'Espagnat, "The quantum theory and reality", *Scientific American* **241**, 128–140 (November 1979).

[146] S.J. Freedman and J.F. Clauser, "Experimental test of local hidden variable theories", *Phys. Rev. Lett.* **28**, 938–941 (1972); S.J. Freedman, thesis, University of California, Berkeley.

[147] J.F. Clauser, "Experimental investigations of a polarization correlation anomaly", *Phys. Rev. Lett.* **36**, 1223 (1976).

[148] E.S. Fry and R.C. Thompson, "Experimental test of local hidden variable theories", *Phys. Rev. Lett.* **37**, 465–468 (1976).

[149] M. Lamehi-Rachti and W. Mittig, "Quantum mechanics and hidden variables: a test of Bell's inequality by the measurement of the spin correlation in low energy proton–proton scattering", *Phys. Rev.* **D 14**, 2543–2555 (1976).

[150] O. Freire, *The Quantum Dissidents; Rebuilding the Foundations of Quantum Mechanics*, Springer (2015).

[151] A. Aspect, P. Grangier, and G. Roger, "Experimental tests of realistic local theories via Bell's theorem", *Phys. Rev. Lett.* **47**, 460–463 (1981).

[152] A. Aspect, P. Grangier, and G. Roger, "Experimental realization of Einstein–Podolsky–Bohm Gedankenexperiment: a new violation of Bell's inequalities", *Phys. Rev. Lett.* **49**, 91–94 (1982).

[153] A. Aspect, J. Dalibard, and G. Roger, "Experimental tests of Bell's inequalities using time varying analyzers", *Phys. Rev. Lett.* **49**, 1804–1807 (1982).

[154] W. Perrie, A.J. Duncan, H.J. Beyer, and H. Kleinpoppen, "Polarization correlations of the two photons emitted by metastable atomic deuterium: a test of Bell's inequality", *Phys. Rev. Lett.* **54**, 1790–1793 (1985).

[155] T.E. Kiess, Y.E. Shih, A.V. Sergienko, and C.O. Alley, "Einstein–Podolsky–Rosen–Bohm experiments using pairs of light quanta produced by type-II parametric down conversion", *Phys. Rev. Lett.* **71**, 3893–3897 (1993).

[156] W. Tittel, J. Brendel, H. Zbinden, and N. Gisin, "Violations of Bell inequalities by photons more than 10 km apart", *Phys. Rev. Lett.* **81**, 3563–3566 (1998).

[157] T. Scheidl, R. Ursin, J. Kofler, S. Ramelow, X.S. Ma, T. Herbst, L. Ratschbacher, A. Fedrizzi, N.K. Langford, T. Jennewein, and A. Zeilinger, "Violations of local realism with freedom of choice", *Proc. Nat. Acad. Sciences* **107**, 19708–19713 (November 16, 2010).

[158] J. Yin, Y. Cao, Y-H. Li, S-K. Liao, L. Zhang, J-G. Ren, W-Q. Cai, W-Y. Liu, B. Li, H. Dai, G-B. Li, Q-M. Lu, Y-H. Gong, Y. Xu, S-L. Li, F-Z. Li, Y-Y. Yin, Z-Q. Jiang, M. Li, J-J. Jia, G. Ren, D. He, Y-L. Zhou, X-X. Zhang, N. Wang, X. Chang, Z-C. Zhu, N-L. Liu, Y-A. Chen, C-Y. Lu, R. Shu, C-Z. Peng, J-Y. Wang,

and J-W. Pan, "Satellite-based entanglement distribution over 1200 kilometers", *Science* **356**, 1140–1144 (2017).

[159] B. Lamine, R. Hervé, M.T. Jaekel, A. Lambrecht, and S. Reynaud, "Large scale EPR correlation and gravitational waves backgrounds", *Eur. Phys. Lett.* **95**, 20004 (2011).

[160] J.C. Howell, A. Lamas-Linares, and D. Bouwmeester, "Experimental violation of a spin-1 Bell inequality using maximally entangled four photon states", *Phys. Rev. Lett.* **88**, 030401 (2002).

[161] B. Hensen, H. Bernien, A.E. Dréau, A. Reiserer, N. Kalb, M.S. Blok, J. Ruitenberg, R.F.L. Vermeulen, R.N. Schouten, C. Abellan, W. Amaya, V. Pruneri, M.W. Mitchell, M. Markham, D.J. Twitchen, D. Elkouss, S. Wehner, T.H. Taminiau, and R. Hanson, "Loophole-free Bell inequality violation using electron spins separated by 1.3 kilometres", *Nature* **526**, 682–686 (2015) or arXiv:1508.05949; B. Hensen, N. Kalb, M.S. Blok, A.E. Dréau, A. Reiserer, R.F.L. Vermeulen, R.B. Schouten, M. Markham, D.J. Twitchen, K Goodenough, D. Elkouss, S. Wehner, T.H. Taminiau, and R. Hanson; "Lopphole-free Bell test using electron spins in diamond: second experiment and additional analysis, *Scient. Rep.* **6**, 30289 (2016).

[162] M. Giustina, M.A.M Versteegh, S. Wengerowsky, J. Handsteiner, A. Hochrainer, K. Phelan, F. Steinlechner, J. Kofler, J.A. Larsson, C. Abellán, W. Amaya, V. Pruneri, M.W. Mitchell, J. Beyer, T. Gerrits, A.E. Lita, L.K. Shalm, S.W. Nam, T. Scheidl, R. Ursin, B. Wittmann, and A. Zeilinger, "Significant-loophole-free test of Bell's theorem with entangled photons", *Phys. Rev. Lett.* **115**, 250401 (2015).

[163] L.K. Shalm, E. Meyer-Scott, B.G. Christensen, P. Bierhorst, M.A. Wayne, M.J. Stevens, T. Gerrits, S. Glancy, D.R. Hamel, M.S. Allman, K.J. Coakley, S. Dyer, C. Hodge, A.E. Lita, V.B. Verma, C. Lambrocco, E. Tortorici, A.L. Migdall, Y. Zhang, D.R. Kumor, W.H. Farr, F. Marsili, M.D. Shaw, J.A. Stern, C. Abellán, W. Amaya, V. Pruneri, T. Jennewein, M.W. Mitchell, P.G. Kwiat, J.C. Bienfang, R.P. Mirin, E. Knill, and S.W. Nam, "Strong Loophole-Free Test of Local Realism", *Phys. Rev. Lett.* **115**, 250402 (2015).

[164] A. Aspect, "Closing the door on Einstein and Bohr's quantum debate", *Physics* **8**, 123 (2015).

[165] A. Fine, "Hidden variables, joint probability, and the Bell inequalities", *Phys. Rev. Lett.* **48**, 291–295 (1982).

[166] J.S. Bell, "Introduction to the hidden variable question", contribution to *Foundations of Quantum Mechanics*, Proceedings of the International School of Physics Enrico Fermi, course II, Academic (1971), p.171; reprinted in pp. 29–39 of [6].

[167] J.F. Clauser and M.A. Horne, "Experimental consequences of objective local theories", *Phys. Rev.* **D 10**, 526–535 (1974).

[168] T. Norsen, "John S. Bell concept of local causality", *Am. J. Phys.* **79**, 1261–1274 (2011); see also "Bell locality and the nonlocal character of Nature", *Found. Physics Lett.* **19**, 633–655 (2006).

[169] J.S. Bell, "La nouvelle cuisine", §24 of the second edition of [6].

[170] J.S. Bell, "The theory of local beables", *Epistemological Letters*, March 1976; reprinted in pp. 52–62 of [6].

[171] O. Oreshkov, F. Costa, and C. Brukner, "Quantum correlations with no causal order", *Nature Comm.*, Article number: 1092 (2012).

[172] M. Araújo, F. Costa, and C. Brukner, "Computational advantage from quantum-controlled ordering of gates", *Phys. Rev. Lett.* **113**, 250402 (2014).

[173] N. Gisin, *L'impensable hasard*, Odile Jacob (2012).

[174] J.P. Jarrett, "On the physical significance of the locality conditions in the Bell arguments", *Nőus* **18**, 569–589 (1984).

[175] L.E. Ballentine and J.P. Jarrett, "Bell's theorem: does quantum mechanics contradict relativity?", *Am. J. Phys.* **55**, 696–701 (1987).

[176] A. Shimony, "Bell's theorem", in Stanford Encyclopedia of Philosophy (2004 and 2009), http://plato.stanford.edu/entries/qm-modal/.

[177] W.M. Dickson, *Chances and Non-Locality*, Cambridge University Press (1998); see in particular §6.2; by the same author, see also the review "Tim Maudlin: Quantum Non-Locality and Relativity, Metaphysical Intimations of Modern Physics review", *Philosophy of Science* **64**, 516–517 (1997).

[178] T. Maudlin, *Quantum Non-locality and Relativity*, Wiley-Blackwell (2011).

[179] A.J. Leggett and A. Garg, "Quantum mechanics versus macroscopic realism: is the flux there when nobody looks?", *Phys. Rev. Lett.* **54**, 857–860 (1985).

[180] A.J. Leggett, "The current status of quantum mechanics at the macroscopic level", *Proceedings 2nd Int. Symp. Foundations of Quantum Mechanics*, Tokyo, 287–297 (1986).

[181] C. Emary, N. Lambert, and F. Nori, "Leggett-Garg inequalities", *Rep. Progr. Phys.* **77**, 016001 (2014).

[182] G.C. Knee, S. Simmons, E.M. Gauger, J.J.L. Morton, H. Riemann, N.V. Abrosimov, P. Becker, H-J. Pohl, K.M. Itoh, M.W. Thewalt, G.A.D. Briggs, and S.C. Benjamin, "Violation of a Leggett-Garg inequality with ideal noninvasive measurements", *Nature Comm.* **3**, 606 (2012).

[183] C. Robens, W. Alt, D. Meschede, C. Emary, and A. Alberti, "Ideal negative measurements in quantum walks disprove theories based on classical trajectories", *Phys. Rev. X* **5**, 011003 (2015).

[184] P. Grangier, M.J. Potasek, and B. Yurke, "Probing the phase coherence of parametrically generated photon pairs: a new test of Bell's inequalities", *Phys. Rev.* **A 38**, 3132–3135 (1988).

[185] J.D. Franson, "Bell inequality for position and time", *Phys. Rev. Lett.* **62**, 2205–2208 (1989).

[186] J.G. Rarity and P.R. Tapster, "Experimental violation of Bell's inequality based on phase and momentum", *Phys. Rev. Lett.* **64**, 2495–2498 (1990).

[187] J. Brendel, E. Mohler, and W. Martienssen, "Experimental test of Bell's inequality for energy and time", *Eur. Phys. Lett.* **20**, 575–580 (1992).

[188] V. Capasso, D. Fortunato, and F. Selleri, "Sensitive observables of quantum mechanics", *Int. J. Theor. Phys.* **7**, 319–326 (1973).

[189] N. Gisin, "Bell's inequality holds for all non-product states", *Phys. Lett.* **A154**, 201–202 (1991).

[190] N. Gisin and A. Peres, "Maximal violation of Bell's inequality for arbitrarily large spin", *Phys. Lett.* **A 162**, 15–17 (1992).

[191] S. Popescu and D. Rohrlich, "Generic quantum non locality", *Phys. Lett.* **A166**, 293–297 (1992).

[192] S.L. Braunstein, A. Mann, and M. Revzen, "Maximal violation of Bell inequalities for mixed states", *Phys. Rev. Lett.* **68**, 3259–3261 (1992).

[193] R.F. Werner, "Quantum states with Einstein–Podolsky–Rosen correlations admitting a hidden variable model", *Phys. Rev.* **A 40**, 4277–4281 (1989).

[194] S. Popescu, "Bell's inequalities and density matrices: revealing 'hidden' nonlocality", *Phys. Rev. Lett.* **74**, 2619–2622 (1995).

[195] A. Peres, "Collective tests for quantum nonlocality", *Phys. Rev.* **A 54**, 2685–2689 (1996).

[196] B. Yurke and D. Stoler, "Bell's-inequality experiments using independent-particle sources", *Phys. Rev.* **A 46**, 2229–2234 (1992).

[197] F. Laloë and W.J. Mullin, "Interferometry with independent Bose–Einstein condensates: parity as an EPR/Bell quantum variable", *Eur. Phys. J.* **B 70**, 377–396 (2009).

[198] S.M. Tan, D.F. Walls, and M.J. Collett, "Nonlocality of a single photon", *Phys. Rev. Lett.* **66**, 252–255 (1991).

[199] L. Hardy, "Nonlocality of a single photon revisited", *Phys. Rev. Lett.* **73**, 2279–2283 (1994).

[200] L. Heaney, A. Cabello, M.F. Santos, and V. Vedral, "Extreme nonlocality with one photon", arXiv:0911.0770v3 [quant-ph] (2009); *New J. Phys.* **13**, 053054 (2011).

[201] B.F. Toner and D. Bacon, "Communication cost of simulating Bell correlations", *Phys. Rev. Lett.* **91**, 187904 (2003).

[202] N. Brunner, D. Cavalcanti, S. Pironio, V. Scarani, and S. Wehner, "Bell nonlocality", *Rev. Mod. Phys.* **86**, 419–478 (2014).

[203] J.A. Larsson, "Loopholes in Bell inequality tests of local realism", *J. Phys. A* **47**, 424003, 1–33 (2014).

[204] J. Kofler, M. Giustina, J-A. Larsson, and M.W. Mitchell, "Requirements for a loophole-free photonic Bell test using imperfect setting generators", *Phys. Rev.* **A 93**, 032115 (2016).

[205] P. Pearle, "Hidden-variable example based upon data rejection", *Phys. Rev.* **D 2**, 1418–1425 (1970).

[206] J.S. Bell, Oral presentation to the EGAS conference in Paris, July 1979 (published in an abridged version in the next reference).

[207] J.S. Bell, "Atomic cascade photons and quantum mechanical nonlocality", *Comments on Atomic and Molecular Physics* **9**, 121 (1980); CERN preprint TH.2053 and TH 2252; Chapter 13 of [6].

[208] A.O. Barut and P. Meystre, "A classical model of EPR experiment with quantum mechanical correlations and Bell inequalities", *Phys. Lett.* **105A**, 458–462 (1984).

[209] N. Gisin, "Hidden quantum nonlocality revealed by local filters", *Phys. Lett.* **A 210**, 151–156 (1996); see in particular §3.

[210] D.S. Tasca, S.P. Walborn, F. Toscano, and P.H. Souto Ribeiro, "Observation of tunable Popescu–Rohrlich correlations through postselection of a Gaussian state", *Phys. Rev.* **A 80**, 030101 (2009).

[211] I. Gerhardt, Q. Liu, A. Lamas-Linares, J. Skaar, V. Scarani, V. Makarov, and C. Kurtsiefer, "Experimentally faking the violation of Bell's inequalities", *Phys. Rev. Lett.* **107**, 170404 (2011).

[212] G. Weihs, T. Jennewein, C. Simon, H. Weinfurter, and A. Zeilinger, "Violation of Bell's inequality under strict Einstein locality conditions", *Phys. Rev. Lett.* **81**, 5039–5043 (1998).

[213] G. 't Hooft, *The Cellular Automaton Interpretation of Quantum Mechanics*, Springer (2016).

[214] C.H. Brans, "Bell's theorem does not eliminate fully causal hidden variables", *Int. J. Theor. Phys.* **27**, 219–226 (1988).

[215] J.S. Bell, "Free variables and local causality", *Epistemological Lett.*, Feb. 1977; Chapter 12 of [6].

[216] C. Abellán, W. Amaya, D. Mitrani, V. Pruneri, and M.W. Mitchell, "Generation of fresh and pure random numbers for loophole-free Bell tests", *Phys. Rev. Lett.* **115**, 250403 (2015).

[217] M.J.W. Hall, "Local deterministic model of singlet state correlations based on relaxing measurement independence", *Phys. Rev. Lett.* **105**, 250404 (2010).

[218] P.H. Eberhard, "Background level and counter efficiencies required for a loophole-free Einstein–Podolsky–Rosen experiment", *Phys. Rev.* **A 47**, R747-R750 (1993).

[219] P.G. Kwiat, P.H. Eberhard, A.M. Steinberg, and R.Y. Chiao, "Proposal for a loophole-free Bell inequality experiment", *Phys. Rev.* **A 49**, 3209–3220 (1994).

[220] E.S. Fry, T. Walther, and S. Li, "Proposal for a loophole-free test of the Bell inequalities", *Phys. Rev.* **A 52**, 4381–4395 (1995).

[221] R. Garcia-Patron, J. Fiurasek, N.J. Cerf, J. Wenger, R. Tualle-Brouri, and P. Grangier, "Proposal for a loophole-free Bell test using homodyne detection", *Phys. Rev. Lett* **93**, 130409 (2004).

[222] J. Wenger, M. Hafezi, F. Grosshans, R. Tualle-Brouri, and P. Grangier, "Maximal violation of Bell inequalities using continuous-variable measurements", *Phys. Rev.* **A 67**, 012105 (2003).

[223] M.A. Rowe, D. Kielpinski, V. Meyer, C.A. Sackett, W.M. Itano, C. Monroe, and D.J. Wineland, "Experimental violation of a Bell's inequality with efficient detection", *Nature* **409**, 791–794 (2001).

[224] C. Simon and W.T.M. Irvine, "Robust long-distance entanglement and a loophole-free Bell test with ions and photons", *Phys. Rev. Lett.* **91**, 110405 (2003).

[225] D.N. Matsukevich, T. Chanelière, S.D. Jenkins, S.Y. Lan, T.A.B. Kennedy, and A. Kuzmich, "Entanglement of remote atomic qubits", *Phys. Rev. Lett.* **96**, 030405 (2006).

[226] D.N. Matsukevich, P. Maunz, D.L. Moehring, S. Olmschenk, and C. Monroe, "Bell inequality violation with two remote atomic qubits", *Phys. Rev. Lett.* **100**, 150404 (2008).

[227] M. Ansmann, H. Wang, R.C. Bialczak, M. Hofheinz, E. Lucero, M. Neeley, A.D. O'Connell, D. Sank, M. Weides, J. Wenner, A.N. Cleland, and J.M. Martinis, "Violation of Bell's inequality in Josephson phase qubits", *Nature* **461**, 504–506 (2009).

[228] H. Bernien, B. Hensen, W. Pfaff, G. Koolstra, M.S. Blok, L. Robledo, T.H. Taminiau, M. Markham, D.J. Twitchen, L. Childress, and R. Hanson, "Heralded entanglement between solid-state qubits separated by three metres", *Nature* **497**, 86–90 (2013); arXiv:1212.6136 [quant-ph].

[229] S.D. Barrett and P. Kok, "Efficient high-fidelity quantum computation using matter qubits and linear optics", *Phys. Rev. A* **71**, 060310 (2005).

[230] W. Rosenfeld, D. Burchardt, R. Garthoff, K. Redeker, N. Ortegel, M. Rau, and H. Weinfurter, "Event-ready Bell test using entangled atoms simultaneously closing detection and locality loopholes", arXiv:1611.04604v1 [quant-ph] (2016).

[231] S. Popescu and D. Rohrlich, "Quantum nonlocality as an axiom", *Found. Phys.* **24**, 379–385 (1994).

[232] H.P. Stapp, "Whiteheadian approach to quantum theory and the generalized Bell's theorem", *Found. Phys.* **9**, 1–25 (1979); "Bell's theorem and the foundations of quantum physics", *Am. J. Phys.* **53**, 306–317 (1985).

[233] H.P. Stapp, "Nonlocal character of quantum theory", *Am. J. Phys.* **65**, 300–304 (1997).

[234] N.D. Mermin, "Nonlocal character of quantum theory?", *Am. J. Phys.* **66**, 920–924 (1998).

[235] M. Redhead, *Incompleteness, Nonlocality and Realism: A Prolegomenon to the Philosophy of Quantum Mechanics*, Chapter 4, Clarendon Press (1988).

[236] A.J. Leggett, "Realism and the physical world", *Rep. Progr. Phys.* **71**, 022001 (2008).

[237] N. Gisin, "Non-realism: deep thought or a soft option?", *Found. Phys.* **42**, 80–85 (2012).

[238] H.P. Stapp, "Meaning of counterfactual statements in quantum physics", *Am. J. Phys.* **66**, 924–926 (1998).

[239] B. d'Espagnat, "Nonseparability and the tentative descriptions of reality", *Phys. Rep.* **110**, 201–264 (1984).

[240] B. d'Espagnat, *Reality and the Physicist*, Cambridge University Press (1989).

[241] R.B. Griffiths, "Consistent quantum counterfactuals", *Phys. Rev.* **A 60**, R5–R8 (1999).

[242] N.D. Mermin, "Bringing home the atomic world: quantum mysteries for anybody", *Am. J. Phys.* **49**, 940–943 (1981).

[243] B. Christensen, K. McCusker, J. Altepeter, B. Calkins, T. Gerrits, A. Lita, A. Miller, L. Shalm, Y. Zhang, S. Nam, N. Brunner, C. Lim, N. Gisin, and P. Kwiat, "Detection-loophole-free test of quantum nonlocality, and applications", *Phys. Rev. Lett.* **111**, 130406 (2013).

[244] M. Giustina, A. Mech, S. Ramelow, B. Wittmann, J. Kofler, J. Beyer, A. Lita, B. Calkins, T. Gerrits, S. Woo Nam, R. Ursin, and A. Zeilinger, "Bell violation using entangled photons without the fair-sampling assumption", *Nature* **497**, 227–230 (2013).

[245] B.S. Cirelson, "Quantum generalizations of Bell's inequality" *Lett. Math. Phys.* **4**, 93–100 (1980).

[246] L.J. Landau, "On the violations of Bell's inequality in quantum theory", *Phys. Lett.* **A 120**, 54–56 (1987).

[247] A. Shimony, *Search for a naturalistic world view*, vol. II, p. 131, Cambridge Universtity Press (1993).

[248] A. Shimony, "Events and processes in the quantum world", in *Quantum Concepts in Space and Time*, R. Penrose and C.J. Isham editors, Oxford University Press (1986), pp. 182–203.

[249] J. Barrett, N. Linden, S. Massar, S. Pironio, S. Popescu, and D. Roberts, "Nonlocal correlations as an information-theoretic resource", *Phys. Rev.* **A 71**, 022101 (2005).

[250] L. Masanes, A. Acin, and N. Gisin, "General properties of nonsignaling theories", *Phys. Rev.* **A 73**, 012112 (2006).

[251] G. Brassard, H. Buhrman, N. Linden, A.A. Méthot, A. Tapp, and F. Unger, "Limit on nonlocality in any world in which communication complexity is not trivial", *Phys. Rev. Lett.* **96**, 250401 (2006).

[252] M. Pawlowski, T. Paterek, D. Kaszlikowski, V. Scarani, A. Winter, and D. Rohrlich, "Information causality as a physical principle", *Nature* **461**, 1101–1104 (2009).

[253] H. Barnum, S. Beigi, S. Boixo, M.B Elliott, and S. Wehner, "Local quantum measurements and no-signaling imply quantum correlations", *Phys. Rev. Lett.* **104**, 140401 (2010).

[254] M.L. Almeida, J.-D. Bancal, N. Brunner, A. Acin, N. Gisin, and S. Pironio, "Guess your neighbor's input: a multipartite nonlocal game with no quantum advantage", *Phys. Rev. Lett.* **104**, 230404 (2010).

[255] D.M. Greenberger, M.A. Horne, and A. Zeilinger, "Going beyond Bell's theorem", in *Bell's Theorem, Quantum Theory, and Conceptions of the Universe*, M. Kafatos editor, Kluwer (1989), pp. 69–72; this reference is not always easy to find, but one can also read the following article, published one year later.

[256] D.M. Greenberger, M.A. Horne, A. Shimony, and A. Zeilinger, "Bell's theorem without inequalities", *Am. J. Phys.* **58**, 1131–1143 (1990).

[257] N.D. Mermin, "Quantum mysteries revisited", *Am. J. Phys.* **58**, 731–733 (1990); see also "What's wrong with these elements of reality?", *Physics Today*, 9–11 (June 1990).

[258] D. Bouwmeester, J.W. Pan, M. Daniell, H. Weinfurter, and A. Zeilinger, "Observation of three-photon Greenberger–Horne–Zeilinger entanglement", *Phys. Rev. Lett.* **82**, 1345–1349 (1999).

[259] J.W. Pan, D. Bouwmeester, M. Daniell, H. Weinfurter, and A. Zeilinger, "Experimental test of quantum nonlocality in three-photon Greenberger–Horne–Zeilinger entanglement", *Nature* **403**, 515–519 (2000).

[260] Z. Zhao, T. Yang, Y.-A Chen, A.-N. Zhang, M. Zukowski, and J.W. Pan, "Experimental violation of local realism by four-photon Greenberger–Horne–Zeilinger entanglement", *Phys. Rev. Lett.* **91**, 180401 (2003).

[261] R. Laflamme, E. Knill, W.H. Zurek, P. Catasi, and S.V.S Mariappan, "NMR GHZ", arXiv:quant-phys/9709025 (1997) and *Phil. Trans. Roy. Soc. Lond.* **A 356**, 1941–1948 (1998).

[262] S. Lloyd, "Microscopic analogs of the Greenberger–Horne–Zeilinger experiment", *Phys. Rev.* **A 57**, R1473–1476 (1998).

[263] U. Sakaguchi, H. Ozawa, C. Amano, and T. Fokumi, "Microscopic analogs of the Greenberger–Horne–Zeilinger experiment on an NMR quantum computer", *Phys. Rev.* **60**, 1906–1911 (1999).

[264] N.D. Mermin, "Extreme quantum entanglement in a superposition of macroscopically distinct states", *Phys. Rev. Lett.* **65**, 1838–1841 (1990).

[265] G. Svetlichny, "Distinguishing three-body from two-body nonseparability by a Bell-type inequality", *Phys. Rev.* **D 35**, 3066–3069 (1987).

[266] J. Acacio de Barros and P. Suppes, "Inequalities for dealing with detector efficiencies in Greenberger–Horne–Zeilinger experiments", *Phys. Rev. Lett.* **84**, 793–797 (2000).

[267] J. Lavoie, R. Kaltenbaek, and K.J. Resch, "Experimental violations of Svetlichny's inequality", *New. J. Physics* **11**, 073051 (2009).

[268] C. Erven, E. Meyer-Scott, K. Fisher, J. Lavoie, B.L. Higgins, Z. Yan, C.J. Pugh, J-P. Bourgoin, R. Prevedel, L.K. Shalm, L. Richards, N. Gigov, R. Laflamme, G. Weihs, T. Jennenwein, and K.J. Resch, "Experimental three-photon quantum nonlocality under strict locality conditions", *Nature Photonics*, **8** 292–296 (2013).

[269] B. Yurke and D. Stoler, "Einstein–Podolsky–Rosen effects from independent particle sources", *Phys. Rev. Lett.* **68**, 1251–1254 (1992).

[270] S. Massar and S. Pironio, "Greenberger–Horne–Zeilinger paradox for continuous variables", *Phys. Rev.* **A 64**, 062108 (2001).

[271] H. J. Bernstein, D.M. Greenberger, M.A. Horne, and A. Zeilinger, "Bell theorem without inequalities for two spinless particles", *Phys. Rev.* **A 47**, 78–84 (1993).

[272] F. Laloë, "Correlating more than two particles in quantum mechanics", *Current Science* **68**, 1026–1035 (1995); http://hal.archives-ouvertes.fr/hal-00001443.

[273] D.J. Wineland, J.J. Bollinger, W.M. Itano, F.L. Moore, and D.J. Heinzen, "Spin squeezing and reduced quantum noise in spectroscopy", *Phys. Rev.* **A 46**, R6797–6800 (1992).

[274] J.J. Bollinger, W.M. Itano, D.J. Wineland, and D.J. Heinzen, "Optimal frequency measurements with maximally correlated states", *Phys. Rev.* **A 54**, R4649–4652 (1996).

[275] J.A. Dunningham, K. Burnett, and S.M. Barnett, "Interferometry below the standard limit with Bose-Einstein condensates lithography", *Phys. Rev. Lett.* **89**, 150401 (2002).

[276] A.N. Boto, P. Kok, D.S. Abrams, S.L. Braunstein, C.P. Williams, and J.P. Dowling, "Quantum interferometric optical lithography: exploiting entanglement to beat the diffraction limit", *Phys. Rev. Lett.* **85**, 2733–2736 (2000).

[277] G. Björk, L.L. Sanchez-Soto, and J. Söderholm, "Entangled state lithography: tailoring any pattern with a single state", *Phys. Rev. Lett.* **86**, 4516–4519 (2001).

[278] M. d'Angelo, M.V. Chekhova, and Y. Shih, "Two-photon diffraction and quantum lithography", *Phys. Rev. Lett.* **87**, 013602 (2001).

[279] A. Zeilinger, M.A. Horne, H. Weinfurter, and M. Zukowski, "Three-particle entanglements from two entangled pairs", *Phys. Rev. Lett.* **78**, 3031–3034 (1997).

[280] K. Mølmer and A. Sorensen, "Multiparticle entanglement of hot trapped ions", *Phys. Rev. Lett.* **82**, 1835–1838 (1999).

[281] C.A. Sackett, D. Klepinski, B.E. King, C. Langer, V. Meyer, C.J. Myatt, M. Rowe, O.A. Turchette, W.M. Itano, D.J Wineland, and C. Monroe, "Experimental entanglement of four particles", *Nature* **404**, 256–259 (2000).

[282] A. Cabello, "Violating Bell's inequalities beyond Cirelson's bound", *Phys. Rev. Lett.* **88**, 060403 (2002).

[283] S. Marcovitch, B. Reznik, and L. Vaidman, "Quantum mechanical realization of a Popescu–Rohrlich box", *Phys. Rev.* **A 75**, 022102 (2007).

[284] N.D. Mermin, "What's wrong with this temptation?", *Phys. Today* **47**, June 1994, pp. 9–11; "Quantum mysteries refined", *Am. J. Phys.* **62**, 880–887 (1994).

[285] D. Boschi, S. Branca, F. De Martini, and L. Hardy, "Ladder proof of nonlocality without inequalities: theoretical and experimental results", *Phys. Rev. Lett.* **79**, 2755–2758 (1997).

[286] S. Goldstein, "Nonlocality without inequalities for almost all entangled states for two particles", *Phys. Rev. Lett.* **72**, 1951–1954 (1994).

[287] G. Ghirardi and L. Marinatto, "Proofs of nonlocality without inequalities revisited", *Phys. Lett.* **A 372**, 1982–1985 (2008).

[288] S. Kochen and E.P. Specker, "The problem of hidden variables in quantum mechanics", *J. Math. Mech.* **17**, 59–87 (1967).

[289] F. Belifante, *Survey of Hidden Variables Theories*, Pergamon Press (1973).

[290] A. Cabello, J.M. Estebaranz, and G. Garcia-Alcaine, "Bell–Kochen–Specker theorem: a proof with 18 vectors", *Phys. Lett.* **A 212**, 183–187 (1996).

[291] A.A. Klyachko, M.A. Can, S. Binicioglu, and A.S. Shumovsky, "Simple tests for hiddden variables in spin-1 systems", *Phys. Rev. Lett.* **101**, 020403 (2008).

[292] A. Peres, "Incompatible results of quantum measurements", *Phys. Lett.* **A 151**, 107–108 (1990).

[293] N.D. Mermin, "Simple unified form for the major no-hidden-variables theorems", *Phys. Rev. Lett.* **65**, 3373 (1990).

[294] G. Tastevin and F. Laloë, "Surrealistic Bohmian trajectories do not occur with macroscopic pointers", arXiv:1802.03783 [quant-ph].

[295] A. Cabello, "Experimentally testable state-independent quantum contextuality", *Phys. Rev. Lett.* **101**, 210401 (2008).

[296] A. Cabello and G. Garcia-Alcaine, "Proposed experimental tests of the Bell–Kochen–Specker theorem", *Phys. Rev. Lett.* **80**, 1797–1799 (1998).

[297] C. Simon, M. Zukowski, H. Weinfurter, and A. Zeilinger, "Feasible Kochen–Specker experiment with single particles", *Phys. Rev. Lett.* **85**, 1783–1786 (2000).

[298] Y.-F. Huang, C.-F. Li, Y.-S. Zhang, J.-W. Pan, and G.-C. Guo, "Experimental test of the Kochen–Specker theorem with single photons", *Phys. Rev. Lett.* **90**, 250401 (2003).

[299] R. Lapkiewicz, P. Li, C. Schaeff, N.K. Langford, S. Ramelow, M. Wieskiak, and A. Zeilinger, "Experimental non-classicality of an indivisible quantum system", *Nature* **474**, 490–493 (2011).

[300] Y. Hasegawa, R. Loidl, G. Badurek, M. Baron, and H. Rauch, "Quantum contextuality in a single-neutron optical experiment", *Phys. Rev. Lett.* **97**, 230401 (2006).

[301] G. Kirchmair, F. Zähringer, R. Gerritsma, M. Kleinmann, O. Gühne, A. Cabello, R. Blatt, and C.F. Roos, "State-independent experimental test of quantum contextuality", *Nature* **460**, 494–497 (2009).

[302] O. Moussa, C.A. Ryan, D.G. Gory, and R. Laflamme, "Testing contextuality on quantum ensembles with one clean qubit", *Phys. Rev. Lett.* **104**, 160501 (2010).

[303] P. Grangier, "Contextual objectivity: a realistic interpretation of quantum mechanics", *Eur. J. Phys.* **23**, 331–337 (2002); arXiv:quant-ph/0012122 (2000), quant-ph/0111154 (2001), quant-ph/0301001 (2003) and quant-ph/0407025 (2004).

[304] N. Harrigan and R.W. Spekkens, "Einstein, incompleteness, and the epistemic view of quantum states", *Found. Phys.* **40**, 125–157 (2010).

[305] M.S. Leifer, "Is the quantum state real? An extended review of ψ-ontology theorems", *Quanta* **3**, 67–155 (2014); arXiv:1409.1570 [quant-ph]; see also: http://mattleifer.info/2011/11/20/can-the-quantum-state-be-interpreted-statistically/.

[306] Y. Aharonov, J. Anandan, and L. Vaidman, "Meaning of the wave function", *Phys. Rev.* **A 47**, 4616–4626 (1993).

[307] W.G. Unruh, "Reality and measurement of the wave function", *Phys. Rev.* **A 50**, 882–887 (1993).

[308] M.F. Pusey, J. Barrett and T. Rudolph, "On the reality of the quantum state", *Nature Physics* **8**, 475–478 (2012); "The quantum state cannot be interpreted statistically", arXiv:1111.3328 [quant-phys] (2011).

[309] R. Colbeck and R. Renner, "Is a system's wave function in one-to-one correspondence with its elements of reality?", *Phys. Rev. Lett.* **108**, 150402 (2012); arXiv:1111.6597 [quant-ph].

[310] M. Schlosshauer and A. Fine, "Implications of the Pusey-Barrett-Rudolph non-go theorem", *Phys. Rev. Lett.* **108**, 260404 (2012).

[311] P.G. Lewis, D. Jennings, J. Barrett, and T. Rudolph, "Distinct quantum states can be compatible with a single state of reality", *Phys. Rev. Lett.* **109**, 150404 (2012).

[312] L. Hardy, "Are quantum states real?", *Int. J. Mod. Phys.* **27**, 1345012 (2013); arXiv:1205.1439 [quant-ph].

[313] J. Barrett, E.G. Cavalcanti, R. Lal, and O.J.E. Maroney, "No ψ-epistemic model can fully explain the indistinguishability of quantum states", *Phys. Rev. Lett.* **112**, 250403 (2014).

[314] C. Branciard, "How ψ-epistemic models fail at explaining the indistinguishability of quantum states", *Phys. Rev. Lett.* **113**, 020409 (2014).

[315] R. Colbeck and R. Renner, "A system's wave function is uniquely determined by its underlying physical state", *New J. Phys.* **19**, 013016 (2017).

[316] D. Nigg, T. Monz, P. Schindler, E.A. Martinez, M. Hennrich, R. Blatt, M.F. Pusey, T. Rudolph, and J. Barrett, "Can different quantum state vectors correspond to the same physical state? An experimental test", *New J. Phys.* **18**, 013007 (2016).

[317] M. Ringbauer, B. Duffus, C. Branciard, E.G. Calvacanti, A.G. White, and A. Fedrizzi, "Measurement of the reality of the wave function", *Nature Physics* **11**, 249–254 (2015).

[318] K.Y. Liao, X.D. Zhang, G.Z. Guo, B.Q. Ai, H. Yan, and S.L. Zhu, "Experimental test of the non-go theorem for continuous ψ-epistemic models", *Nature, scientific reports* **6**, 26519 (2016).

[319] E. Schrödinger, "Discussion of probability relations between separated systems", *Proc. Cambridge Phil. Soc.* **31**, 555 (1935); "Probability relations between separated systems", *Proc. Cambridge Phil. Soc.* **32**, 446 (1936).

[320] http://en.wikiquote.org/wiki/Werner_Heisenberg.

[321] M. Horodecki, P. Horodecki, and R. Horodecki, "Limits for entanglement measures", *Phys. Rev. Lett.* **84**, 2014–2017 (2000).

[322] M. B. Plenio and S. Virmani, "An introduction to entanglement measures", quant-ph/0504163 (2006); *Quant. Info. Comput.* **7**, 1–51 (2007).

[323] A. Méthot and V. Scarani, "An anomaly of nonlocality", quant-ph/0601210 (2006); *Quant. Info. Comput.* **7**, 157–170 (2007).

[324] V. Coffman, J. Kundu, and W.K. Wootters, "Distributed entanglement", *Phys. Rev.* **A 61**, 052306 (2000).

[325] T.J. Osborne and F. Verstraete, "General monogamy inequality for bipartite qubit entanglement", *Phys. Rev. Lett.* **96**, 220503 (2006).

[326] B. Toner, "Monogamy of nonlocal quantum correlations", *Proc. Roy. Soc.* **A 465**, 59–68 (2009).

[327] B. Toner and F. Verstraete, "Monogamy of Bell correlations and Tsirelson's bound", arXiv:quant-ph/0611001 (2006).

[328] A. Peres, "Separability criterion for density matrices", *Phys. Rev. Lett.* **77**, 1413–1415 (1996).

[329] M. Horodecki, P. Horodecki, and R. Horodecki, "Separability of mixed states: necessary and sufficient conditions", *Phys. Lett.* **A 223**, 1–8 (1996).

[330] E. Hagley, X. Maître, G. Nogues, C. Wunderlich, M. Brune, J.M. Raimond, and S. Haroche, "Generation of Einstein–Podolsky–Rosen pairs of atoms", *Phys. Rev. Lett.* **79**, 1–5 (1997).

[331] Q.A. Turchette, C.S. Wood, B.E. King, C.J. Myatt, D. Leibfried, W.M. Itano, C. Monroe, and D.J. Wineland, "Deterministic entanglement of two trapped ions", *Phys. Rev. Lett.* **81**, 3631–4 (1998).

[332] J.I. Cirac and P. Zoller, "Quantum computations with cold trapped ions", *Phys. Rev. Lett.* **74**, 4091–4094 (1995).

[333] R. Blatt and D. Wineland, "Entangled states of trapped atomic ions", *Nature* **453**, 1008–1015 (2008).

[334] M. Steffen, M.A. Ansmann, R.C. Bialczak, N. Katz, E. Lucero, R. McDermott, M. Neeley, E.M. Weig, A.N. Cleland, and J.M. Martinis, "Measurement of the entanglement of two superconducting qubits via state tomography", *Science* **313**, 1423–1425 (2006).

[335] M. Zukowski, A. Zeilinger, M.A. Horne, and A.K. Ekert, "Event-ready-detectors Bell experiment via entanglement swapping", *Phys. Rev. Lett.* **71**, 4287–4290 (1993).

[336] J.W. Pan, D. Bouwmeester, H. Weinfurter, and A. Zeilinger, "Experimental entanglement swapping: entangling photons that never interacted", *Phys. Rev. Lett.* **80**, 3891–3894 (1998).

[337] D. Leibfried, E. Knill, S. Seidelin, J. Britton, R.B. Blakestad, J. Chiaverini, D.B. Hume, W.M. Itano, J.D. Jost, C. Langer, R. Ozeri, R. Reichle, and D.J. Wineland, "Creation of a six-atom 'Schrödinger cat' state", *Nature* **438**, 639–642 (2005).

[338] H. Häffner, W. Hänsel, C.F. Roos, J. Benhelm, D. Chek-al-kar, M. Chwalla, T. Körber, U.D. Rapol, M. Riebe, P.O. Schmidt, C. Becher, O. Gühne, W. Dür, and R. Blatt, "Scalable multiparticle entanglement of trapped ions", *Nature* **438**, 643–646 (2005).

[339] M. Radmark, M. Zukowski, and M. Bourennane, "Experimental tests of fidelity limits in six-photon interferometry and of rotational invariance properties of the photonic six-qubit entanglement singlet state", *Phys. Rev. Lett.* **103**, 150501 (2009).

[340] M. Radmark, M. Wiesniak, M. Zukowski, and M. Bourennane, "Experimental filtering of two-, four-, and six-photon singlets from a single parametric down-conversion source", *Phys. Rev.* **A 80**, 040302(R) (2009).

[341] T. Wilk, A. Gaëtan, C. Evellin, J. Wolters Y. Miroshnychenko, P. Grangier, and A. Browaeys, "Entanglement of two individual neutral atoms using Rydberg blockade", *Phys. Rev. Lett.* **104**, 010502 (2010).

[342] L. Isenhower, E. Urban, X.L. Zhang, A.T. Gill, T. Henage, T.A. Hohnson, T.G. Walker, and M. Saffman, "Demonstration of a neutral atom controlled-NOT quantum gate", *Phys. Rev. Lett.* **104**, 010503 (2010).

[343] W. Chen, J. Hu, Y. Duan, B. Braveman, H. Zhang, and V. Vuletić, "Carving complex many-atom entangled states by single photon detection", *Phys. Rev. Lett.* **115**, 250502 (2015).

[344] S. Welte, B. Hacker, S. Daiss, S. Ritter, and G. Rempe, "Cavity carving of atomic Bell states", *Phys. Rev. Lett.* **118**, 210503 (2017).

[345] The use of these words was suggested by Roger Balian in a private conversation.

[346] M. Schlosshauer, "Decoherence, the measurement problem, and interpretations of quantum mechanics", *Rev. Mod. Phys.* **76**, 1267–1305 (2005).

[347] J.F. Poyatos, J.I. Cirac, and P. Zoller, "Quantum reservoir engineering with laser cooled trapped ions", *Phys. Rev. Lett.* **77**, 4728–4731 (1996).

[348] S. Diehl, A. Micheli, A. Kantian, B. Kraus, H.P. Büchler, and P. Zoller, "Quantum states and phases in driven open quantum systems with cold atoms", *Nature Physics* **4**, 878–883 (2008).

[349] B. Kraus, H.P. Büchler, S. Diehl, A. Kantian, A. Micheli, and P. Zoller, "Preparation of entangled states by quantum Markov processes", *Phys. Rev.* **A 78**, 042307 (2008).

[350] C. Cohen-Tannoudji and A. Kastler, "Optical pumping", *Progress in Optics* **5**, 3–81 (1966).

[351] W. Happer, "Optical pumping", *Rev. Mod. Phys.* **44**, 169–250 (1966).

[352] A. Carvalho, P. Milman, R de Matos Filho, and L. Davidovich, "Decoherence, pointer engineering, and quantum state protection", *Phys. Rev. Lett.* **86**, 4988–4991 (2001).

[353] F. Verstraete, M. Wolf, and I. Cirac, "Quantum computation and quantum state engineering driven by dissipation", *Nature Physics* **5**, 633–636 (2009).

[354] S. Pielawa, L. Davidovich, D. Vitali, and G. Morigi, "Engineering atomic quantum reservoirs for photons", *Phys. Rev.* **A 81**, 043802 (2010).

[355] M. Müller, S. Diehl, G. Pupillo, and P. Zoller, "Engineered open systems and quantum simulations with atoms and ions", *Advances in atomic, molecular and optical physics* **61**, 1–80 (2012).

[356] J. T. Barreiro, M. Mller, P. Schindler, D. Nigg, T. Monz, M. Chwalla, M. Hennrich, C.F. Roos, P. Zoller, and R. Blatt, "An open-system quantum simulator with trapped ions", *Nature* **470**, 486-491 (2011).

[357] R.P. Feynman and F.L. Vernon, "The theory of a general quantum system interacting with a linear dissipative system", *Ann. Phys.* **24**, 181–173 (1963).

[358] A.0. Caldeira and A.J. Leggett, "Influence of dissipation on quantum tunneling in macroscopic systems", *Phys. Rev. Lett.* **46**, 211–214 (1981); "Quantum tunneling in a dissipative system", *Annals of Physics* **149**, 374–456 (1983).

[359] A.J. Leggett, S. Chakravarty, A.T. Dorsey, M.P.A. Fisher, A. Garg, and W. Zwerger, "Dynamics of the dissipative two state system", *Rev. Mod. Phys.* **59**, 1–86 (1987).

[360] N.V. Prokof'ev and P.C.E. Stamp, "Theory of the spin bath", *Rep. Prog. Phys.* **63**, 669–726 (2000).

[361] P.C.E. Stamp, "The decoherence puzzle", *Studies Hist. Phil. Mod. Phys.* **37**, 467–497 (2006).

[362] A. Hagar, "Decoherence: the view from the history and philosophy of science", *Philo. Trans. Royal Soc.* **A 270**, 4594–4609 (2012).

[363] M. Brune, E. Hagley, J. Dreyer, X. Maître, A. Maali, C. Wunderlich, J.M. Raimond, and S. Haroche, "Observing the progressive decoherence of the 'meter' in a quantum measurement", *Phys. Rev. Lett.* **77**, 4887–4890 (1996).

[364] C.H. van der Wal, A.C.J. ter Haar, F.K. Wilhelm, R.N. Schouten, C.J.P.M. Harmans, T.P Orlando, S. Lloyd, and J.E. Mooij, "Quantum superposition of macroscopic persistent-current states", *Science* **290**, 773–777 (2000).

[365] I. Chiorescu, Y. Nakamura, C.J.P. M. Harmans, and J.E. Mooij, "Coherent quantum dynamics of a superconducting flux qubit", *Science* **299**, 1869–1871 (2003).

[366] S. Takahashi, I.S. Tupitsyn, J. van Tol, C.C. Beedle, D.N. Hendrickson, and P.C.E. Stamp, "Decoherence in crystals of quantum molecular magnets", *Nature* **476**, 76–79 (2011).

[367] R.P. Feynman, F.B. Morinigo, and W.G. Wagner, *Feynman lectures on gravitation*, Westview Press (2003) and CRC Press (2018).

[368] B. Lamine, R. Hervé, A. Lambrecht, and S. Reynaud"Ultimate decoherence border for matter-wave interferometry", *Phys. Rev. Lett.* **96**, 050405 (2006).

[369] S. Reynaud, P.A. Maia Neto, A. Lambrecht, and M.T. Jaekel, "Gravitational decoherence in planetary motion", *Europhys. Lett.* **54**, 135–140 (2001).

[370] C.H. Bennett, G. Brassard, S. Popescu, B. Schumacher, J.A. Smolin and W.K. Wootters, "Purification of noisy entanglement and faithful teleportation via noisy channels", *Phys. Rev. Lett.* **76**, 722–725 (1996).

[371] C.H. Bennett, H. Bernstein, S. Popescu, and B. Schumacher, "Concentrating partial entanglement by local operations", *Phys. Rev.* **A 53**, 2046–2052 (1996).

[372] C.H. Bennett, D.P. DiVincenzo, J.A. Smolin, and W.K. Wootters, "Mixed-state entanglement and quantum error correction", *Phys. Rev.* **A 54**, 3824–3851 (1996).

[373] J.W. Pan, C. Simon, C. Brukner, and A. Zeilinger, "Entanglement purification for quantum communication", *Nature* **410**, 1067–1070 (2001).

[374] S. Haroche and J.-M. Raimond, *Exploring the Quantum; Atoms, Cavities and Photons*, Oxford University Press (2008).

[375] C. Cohen-Tannoudji, J. Dupont-Roc, and G. Gryndberg, *Atom–Photon Interactions*, Wiley (1992).

[376] W.K. Wootters and W.H. Zurek, "A single quantum cannot be cloned", *Nature* **299**, 802–803 (1982).

[377] D. Dieks, "Communication by EPR devices", *Phys. Lett.* **A 92**, 271–272 (1982).

[378] D.T. Smithey, M. Beck, M.G. Raymer, and A. Faridani, "Measurement of the Wigner distribution and the density matrix of a light mode using optical homodyne tomography: application to squeezed states and the vacuum", *Phys. Rev. Lett.* **70**, 1244–1247 (1993).

[379] D.T. Smithey, M. Beck, J. Cooper, and M.G. Raymer, "Measurement of number-phase uncertainty relations of optical fields", *Phys. Rev.* **A 48**, 3159–3167 (1993).

[380] U. Leonhardt, *Measuring the Quantum State of Light*, Cambridge University Press (1997).

[381] Y. Aharonov and D. Rohrlich, *Quantum Paradoxes; Quantum Theory for the Perplexed*, Wiley-VCH (2005).

[382] Y. Aharonov, D.Z. Albert, and L. Vaidman, "How the result of a measurement of a component of the spin of a spin-1/2 particle can turn out to be 100", *Phys. Rev. Lett.* **60**, 1351–1354 (1988).

[383] Y. Aharonov and L. Vaidman, "Properties of a quantum system during the time interval between two measurements", *Phys. Rev.* **A 41**, 11–20 (1990).

[384] Y. Aharonov, S. Popescu, and J. Tollaksen, "A time-symmetric formulation of quantum mechanics", *Physics Today* (November 2010), 27–32.

[385] J.S. Lundeen, B. Sutherland, A. Patel, C. Stewart, and C. Bamber, "Direct measurement of the quantum wavefunction", *Nature* **474**, 188–191 (2011).

[386] A.E. Allahverdyan, R. Balian and Th. M. Nieuwenhuizen, "Determining a quantum state by means of a single apparatus", *Phys. Rev. Lett.* **92**, 120402 (2004).

[387] A. Peres, "How the no-cloning theorem got its name", *Fortschritte der Phys.* **51**, 458–461 (2003).

[388] N. Gisin and S. Massar, "Optimal quantum cloning machines", *Phys. Rev. Lett.* **79**, 2153–2156 (1997).

[389] C.H. Bennett and G. Brassard, "Quantum cryptography: public key distribution and coin tossing", in Proceedings of the IEEE International Conference on Computers Systems and Signal Processing, Bangalore India (1984), pp. 175–179.

[390] A.K. Ekert, "Quantum cryptography based on Bell's theorem", *Phys. Rev. Lett.* **67**, 661–663 (1991).

[391] C.H. Bennett, G. Brassard, and N.D. Mermin, "Quantum cryptography without Bell's theorem", *Phys. Rev. Lett.* **68**, 557–559 (1992).

[392] C.H. Bennett, G. Brassard, and A.K. Ekert, "Quantum cryptography", *Scientific American* **267**, 50–57 (October 1992).

[393] N. Gisin, G. Ribordy, W. Tittel, and H. Zbinden, "Quantum cryptography", *Rev. Mod. Phys.* **74**, 145–195 (2002).

[394] C.H. Bennett, "Quantum cryptography using any two nonorthogonal states", *Phys. Rev. Lett.* **68**, 3121–3124 (1992).

[395] P. Townsend, J.G. Rarity, and P.R. Tapster, "Single photon interference in a 10 km-long optical fiber interferometer", *Electron. Lett.* **29**, 634–635 (1993).

[396] C.H. Bennett, G. Brassard, C. Crépeau, R. Jozsa, A. Peres, and W.L. Wootters, "Teleporting an unknown quantum state via dual classical and Einstein–Podolsky–Rosen channels", *Phys. Rev. Lett.* **70**, 1895–1898 (1993).

[397] D. Bouwmeester, J.W. Pan, K. Mattle, M. Eibl, H. Weinfurter, and A. Zeilinger, "Experimental quantum teleportation", *Nature* **390**, 575–579 (1997).

[398] A. Peres, *Quantum Theory: Concepts and Methods*, Kluwer (1993); see also [118].

[399] M. Le Bellac, *Physique Quantique*, 2nd edition, CNRS Editions and EDP Sciences (2007).

[400] S. Massar and S. Popescu, "Optimal extraction of information from finite quantum ensembles", *Phys. Rev. Lett.* **74**, 1259–1263 (1995).

[401] S. Popescu, "Bell's inequalities versus teleportation: what is nonlocality?", *Phys. Rev. Lett.* **72**, 797–800 (1994); arXiv:quant-ph/9501020 (1995).

[402] T. Sudbery, "The fastest way from A to B", *Nature* **390**, 551–552 (1997).

[403] G.P. Collins, "Quantum teleportation channels opened in Rome and Innsbruck", *Phys. Today* **51**, 18–21 (February 1998).

[404] Y. Xia, J. Song, P.-M. Lu, and H-S. Song, "Teleportation of an N-photon Greenberger–Horne–Zeilinger (GHZ) polarization-entangled state using linear elements", *J. Opt. Soc. Am.* **B 27**, A1–A6 (2010).

[405] X.M. Jin, J.G. Ren, B. Yang, Z.H. Yi, F. Zhou, X.F. Xu, S.K. Wang, D. Yang, Y.F. Hu, S. Jiang, T. Yang, H. Yin, K. Chen, C.Z. Peng, and J.W. Pan, "Experimental free-space quantum teleportation", *Nature Photonics* **4**, 376–381 (2010).

[406] C.H. Bennett, "Quantum information and computation", *Phys. Today* **48**, 24–30 (October 1995).

[407] D.P. DiVincenzo, "Quantum computation", *Science* **270**, 255–261 (October 1995).

[408] C.H. Bennett and D.P. DiVincenzo, "Quantum information and computation", *Science* **404**, 247–255 (2000).

[409] D. Bouwmeester, A.K. Ekert, and A. Zeilinger editors, *The Physics of Quantum Information: Quantum Cryptography, Quantum Teleportation, Quantum Computation*, Springer (2000).

[410] N.D. Mermin, *Quantum Computer Science: An Introduction*, Cambridge University Press (2007).

[411] S.M. Barnett, *Quantum Information*, Oxford University Press (2009).

[412] M.A. Nielsen and I.L Chuang, *Quantum Computation and Quantum Information*, Cambridge University Press (2011).

[413] D. Deutsch, "Quantum theory, the Church–Turing principle and the universal quantum computer", *Proc. Roy. Soc.* **A 400**, 97–117 (1985).

[414] http://en.wikipedia.org/wiki/History_of_quantum_computing

[415] M. Le Bellac, *Le monde quantique*, EDP Sciences (2010).

[416] P. Shor, *Proceedings of the 55th Annual Symposium on the Foundations of Computer Science*, IEEE Computer Society Press, Los Alamitos, California (1994), pp. 124–133.

[417] N.D. Mermin, "What has quantum mechanics to do with factoring?", *Phys. Today* **60**, 8–9 (April 2007); "Some curious facts about quantum factoring", *Phys. Today* **60**, 10–11 (October 2007).

[418] L.K. Grover, "A fast quantum mechanical algorithm for database search", Proceedings, 28th Annual ACM Symposium on the Theory of Computing (May 1996), p. 212; "From Schrödinger's equation to quantum search algorithm", *Am. J. Phys.* **69**, 769–777 (2001).

[419] D. Deutsch and R. Jozsa, "Rapid solution of problems by quantum computation", *Proceedings of the Royal Society of London A* **439**, 553–558 (1992).

[420] D.S. Abrams and S. Lloyd, "Simulation of many-body Fermi systems on a universal quantum computer", *Phys. Rev. Lett.* **79**, 2586–2589 (1997).

[421] A.W. Harrow, A. Hassidim, and S. Lloyd, "Quantum algorithm for linear systems of equations", *Phys. Rev. Lett.* **103**, 150502 (2009).

[422] L.M.K. Vandersypen, M. Steffen, G. Breyta, C.S. Yannoni, M.H. Sherwood, and I.L. Chuang, "Experimental realization of quantum Shor's factoring algorithm using nuclear magnetic resonance", *Nature* **414**, 883–887 (2001).

[423] E. Martin-Lopez, A. Laing, T. Lawson, R. Alvarez, X. Zhou, and J.L. O'Brien, "Experimental realisation of Shor's quantum factoring algorithm using qubit recycling", *Nature Photonics* **6**, 773–776 (2012).

[424] N. Xu, J. Zhu, D. Lu, X. Zhou, X. Peng, and J. Du, "Quantum factorization of 143 on a dipolar-coupling nuclear magnetic resonance system", *Phys. Rev. Lett.* **108**, 130501 (2012); see also *Phys. Rev. Lett.* **109**, 269902 (2012).

[425] S. Haroche and J.M. Raimond, "Quantum computing: dream or nightmare?", *Phys. Today* **49**, 51–52 (August 1996).

[426] P.W. Shor, "Scheme for reducing decoherence in quantum computer memory", *Phys. Rev.* **A 52**, R2493–R2496 (1995).

[427] A.M. Steane, "Error correcting codes in quantum theory", *Phys. Rev. Lett.* **77**, 793–796 (1996).

[428] J. Preskill, "Battling decoherence: the fault-tolerant quantum computer", *Phys. Today* **52**, 24–30 (June 1999); "Reliable quantum computers", *Proc. Roy. Soc. Lond.* **A 454**, 385–410 (1998) or arXiv:quant-ph/9705031v3.

[429] C.H. Bennett, G. Brassard, S. Popescu, B. Schumacher, J.A. Smolin, and W.K. Wootters, "Purification of noisy entanglement and faithful teleportation via noisy channels", *Phys. Rev. Lett.* **76**, 722–725 (1996).

[430] S.J. Devitt, W.J. Munro, and K. Nemoto, "Quantum error correction for beginners", *Rep. Progr. Phys.* **76**, 07001 (35 pages) (2013).

[431] B.M. Terhal, "Quantum error correction for quantum memories", *Rev. Mod. Phys.* **87**, 307–343 (2015).

[432] D. Gottesman, "An Introduction to Quantum Error Correction and Fault-Tolerant Quantum Computation", in "Quantum Information Science and Its Contributions to Mathematics", *Proceedings of Symposia in Applied Mathematics* **68**, 13–58 (Amer. Math. Soc., Providence, Rhode Island, 2010); or arXiv:0904.2557 [quant-ph].

[433] J. Kempe, "Approaches to quantum error correction", in "Quantum Decoherence", Poincaré seminar 2005, Progress in Mathematical Physics series, 85–123 (2006); arXiv:quant-ph/0612185. J. Kempe, O. Regev, F. Unger, and R. de Wolf, "Upper bounds on the noise threshold for fault-tolerant quantum computing", arXiv:0802.1464 [quant-ph] (2008).

[434] C.H. Bennett, D.P. DiVincenzo, J.A. Smolin, and W.A. Wootters, "Mixed-state entanglement and quantum error correction", *Phys. Rev.* **A 54**, 3824–3851 (1996).

[435] H.J. Briegel, W. Dür, J.I. Cirac, and P. Zoller, "Quantum repeaters: the role of imperfect local operations in quantum communication", *Phys. Rev. Lett.* **81**, 5932–5935 (1998).

[436] R.B. Griffiths and Chi-Sheng Niu, "Semiclassical Fourier transform for quantum computation", *Phys. Rev. Lett.* **76**, 3228–3231 (1996).

[437] F. Verstraete, M.M. Wolf, and J.I Cirac, "Quantum computation and quantum-state engineering driven by dissipation", *Nature Physics* **5**, 633–636 (2009).

[438] I. Cirac and P. Zoller, "Goals and opportunities in quantum simulation", *Nature Physics* **8**, 264–266 (2012).

[439] I. Buluta and F. Nori, "Quantum simulators", *Science* **326**, 108–111 (2009).

[440] J.D. Biamonte, V. Bergholm, J.D. Whitfield, J. Fitzsimons, and A. Aspuru-Guzik, "Adiabatic quantum simulators", *AIP Advances* **1**, 022126 (2011).

[441] I.M. Georgescu, S. Ashab, and F. Nori, "Quantum simulation", *Rev. Mod. Phys.* **86**, 153–185 (2014).

[442] P. Hauke, F.M. Cucchiette, L. Tagliacozzo, I. Deutsch, and M. Lewenstein, "Can one trust quantum simulators?", *Rep. Progr. Phys.* **75**, 082401 (2012).

[443] M. Greiner, O. Mandel, T. Esslinger, T.W. Hänsch, and I. Bloch, "Quantum phase transition from a superfluid to a Mott insulator in a gas of ultracold atoms", *Nature* **415**, 39–44 (2002).

[444] X. Peng, J. Zhang, J. Du, and D. Suter, "Quantum simulation of a system with competing two- and three-body interactions", *Phys. Rev. Lett.* **103**, 140501 (2009).

[445] J.Q. You and F. Nori, "Quantum information", *Physics Today* **58**, 42–47 (2005); "Atomic physics and quantum optics using superconducting circuits", *Nature* **474**, 589-597 (2011).

[446] B.P. Lanyon, J.D. Whitfield, G.G. Gillett, M.E. Goggin, M.P. Almeida, I.Kassal, J.D. Biamonte, M. Mohseni, B.J. Powell, M. Barbieri, A. Aspuru-Guzik and A.G.

White, "Towards quantum chemistry on a quantum computer", *Nature Chemistry* **2**, 106–111 (2010).

[447] J. Dalibard, F. Gerbier, G. Juzeliunas, and P. Öhberg, "Artificial gauge potentials for neutral atoms", *Rev. Mod. Phys.* **83**, 1523–1543 (2011).

[448] P. Grangier, J.A. Levenson, and J.P. Poizat, "Quantum nondemolition measurements in optics", *Nature* **396**, 537–542 (1998).

[449] H.D. Zeh, "On the interpretation of measurement in quantum theory", *Found. Phys.* **I**, 69–76 (1970).

[450] W.H. Zurek, "Pointer basis of quantum apparatus: into what mixture does the wave packet collapse?", *Phys. Rev.* **D 24**, 1516–1525 (1981); "Environment-induced superselection rules", *Phys. Rev.* **D 26**, 1862–1880 (1982).

[451] W.H. Zurek, "Decoherence, einselection and the quantum origin of the classical", *Rev. Mod. Phys.* **75**, 715–775 (2003).

[452] K. Hepp, "Quantum theory of measurement and macroscopic observables", *Helv. Phys. Acta* **45**, 237–248 (1972).

[453] J.S. Bell, "On wave packet reduction in the Coleman-Hepp model", *Helv. Phys. Acta* **48**, 93–98 (1975); reprinted in [6].

[454] W.H. Zurek, "Preferred states, predictability, classicality and the environment-induced decoherence", *Progr. Theor. Phys.* **89**, 281–312 (1993); a shorter version is available in "Decoherence and the transition from quantum to classical", *Phys. Today* **44**, 36–44 (October 1991).

[455] M. Simonius, "Spontaneous symmetry breaking and blocking of metastable states", *Phys. Rev. Lett.* **40**, 980–983 (1978).

[456] W.H. Zurek, "Environment-assisted invariance, entanglement and probabilities in quantum physics", *Phys. Rev. Lett.* **90**, 120404 (2003).

[457] F. Hund, "Zur Deutung der Molekelspektren III", *Zeit. Phys.* **43**, 805–826 (1927).

[458] J. Trost and K. Hornberger, "Hund's paradox and the collisional stabilization of chiral molecules", *Phys. Rev. Lett.* **103**, 023202 (2009).

[459] E.P. Wigner, "Die Messung quantenmechanischer Operatoren", *Z. Phys.* **131**, 101–108 (1952).

[460] P. Busch, "Translation of 'Die Messung quantenmechanischer Operatoren' by E.P. Wigner", arXiv:1012.4372v1 [quant-ph] (2010).

[461] H. Araki and M. Yanase, "Measurement of quantum mechanical operators", *Phys. Rev.* **120**, 622–626 (1960).

[462] M. Yanase, "Optimal measuring apparatus", *Phys. Rev.* **123**, 666–668 (1961).

[463] T. Ohira and P. Pearle, "Perfect disturbing measurements", *Am. J. Phys.* **56**, 692–695 (1988).

[464] G.C. Ghirardi, F. Miglietta, A. Rimini, and T. Weber, "Determination of the minimal amount of non-ideality and identification of the optimal measuring apparatuses", *Phys. Rev.* **D 24**, 347–352 (1981); "Analysis of a model example", Phys. Rev. **D 24**, 353–358 (1981).

[465] M. Burgos, "Contradiction between conservation laws and orthodox quantum mechanics", *J. Modern Phys.* **1**, 137–142 (2010).

[466] L. Loveridge and P. Busch, "Measurement of quantum mechanical operators revisited", *Europ. Phys. J.* **62**, 297–307 (2011).

[467] Y. Aharonov, J. Anandan, S. Popescu, and L. Vaidman, "Superpositions of time evolutions of a quantum system and a quantum time-translation machine", *Phys. Rev. Lett.* **64**, 2965–2968 (1990).

[468] C. Ferries and J. Combes, "How the result of a single coin toss can turn out to be 100 heads", *Phys. Rev. Lett.* **113**, 120404 (2014).

[469] L. Vaidman, "Comment on 'How the result of a single coin toss can turn out to be 100 heads", arXiv:1409.5386v1 (2014).

[470] L. Vaidman, "Weak value controversy", *Phil. Trans. Roy. Soc.* **A 375** (2017).

[471] S. Tanaka and N. Yamamoto, "Information amplification via postselection: a parameter-estimation perspective ", *Phys. Rev.* **A 88**, 042116 (2013).

[472] C. Ferries and J. Combes, "Weak value amplification is suboptimal for estimation and detection", *Phys. Rev. Lett.* **112**, 040406 (2014).

[473] G.C. Knee and E.M. Gauger, "When amplification with weak values fails to suppress technical noise", *Phys. Rev.* **X 4**, 011032 (2014).

[474] A.N. Jordan, J. Martinez-Rincón, and J.C.Howell, "Technical advantages for weak-value amplification: when less is more", *Phys. Rev.* **X 4**, 011031 (2014).

[475] G.I. Viza, J. Martinez-Rincón, G.B. Alves, A.N. Jordan, and J.C.Howell, "Experimentally quantifying the advantages of weak-value-based metrology", *Phys. Rev.* **A 92**, 0312127 (2015).

[476] J. Sinclair, M. Hallaji, A. Steinberg, J. Tollaksen, and A. Jordan, "Weak-value amplification and optimal parameter estimation in the presence of correlated noise", *Physical Review* **A 96**, 052128 (2017).

[477] N.W.M. Richtie, J.G. Story, and R. G. Hulet, "Realization of a measurement of a weak value", *Phys. Rev. Lett.* **66**, 1107–1110 (1991).

[478] D.R. Solli, C.F. McCormick, R.Y. Chiao, S. Popescu, and J.M. Hickmann, "Fast light, slow light, and phase singularities: a connection to generalized weak values", *Phys. Rev. Lett.* **92**, 043601 (2004).

[479] N. Brunner, V. Scarani, M. Wegmüller, M. Legré, and N. Gisin, "Direct measurement of superluminal group velocity and signal velocity in an optical fiber", *Phys. Rev. Lett.* **93**, 203902 (2004).

[480] G.J. Pryde, J.L. O'Brien, A.G. White, T.C. Ralph, and H.M. Wiseman, "Measurement of quantum weak values of photon polarization", *Phys. Rev. Lett.* **94**, 220405 (2005).

[481] R. Mir, J.S. Lundeen, M.W. Mitchell, A.M. Steinberg, J.L. Garretson, and H.M. Wiseman, "A double slit 'which way' experiment on the complementarity-uncertainty debate", *New. J. Phys.* **9**, 287–297 (2007).

[482] O. Hosten and P. Kwiat, "Observation of the spin Hall effect of light via weak measurements", *Science* **319**, 787–790 (2008).

[483] J.S. Lundeen and A.M. Steinberg, "Experimental joint weak measurement on a photon pair as a probe of Hardy's paradox", *Phys. Rev. Lett.* **102**, 020404 (2009).

[484] K. Yokota, T. Yamamoto, M. Koashi, and N. Imoto, "Direct observation of Hardy's paradox by joint weak measurement with an entangled photon pair", *New. J. Phys.* **11**, 033011 (2009).

[485] P. Ben Dixon, D.J. Starling, A.N. Jordan, and J.C. Howell, "Ultrasensitive beam deflection measurement via interferometric weak value amplification", *Phys. Rev. Lett.* **102**, 173601 (2009). D.J. Starling, P. Ben Dixon, A.N. Jordan, and J.C. Howell, "Optimizing the signal-to-noise ratio of a beam-deflection measurement with interferometric weak values", *Phys. Rev.* **A80**, 041803 (2009).

[486] N. Brunner and C. Simon, "Measuring small longitudinal phase shifts: weak measurements or standard interferometry", *Phys. Rev. Lett.* **105**, 010405 (2010).

[487] A. Feizpour, M. Hallaji, G. Dmochowski, and A. Steinberg, "Observation of the nonlinear phase shift due to single postselected photons", *Nature Phys.* **11** DOI:10.1038/NPHYS3433.

[488] N.S. Williams and A.N. Jordan, "Weak values and the Leggett-Garg inequality in solid-state qubits", *Phys. Rev. Lett.* **100**, 026804 (2008).

[489] D.T. Gillepsie, "The mathematics of Brownian motion and Johnson noise", *Am. J. Phys.* **64**, 225–240 (1995).

[490] H.P. McKean, *Stochastic Integrals*, AMS Chelsea Publishing, Providence (1969).

[491] N. Gisin, "A simple nonlinear dissipative quantum evolution equation", *J. Phys.* **A 14**, 2259–2267 (1981).

[492] N. Gisin, "Irreversible quantum dynamics and the Hilbert space structure of quantum dynamics", *J. Math. Phys.* **24**, 1779–1782 (1983).

[493] N. Gisin, "Quantum measurements and stochastic processes", *Phys. Rev. Lett.* **52**, 1657–1660 (1984).

[494] T.A. Brun, "A simple model of quantum trajectories", *Am. J. Phys.* **70**, 719–737 (2002).

[495] K. Jacobs and D.A. Steck, "A straightforward introduction to continuous quantum measurement", *Contemp. Phys.* **47**, 279–303 (2007), arXiv:quant-ph/0611067 (2006).

[496] V.P. Belavkin, "Nondemolition measurement and control in quantum dynamical systems", *Proc. of CISM Seminars on Information Complexity and Control in Quantum Systems*, A. Blaquière, S. Diner, and G. Lochak editors, Springer Verlag (1987), pp. 311–329.

[497] N.F. Mott, "The wave mechanics of α-ray tracks", *Proc. Royal Soc.* **A 126**, 79–84 (1929); reprinted in "Quantum theory of measurement", J.A. Wheeler and W.H. Zurek editors, Princeton University Press (1983), pp. 129–134.

[498] W. Nagourney, J. Sandberg and H. Dehmelt, "Shelved optical electron amplifier: observation of quantum jumps", *Phys. Rev. Lett.* **56**, 2797–2799 (1986); H. Dehmelt, "Experiments with an isolated subatomic particle at rest", *Rev. Mod. Phys.* **62**, 525–530 (1990).

[499] T. Sauter, W. Neuhauser, R. Blatt, and P.E. Toschek, "Observation of quantum jumps", *Phys. Rev. Lett.* **57**, 1696–1698 (1986).

[500] J.C. Bergquist, R.G. Hulet, W.M. Itano, and D.J. Wineland, "Observation of quantum jumps in a single atom", *Phys. Rev. Lett.* **57**, 1699–1702 (1986).

[501] W.M. Itano, J.C. Bergquist, R.G. Hulet, and D.J. Wineland, "Radiative decay rates in Hg$^+$ from observation of quantum jumps in a single ion", *Phys. Rev. Lett.* **59**, 2732–2735 (1987).

[502] E. Schrödinger, "Are there quantum jumps?", *British J. Phil. Sci.* **3**, 109–123 and 233–242 (1952).

[503] G. Greenstein and A.G. Zajonc, "Do quantum jumps occcur at well-defined moments of time?", *Am. J. Phys.* **63**, 743–745 (1995).

[504] C. Cohen-Tannoudji and J. Dalibard, "Single-atom laser spectroscopy looking for dark periods in fluorescence light", *Europhys. Lett.* **1**, 441–448 (1986).

[505] M. Porrati and S. Puttermann, "Wave-function collapse due to null measurements: the origin of intermittent atomic fluorescence", *Phys. Rev.* **A 36**, 929–932 (1987).

[506] S. Peil and G. Gabrielse, "Observing the quantum limit of an electron cyclotron: QND measurements of quantum jumps between Fock states", *Phys. Rev. Lett.* **83**, 1287–1290 (1999).

[507] D. Hanneke, S. Fogwell, and G. Gabrielse, "New measurement of the electron magnetic moment and the fine structure constant", *Phys. Rev. Lett.* **100**, 120801 (2008).

[508] M. Brune, S. Haroche, V. Lefevre, J.M. Raimond, and N. Zagury, "Quantum nondemolition measurement of small photon numbers by Rydberg-atom phase-sensitive detection", *Phys. Rev. Lett.* **65**, 976–979 (1990).

[509] S. Gleyzes, S. Kuhr, C. Guerlin, J. Bernu, S. Deleglise, U.B. Hoff, M. Brune, J.-M. Raimond, and S. Haroche, "Quantum jumps of light recording the birth and death of a photon in a cavity", *Nature* **446**, 297–300 (2007); C. Guerlin, J. Bernu, S. Deleglise, C. Sayrin, S. Gleyzes, S. Kuhr, M. Brune, J.M. Raimond, and S. Haroche, "Progressive state collapse and quantum nondemolition photon counting", *Nature* **448**, 889–893 (2007).

[510] J. Javanainen and S.M Yoo, "Quantum phase of a Bose–Einstein condensate with arbitrary number of atoms", *Phys. Rev. Lett.* **76**, 161–164 (1996).

[511] M.R. Andrews, C.G. Townsend, H.J. Miesner, D.S. Durfee, D.M. Kurn, and W. Ketterle, "Observation of interference between two Bose condensates", *Science* **275**, 637–641 (1997).

[512] A.J. Leggett and F. Sols, "On the concept of spontaneously broken gauge symmetry in condensed matter physics", *Found. Phys.* **21**, 353–364 (1991).

[513] E.P. Wigner, "Interpretation of quantum mechanics", lectures given in 1976 at Princeton University, later published in *Quantum Theory of Measurement*, J.A. Wheeler and W.H. Zurek editors, Princeton University Press (1983), pp. 260–314; see also Wigner's contribution in "Foundations of quantum mechanics", *Proc. Enrico Fermi Int. Summer School*, B. d'Espagnat editor, Academic Press (1971).

[514] N.D. Mermin, "What is quantum mechanics trying to tell us?", *Am. J. Phys.* **66**, 753–767 (1998).

[515] B. Misra and E.C.G. Sudarshan, "The Zeno's paradox in quantum theory", *J. Math. Phys.* (NY) **18**, 756–763 (1977).

[516] A. Zeilinger, "A foundational principle for quantum mechanics", *Found. Phys.* **29**, 631–643 (1999).

[517] C. Brukner and A. Zeilinger, "Operationally invariant information in quantum measurements", *Phys. Rev. Lett.* **83**, 3354–3357 (1999).

[518] C.A. Fuchs, "Quantum foundations in the light of quantum information", arXiv:quant-ph/0106166 (2001).

[519] C.A. Fuchs, "Quantum mechanics as quantum information (and only a little more)", arXiv:quant-ph/0205039 (2002).

[520] I. Pitowsky, "Betting on the outcomes of measurements: a Bayesian theory of quantum probability", *Studies in History and Philosophy of Modern Physics* **34**, 395–414 (2003).

[521] G. Auletta, *Foundations and interpretations of quantum mechanics*, World Scientific (2001); "Quantum information as a general paradigm", *Found. Phys.* **35**, 787–815 (2005).

[522] G. Auletta, M. Fortunato, G. Parisi, *Quantum mechanics*, Cambridge Univerity Press (2014).

[523] J. Bub, "Quantum probabilities: an information-theoretic interpretation", in *Probabilities in Physics*, C. Beisbart and S. Hartmann editors, Oxford University Press (2011).

[524] D. Deutsch and P. Hayden, "Information flow in entangled quantum systems", *Proc. Royal Soc.* **A 456**, 1759–1774 (2000).

[525] L.E. Ballentine, "The statistical interpretation of quantum mechanics", *Rev. Mod. Phys.* **42**, 358–381 (1970).

[526] A.J. Leggett, "Probing quantum mechanics towards the everyday world: where do we stand?", *Physica Scripta* **T102**, 69–73 (2002).

[527] A.E. Allahverdyan, R. Balian, and T.M. Nieuwenhuizen, "A subensemble theory of ideal quantum measurement processes", *Annals of Physics* **376**, 324–352 (2017); arXiv:1303.7257v4 [quant-ph].

[528] C. Rovelli, "Relational quantum mechanics", *Int. J. Theor. Phys.* **35**, 1637–1678 (1996); ArXiv:quant-ph/9609002v2 (2008).

[529] F. Laudisa and C. Rovelli, "Relational quantum mechanics", *Stanford Encyclopedia of Philosophy* (2008), http://plato.stanford.edu/entries/qm-relational/

[530] M. Smerlak and C. Rovelli, "Relational EPR", *Found. Phys.* **37**, 427–445 (2007).

[531] C.M. Caves, C.A. Fuchs, and R. Schack, "Subjective probability and quantum certainty", *Stud. Hist. Phil. Mod. Phys.* **38**, 255–274 (2007).

[532] C.A. Fuchs, "QBism, the perimeter of quantum Bayesianism", arXiv:1003.5290v1 [quant-ph](2010).

[533] R. Healey, "Quantum-Bayesian and pragmatist views of quantum theory", *Stanford Encyclopedia of Physics* (2017) https://plato.stanford.edu/entries/quantum-bayesian/

[534] C.A. Fuchs, N.D. Mermin, and R. Schack, "An introduction to QBism with an application to the locality in quantum mechanics", arXiv:1311.5253v1 [quant-ph](2013).

[535] F. Zwicky, "On a new type of reasoning and some of its possible consequences", *Phys. Rev.* **43**, 1031–1033 (1933).

[536] G. Birkhoff and J. von Neumann, "The logic of quantum mechanics", *Ann. Math.* **37**, 823–843 (1936).

[537] M. Strauss, "Grundlagen der modernen Physik", in *Mikrokosmos-Makrokosmos: Philosophish-theoretische Probleme der Naturwissenchaften, Technik und Medizin*, Akademie Verlag, Berlin (1967).

[538] K.R. Popper, "Birkhoff and von Neumann's interpretation of quantum mechanics", *Nature* **219**, 682–685 (1968).

[539] P. Jordan, "Zur Quanten-Logik", *Archiv der Mathematik* **2**, 166–171 (1949).

[540] F. David, *The formalisms of quantum mechanics: an introduction*, Spinger (2015).

[541] R. Hughes, "La logique quantique", *Pour la Science* December 1981, 36–49.

[542] P. Mittelstaedt, *Quantum Logic*, Kluwer Academic Publishers (1978).

[543] E.G. Beltrametti and G. Cassinelli, *The Logic of Quantum Mechanics*, Cambridge University Press (1984).

[544] A. Grinbaum, "The significance of information in quantum theory", Ph.D. thesis, Ecole polytechnique (2004), arXiv:quant-ph/0410071 (2004). "Reconstruction of quantum theory", *Brit. J. Phil. Sci* **58**, 387–408 (2007).

[545] C. de Ronde, G. Domenech, and H. Freytes, "Quantum logic in historical and philosophical perspective", Internet Encyclopedia of Philosophy, http://www.iep.utm.edu/qu-logic/

[546] J.S. Bell, "A new approach to quantum logic", *Brit. J. Phil. Sci.* **37** 83–99 (1986).

[547] H. Reichenbach, *Philosophic Foundations of Quantum Mechanics*, University of California Press (1965).

[548] C.F. von Weizsäcker, *Göttingische Gelehrte Anzeigen*, **208**, 117–136 (1954).

[549] R. Haag, *Local quantum physics: Fields, particles, algebras*, Springer (1996).

[550] P. Jordan, J. von Neumann, and E. Wigner, "An algebraic generalization of the quantum mechanical formalism", *Ann. Math.* **35**, 29–64 (1934).

[551] I.M. Gelfand and M.A. Naimark, "On the embedding of normed rings into the ring of operators in Hilbert space", *Mat. Sbornik* **12**, 197–213 (1943).

[552] I.E. Segal, "Irreductible representations of operator algebras", *Bull. Amer. Math. Soc.* **61**, 69–105 (1947); "Postulates for general quantum mechanics", *Ann. Math.* **48**, 930–948 (1947).

[553] R. Haag and D. Kastler, "An algebraic approach to quantum field theory", *J. Math. Phys.* **7**, 848–861 (1964).

[554] F.J. Murray and J. von Neumann, "On rings of operators", *Ann. Math.* **37**, 116–229 (1936).

[555] A. Connes, "Une classification des facteurs du type III", *Ann. Sci. Ecole Norm. Sup.* **6**, 133–252 (1973).

[556] A. Connes, *Noncommutative Geometry*, Academic Press (1994).

[557] G. Mackey, *Mathematical Foundations of Quantum Mechanics*, Benjamin, New York (1963).

[558] C. Piron, "Axiomatique quantique", *Helv. Phys. Acta* **37**, 439–468 (1964).

[559] J.M. Jauch and C. Piron, "On the structure of quantal proposition systems", *Helv. Phys. Acta* **42**, 842–848 (1969).

[560] M.P. Solèr, "Characterization of Hilbert spaces by orthomodular spaces", *Comm. Algebra* **23**, 219–243 (1995).

[561] B. Coecke and E.O. Paquette, "Categories for the practicing physicist", arXiv:0905.3010v2 [quant-ph]; in *New Structures for Physics*, Springer (2011), pp. 173–286.

[562] B. Coecke, "Quantum picturalism", *Contemp. Phys.* **51**, 59–83 (2010).

[563] H. Barnum and A. Wilce, "Information processing in convex operational theories", *Electronic Notes in Theor. Computer Sci.* **12**, 3–15 (2011).

[564] A. Wilce, "Quantum logic and probability theory", *Stanford encyclopedia of philosophy* (2008), http://plato.stanford.edu/entries/qt-quantlog/.

[565] A.M. Gleason, "Measures on the closed subspaces of a Hilbert space", *J. Math. and Mech.* **6**, 885–893 (1957).

[566] P. Bush, "Quantum states and generalized observables: a simple proof of Gleason's theorem", *Phys. Rev. Lett.* **91**, 120403 (2016);

[567] A. Auffèves and P. Grangier, "Contexts, systems and modalities: a new ontology for quantum mechanics", *Found. Phys.* **46**, 121–137 (2015); "Recovering the quantum formalism from physically realist axioms", arXiv:1610.06164v2.

[568] R. Omnès, "Logical reformulation of quantum mechanics", *J. Stat. Phys.* **53**, "I: Foundations", 893–932; "II: Interferences and the EPR experiments", 933–955; "III: Classical limit and irreversibility", 957–975 (1988).

[569] M. Gell-Mann and J.B. Hartle, "Classical equations for quantum systems", *Phys. Rev.* **D 47**, 3345–3382 (1993).

[570] R. Omnès *The Interpretation of Quantum Mechanics*, Princeton University Press (1994); *Understanding Quantum Mechanics*, Princeton University Press (1999).

[571] R.B. Griffiths and R. Omnès, "Consistent histories and quantum measurements", *Phys. Today* **52**, 26–31 (August 1999).

[572] P.C. Hohenberg, "*Colloquium:* An introduction to consistent quantum theory", *Rev. Mod. Phys.* **82**, 2835–2844 (2010).

[573] Y. Aharonov, P.G. Bergmann, and J.L. Lebowitz, "Time symmetry in the quantum process of measurement", *Phys. Rev.* **B 134**, 1410–1416 (1964).

[574] R.B. Griffiths, "Consistent histories and quantum reasoning", *Phys. Rev.* **A 54**, 2759–2774 (1996).

[575] R.B. Griffiths, "Choice of consistent family and quantum incompatibility", *Phys. Rev.* **A 57**, 1604–1618 (1998).

[576] "Observant readers take the measure of novel approaches to quantum theory: some get Bohmed", *Phys. Today* **52**, 11–15 and 89–92 (February 1999).

[577] R.B. Griffiths, "Correlations in separated quantum systems: a consistent history analysis of the EPR problem", *Am. J. Phys.* **55**, 11–17 (1987).

[578] F. Dowker and A. Kent, "Properties of consistent histories", *Phys. Rev. Lett.* **75**, 3038–3041 (1995); "On the consistent histories approach to quantum mechanics", *J. Stat. Phys.* **82**, 1575–1646 (1996).

[579] A. Kent, "Quasiclassical dynamics in a closed quantum system", *Phys. Rev.* **A 54**, 4670–4675 (1996).

[580] T.A. Brun, "Continuous measurements, quantum trajectories and decoherent histories", *Phys. Rev.* **A 61**, 042107 (2000).

[581] J.S. Bell, "Are there quantum jumps?", in *Schrödinger–Centenary Celebration of a Polymath*, C.W. Kilmister editor, Cambridge University Press (1987), p. 41; see also Chapter 22 of [6].

[582] J.S. Bell, "Beables for quantum field theory", CERN-TH.4035/84 (August 2 1984); *Phys. Rep.* **137**, 49–54 (1986); Chapter 19 of [6].

[583] L. de Broglie, "La mécanique ondulatoire et la structure atomique de la matière et du rayonnement", *J. Physique et le Radium*, série VI, tome VIII, 225–241 (1927); "Interpretation of quantum mechanics by the double solution theory", *Ann. Fond. Louis de Broglie* **12**, Nr 4 (1987).

[584] L. de Broglie, *Tentative d'Interprétation Causale et Non-linéaire de la Mécanique Ondulatoire*, Gauthier-Villars, Paris (1956).

[585] L. de Broglie, *Les Incertitudes d'Heisenberg et l'Interprétation Probabiliste de la Mécanique Ondulatoire*, Gauthier-Villars and Bordas, Paris (1982).

[586] J.T. Cushing, *Quantum Mechanics*, The University of Chicago Press (1994).

[587] D. Bohm, "Proof that probability density approaches $|\Psi|^2$ in causal interpretation of quantum theory", *Phys. Rev.* **89**, 458–466 (1953).

[588] P.R. Holland, *The Quantum Theory of Motion*, Cambridge University Press (1993).

[589] S. Goldstein, "Bohmian mechanics", Stanford Encyclopedia of Philosophy, https://plato.stanford.edu/entries/qm-bohm/ (2001 and 2013).

[590] D. Dürr and S. Teufel, *Bohmian mechanics*, Springer (2009).

[591] X. Oriols and J. Mompart, *Applied Bohmian mechanics: from nanoscale systems to cosmology*, Editorial Pan Stanford Publishing Pte. Ltd (2012); see chapter 1 "Overview of Bohmian mechanics", or ArXiv:1206.1084v2 [quant-ph].

[592] J. Bricmont, *Making Sense of Quantum Mechanics*, Springer (2016).

[593] E. Madelung, "Quantentheorie in hydrodynamische Form", *Z. Phys.* **40**, 322–326 (1927).

[594] D. Dürr, S. Goldstein, and N. Zanghì, "Quantum equilibrium and the origin of absolute uncertainty", *J. Stat. Phys.* **67**, 843–907 (1992).

[595] A. Valentini, "Signal-locality in hidden-variables theories", *Phys. Lett.* **A 297**, 273–278 (2002); "Beyond the quantum", *Physics World* 32–37 (November 2009).

[596] J.S. Bell, chapter 18 of [55] (p. 128 of [6]).

[597] D. Dürr, S. Goldstein, and N. Zanghi, *Quantum Physics Without Quantum Philosophy*, Springer (2012), chapter 12; see also S. Goldstein and N. Zanghi, "Reality and the role of the wavefunction in quantum theory", arxiv:1101.4575v1.

[598] P. Holland, "Hamiltonian theory of wave and particle in quantum mechanics II: Hamilton–Jacobi theory and particle back-reaction", *Nuov. Cim.* **B 116**, 1143–1172 (2001).

[599] C. Philippidis, C. Dewdney, and B.J. Hiley, "Quantum interference and the quantum potential", *Nuov. Cim.* **52 B**, 15–23 (1979).

[600] J.S. Bell, "De Broglie–Bohm, delayed choice double slit experiment, and density matrix", *International Journal of Quantum Chemistry* **18**, supplement symposium 14, 155–159 (1980); chapter 14 of [6].

[601] B.J. Hiley, "Welcher Weg experiments from the Bohm perspective", contribution to the Växjö conference (2005),
       http://www.bbk.ac.uk/tpru/BasilHiley/WelcherWegBohmBJH2.pdf.

[602] D. Greenberger, M. Horne, and A. Zeilinger, "Multiparticle interferometry and the superposition principle", *Phys. Today* **46**, 22–29 (1993).

[603] K. Gottfried, "Two particle interference", *Am. J. Phys.* **68**, 143–147 (2000).

[604] E. Guay and L. Marchildon, "Two-particle interference in standard and Bohmian quantum mechanics", *J. Phys.* **A 36**, 5617–5624 (2003).

[605] L. Vaidman, "The reality of Bohmian quantum mechanics or Can you kill with an empty wave bullet?", *Found. Phys.* **35**, 299–312 (2005).

[606] E. Deotto and G.C. Ghirardi, "Bohmian mechanics revisited", *Found. Phys.* **28**, 1–30 (1998).

[607] P. Holland, "Uniqueness of paths in quantum mechanics", *Phys. Rev.* **A 60**, 4326–4330 (1999); "Uniqueness of conserved currents in quantum mechanics", *Ann. Phys. (Leipzig)* **12**, 446–462 (2003).

[608] P. Holland and C. Philippidis, "Implications of Lorentz covariance for the guidance equation in two-slit quantum interference", **A 67**, 062105 (2003).

[609] H.M. Wiseman, "Grounding Bohmian mechanics in weak values and bayesianism", *New J. Phys.* **9**, 165 (2007).

[610] T. Nikuni and J.E. Williams, "Kinetic theory of a spin 1/2 Bose-condensed gas", *J. Low Temperature Phys.* **133**, 323–374 (2003).

[611] B.G. Englert, M.O. Scully, G. Süssmann, and H. Walther, "Surrealistic Bohm trajectories", *Z. Naturforschung* **47a**, 1175–1186 (1992).

[612] C. Dewdney, P.R. Holland, and A. Kyprianidis, "What happens in a spin measurement?", *Phys. Lett.* **A 119**, 259–267 (1986).

[613] E.P. Wigner, "Rejoinder", *Am. J. Phys.* **39**, 1097 (1971).

[614] J. Clauser, "von Neumann's informal hidden-variable argument", *Am. J. Phys.* **39**, 1095 (1971); "Reply to Dr Wigner's objections", *Am. J. Phys.* **39**, 1098 (1971).

[615] C. Dewdney, P.R. Holland, and A. Kyprianidis, "A causal account of nonlocal Einstein-Podolsky-Rosen spin correlations", *J. Phys. Math. Gen.* **20**, 4717–4732 (1987).

[616] C. Dewdney, "Nonlocally correlated trajectories in two-particle quantum mechanics", *Found. of Phys.* **18**, 867–886 (1988).

[617] J.S. Bell, "De Broglie-Bohm, delayed-choice double-slit experiment, and density matrix", *Int. J. Quant. Chem.*, Quantum Chemistry Symposium **14**, 155–159 (1980); reprinted in [6].

[618] J. des Cloizeaux, "A reformulation of Schrödinger and Dirac equations in terms of observable local densities and electromagnetic fields: a step towards a new interpretation of quantum mechanics?", *J. Physique* **44**, 885–908 (1983).

[619] S. Colin and W. Struyve, "A Dirac sea pilot-wave model for quantum field theory", *J. Phys. A Math. Theor.* **40**, 7309–7341 (2007).

[620] D. Dürr, S. Goldstein, R. Tumulka, and N. Zanghì, "Trajectories and particle creation and annihilation in quantum field theory", *J. Phys. A Math. Gen.* **36**, 4143–4149 (2003).

[621] D. Dürr, S. Goldstein, R. Tumulka and N. Zanghì, "Bohmian mechanics and quantum field theory", *Phys. Rev. Lett.* **93**, 090402 (2004).

[622] W. Struyve, "Field beables for quantum field theory", *Rept. Prog. Phys.* **73**, 106001 (2010); arXiv:0707.3685v2 [quant-ph] (2007).

[623] K. Berndl, D. Dürr, S. Goldstein, and N. Zhanghi, "Nonlocality, Lorentz invariance, and Bohmian quantum theory", *Phys. Rev.* **A 53**, 2062–2073 (1996).

[624] G. Horton and C. Dewdney, "A nonlocal, Lorentz-invadiant, hidden-variable interpretation of relativistic quantum mechanics based on particle trajectories", *J. Phys. A Math. Gen.* **34**, 9871–9878 (2001); "A relativistically covariant version of Bohm's quantum field theory for the scalar field", *J. Phys. A Math. Gen.* **37**, 11935–11943 (2004).

[625] H. Nikolić, "Relativistic quantum mechanics and the Bohmian interpretation", *Found. Phys. Lett.* **18**, 549–561 (2005); "QFT as pilot-wave theory of particle creation and destruction", *J. Mod. Phys.* **A 25**, 1477–1505 (2010); "Bohmian mechanics in relativistic quantum mechanics, quantum field theory and string theory ", *J. Phys. Conference Series* **67**, 012035 (2007).

[626] C.L. Lopreone and R.E. Wyatt, "Quantum wave packet dynamics with trajectories", *Phys. Rev. Lett.* **82**, 5190–5193 (1999).

[627] I.P. Christov, "Time-dependent quantum Monte Carlo: preparation of the ground state", *New Journal Phys.* **9**, 70 (2007); "Polynomial-time-scaling quantum dynamics with time-dependent quantum Monte Carlo", *J. Chem. Phys.* **A 113**, 6016–6021 (2009).

[628] L. Shifren, R. Akis, and D.K. Ferry, "Correspondence between quantum and classical motion: comparing Bohmian mechanics with a smoothed effective potential approach", *Phys. Lett.* **A 274**, 75–83 (2000).

[629] G. Allaberda, D. Marian, A. Benali, S. Yaro, N. Zanghi, and X. Oriols, "Time-resolved transport with quantum trajectories", *J. Comput. Electron.* 12:405–419 (2013).

[630] A. Benseny, G. Albareda, A. Sanz, J. Mompart, and W. Oriols, "Applied Bohmian mechanics", *Eur. Phys. J.* 68:286 (2014).

[631] P. Peter, E. Pinho, and N. Pinto-Neto, "Tensor perturbations in quantum cosmological backgrounds", *JCAP* **07**, 014 (2005), "Gravitational wave background in perfect fluid quantum cosmologies", *Phys. Rev.* **D73**, 104017 (2006).

[632] E. Pinho and N. Pinto-Neto, "Scalar and vector perturbations in quantum cosmological backgrounds ", *Phys. Rev.* **D76**, 023506 (2007).

[633] J. Acacio de Barros, N. Pinto-Neto, and M.A. Sagioro-Leal, "The causal interpretation of dust and radiation fluid nonsingular quantum cosmologies", *Phys. Lett.* **A241**, 229–239 (1998).

[634] D. Bohm and J.P. Vigier, "Model of the causal interpretation of quantum theory in terms of a fluid with irregular fluctuations", *Phys. Rev.* **96**, 208–216 (1954).

[635] A. Valentini, "Signal-locality, uncertainty, and the subquantum H-theorem" I, *Phys. Lett.* **A 156**, 5–11 (1991); II, *Phys. Lett.* **A 158**, 1–8 (1991) .

[636] A. Valentini and H. Westman, "Dynamical origin of quantum probabilities", *Proc. Roy. Soc.* **A 461**, 253–272 (2004).

[637] M.D. Towler, N.J. Russell, and A. Valentini, "Time scales for dynamical relaxation to the Born rule", *Proc. Royal Soc.* **A 468**, 990–1013 (2015).

[638] G. Garcia de Polavieja, "A causal quantum theory in phase space", *Phys. Lett.* **A 220**, 303–314 (1996).

[639] M.O. Scully, "Do Bohm trajectories always provide a trustworthy physical picture of particle motion?", *Phys. Scripta* **T 76**, 41–46 (1998).

[640] C. Dewdney, L. Hardy, and E.J. Squires, "How late measurements of quantum trajectories can fool a detector", *Phys. Lett.* **A 184**, 6–11 (1993).

[641] N. Gisin, "Why Bohmian mechanics? One and two-time measurements, Bell inequalities, philosophy and physics", arXiv:1509.00767 [quant-ph].

[642] R.B. Griffiths, "Bohmian mechanics and consistent histories", *Phys. Lett.* **A 261**, 227–234 (1999).

[643] Y. Aharonov and L. Vaidman, "About position measurements which do not show the Bohmian particle position", in J.T. Cushing *et al.* eds *Bohmian theory, an appraisal*, Kluwer (1996); arXiv:quant-ph/9511005 (1995).

[644] Y. Aharonov, B-G. Englert, and M.O. Scully, "Protective measurements and Bohm trajectories", *Phys. Lett.* **A 263**, 137–146 (1999).

[645] C. Cohen-Tannoudji, B. Diu, and F. Laloë, *Mécanique quantique*, Hermann (1973 and 1977); *Quantum Mechanics*, Wiley (1977).

[646] N. Bohr, "Discussions with Einstein on epistemological problems in atomic physics", in [2], 200–241; reprinted in *Quantum Theory and Measurement*, J.A. Wheeler and W.H. Zurek editors, Princeton University Press (1983), pp. 9–49.

[647] M. Correggi and G. Morchio, "Quantum mechanics and stochastic mechanics for compatible observables at different times", *Ann. Physics* **296**, 371–389 (2002).

[648] A. Neumaier, "Bohmian mechanics contradicts quantum mechanics", arXiv:quant-ph/0001011 (2000).

[649] G. Brida, E. Cagliero, G. Falzetta, M. Genovese, M. Gramegna, and C. Novero, "Experimental realization of a first test of de Broglie–Bohm theory", *J. Phys. B* **35**, 4751–4756 (2002); "A first experimental test of the de Broglie–Bohm theory against standard quantum mechanics", arXiv:quant-ph/0206196 (2002).

[650] G. Brida, E. Cagliero, G. Falzetta, M. Genovese, M. Gramegna, and E. Prerdazzi, "Biphoton double slit experiment", *Phys. Rev.* **A 68**, 033803 (2003).

[651] P. Ghose, "An experiment to distinguish between de Broglie–Bohm and standard quantum mechanics", arXiv:quant-ph/0003037 (2003).

[652] D. Dürr, S. Goldstein, R. Tumulka, and N. Zanghi, "Bell-type quantum field theories", *J. Phys. A Math. Gen.* **38**, R1 (2005); arXiv:quant-ph/0407116v1 (2004).

[653] A. Sudbery, "Objective interpretations of quantum mechanics and the possibility of a deterministic limit", *J. Phys.* **A 20**, 1743–1750 (1987); "Single-world theory of the extended Wigner's friend experiment", *Found. Phys.* **47**, 658–669 (2017); arXiv:1608.05373v3 (2017).

[654] A. Oldofredi, "Stochasticity and Bell-type quantum field theory", *Synthese*, 1–20 (2018); arXiv:1802.01898v1 (2018).

[655] I. Fényes, "Eine wahrscheinlichkeitstheoretische Begründung und Interpretation der Quantenmechanik", *Zeit. Physik* **132**, 81–106 (1952).

[656] E. Nelson, "Derivation of the Schrödinger equation from Newtonian mechanics", *Phys. Rev.* **150**, 1079–1085 (1966).

[657] R. Werner, "A generalization of stochastic mechanics and its relation to quantum mechanics", *Phys. Rev.* **D 34**, 463–469 (1986).

[658] T.C. Wallstrom, "Inequivalence between the Schrödinger equation and the Madelung hydrodynamic equations", *Phys. Rev.* **A 49**, 1613–1617 (1994).

[659] P. Damgaard and H. Hüffel editors, *Stochastic Quantization*, World Scientific (1988).

[660] M. Masujima, *Path Integral Quantization and Stochastic Quantization*, Springer Verlag (2000 and 2009).

[661] G. Parisi and Y-S. Wu, "Perturbation theory without gauge fixing", *Sci. Sin.* **24**, 483–496 (1981).

[662] E. Gozzi, "Functional-integral approach to Parisi-Wu stochastic quantization: scalar theory", *Phys. Rev.* **D 28**, 1922–1930 (1983).

[663] M. Dickson and D. Dieks, "Modal interpretation of quantum mechanics", *Stanford Encyclopedia of Philosophy* (2007): http://plato.stanford.edu/entries/qm-modal/

(now replaced by the next reference, but still accessible on the site of the Encyclopedia).

[664] O. Lombardi and D. Dieks, "Modal interpretations of quantum mechanics", *Stanford Encyclopedia of Philosophy* (2012):
https://plato.stanford.edu/entries/qm-modal/

[665] B.C. van Fraassen, "A formal approach to the philosophy of science", in *Paradigms and Paradoxes: The Philosophical Challenge of the Quantum Domain*, R. Colodny editor, University of Pittsburg Press (1972), pp. 303–366; "The Einstein–Podolsky–Rosen paradox", *Synthese*, **29**, 291–309 (1974); *Quantum Mechanics: An Empiricist View*, Oxford, Clarendon Press (1991).

[666] S. Kochen, "A new interpretation of quantum mechanics", in *Symposium on the Foundations of Modern Physics*, P. Mittelstaedt and P. Lahti editors, World Scientific (1985), pp. 151–169.

[667] D. Dieks, "The formalism of quantum theory: an objective description of reality?", *Annalen der Physik* **500**, 174–190 (1988); "Quantum mechanics without the projection postulate and its realistic interpretation", *Found. Phys.* **19**, 1397–1423 (1989); "Resolution of the measurement problem through decoherence of the quantum state", *Phys. Lett.* **A 142**, 439–446 (1989); "Modal interpretation of quantum mechanics, measurements, and macroscopic behaviour", *Phys. Rev.* **A 49**, 2290–2300 (1994).

[668] R. Healey, *The Philosophy of Quantum Mechanics: An Interactive Interpretation*, Cambridge University Press (1989); "Measurement and quantum indeterminateness", *Found. Phys. Lett.* **6**, 307–316 (1993).

[669] G. Bacciagaluppi, "Topics in the modal interpretation of quantum mechanics", dissertation, Cambridge University (1996); "Delocalized properties in the modal interpretation of a continuous model of decoherence", *Found. Phys.* **30**, 1431–1444 (2000).

[670] M. Dickson, "Wavefunction tails in the modal interpretation", *Proceedings of the Philosophy of Science Association 1994*, D. Hull, M. Forbes, and R. Burian editors, **1**, 366–376 (1994).

[671] J. Berkovitz and M. Hemmo, "Modal interpretations of quantum mechanics and relativity: a reconsideration", *Found. Phys.* **35**, 373–397 (2005).

[672] R. Healey, "Modal interpretation, decoherence, and the quantum measurement problem", in *Quantum Measurement: Beyond Paradox*, R. Healey and G. Hellmann editors, *Minnesota Studies in the Philosophy of Science* **17**(1998), pp. 52–86.

[673] W. Myrvold, "Modal interpretation and relativity", *Found. Phys.* **32**, 1173–1784 (2002).

[674] R. Clifton, "The modal interpretation of algebraic quantum field theory", *Phys. Lett.* **A 271**, 167–177 (2000).

[675] L. Diosi, "Quantum stochastic processes as models for state vector reduction", *J. Phys.* **A 21**, 2885–2898 (1988).

[676] R. Haag, "Fundamental irreversibility and the concept of events", *Comm. Math. Phys.* **132**, 245–251 (1990); "An evolutionary picture for quantum physics", *Comm. Math. Phys.* **180**, 733–743 (1996).

[677] A. Jadczyk, "On quantum jumps, events, and spontaneous localization models", *Found. Phys.* **25**, 743–762 (1995).

[678] P. Pearle, "How stands collapse I", *J. Phys. A: Math. Theor.* **40**, 3189–3204 (2007).

[679] P. Pearle, "On the time it takes a state vector to reduce", *J. Stat. Phys.* **41**, 719–727 (1985).

[680] A. Barchielli, L. Lanz, and G.M. Prosperi, "A model for the macroscopic description and continual observations in quantum mechanics", *Nuov. Cim.* **42 B**, 79–121 (1982).

[681] A. Barchielli, "Continual measurements for quantum open systems", *Nuov. Cim.* **74 B**, 113–138 (1983); "Measurement theory and stochastic differential equations in quantum mechanics", *Phys. Rev.* **A 34**, 1642–1648 (1986).

[682] F. Benatti, G.C. Ghirardi, A. Rimini, and T. Weber, "Quantum mechanics with spontaneous localization and the quantum theory of measurement", *Nuov. Cim.* **100 B**, 27–41 (1987).

[683] F. Benatti, G.C. Ghirardi, A. Rimini, and T. Weber, "Operations involving momentum variables in non-Hamiltonian evolution equations", *Nuov. Cim.* **101 B**, 333–355 (1988).

[684] P. Blanchard, A. Jadczyk, and A. Ruschhaupt, "How events come into being: EEQT, particle tracks, quantum chaos and tunneling time", in *Mysteries, Puzzles and Paradoxes in Quantum Mechanics*, R. Bonifacio editor, American Institute of Physics, AIP Conference Proceedings, n^r 461 (1999); *J. Mod. Optics* **47**, 2247–2263 (2000).

[685] P. Pearle, "Combining stochastic dynamical state-vector reduction with spontaneous localization", *Phys. Rev.* **A 39**, 2277–2289 (1989).

[686] G.C. Ghirardi, P. Pearle, and A. Rimini, "Markov processes in Hilbert space and continuous spontaneous localization of systems of identical particles", *Phys. Rev.* **A 42**, 78–89 (1990).

[687] P. Pearle, "Cosmogenesis and collapse", arXiv:1003.5582v2 [gr-qc] (2010); *Found. Phys.* **42**, 4–18 (2012).

[688] *Experimental Metaphysics: Quantum Mechanical Studies for Abner Shimony*, Festschrift volumes 1 and 2, R.S. Cohen, M.A. Horne, and J.J. Stachel editors, *Boston Studies in the Philosophy of Science*, volumes 193 and 194, Kluwer Academic Publishers (1997); P. Pearle, volume 1, p. 143; G. Ghirardi and T. Weber, volume 2, p. 89.

[689] A. Shimony, "Desiderata for a modified quantum dynamics", pp. 49–59 in *"PSA 1990* vol. 2, *Proceedings of the 1990 Biennial Meeting of the Philosophy of Science Association*, A. Fine, M. Forbes, and L. Wessel editors, Philosophy of Science Association.

[690] P. Pearle, "How stands collapse II", in *Quantum Reality, Relativistic Causality and Closing the Epistemic Circle: Essays in Honour of Abner Shimony*, W. Mryvold and J. Christian editors, Springer (2009), pp. 257–292.

[691] L.F. Santos and C.O. Escobar, "A proposed solution to the tail problem of dynamical reduction models", *Phys. Lett.* **A 278**, 315–318 (2001).

[692] L. Diosi, "Continuous quantum measurement and Itô formalism", *Phys. Lett.* **129 A**, 419–423 (1988).

[693] L. Diosi, "Models for universal reduction of macroscopic quantum fluctuations", *Phys. Rev.* **A 40**, 1165–1174 (1989).

[694] G.C. Ghirardi, R. Grassi, and A. Rimini, "Continuous-spontaneous-reduction model involving gravity", *Phys. Rev.* **A 42**, 1057–1064 (1990).

[695] R. Penrose, *The Emperor's New Mind*, Oxford University Press (1989); *Shadows of the Mind*, Oxford University Press (1994).

[696] R. Penrose, "On gravity's role in quantum state reduction", *General Relativity and Gravitation* **28**, 581–600 (1996).

[697] N. Gisin, "Stochastic quantum dynamics and relativity", *Helv. Phys. Acta* **62**, 363–371 (1989).

[698] G.C. Ghirardi, R. Grassi, and P. Pearle, "Relativistic dynamical reduction models: general framework and examples", *Found. Phys.* **20**, 1271–1316 (1990).

[699] P. Pearle, "Completely quantized collapse and consequences", *Phys. Rev.* **A 72**, 022112 (2005).

[700] D.J. Bedingham, "Relativistic state reduction dynamics", *Found. Phys.* **41**, 686–704 (2011); arXiv:1003.2774v2 [quant-ph] (2010). "Relativistic state reduction model", *J. Phys. Conf. Series* **306**, 012034 (2011).

[701] D.J. Bedingham, D. Dürr, G. Ghirardi, S. Goldstein, R. Tumulka, and N. Zanghì, "Matter density and relativistic models of wave function collapse", *J. Stat. Phys.* **154**, 623–631 (2014); arXiv:1111.1425v2 [quant-ph] (2011).

[702] N. Gisin, "Weinberg's nonlinear quantum mechanics and supraluminal communications", *Phys. Lett.* **A 143**, 1–2 (1990).

[703] J. Polchinski, "Weinberg's nonlinear quantum mechanics and the Einstein–Podolsky-Rosen paradox", *Phys. Rev. Lett.* **66**, 397–400 (1991).

[704] R. Tumulka, "On spontaneous wave function collapse and quantum field theory", *Proc. Roy. Soc.* **A 462**, 1897–1908 (2006); "A relativistic version of the Ghirardi–Rimini–Weber model", *J. Stat. Phys* **125**, 821–840 (2006); "Collapse and relativity", arXiv:quant-ph/0602208 (2006).

[705] A. Bassi and G. Ghirardi, "Dynamical reduction models", *Phys. Rep.* **379**, 257–426 (2003).

[706] S. Weinberg, "Precision tests of quantum mechanics", *Phys. Rev. Lett.* **62**, 485–488 (1989); "Testing quantum mechanics", *Ann. of Phys.* **194**, 336–386 (1989).

[707] K. Wódkiewicz and M.O. Scully, "Weinberg's nonlinear wave mechanics", *Phys. Rev.* **A 42**, 5111–5116 (1990).

[708] P. Pearle, J. Ring, J.I Collar, and F.T. Avignone, "The CSL collapse model and spontaneous radiation: an update", *Found. Phys.* **29**, 465–80 (1998).

[709] H.S. Miley, F.T. Avignone, R.L. Brodzinski, J.I. Collar, and J.H. Reeves, "Suggestive evidence for the two-neutrino double-β decay of ^{76}Ge", *Phys. Rev. Lett.* **65**, 3092–3095 (1990).

[710] F. Laloë, W. Mullin, and P. Pearle, "Heating of trapped ultracold atoms by collapse dynamics", *Phys. Rev.* **A 90**, 052119 (2014).

[711] K. Hornberger, S. Gerlich, P. Haslinger, S. Nimmrichter, and M. Arndt, "Quantum interference of clusters and molecules", *Rev. Mod. Phys.* **84**, 157–173 (2012).

[712] A. Bassi, K. Lochan, S. Satin, T.P. Singh, and H. Ulbricht, "Models of wave-function collapse, underlying theories and experimental tests", *Rev. Mod. Phys.* **85**, 471–527 (2013).

[713] G.C. Ghirardi, "Quantum superpositions and definite perceptions: envisaging new feasible tests", *Phys. Lett.* **A 262**, 1–14 (1999).

[714] J. Dalibard, Y. Castin, and K. Mølmer, "Wave function approach to dissipative processes in quantum optics", *Phys. Rev. Lett.* **68**, 580–583 (1992).

[715] K. Mølmer, Y. Castin, and J. Dalibard, "Monte Carlo wave-function method in quantum optics", *Journ. Optical. Soc. Am. B* **10**, 524–538 (1993).

[716] H.J. Carmichael, *An Open System Approach to Quantum Optics*, Lectures notes in Physics, monograph 18, Springer-Verlag (1993).

[717] M.B. Plenio and P.L. Knight, "The quantum-jump approach to dissipative dynamics in quantum optics", *Rev. Mod. Phys.* **70**, 101–141 (1998).

[718] N. Gisin and I.C. Percival, "The quantum-state diffusion model applied to open systems", *J. Phys.* **A 25**, 5677–5691 (1992); "Quantum state diffusion, localization and quantum dispersion entropy", **26**, 2233–2243 (1993); "The quantum state diffusion picture of physical processes", **26**, 2245–2260 (1993).

[719] I.C. Percival, *Quantum State Diffusion*, Cambridge University Press (1998).

[720] F. Laloë, "Modified Schrödinger dynamics with attractive densities", *Eur. Phys. J.* **D 69**, 162 (2015).

[721] J.G. Cramer, "The transactional interpretation of quantum mechanics", *Rev. Mod. Phys.* **58**, 647–687 (1986); in an appendix, this article contains a review of the various interpretations of quantum mechanics.

[722] J.G. Cramer, "Generalized absorber theory and the Einstein–Podolsky–Rosen paradox", *Phys. Rev.* **D 22**, 362–376 (1980).

[723] H. Price and K. Wharton, "Does time-symmetry imply retrocausality? How the quantum world says 'maybe'", ArXiv:1002:0906v3 (2011).

[724] H. Price and K. Wharton, "Disentangling the quantum world", *Entropy* **17**, 7752–7767 (2015).

[725] M.S. Leifer and M.F. Pusey, "Is a time symmetric interpretation of quantum theory possible without retrocausality?", ArXiv:1607:0787v2 (2017).

[726] H. Everett III, "Relative state formulation of quantum mechanics", *Rev. Mod. Phys.* **29**, 454–462 (1957); reprinted in *Quantum Theory and Measurement*, J.A. Wheeler and W.H. Zurek editors, Princeton University Press (1983), pp. 315–323.

[727] B.S. DeWitt and N. Graham *The Many-Worlds Interpretation of Quantum Mechanics*, Princeton Series in Physics, Princeton University Press (1973).

[728] H. Everett III, Letter to L.D. Raub dated April 7 (1983),
http://dspace.nacs.uci.edu/xmlui/handle/10575/1205.

[729] D. Deutsch, "The structure of the multiverse", *Proc. Roy. Soc. London* **A 458**, 2911–2923 (2002).

[730] A. Kent, "Against many world interpretations", *Int. Journ. Mod. Phys* **A 5**, 1745–1762 (1990).

[731] P. Van Esch, "On the Born rule and the Everett programme", *Ann. Fond. Louis de Broglie* **32**, 51–59 (2007).

[732] D. Deutsch, "Quantum theory of probability and decisions", *Proc. Roy. Soc. London* **A 455**, 3129–3137 (1999).

[733] M.A. Rubin, "Relative frequency and probability in the Everett interpretation of Heisenberg–picture quantum mechanics", *Found. Phys.* **33**, 379–405 (2002).

[734] D. Wallace, "Everettian rationality: defending Deutsch's approach to probability in the Everett interpretation", *Stud. Hist. Phil. Mod. Phys.* **34**, 415–438 (2003).

[735] S. Saunders, "Derivation of the Born rule from operational assumptions", *Proc. Roy. Soc. London* **A 460**, 1771–1788 (2004).

[736] W.H. Zurek, "Probabilities from entanglement, Born's rule $p_k = |\Psi_k|^2$ from envariance", *Phys. Rev.* **A 71**, 052105 (2005).

[737] D. Wallace, "Quantum probability from subjective likelihood: improving on Deutsch's proof of the probability rule", *Studies in History and Philosophy of Modern Physics* **38**, 311–332 (2007).

[738] H. Price, "Probability in the Everett world: comments on Wallace and Greaves", arXiv:quant-ph/0604191 (2006); "Decisions, decisions, decisions: can Savage salvage the Everettian probability?", arXiv:quant-ph/0802.1390 (2008).

[739] H.D. Zeh, "Roots and fruits of decoherence", *Séminaire Poincaré* **1**, 115–129 (2005); available at http://www.bourbaphy.fr/.

[740] D. Deutsch and P. Hayden, "Information flow in entangled quantum systems", *Proc. Roy. Soc. London* **A 456**, 1759–1774 (2000).

[741] B. DeWitt, *The Global Approach to Quantum Field Theory*, vol. 1, Clarendon Press (2003), p. 144.

[742] M. Tegmark, "Parallel universes", in *Science and Ultimate reality: From Quantum to Cosmos*, J.D. Barrow, P.C.W. Davies, and C.L. Harper editors, Cambridge University Press (2003); "Many worlds in context", arXiv:0905.2182v2 [quant-ph] (2010); also in *Many Worlds? Everett, Quantum Theory and Reality*, S. Saunders, J. Barrett, A. Kent and D. Wallace editors, Oxford University Press (2010).

[743] T. Damour, "Einstein 1905–1955: son approche de la physique", *Séminaire Poincaré* **1**, 1–25 (2005); available at http://www.bourbaphy.fr/.

[744] J.S. Bell, "The measurement theory of Everett and de Broglie's pilot wave", in *Quantum Mechanics, Determinism, Causality, and Particles*, M. Flato *et al.* editors, Dordrecht-Holland, D. Reidel (1976), pp. 11–17; Chapter 11 of 2004 edition of [6].

[745] A. Einstein, "Quantentheorie des einatomigen idealen Gases", *Sitzungsberichte der Preussischen Akademie der Wissenschaften* **1**, 3–14 (1925).

[746] M.H. Anderson, J.R. Ensher, M.R. Matthews, C.E. Wieman, and E.A. Cornell, "Observation of Bose–Einstein condensation in a dilute atomic vapor", *Science* **269**, 198–201 (1995).

[747] K.B. Davis, M.-O. Mewes, M.R. Andrews, N.J. van Druten, D.S. Durfee, D.M. Kurn, and W. Ketterle, "Bose–Einstein condensation in a gas of sodium atoms", *Phys. Rev. Lett.* **75**, 3969–3973 (1995).

[748] H. Hertz, *Miscellaneous Papers*, translated from first German edition (1895) by D.E. Jones and G.A. Schott, MacMillan, volume 1, p. 318.

Index